2017 IEEE International Symposium on Nanoelectronic and Information Systems (iNIS 2017)

Bhopal, India
18-20 December 2017

IEEE Catalog Number: CFP17C48-POD
ISBN: 978-1-5386-1357-3

**Copyright © 2017 by the Institute of Electrical and Electronics Engineers, Inc.
All Rights Reserved**

Copyright and Reprint Permissions: Abstracting is permitted with credit to the source. Libraries are permitted to photocopy beyond the limit of U.S. copyright law for private use of patrons those articles in this volume that carry a code at the bottom of the first page, provided the per-copy fee indicated in the code is paid through Copyright Clearance Center, 222 Rosewood Drive, Danvers, MA 01923.

For other copying, reprint or republication permission, write to IEEE Copyrights Manager, IEEE Service Center, 445 Hoes Lane, Piscataway, NJ 08854. All rights reserved.

*** *This is a print representation of what appears in the IEEE Digital Library. Some format issues inherent in the e-media version may also appear in this print version.*

IEEE Catalog Number: CFP17C48-POD
ISBN (Print-On-Demand): 978-1-5386-1357-3
ISBN (Online): 978-1-5386-1356-6

Additional Copies of This Publication Are Available From:

Curran Associates, Inc
57 Morehouse Lane
Red Hook, NY 12571 USA
Phone: (845) 758-0400
Fax: (845) 758-2633
E-mail: curran@proceedings.com
Web: www.proceedings.com

Proceedings

2017 IEEE International Symposium on Nanoelectronic and Information Systems

18–20 December 2017
Bhopal, India

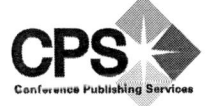

Los Alamitos, California

Washington • Tokyo

Proceedings

2017 IEEE International Symposium on Nanoelectronic and Information Systems

iNIS 2017

2017 IEEE International Symposium on Nanoelectronic and Information Systems

iNIS 2017

Table of Contents

Message from the General Chairs...xi

Message from the Technical Program Chairs...xii

Organizing Committee..xiii

Technical Program Committee..xv

Steering Committee..xix

Keynotes...xx

Session 01: Cyber Physical Systems

Design Optimization of DSP for Wearable Biomedical Device 1

    M. Naga Sasikanth, Sashank Gambhira, and Mrigank Sharad

Embedded Hardware Prototype for Gas Detection and Monitoring System in Android
Mobile Platform 6

    Rahul Kurzekar, Hardik Arora, and Rahul Shrestha

Session 02 (Special): Reliability and Performance Aware SoC solutions for IoT Framework - Part 1

Reliability and Threat Analysis of NBTI Stress on DSP Cores 11

    Anirban Sengupta, Deepak Kachave, Shuba Neema, and Sri Harsha Panugothu

A Firefly Algorithm Driven Approach for High Level Synthesis 15

    Pallabi Sarkar, Anirban Sengupta, Santosh Rathlavat, and Mrinal Kanti Naskar

Mathematical Validation of HWT Based Lossless Image Compression 20

    Anirban Sengupta and Dipanjan Roy

Session 03: Low Power Quantum Computing

Lookup Table-Based Low-Power Implementation of Multi-channel Filters
for Software Defined Radio 23
  Subhendu Kumar Sahoo and P. K. Meher

Design of Practical Parity Generator and Parity Checker Circuits in QCA 28
  Dharmendra Kumar, Chintoo Kumar, Shipra Gautam, and Debasis Mitra

Session 04: Emerging Technologies I

Design and Optimization of Single Electron Transistor Based 4-Bit Arithmetic
and Logic Unit at Room Temperature Operation 34
  Rathin Joshi, Rutu Parekh, and Yash Agrawal

Realizing All Logic Operations Using mRNA-Ribosome System as a Post Si
Alternative 40
  Pratima Chatterjee and Prasun Ghosal

Design of a High Performance Carry Generation Circuit for Ternary Full Adder Using
CNTFET 46
  Subhendu Kumar Sahoo, Gangishetty Akhilesh, and Rasmita Sahoo

Session 05: NoCs and Memory

Energy Efficient NoC Router for High Throughput Applications in Many-Core GPUs 50
  Shrestha Bansal, Hemanta Kumar Mondal, Sri Harsha Gade, and Sujay Deb

Routing Algorithm for Application-Specific Network-on-Chip with Irregular Core
Sizes 56
  Grandhi Sai Anirudh and Soumya J

STT-MRAM for Low Power Access for Read-Intensive Parallel Deep-Learning
Architectures 61
  Saranyu Chattopadhyay, Kaustav Brahma, Arkaprova Ray, and Mrigank Sharad

Session 06 (Special): Reliability and Performance Aware SoC solutions for IoT Framework - Part 2

Comprehensive Operation Chaining Based Schedule Delay Estimation During High
Level Synthesis 66
  Vipul Kumar Mishra and Anirban Sengupta

Cost Aware Majority Logic Synthesis for Emerging Technologies 69
  Vipul Kumar Mishra

vi

Session 07: Emerging Technologies II

Implementation of a 6 GHz MEMS Switch 74
Saurabh Chaturvedi, Mladen Božanić, and Saurabh Sinha

Towards the Approximation of Cell Wise Switching Time in Quantum-Dot Cellular
Automata 78
Soudip Sinha Roy

Fault Tolerance and Temperature Stability: The Dynamic Error Estimation
in Quantum-Dot Cellular Automata 84
Soudip Sinha Roy

Session 08 (Special): Emerging Nanoscale Transistors for Biosensing and IoT applications

Tunneling Field Effect Transistors for Energy Efficient Logic, Sensor Interface and 3D
IC Circuits for IoT Platforms 90
Japa Aditya, T. Nagateja, and Ramesh Vaddi

SiGe Source Charge Plasma TFET for Biosensing Applications 93
Nawaz Shafi, Chitrakant Sahu, C. Periasamy, and Jawar Singh

Session 09: Security and Privacy

Neutralization of the Effect of Hardware Trojan in SCADA System Using Selectively
Placed TMR 99
Nagendra Babu Gunti and Karthikeyan Lingasubramanian

Fault Sensitive Neutralization of Hardware Trojans Using Multi-level Triple Modular
Redundancy Scheme 105
Nagendra Babu Gunti and Karthikeyan Lingasubramanian

A Novel Intrusion Detection Algorithm: An AODV Routing Protocol Case Study 111
Gurveen Vaseer, Garima Ghai, and Pushpinder Singh Patheja

Session 10 (Special): Circuit Design in Rebooting Computing

Security Evaluation of MTJ/CMOS Circuits Against Power Analysis Attacks 117
S. Dinesh Kumar and Himanshu Thapliyal

Quantum Circuit Designs of Integer Division Optimizing T-Count and T-Depth 123
*Himanshu Thapliyal, T. S. S. Varun, Edgard Munoz-Coreas, Keith A. Britt,
and Travis S. Humble*

Session 11: ADC and Amplifiers

High Performance Sense Amplifier Based Flip Flop for Driver Applications 129

    Anoop D, Nithin Kumar Y. B., and Vasantha M. H.

A Novel Low Power High Speed BEC for 2GHz Sampling Rate Flash ADC in 45nm
Technology 133

    Sarfraz Hussain, Rajesh Kumar, and Gaurav Trivedi

Comparison and Design of Dynamic Comparator in 180nm SCL Technology for Low
Power and High Speed Flash ADC 139

    Sarfraz Hussain, Rajesh Kumar, and Gaurav Trivedi

Session 12 (Special): Applications of Big Data Analytics and IoT for Tomorrow's Smart City

An Automated Game Theoretic Approach for Cooperative Road Traffic Management
in Disaster 145

    Samya Muhuri, Debasree Das, and Susanta Chakraborty

Security Enhancements to System on Chip Devices for IoT Perception Layer 151

    Sudeendra Kumar K, Sauvagya Sahoo, Abhishek Mahapatra, Ayas Kanta Swain,
    and K.K. Mahapatra

An Efficient MapReduce-Based Adaptive K-Means Clustering for Large Dataset 157

    Tapan Chowdhury, Arijit Mukherjee, and Susanta Chakraborty

Session 13: MOSFET Design

Enhanced Look-Up Table Approach for Modeling of Floating Body SOI MOSFET 163

    Sitansusekhar Roymohapatra, Ganesh R Gore, Akanksha Yadav, Mahesh B. Patil,
    Krishnan S Rengrajan, and Maryam Shojaei Baghini

Analysis of Barrier Layer Thickness on Performance of In1-xGaxAs Based Gate Stack
Cylindrical Gate Nanowire MOSFET .... N/A

    Sanjeev Kumar Sharma, Balwinder Raj, Mamta Khosla, and Jeetendra Singh

Basic CMOS Gate Design by Mixed-Mode Analysis of Step-Channel
TMDG-MOSFET 173

    Pankaj Kumar, Syed Samsuz Zaman, Mansh Pratim Sarma, Ashok Ray, and Gaurav Trivedi

Modeling of Threshold Voltage and Subthreshold Current for P-Channel Symmetric
Double-Gate MOSFET in Nanoscale Regime 179

    Rekib Uddin Ahmed and Prabir Saha

Session 14: Image and Signal Processing

Digital Video Stabilization- Review with a Perspective of Real Time Implementation 184
Mohammed Ahmed

Session 15: Hardware Devices and Layout

Gate Metal Work Function Engineering for the Improvement of Electrostatic
Behaviour of Doped Tunnel Field Effect Transistor 190
Deepak Soni, Dheeraj Sharma, Shivendra Yadav, Mohd. Aslam,
Dharmendra Singh Yadav, and Neeraj Sharma

A Comparative Study of GaP/SiGe Hetero Junction Double Gate Tunnel Field Effect
Transistor 195
Dharmendra Singh Yadav, Dheeraj Sharma, Sukeshni Tirkey, Deepak Soni,
Deepak G. Sharma, Shriya Bajpai, and Neeraj Sharma

Design and Simulation of SF-FinFET and SD-FinFET and Their Performance
in Analog, RF and Digital Applications 200
Syed Samsuz Zaman, Pankaj Kumar, Manash Pratim Sarma, Ashok Ray,
and Gaurav Trivedi

Session 16 (Special): MOS Current Mode Logic: An Alternative for High Speed Signaling

A Design Methodology for MOS Current Mode Logic VCO 206
Abir J Mondal, Alak Majudmer, and Bidyut K Bhattacharyya

A 90nm Novel MUX-Dual Latch Design Approach for Gigascale Serializer
Application 210
Monalisa Das, Alak Majumder, Abir J Mondal, and Bidyut K Bhattacharyya

Session 17: Low Power and Reliability

Leakage Reduction in DT8T SRAM Cell Using Body Biasing Technique 215
Rajani Suthar, Kirti S. Pande, and Murty N.S.

A Single-Ended Read Decoupled 9T SRAM Cell for Low Power Applications 220
S. R. Mansore, R. S. Gamad, and D. K. Mishra

Session 18 (Special): Gated Clock Distribution for Silicon Chips

Current Profile Generated by Gating Logic Reduces Power Supply Noise of Integrated
CPU Chip 224
Alak Majumder and Pritam Bhattacharjee

Binary Counter Based Gated Clock Tree for Integrated CPU Chip 229
Bipasha Nath and Alak Majumder

Session 19: Student Research Forum

Design and Analysis of Schmitt Trigger Based 10T SRAM in 32 nm Technology 234
Amit Singh Rajput, Manisha Pattanaik, and R.K. Tiwari

An Efficient & Effective Approach of Chip Power Calculation by Unified Power
Format (UPF) ....N/A
Arti Devulapalli, Akhilesh Chandra Mishra, and D. Prem Prasad

Session 20 (Special): Networks-on-Chip Architecture

A Power, Thermal and Reliability-Aware Network-on-Chip 243
*Ashish Sharma, Yogendra Gupta, Sonal Yadav, Lava Bhargava, Manoj Singh Gaur,
and Vijay Laxmi*

Session 21 (Special): Secured High Performance Nanoscale System Design for Tomorrow

Microprocessor Based Physical Unclonable Function 246
*Sudeendra kumar K, Sauvagya Sahoo, Abhishek Mahapatra, Ayas Kanta Swain,
and K.K. Mahapatra*

On-chip RO-Sensor for Recycled IC Detection 252
Sauvagya Ranjan Sahoo, Sudeendra K, A. Mahapatra, A.K. Swain, and K.K. Mahapatra

MSM: Performance Enhancing Area and Congestion Aware Network-on-Chip
Architecture 257
Tuhin Subhra Das and Prasun Ghosal

Session 22 (Special): IoT and Smart Cities

Rapid Prototyping IoT End Applications Using Software Development Kits and Add
on Plugins 263
Manoj R and Adrian Fernandez

Author Index 268

Message from the General Chairs

It is a distinct privilege to welcome all the participants to the city of Bhopal, Madhya Pradesh (MP), India. We warmly welcome you to the 3rd IEEE International Symposium on Nanoelectronic and Information Systems (IEEE-iNIS, http://www.ieee-inis.org/). The 1st and 2nd iNIS meetings that took place at Indore and Gwalior, India respectively were immensely successful. Those were very well attended by researchers from industry and academia. Proceedings of both iNIS 2015 and 2016 are available in IEEE Xplore. They have been produced through IEEE Computer Society Conference Publishing Services (CPS). The main goal of iNIS is to provide a platform for both hardware and software researchers to interact under one roof for research and development which may lead to realization of efficient, robust, and secure information processing circuits and systems. iNIS has been initiated as a sponsored meeting of Technical Committee on VLSI (TCVLSI, http://www.ieee-tcvlsi.org/), of IEEE Computer Society (IEEE-CS). TCVLSI is one among two-dozen technical committees of IEEE-CS which endorse different meetings in the scope of IEEE-CS. TCVLSI endorses a league of successful meetings including ARITH, ASAP, ISVLSI, IWLS, and SLIP. iNIS right from the birth is among the league of many immensely successful "Sister Conferences". iNIS 2017 is sponsored by IEEE-CS through TCVLSI and technically co-sponsored by IEEE-CEDA. iNIS 2017 proceeding is also published by IEEE-CS Conference Publication Services (CPS). iNIS 2017 has continued attracting attendees from all over the globe. We hope that iNIS will continue attracting researchers from various parts of the globe and continue serving the community in years to come.

The city of Bhopal in Madhya Pradesh has been selected as a venue of iNIS 2017 which is located in the central region of Madhya Pradesh. Bhopal is 13-14 hours' drive from New Delhi, capital of India. Bhopal is the capital of the state and is known as "City of Lakes". It has always been an important center of the state. There are many tourist attractions in Bhopal including Birla Mandir, Bharat Bhawan, Upper Lake, Archeological Museum, Bhojpur and Bhimbethika. The world famous Taj Mahal is at just 10 hours driving distance from Bhopal. This is a must visit for the delegates.

The general chairs would like to thank the steering committee chair, Saraju Mohanty for his support throughout. We would like to thank Oriental Group of Institutions and other sponsors in helping iNIS 2017. The standard of the symposium is measured by the quality of the papers submission, reviews and the acceptance rate and iNIS 2017 exceeds the standards. iNIS 2017 has 6 keynotes from renowned researchers and around 23 sessions from high quality researchers around the globe. We would like to specifically thank the keynote speakers for their support and exciting talks to the iNIS 2017 attendee. We would like to specifically thank the program chairs, Sudeep Pasricha, and Anirban Sengupta, who did an excellent job in selecting quality papers for presentations. We would sincerely thank the special session chairs, publication chairs, finance chairs, web chairs, publicity chairs, local arrangement chairs, student research forum chairs, registration chairs, and all other active volunteers, for their fantastic job in running the symposium. We would like to thank the sponsors for supporting the iNIS 2017.

Dhruva Ghai
General Co-Chair,
Oriental University, Indore, India
dhruvaghai@orientaluniversity.in

Xin Li
General Co-Chair, iNIS 2017
Duke University, USA
xinli.ece@duke.edu

Prasun Ghosal
General Co-Chair, iNIS 2017
IIEST, Shibpur, India
p_ghosal@it.iiests.ac.in

Message from the Technical Program Chairs

It is with distinct pleasure that we welcome you to the 3rd IEEE International Symposium on Nanoelectronic and Information Systems (IEEE-iNIS) to be held in Bhopal, India from 18 – 20th December 2017. Over the last few years, iNIS has evolved into a leading research forum for academic and industrial community to share their innovative ideas and research results. The primary objective of IEEE-iNIS is to provide a platform for both hardware and software researchers to interact under one umbrella.

This year, iNIS consists of six tracks: Nanoelectronic VLSI and Sensor Systems (NVS), Energy-Efficient, Reliable VLSI Systems (ERS), Hardware/Software for Internet of Things and Consumer Electronics (IoT), Hardware for Secure Information Processing (SIP), Hardware/Software Solutions for Big Data (SBD), and Cyber Physical Systems and Social Networks (CSN). iNIS 2017 attracted high quality papers from across the globe. After a thorough review process, the top 69 papers were selected for presentation. The reviews were conducted by a strong technical program committee and reviewers who were among the leading researchers in their fields. The iNIS 2017 proceedings will be published and indexed at IEEE Xplore. The papers from previous iterations of this conference have been extended and published in top archival journals in the past. This year as well, selected iNIS 2017 papers will be invited to extend their contributions for consideration in special issues in leading journals such as IEEE Consumer Electronics and IEEE Transactions on Nanotechnology. iNIS 2017 is a truly global conference, with participation of authors and program committee members from countries such as USA, India, Australia, Germany, Italy, France, Korea, Greece, UK, Singapore and China.

We would like to express our sincere gratitude to the tracks chairs for their tireless efforts in managing the tracks by forming a strong program committee, assigning reviews, consolidating the scores, and finally making the recommendations. We would like to thank the technical program committee members for their time and efforts in providing detailed and timely reviews. We would also like to thank all of the authors without whose contributions, the technical program would not have been possible. We would like to thank the Steering Committee and the Organizing Committee for their support and cooperation in every step of the process.

We hope you a very productive iNIS 2017 and wish you a very pleasant stay in Bhopal.

Program Chair
Sudeep Pasricha
Colorado State University
Fort Collins, Colorado, USA
sudeep@colostate.edu

Program Chair
Anirban Sengupta
Indian Institute of Technology
Indore, India
asengupt@iiti.ac.in

Organizing Committee

General Chairs
Dhruva Ghai, *Oriental University, India*
Xin Li, *Duke University, USA*
Prasun Ghosal, *IIEST, India*

Program Chairs
Sudeep Pasricha, *Colorado State University, USA*
Anirban Sengupta, *IIT Indore, India*

Publication Chairs
Karthikeyan Lingasubramanian, *University of Alabama at Birmingham, USA*
Shubhajit Roy Chowdhury, *IIT Mandi, India*

Web Chair
Mike Borowczak, *University of Wyoming, USA*

Publicity Chairs
Volkan Kursun, *HKUST, Hong Kong*
Alak Majumdar, *NIT Arunachal Pradesh, India*
Gargi Khanna, *NIT Hamirpur, India*
Tripta Thakur, *MANIT, India*
Umar Albalawi, *University of Tabuk, Saudi Arabia*
Saumya Kanti Datta, *Eurecom, France*

Local Arrangement Chairs
Deepika Masand, *OIST, India*
Taruna Jain, *Barkatullah UIT, India*
Deepak Verma, *MANIT, India*

Special Session Chairs
Himanshu Thapliyal, *University of Kentucky, USA*
Hai (Helen) Li, *Duke University, USA*

Student Research Forum Chair
Susanta Chakraborty, *IIEST, Shibpur, India*

Finance Chairs
Garima Ghai, *Oriental University, India*
Sunil Singh, *OIST, India*

Registration Chairs

Brajesh K Kaushik, *IIT Roorkee, India*
Rahul Dubey, *OIST, India*
Bhoopendra Singh, *RGPV, India*

Industry Liaison Chairs

Santosh K. Vishvakarma, *IIT Indore, India*
Nagi Naganathan, *Avago Technologies, USA*
Dheeraj Agarwal, *MANIT, India*

Steering Committee Chairs

Saraju P. Mohanty, *University of North Texas, USA*

Technical Program Committee

CSN (Cyber Physical Systems and Social Networks) Track Chairs & Members
Track Chairs
Tie Qiu, *Dalian University of Technology (DUT), China*
Mehdi Maasoumy, *University of California at Berkeley*

Navin Agrawal, *Oriental College of Technology, Bhopal, India*
Prasun Ghosal, *Indian Institute of Engineering Science and Technology, Shibpur, India*
Antonio Iannopollo, *University of California, Berkeley*, USA
Baihong Jin, *University of California, Berkeley, USA*
Amey Kulkarni, *Velodyne LiDAR, Inc., USA*
Chung-Wei Lin, *Toyota InfoTechnology Center, USA*
Mohammad M.R. Mozumdar, *CSU Long Beach, USA*
Deepak Puthal, *University of Technology Sydney, Australia*
Sir Researcher, *TUM CREATE, Singapore*
Bibhudutta Rout, *UNT, USA*
Fatemeh Tehranipoor, *San Francisco State University, USA*
Upasna Vishnoi, *Marvell Semiconductors, USA*
Hui Zhao, *UNT , USA*

ERS (Energy – Efficient, Reliable VLSI Systems) Track Chairs & Members
Track Chairs

Sujay Deb, *Indraprastha Institute of Information Technology Delhi, India*
Saket Srivastava, *University of Lincoln, United Kingdom*

Vijit Gadi, *Synopsys India*
Chandan Giri, *Indian Institute of Engineering Science & Technology Shibpur, India*
Mohammad Hashmi, *IIITD, India*
Karthikeyan Lingasubramanian, *UAB, USA*
Sudip Roy, *Indian Institute of Technology Roorkee, India*
Sandeep Saini, LNM IIT Jaipur, India
Jaspreet Singh, *Synopsys India*
Baris Taskin, *Drexel University, USA*

IoT (Hardware / Software for Internet of Things and Consumer Electronics) Track Chairs & Members
Track Chairs
Vaskar Raychoudhury, *Indian Institute of Technology Roorkee, India*
Abhishek Roy, *SAMSUNG ELECTRONICS CO. LTD, South Korea*

Christian Becker, *University of Mannheim, Germany*
Aniello Castiglione, *University of Salerno, Italy*
Bin Guo, *Institut Telecom SudParis, France*
Yu Hua, *Huazhong University of Science and Technology, China*
Sushanta Karmakar, *Indian Institute of Technology Guwahati, India*
Sateesh Kumar Peddoju, *INDIAN INSTITUTE OF TECHNOLOGY ROORKEE, India*
Nidhi Rajshree, *IBM India Ltd., India*
Bharat Sahu, *Sungkyunkwan University, India*
Divya Saxena, JIIT Noida, India
Navrati Saxena, *Sungkyunkwan University, South Korea*
Sukhdeep Singh, *Samsung Research Institute Bangalore, India*
Weigang Wu, *Sun Yat-sen University, China*

NVS (Nanoelectronic VLSI and Sensor Systems) Track Chairs & Members
Track Chairs
Maryam Shojaei Baghini, *Indian Institute of Technology Bombay, India*
Yiyu Shi, *University of Notre Dame, USA*

Mahima Arrawatia, *IIT Jodhpur, India*
K V Arya, *ABV-IIITM Gwalior, India*
Ganesh Balakrishnan, *University of New Mexico, USA*
Yaser Banadaki, *Southern University and A&M College - Baton Rouge, USA*
Neena Gilda, *Indian Institute of Technology Bombay, India*
Brajesh Kumar Kaushik, *Indian Institute of Technology-Roorkee, India*
Elias Kougianos, *University of North Texas, USA*
Amey Kulkarni, *Velodyne LiDAR, Inc., USA*
Jagadesh Kumar, *IIT-Delhi, India*
Joycee Mekie, *IIT Gandhinagar, India*
Durga Misra, *NJIT, USA*
Pragya Nema, *Oriental University Indore, India*
Vikram Palodiya, *Oriental University, India*
Shilpa Pendyala, *Intel, USA*
Madhav Rao, *IIIT-B, India*
Bibhudutta Rout, *University of North Texas, USA*
Chitrakant Sahu, *MNIT Jaipur, India*
Sergio Saponara, *University of Pisa, Italy*
Pankaj Shrivastava, *N/A, USA*

Nozar Tabrizi, *Kettering University, USA*
Upasna Vishnoi, *Marvell Semiconductor Inc., USA*
Santosh Vishvakarma, *IIT Indore, India*
Hui Zhao, *UNT , USA*

SBD (Hardware / Software Solutions for Big Data) Track Chairs & Members
Track Chairs
Karan Mitra, *Luleå University of Technology, Sweden*
Theocharis Theocharides, University of Cyprus, Cyprus

Malay Bhattacharyya, *Indian Institute of Engineering Science and Technology, Shibpur, India*
Cornelia Caragea, *Kansas State University, USA*
Susanta Chakraborty, *Indian Institute of Engineering Science and Technology, Shibpur, India*
Prem Prakash Jayaraman, *Swinburne University of Technology, Australia*
Debajyoti Mukhopadhyay, *Maharashtra Institute of Technology, India*
Madhu Mutyam, *Indian Institute of Technology, Madras, India*
Dhavalkumar Thakker, *University of Bradford, United Kingdom*
Laurence T. Yang, *St Francis Xavier University, Canada*

SIP (Hardware for Secure Information Processing) Track Chairs & Members
Track Chairs
Apostolos Fournaris, *University of Patras, Greece*
Kamalakanta Mahapatra, NIT ROURKELA, India

Amit Acharyya, *Indian Institute of Technology Hyderabad, India*
Mridul Sankar Barik, *Jadavpur University, India*
Bibhas Chandra Dhara, *Dept of IT, India*
Jaya Dofe, *UNH, USA*
Yier Jin, *University of Florida, USA*
Odysseas Koufopavlou, *Department of Electrical and Computer Engineering, Greece*
Christos Kyrkou, *University of Cyprus, Cyprus*
Ashis Kumar Mal, *NIT Durgapur, India*
Kailash Chandra Ray, *Indian Institute of Technology Patna, India*
Kishor Sarawadekar, *Indian Institute of Technology, India*
Nicolas Sklavos, *University of Patras, Greece*
Kunlin Tsai, *Tunghai University Taichung, Taiwan R.O.C, Taiwan*

SRF (Student Research Forum) Track Chair & Members
Track Chair
Susanta Chakraborty, *IIEST, Shibpur, India*

Navin Agrawal, *Oriental College of Technology, Bhopal, India*
Siddhartha Bhattarcharyya, *RCC IIIT, India*
Kavita Burse, *Oriental Group of Institutes, Bhopal, India*
Amey Kulkarni, *Velodyne LiDAR, Inc., USA*
Pralay Mitra, *Indian Institute of Technology, Kharagpur, India*
Bikromadittya Mondal, *BP Poddar Institute of Management and Technology, India*
Sarbani Palit, *Indian Statistical Institute Kolkata, India*
Shilpa Pendyala, *Intel, USA*
Sarbani Roy, *Jadavpur University, India*
Sanjit Kumar Setua, *University of Calcutta, India*

SSP (Special Sessions and Panels) Track Chairs & Members
Track Chair
Helen Li, *Duke University, USA*
Himanshu Thapliyal, *University of Kentucky, USA*

Vijay Holimath, *VividSparks IT Solution Pvt Ltd, India*
Vipul Mishra, *Bennett University, India*
Deepak Puthal, *University of Technology Sydney, Australia*
Fatemeh Tehranipoor, *San Francisco State University, USA*
UpasnaVishnoi, *Marvell Semiconductors, USA*
Dhruva Ghai, *Oriental University, India*
Saraju Mohanty, *University of North Texas, USA*
Sudeep Pasricha, *Colorado State University, USA*
Anirban Sengupta, *Indian Institute of Technology (IIT) Indore, India*

Steering Committee

Chair
Saraju P. Mohanty, *University of North Texas, USA*

Vice-Chair
Dhruva Ghai, *Oriental University, India*

Steering Committee Members
Aida Todri-Sanial, *CNRS-LIRMM, France*
Anirban Sengupta, *Indian Institute of Technology, Indore*
Ashok Srivastava, *Louisiana State University, USA*
Hai (Helen) Li, *Duke University, USA*
Himanshu Thapliyal, *University of Kentucky, USA*
Jia Di, *University of Arkansas, USA*
Nabanita Das, *Indian Statistical Institute, Kolkata, India*
Prasun Ghosal, *IIEST, Shibpur, India*
Sudeep Pasricha, *Colorado State University, USA*
Xin Li, *Duke University, USA*

Keynotes

ThirdEye: Visual Assist for Grocery Shopping
Dr. Vijay Krishnan Narayan
Department of Computer Science and Engineering
Pennsylvania State University

Abstract

Shopping is widely considered as a relaxing leisure activity. However, grocery shopping can be a frustrating experience for those with visual impairment. While getting to a grocery shop itself is not as much of a challenge for them, locating and picking the items in the grocery shelf becomes a task as challenging as picking a needle from the haystack. Imagine picking up five items for your dinner recipe from a typical grocery store in the US that carries around 35,000 unique items and can have more than 30 aisles spanning 45,000 square meters. This talk will showcase synergistic advances in algorithms, architectures and interface design for assisting those with visual impairment to do shopping.

Biography

Vijay Narayanan is a Distinguished Professor of Computer Science and Engineering and Electrical Engineering at The Pennsylvania State University. He is the director of the NSF Expeditions-in-Computing Program on Visual Cortex on Silicon and a thrust leader for the DARPA-MARCO LEAST Center. He has published more than 400 papers and won several awards in recognition of his research in power-aware systems, embedded systems and computer architecture. He is a fellow of IEEE and ACM.

Challenges of Converging Nanoelectronics and Nanotechnology for Internet of Things

Dr. Durgamadhab (Durga) Mishra

Professor & Associate Chair for Graduate Program, FELLOW of The Electrochemical Society
Department of Electrical and Computer Engineering, New Jersey Institute of Technology
Newark, NJ, USA

Abstract

Current trends in Internet of Things (IoT) require the convergence of Nanoelectronics, Nanotechnology, Communication Technology and Information Technology. Sensor systems monitoring environment, health care, water quality, vehicle traffic, smart cities are becoming the norm. Despite extended range of applications low power requirement is the key to these nanosystems. Incorporation of different nanodevices into these nanosystems with functionalities that do not necessarily scale according to "Moore's Law," but provide additional value in different ways (more than Moore), is necessary. It is therefore important to get exposed to the current trend in chip fabrication, device structures and fabrication (gate stack design and fabrication), device and circuit relationship and design, reliability of new devices and processes. Furthermore, nanoelectronic devices with extremely low power consumption depends on the next generation high-k deposition process, precise selection of deposition parameters, pre-deposition surface treatments and subsequent annealing temperatures. In this talk, some of the recent developments in device fabrication for electronics devices and IoT devices will be outlined.

Biography

Durga Misra is a Professor in the Department of Electrical and Computer Engineering, New Jersey Institute of Technology, Newark, USA. He served as the Director of Microelectronics Research Center at NJIT andhad a short-term appointment at Bell Laboratories, Murray Hill, NJ, in 1997. His current research interests are in the areas of nanoelectronic/optoelectronic devices and circuits; especially in the area of nanometer CMOS gate stacks and device reliability. He is currently a Distinguished Lecturer of IEEE Electron Devices Society (EDS) and serving in the IEE EDS Board of Governors. He is a Fellow of the Electrochemical Society (ECS) and served in the ECS Board as a Board Member (2008-10). He received the Thomas Collinan Award from the Dielectric Science & Technology Division and Electronic and Photonic Division Award from ECS. He edited and co-edited more than 40 books and conference proceedings in his field of research. He has published more than 95 technical articles in peer reviewed Journals and more than 160 articles in International Conference proceedings including 75 Invited Talks. He has graduated 15 PhD students and 35 MS students. He received the M.S. and Ph.D. degrees in electrical engineering from the University of Waterloo, Waterloo, ON, Canada.

Considerations for Designing Secure and Efficient Nanoelectronic Computer Architectures

Dr. Garrett S. Rose

Associate Professor, Department of Electrical Engineering and Computer Science,
The University of Tennessee, Knoxville, TN 37996-2250 USA

Abstract

In the integrated circuits industry today, electronic devices are being scaled to the point where feature sizes are on the order of tens or even a few nanometers such that defect rates are increased, leakage is non-negligible and quantum effects have begun to dominate. With so many potential issues facing conventional microelectronic technologies, novel approaches to circuit design and even non-classical devices warrant exploration. Add to this mix of challenges the reality of security vulnerabilities that must be addressed at early design stages such that emerging nanoelectronic systems must be both efficient and trustworthy. As a specific case study, this talk will focus on memristor based systems. Given their low power operation and small footprint, memristors have emerged as excellent candidates for future memory and logic. However, the non-volatility of memristors also presents certain security challenges whereby sensitive data may be vulnerable. At the same time, memristors also show promise for effective security primitives such as physical unclonable functions and random number generators.

In this talk we will consider the security pros and cons of nanoelectronic systems and also discuss design techniques that best balance security concerns with performance needs. In addition to addressing energy-efficiency and security in conventional systems, nanoelectronic technology should also be considered as a means of enabling truly novel forms of computer architectures. To this end, nano-enabled neuromorphic systems offer exciting opportunities for future computing applications. Again, as such novel technologies and novel approaches to computing continue to emerge, careful attention must be paid to balanging security issues against traditional energy performance metrics.

Biography

Garrett S. Rose received the B.S. degree in computer engineering from Virginia Polytechnic Institute and State University (Virginia Tech), Blacksburg, in 2001 and the M.S. and Ph.D. degrees in electrical engineering from the University of Virginia, Charlottesville, in 2003 and 2006, respectively. His Ph.D. dissertation was on the topic of circuit design methodologies for molecular electronic circuits and computing architectures.

Presently, he is an Associate Professor in the Department of Electrical Engineering and Computer Science at the University of Tennessee, Knoxville where his work is focused on research in the areas of nanoelectronic circuit design, neuromorphic computing and hardware security. Prior to that, from June 2011 to July 2014, he was with the Air Force Research Laboratory, Information Directorate, Rome, NY. From August 2006 to May 2011, he was an Assistant Professor in the Department of Electrical and Computer Engineering at the Polytechnic Institute of New York University, Brooklyn, NY. From May 2004 to August 2005 he was with the MITRE Corporation, McLean, VA, involved in the design and simulation of nanoscale circuits and systems. His research interests include low-power circuits, system-on-chip design, trusted hardware, and developing VLSI design methodologies for novel nanoelectronic technologies.

Dr. Rose is a member of the Association of Computing Machinery, IEEE Circuits and Systems Society and IEEE Computer Society. He serves and has served on Technical Program Committees for several IEEE conferences (including ISVLSI, GLSVLSI, NANOARCH) and workshops in the area of VLSI design. In 2010, he was a guest editor for a special issue of the *ACM Journal of Emerging Technologies in Computing Systems* that presented key papers from the *IEEE/ACM International Symposium on Nanoscale Architectures (NANOARCH'09)*. From April 2014 through March 2017 he was an associate editor for *IEEE Transactions on Nanotechnology*.

QUADSEAL: A Hardware Countermeasure against Side channel Attacks on AES

Dr. Sri Parameswaran
Professor & Program Director for Computer Engineering
School of Computer Science and Engineering
The University of New South Wales, Australia

Abstract

Deep devastation is felt when privacy is breached, personal information is lost, or property is stolen. Now imagine when all of this happens at once, and the victim is unaware of its occurrence until much later. This is the reality, as increasing amount of electronic devices are used as keys, wallets and files. Security attacks targeting embedded systems illegally gain access to information or destroy information. Advanced Encryption Standard (AES) is used to protect many of these embedded systems. While mathematically shown to be quite secure, it is now well known that AES circuits and software implementations are vulnerable to side channel attacks. Side-channel attacks are performed by observing properties of the system (such as power consumption, electromagnetic emission, etc.) while the system performs cryptographic operations. In this talk, differing power based attacks are described, and various countermeasures are explained. In particular, a countermeasure titled Algorithmic Balancing is described in detail. Implementation of this countermeasure in hardware and software is described. Since process variation impairs countermeasures, we show how this countermeasure can be made to overcome process variations.

Biography

Sri Parameswaran is a Professor in the School of Computer Science and Engineering at the University of New South Wales. He also serves as the Postgraduate Research and Scholarships coordinator at the same school. Prof. Parameswaran received his B. Eng. Degree from Monash University and his Ph.D. from the University of Queensland in Australia. He has held visiting appointments at University of California, Kyushu University and Australian National University. He has also worked as a consultant to the NEC Research laboratories at Princeton, USA and to the Asian Development Bank in Philippines. His research interests are in System Level Synthesis, Low power systems, High Level Systems, Network on Chips and Secure and Reliable Processor Architectures. He is the Editor-in-Chief of *IEEE Embedded Systems Letters*. He serves or has served on the editorial boards of *IEEE Transactions on Computer Aided Design, ACM Transactions on Embedded Computing Systems,* the *EURASIP Journal on Embedded Systems* and the *Design Automation of Embedded Systems*. He has served on the Program Committees of Design Automation Conference (DAC), Design and Test in Europe (DATE), the International Conference on Computer Aided Design (ICCAD), the International Conference on Hardware/Software Codesign and

System Synthesis (CODES-ISSS), and the International Conference on Compilers, Architectures and Synthesis for Embedded Systems (CASES).

Reduced Dimension-based Emerging Novel Switching Transistors and Interconnects for Post-CMOS Electronics

Dr. Ashok Srivastava

Wilbur D. and Camille V. Fugler, Jr., Professor of Engineering, Professor of Electrical & Computer Engineering,
School of Electrical Engineering & Computer Science
Louisiana State University, Baton Rouge, LA 70803

Abstract

Focus of this presentation will be on reduced dimension materials such as carbon nanotubes, graphene and other than graphene-based emerging novel switching transistors and interconnects for future integrated circuit design and wide-ranging other applications. Addressing the needs of post-CMOS electronics, development of current transport models of transistors based on these materials will be presented which can be used for the design of ultra-low power and high frequency nanoscale integrated electronic circuits. Use of carbon nanotubes, graphene and graphene-copper hybrid material as a possible solution for the replacement of copper interconnect in nanometer CMOS technology nodes will also be included in presentation.

Biography

Dr. Ashok Srivastava obtained M. Tech. and Ph.D. degrees in Solid State Physics and Semiconductor Electronics area from Indian Institute of Technology, Delhi in 1970 and 1975, respectively. He joined the Department of Electrical & Computer Engineering of Louisiana State University, Baton Rouge in 1990 and is Wilbur D. and Camille V. Fugler, Jr., Professor of Engineering in the School of Electrical Engineering & Computer Science. In year 2011, he held visiting appointments at the Institute of Electrical Engineering NanoLab, Swiss Federal Institute of Technology (EPFL), Lausanne, Switzerland; Katholiek Universiteit/Inter-university Microelectronics Center (IMEC), Leuven, Belgium; Indian Institute of Information Technology (IIIT), Allahabad; and in year 2001 at the Philips Research Laboratory, Eindhoven, The Netherlands. His other past appointments include Central Electronics Engineering Research Institute, Pilani, India (1975-84); Birla Institute of Technology and Science, Pilani, India (1975); North Carolina State University, Raleigh (1985-86); State University of New York, New Paltz (1986-90); University of Cincinnati, Cincinnati (1979); University of Arizona, Tucson (1979-80); Kirtland Air Force Base, New Mexico (Summer 1996); and Jet Propulsion Laboratory/California Institute of Technology, Pasadena (Summer 2004).

Time in Cyber-Physical Systems

Dr. Aviral Shrivastava
Associate professor, School of Computing, informatics, and Decision Systems Engineering,
School of Electrical, Computer and Energy Engineering
Arizona State University, USA

Abstract

Cyber-Physical systems are those that tightly integrate physical and computational systems. One of the big challenges in distributed cyber-physical systems is establishing a common notion of time between the physical world and the computational system. Many modern CPS, especially industrial automation systems, require the actions of different computational systems to be synchronized at much higher rates than is possible through ad hoc designs. Fundamental research is needed in synchronizing clocks of computing systems to a higher degree, and even if the clocks are synchronized, designing CPS nodes so that they can perform actions in a synchronized manner is challenging. We need to find ways to specify distributed CPS applications, ways to specify and verify timing requirements on distributed CPS, confident top-down design methodologies that can ensure the system meets its timing requirements in the first go, dynamically creating and dissolving timing domains using differently build components, and much more.

In this talk, I will present some of the work that we have done, and some of the ideas that we want to pursue in order to solve the challenge of confident and simplified CPS design (from the timing perspective). We believe that confident CPS design is possible only when the timing requirements of CPS are specified in the application itself, and not as a separate document. It should not be a list of separate requirements, but must be married to the application specification in as natural way as possible. Second, we need techniques to design the CPS in one-shot. Provably correct by construction is very good, but even design methodologies that improve the confidence in design are also very important. Finally, there should be automated methods to test the CPS.

Biography

Prof. Aviral Shrivastava is Associate Professor in the School of Computing Informatics and Decision Systems Engineering at the Arizona State University, where he has established and heads the Compiler and Microarchitecture Labs (CML) (http://aviral.lab.asu.edu/). He received his Ph.D. and Masters in Information and Computer Science from University of California, Irvine, and bachelors in Computer Science and Engineering from Indian Institute of Technology, Delhi. He is a 2011 NSF CAREER Award

Recipient, and recipient of 2012 Outstanding Junior Researcher in CSE at ASU. His 2 students have received the outstanding MS thesis award in CSE at ASU. His papers have been the best paper candidate at DAC 2017, ASPDAC 2008, and won the best student paper award at VLSI 2016. His research lies at the intersection of compilers and architectures of embedded and multi-core systems, with the goal of improving power, predictability, performance, temperature, energy, reliability and robustness. NSF and several industries including Microsoft, Raytheon Missile Systems, Intel, Nvidia, etc fund his research. He serves on organizing and program committees of several premier embedded system conferences, including DAC, ICCAD, ISLPED, CODES+ISSS, EMSOFT, CASES and LCTES, and regularly serves on NSF and DOE review panels.

Design Optimization of DSP for Wearable Biomedical Device

M.Naga Sasikanth[1], Sashank Gambhira[2], Mrigank Sharad[3]

Department of Electronics and Electrical Communication engineering

IIT Kharagpur

Kharagpur, West Bengal-721302, India

[1]kanthsasi66@gmail.com, [2]sashankgambhira@g.ucla.edu, [3]mrigank@ece.iitkgp.ernet.in

Abstract—Recent trends towards low power and miniaturized wearable healthcare devices require careful selection of computing modules to be integrated. On-device computing can help extract and transmit only the essential features of the target biomedical signals, thereby leading to reduced power consumption due to wireless communication. However, prior to such information extraction, tackling signal non-idealities, like noise, offset and motion-artifact becomes crucial. While, noise and offset considerations may be addressed in the analog domain, the residual motion artifact often mandates digital domain processing. In this work, we present the system design for photo-plethysmographic (PPG) signal acquisition and processing, where low power custom DSP is designed to minimize the motion artifact (MA), prior to on-device feature extraction and classification. In order to retain the benefits of on-device information extraction, the DSP of MA-removal must be power constrained. We propose a computationally efficient method for MA removal that employs a low power noise-estimation block and adaptive filter for fast MA tracking and removal (MAR). Our proposed algorithm has been implemented in Verilog and synthesized using synopsys in 180nm CMOS technology. This custom DSP draws 750nW of power from a supply of 2V.

keywords—PPG, Photoplethysmogram, Adaptive filter, Motion Artifact removal, Fundamental frequency estimation, Notch filter, DSP, Motion artifact.

I. INTRODUCTION

Wearable devices are becoming increasingly popular to monitor vital signals for health care applications. With smart phones becoming more and more accessible everyday, it can be paired up with a reliable wearable technology for remote monitoring of patient. Through recent advances in VLSI miniaturization, we have fostered the growth of wearable devices, but the technology needs further improvement in terms of battery life while not compromising on reliability. Susceptibility of sensors present in the wearable devices to motion artifacts and other forms of noise often affect the reliability of these devices and could restrict their utility. In order to improve the signal integrity, substantial research has been going on in the recent years and newer methods are constantly being explored to revolutionize the wearable health device sector. Through this paper, we address the same issue and present a novel method for motion artifact removal in PPG signals

Pulse oximeter is commonly used for measuring the PPG signal. Light illuminated by pulse oximeter is reflected back from the skin and the changes in absorption is measured

as PPG signal. PPG signal measured from pulse oximeters essentially have two constituents; one due to continuous absorption into the skin which represents the DC part of the signal and other due to pulsating arteries pumping blood as a consequence of heart beat which represents AC part of signal. The integration of pulse oximeter sensors into wearable devices make PPG signal a strong contender for health care applications. However the downside of using PPG signal is its vulnerability in presence of motion artifacts. Wearable bio-devices are predominantly used by people during physical activity, whose movements significantly affect the signal measured from pulse oximeter. Overcoming motion artifacts has always been a very challenging issue and attracted the interest of several research groups to design novel and more efficient methods to tackle the issue. Moving Average method as utilized in [10] can be seen as one of the traditional, commonly used methods. However, this method is limited by artifact range. Although moving average filter works adequately in low noise scenarios, it is effected to a great degree in presence of sudden peaking of noise. As a solution to this issue, adaptive filters were later proposed considering their efficacy in dealing with the in-band noise. However, these methods necessitate a reference signal which needs to be uncorrelated with one signal(signal/artifact) while being strongly correlated with the another(artifact/signal). Several ideas using adaptive filters demand additional hardware like accelerometers[11][12] or additional transducers. Exploiting the non stationary of PPG signals led to researchers proposing methods involving Wavelet transform [14] and Wigner-Ville distribution[15] applications. Other methods include [5] in which ICA, SVD and FFT are used to obtain noise references but hardware realization is expensive. In [1], two stage adaptive noise cancellation is introduced. Newer methods [6],[8] have been based on estimating the fundamental frequency, using this data produce reference signal followed by an adaptive filter to get a clean signal. These algorithms have succeeded in improving the performance but usage of autocorrelation of input data to compute fundamental frequency involves more hardware and consume higher power during operation. Our focus, thus, has been to design a computationally efficient, low power system to generate a cleaner PPG signal.

We have proposed a newer and computationally efficient method to estimate the fundamental frequency. This, followed

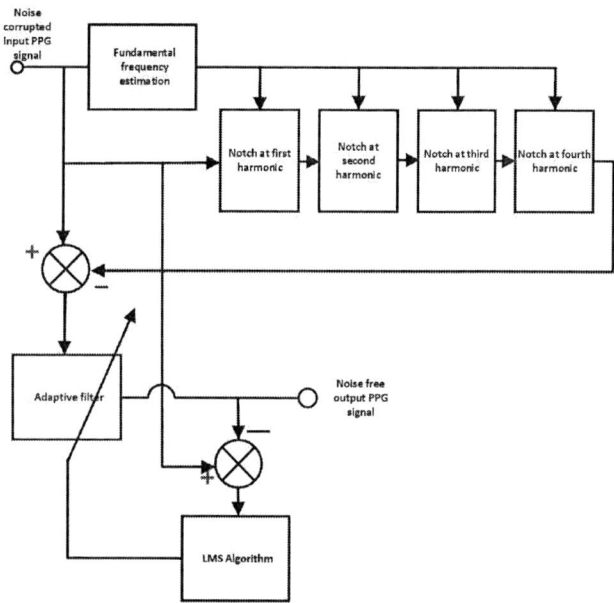

Fig. 1. Proposed System

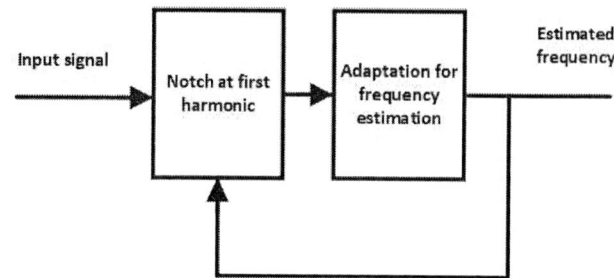

Fig. 2. Adaptive notch for fundamental frequency estimation

by a cascaded notch filter stage and LMS adaptive filtering stage complete the design and the system efficiently generates PPG signal without motion artifacts. The rest of the paper is organized as follows. In Part A of section-II we propose a novel method for fundamental frequency estimation of PPG signal corrupted with noise. In part B we explain use of cascaded notch filter for reference noise generation. In part C adaptive filter for motion artifact removal is explained. Section-III contains Results of our proposed algorithm. Section-IV discusses Hardware Implementation of proposed algorithm and Section-V provides a conclusion.

II. PROPOSED METHOD

The algorithm we proposed involves three stages in essence as shown in fig.1.

A. Fundamental Frequency Estimation

1) Adaptive LMS Algorithm for Fundamental frequency detection:: With emphasis on reducing the hardware requirement for frequency detection, we have employed a LMS algorithm based approach for frequency detection to alter the notch frequency according to input as shown in fig.2.

Algorithm: For a notch filter at a frequency of f_c and sampling frequency of f_s input and output are related by

$$\frac{Y(z)}{X(z)} = \frac{1 - 2cos(\theta)z^{-1} + z^{-2}}{1 - 2rcos(\theta)z^{-1} + z^{-2}}$$

Where $\theta = 2\pi f_c/f_s$.
Now y(n) and x(n) are related as

$$y(n) = x(n) - 2cos(\theta)x(n-1) + x(n-2) + 2rcos(\theta)y(n-1) - y(n-2)$$

Now using LMS algorithm to vary the value of theta

$$\theta(n+1) = \theta(n) - \mu\frac{\partial|e(n)|^2}{\partial\theta}$$

Where e(n) is same as y(n), as our objective is to minimize the output by removing the fundamental component. On simplification, we obtain

$$\frac{\partial e(n)}{\partial \theta} = -2rsin(\theta)y(n-1) + 2sin(\theta)x(n-1)$$

The final expression for update equation can be re-written as

$$\theta(n+1) = \theta(n) - 2\mu(-2rsin(\theta)y(n-1) + 2sin(\theta)x(n-1))y(n).$$

The value of step size for LMS algorithm is kept as 5e-3 and the value of r of notch is taken as 0.99 for a sharp filter.

From the above frequency estimation stage, $f_c = f_s * \theta/2\pi$ is used as input to cascaded notch filter for reference noise generation.

B. Reference PPG Generation

Quasi periodic nature of PPG signal makes it's energy mainly concentrated at fundamental frequency and it's harmonics. Hence comb filter or notch filter can be used to generate noise reference signal by eliminating it's harmonics. But comb filter of the form $(1 - z^{-p})/2$ or $(1 - z^{-p})/2$ $(1 - az^{-p})/2$ where $p = f_s/f_0$, f_s = sampling frequency, f_0 = fundamental frequency of PPG signal, this can be hardware realizable only if p is an integer, hence we used a notch filter for reference noise generation. From the spectrum (fig.8,9) of PPG signal it can be observed that a large part of it's energy is contained in it's first four harmonics. So we used a cascaded 4 stage notch filter for generation of reference noise signal, where each notch filter is of the form,

$$H(z) = \frac{1 - 2cos(\theta)z^{-1} + z^{-2}}{1 - 2rcos(\theta)z^{-1} + z^{-2}}$$

The value of θ is obtained from fundamental frequency estimation stage each of these notch filters operating at center frequencies of f_0, $2f_0$, $3f_0$, $4f_0$. The signal obtained from the cascaded The value of r used in this stage is 0.99 for each of the four notch filters in cascade This signal is used in generation of reference PPG signal for adaptive stage, which refines the input PPG signal thereby making it nearly free of motion artifacts.

978-1-5386-1357-3/17 $31.00 © 2017 IEEE

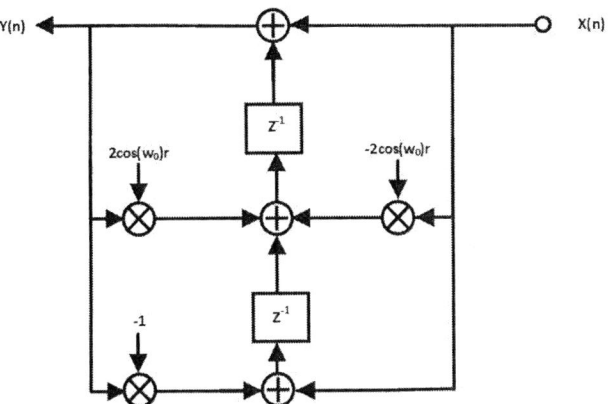

Fig. 3. Notch filter at w_0

C. Adaptive Filtering Stage

The modified PPG signal from cascaded notch filter stage is used as input to adaptive filter. s(n) corresponds to PPG signals and n_0(n) corresponds to motion artifact corrupted with the PPG signal. The output from four stage notch filter is subtracted from s(n) to get a reference PPG signal that is

$$s_{ref}(n) = s(n) - y(n)$$

where y(n) is the output of 4 stage notch filter. s(n)+n_0(n) denotes motion corrupted PPG signal, s_{ref}(n) denotes input to adaptive filter and s_{out}(n) denotes refined output PPG signal generated by adaptive filter. If we assume that motion artifact is uncorrelated to PPG signal, the energy of error signal e(n) from adaptive stage can be written as

$$e(n) = (s(n) + n_0(n)) - s_{out}(n)$$

$$E(e(n)^2) = E((s(n) + n_0(n) - s_{out}(n))^2)$$

$$E(e(n)^2) = E(n_0(n)^2) + E((s(n) - s_{out}(n))^2)$$

Hence, if motion artifact is uncorrelated to input signal, the adaptive filter makes error as low as possible, thus making $s_{out}(n)$ = s(n) eventually.

Equations for LMS algorithm:

w corresponding to the weights of FIR filter is initialized as p sized zero row vector.The equations for LMS algorithm are shown below.

$$x = s_{ref}$$

$$X = [x(i)x(i-1)....x(i-p+1)]^T$$

$$e(n) = s(n) + n_0(n) - s_{out}(n)$$

$$s_{out}(n) = w^T X$$

$$w(n+1) = w(n) + 2\mu e(n)X$$

μ = step size of LMS algorithm For the above algorithm the size of filter(p) is 10.

III. RESULTS

The proposed algorithm was simulated for various input data from MIT-BIH database in physionet.org. The sampling frequency of input data from the database is 125Hz and dataset we have tested is of 1hr duration which corresponds to 450000 samples. The PPG signal from a dataset in MIT-BHU database is shown in Fig.4. An example for Noisy portion in a data signal from MIT-BHU database is shown in Fig.5. Fig.6 shows the output from our motion artifact removal algorithm while Fig.7 shows reference noise which essentially is the motion artifact component. Different types of input signals considered for testing the robustness of our algorithm include ones with high input motion artifact noise as in Fig.8. and the one with low input motion artifact noise as in Fig.9. The first image in each of the two result set corresponds to spectrum of input signal (FFT), the second image corresponds to spectrum of output signal (FFT) and the third image corresponds to Fundamental Frequency estimated from Fundamental Frequency estimator algorithm we proposed. The proposed algorithm for fundamental frequency estimation is robust as seen in each of the figure. It can be seen that the output spectrum has significantly improved in comparison to input spectrum indicating that the output is free from motion artifact components corresponding to low frequency for each of the two considered cases.

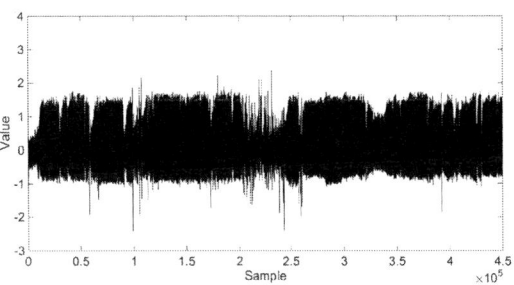

Fig. 4. PPG Signal From a database

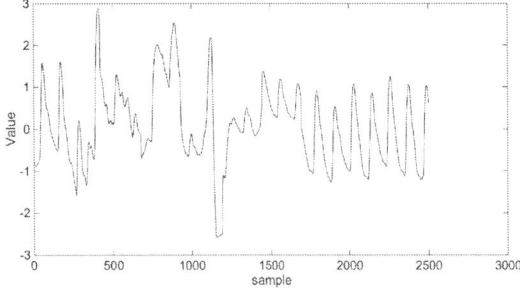

Fig. 5. PPG signal with noise

IV. HARDWARE IMPLEMENTATION

The algorithm has been implemented in verilog and synthesized in 180nm CMOS technology using synopsys design

978-1-5386-1357-3/17 $31.00 © 2017 IEEE

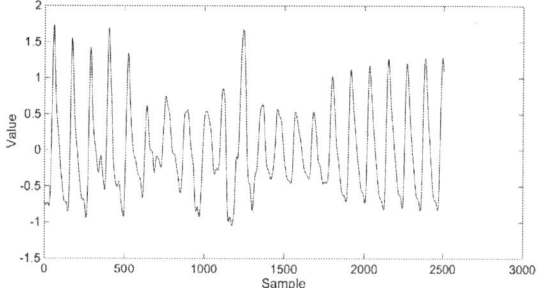

Fig. 6. Output PPG signal for the above signal

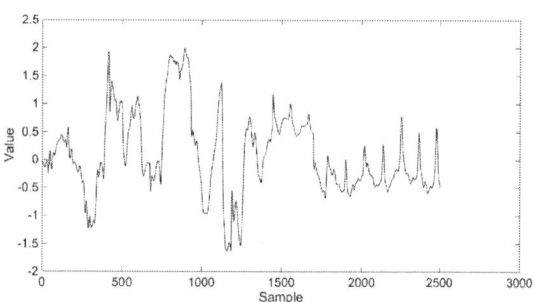

Fig. 7. Reference noise generated by system

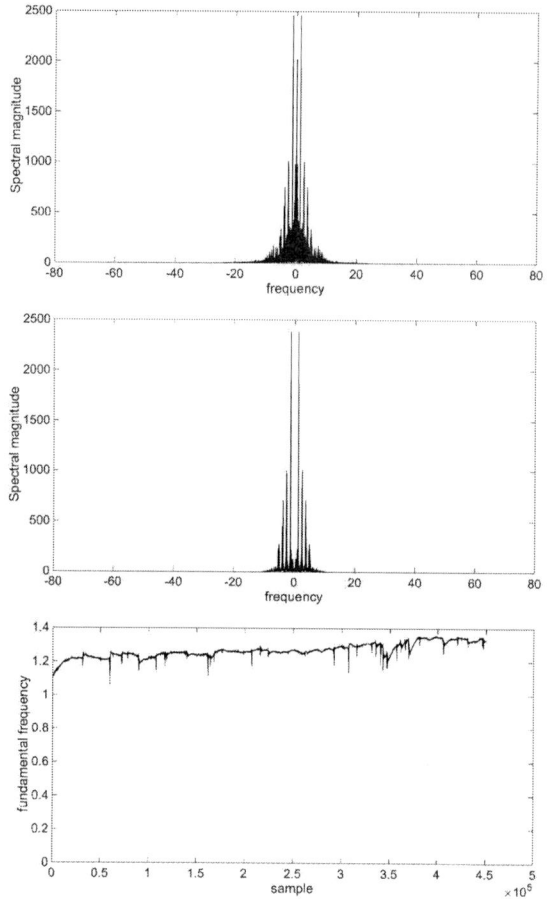

Fig. 8. FFT of input, FFT of output, Fundamental Frequency estimated for high noise data

compiler. The synthesized design occupies an area of $0.8mm^2$ and showed superior results when compared to [6] in terms of power. Table - I compares power consumption (from supply of 2V and sample frequency of 500Hz) of our design with [6], which similarly did not use any additional hardware (accelerator) for reference noise. We have synthesized the algorithm in [6] which uses autocorrelation based frequency estimation with a window size of 20 (much lesser than 2000 that is suggested in reference). As it can be seen our design consumes lesser power in comparison to [6] even with a window size of 20. This is due to the requirement of large number of multipliers for realizing autocorrelation operation in [6].

For the hardware design, cos and sin in notch filter and fundamental frequency estimation blocks are realized through cordic blocks[9]. Adaptive filter is realized as direct form FIR filter. Notch and fundamental frequency estimation blocks are realized as direct form-I IIR filters shown in fig.3 .

V. CONCLUSION

A computationally efficient method has been proposed for motion artifact removal in PPG signals which includes a fundamental frequency detector that serves as input to 4-stage notch filter. This 4 stage notch is used to produce reference noise for adaptive filter. Adaptive filter which is based on LMS algorithm, produces a refined PPG signal free of motion artifact. When compared to autocorrelation based method [6],[8] that require a large number of multiply operations (more power), our design is less power hungry.

TABLE I
POWER COMPARISON BETWEEN REF[6] AND THIS WORK

Ref	Dynamic Power	Leakage Power	Total Power
[6]	$5.6072\mu W$	$0.276\mu W$	$5.8824\mu W$
This work	$0.7509\mu W$	$0.0996\mu W$	$0.85058\mu W$

REFERENCES

[1] H. Kim, S. Kim, N. Van Helleputte, T. Berset, R. Di Geng, I. Romero, J. Penders, C. Van Hoof, R. F. Yazicioglu, "Motion artifact removal using cascade adaptive filtering for ambulatory ECG monitoring system", Proc IEEE Biomedical Circuits and Systems Conf., pp. 160-163, 2012-Nov.-2830. 2

[2] Ram, M.R.; Madhav, K.V.; Krishna, E.H.; Komalla, N.R.; Reddy, K.A. "A novel approach for artifact reduction in PPG signals based on AS-LMS adaptive filter", IEEE Instrum. Meas. 2012, 61, 14451457. 3

[3] M. A. F. M. R. Hasan and T. Shimamura, "A fundamental frequency extraction method based on windowless and normalized autocorrelation functions", in Proc. 6th WSEAS Int. Conf. Circuits, Systems, Signal and Telecommunications, Cambridge, 2012, pp. 305-309. 4

[4] Wood LB, Asada HH, "Low Variance Adaptive Filter for Cancelling Motion Artifact in Wearable Photoplethysmogram Sensor Signals", Engineering in Medicine and Biology Society, 2007 EMBS 2007 29th Annual International Conference of the IEEE; 22-26 Aug. 2007 2007, 652-655. 5

978-1-5386-1357-3/17 $31.00 © 2017 IEEE

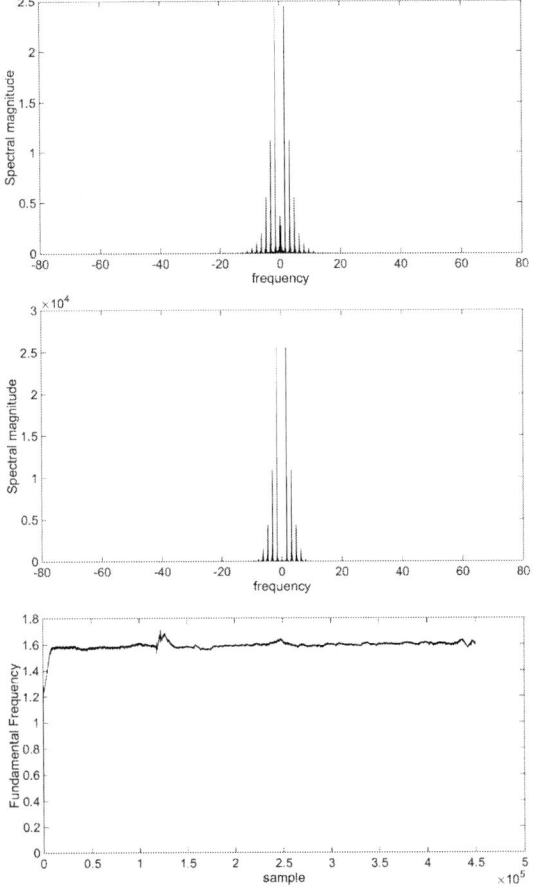

Fig. 9. FFT of input, FFT of output, Fundamental Frequency estimated for low noise data

Conf EMBSIBMES, Houston, USA, Oct. 23-26, 2002, pp. 1769-1770. 13

[13] P. F. Stetson, "Independent component analysis of pulse oximetry signals", Proc. 26th Annu. Int. Conf. IEEE Eng. Med. Biol. Soc., pp. 231-234, 2004-Sep.-15. 14

[14] C. M. Lee, Y. T. Zhang, "Reduction of motion artifacts from photo-plethysmographic recordings using a wavelet denoising approach", Proc. IEEE EMBS Asian-Pacific Conf. Biomed. Eng., pp. 194-195, 2003. 15

[15] Y. sheng Yan, C. C. Poon, Y. ting Zhang, "Reduction of motion artifact in pulse oximetry by smoothed pseudo Wigner-Ville distribution", J. NeuroEng. Rehab., vol. 2, no. 3, 2005. 16

[16] Ying He et al., "The Applications and Simulation of Adaptive Filter in Noise Canceling", International Conference on Computer Science and Software Engineering, 2008.

[5] Ram, M.R.; Madhav, K.V.; Krishna, E.H.; Komalla, N.R.; Reddy, K.A. " A novel approach for artifact reduction in PPG signals based on AS-LMS adaptive filter" . IEEE Instrum. Meas. 2012, 61, 14451457. 6

[6] R. Yousefi, M. Nourani, S. Ostadabbas and I. Panahi, "A Motion-Tolerant Adaptive Algorithm for Wearable Photoplethysmographic Biosensors", IEEE Journal of Biomedical and Health Informatics, vol. 18, no. 2, pp. 670-681, 2014. 7

[7] M. R. Ram, K. V. Madhav, E. H. Krishna, K. N. Reddy, K. A. Reddy, "Adaptive reduction of motion artifacts from PPG signals using a synthetic noise reference signal", Proc. IEEE EMBS Conf. Biomed. Eng. Sci., pp. 315-319, 2010-Nov./Dec. 8

[8] Rasoul Yousefi, Mehrdad Nourani, Issa Panahi, "Adaptive cancellation of motion artifact in wearable biosensors", Engineering in Medicine and Biology Society (EMBC) 2012 annual International Conference of the IEEE, pp. 2004-2008, 2012. 9

[9] K. Maharatna, S. Banerjee, E. Grass, M. Krstic, A. Troya, "Modified virtually scaling-free adaptive CORDIC rotator algorithm and architecture", IEEE Trans. Circuits Syst. Video Technol., vol. 11, no. 11, pp. 1463-1474, Nov. 2005. 10

[10] H. W. Lee, J. W. Lee, W. G. Jung, G. K. Lee, "The periodic moving average filter for removing motion artifacts from PPG signals", Int. J. Control Autom. Syst., vol. 5, pp. 701-706, Dec. 2007. 11

[11] A. B. Barreto, L. M. Vicente and I. K. Persad, " Adaptive Cancellation of Motion Artifact in Photoplethysmographic Blood Volume Pulse Measurements for Exercise Evaluation," in Proc IEEE-EMBC/CMBEC, Sept. 20-23,1995, vol. 2, pp. 983-984. 12

[12] A. R. Relente and L. G. Sison, "Characterization and adaptive filtering of motion artifacts in pulse oximetry using accelerometers ", in Proc.

Embedded Hardware Prototype for Gas Detection and Monitoring System in Android Mobile Platform

Rahul Kurzekar and Hardik Arora
Center for VLSI and Embedded System Technologies
IIIT-Hyderabad
Hyderabad, India
e-mail: rahul.bhagwan, hardik.arora@students.iiit.ac.in

Rahul Shrestha, *Member, IEEE*
School of Computing and Electrical Engineering
IIT-Mandi
Mandi, India
email: rahul_shrestha@iitmandi.ac.in

Abstract—**This paper presents the hardware implementation of liquefied petroleum gas (LPG) detection and monitoring system using microcontroller and android application. This system can effectively monitor the level of LPG in a room and transmits the data to the mobile device through 2.4 GHz ZigBee based bluetooth module. The MQ-6 LPG sensor is used for monitoring the LPG level and it transfers the data to ATmega16 microcontroller. It forwards this data to the mobile device (smartphone) of the user via HC-05 bluetooth module. The smartphone is loaded with an android application which has been specifically designed for this purpose. It displays the gas level in real time and additionally provides with the options to control the exhaust fan and electrical switch.**

Keywords- LPG, ATmega16, Bluetooth, Android.

I.   INTRODUCTION

In today's age, where everything is digitized, human safety has become a prime objective. Every now and then, we see incidents of explosions and fires in homes as well as in laboratories. Most of the time these accidents are due to combustible gases such as LPG, Butane and Methane. In our domestic environment such as house, car, storage as well as in laboratories, use of LPG is very important. So to ensure human safety, we need a system that can constantly monitor the LPG levels in these environments and provide an easy interface to control it in case if there is any leakage.

As smartphones have become an integral part of our lives, it is not only fitting but also convenient to use this technology for such a monitoring purpose. The current smartphones are embedded with various technological features that can ease our purpose and the android system enables us to use these features with great ease.

In the proposed system, we are using an MQ-6 gas sensor to monitor the gas level. The user can monitor these levels using an android application. This application interacts with the controller through HC-05 Bluetooth module. If there is any LPG leakage in the environment, the sensor will send elevated level data to the user. On receiving these, the user can then switch ON the exhaust to get the LPG out of the room. Also to a avoid sparks through electrical switches, that can lead to ignition of the LPG gas, the user can also turn OFF the electricity coming into the room through the main line .The exhaust, however, will have to be connected to a different electrical line.

So as we can see, the system not only provides a simple way to monitor the real-time data, but it also converts the smartphone into a remote controller that can control various appliances.

II.   PRINCIPLE OF OPERATION

The structure of the system can be easily understood with the help of the block diagram shown in Figure 1.

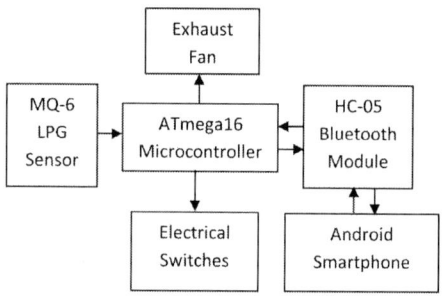

Figure 1. Block Diagram of LPG Detection and Monitoring System

As we can see, the heart of the system is the microcontroller. The sensor, exhaust fan, relay switch, and the Bluetooth module are all connected to the microcontroller. The sensor we are using is an MQ6 gas sensor. It can detect LPG, Butane, Methane and also alcohol and smoke. Inside its cylindrical chamber, there is a SnO_2 which reacts with the combustible compounds present in these gases. So as soon as one of these gases comes into contact with this material, its conductivity changes. Higher the level of the gas in air, higher will be the change in conductivity. In this way, the MQ-6 sensor will detect the LPG level in the room and generate a corresponding voltage at its output.

The controller used here is the ATmega16 microcontroller. It has a built-in ADC and USART module. The ADC module will convert the sensor voltage into corresponding digital value. The ADC module inside ATmega16 is a 10 bit ADC. So the converted value will be in the range of 0-1023. Now this data is then sent to the Bluetooth module HC-05 using the USART of ATmega16. The communication takes place serially with the help of two wires viz. Rx and Tx. Rx of the controller is connected to the

Tx of the HC-05. Similarly, Tx of the controller is connected to the Rx of the HC-05.

The Bluetooth module HC-05 uses RF channel to communicate serially with other Bluetooth devices. It receives the data sent by the controller through Rx pin and then transmits it serially over the RF channel. Of course, a Bluetooth device needs to be paired and connected to the HC-05 if it wants to receive or send any data to it. The HC-05 can be used in both master and slave mode. For our purpose, we have kept it in slave mode always. The user part of the system is the android application. This application can be loaded on any android based smart phone having Bluetooth feature. The application is very simple and the interface is very easy to use. It allows the user to search for the Bluetooth module. Once we select the desired Bluetooth module, it continuously displays the gas level sensed by the sensor. As stated above, the application allows us to search and connect with different Bluetooth devices. So we can have multiple LPG detectors set up across different environments, such as house and car, and still be able to monitor them using a single smartphone.

Apart from receiving and displaying the gas levels, the application also provides few buttons. When these buttons are pressed, the application sends some predefined strings to the Bluetooth module. The Bluetooth module passes these strings to the controller through Tx pin. The controller decodes these strings and carries out the predetermined functions. Thus by using the buttons on the application, we can control the exhaust fan and mains switch.

III. SYSTEM HARDWARE

A. LPG sensor

Figure 2. MQ-6 gas sensor used for LPG detection

Features:
- Highly sensitive to LPG, Butane, Methane
- Can also detect alcohol and smoke
- Robust structure and long life
- Simple drive circuit
- Works on 5V supply

Figure 3 shows the sensitivity list for MQ-6 gas sensor. As we can see, it is suitable for a wide range of LPG concentration. Furthermore, being cheap makes it to be more suitable for low cost system development.

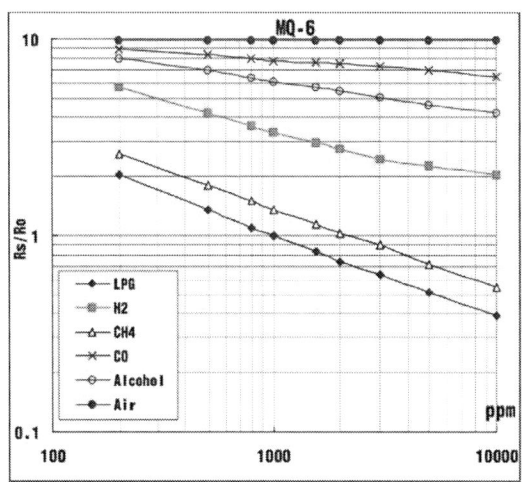

Figure 3. MQ-6 Sensitivity Graph

B. ATmega16 microcontroller

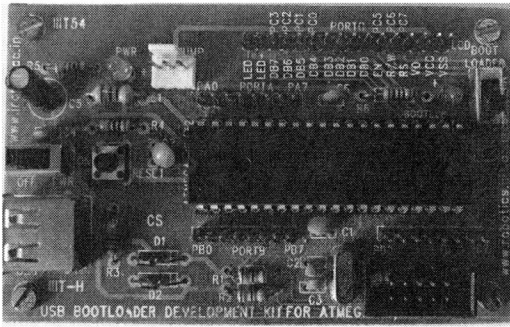

Figure 4. ATmega16 embedded platform used for LPG detection

Features:
- Works on 5V. Suitable for low power application.
- Built-in ADC module
- Multiplexed ADC pins for multiple sensors.
- Supports USART communication
- Configurable BAUD rate
- 16 KB of Flash memory
- Supports internal as well as external oscillators

C. HC-05 Bluetooth module

Figure 5. HC-05 Bluetooth Module

Features:

- Works on 3.3V. Suitable for low power application
- GFSK modulation with 2.4GHz frequency
- Configurable BAUD rate
- Supports both master and slave mode
- Can connect to multiple devices.
- Configurable pairing key provides a level of security
- Can remember the device address so as to deny unwanted connections

D. Android based Smart Phone

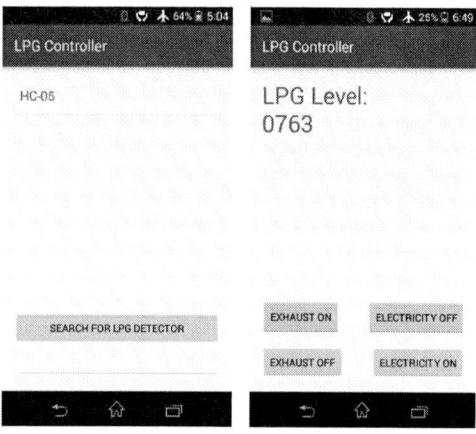

Figure 6. Android Application Interface

Features:

- Embedded Bluetooth module
- Android framework 4.3 or higher
- Simple and user friendly GUI interface
- Can connect to multiple LPG detectors

IV. SYSTEM SOFTWARE

A. Microcontroller program

The ATmega16 has 16KB of flash memory. So we can store a program in it that will be read each time the system is reset. For the proposed system, we have used embedded C as the programming language and Atmel Studio 6.0 as the IDE tool. To access the specific features of ATmega16 such as ADC and USART, certain registers have to be configured. Each module has dedicated register. By reading and writing them, we can use these features of the controller. e.g. To use ADC module, ADMUX, ADC and ADCSRA registers are used. To use USART module, UCSRC, UCSRB, UCSRA, UBBR , UDR registers are used. Some of them serve as input or output buffers while others set the mode of operation like baud rate, frame size, parity bits, conversion depth, etc.

The general program flow of the microcontroller operation can be explained from the flowchart in Figure 7.

B. Android Application

The android application is the user part of the system. It is a GUI based application that provides a simple and easy to use interface. The app is designed in JAVA using the Android Studio as the IDE tool. It uses several android libraries that provide the functions to access the hardware of the smartphone. The application is designed to work on the android framework of 4.3 or higher.

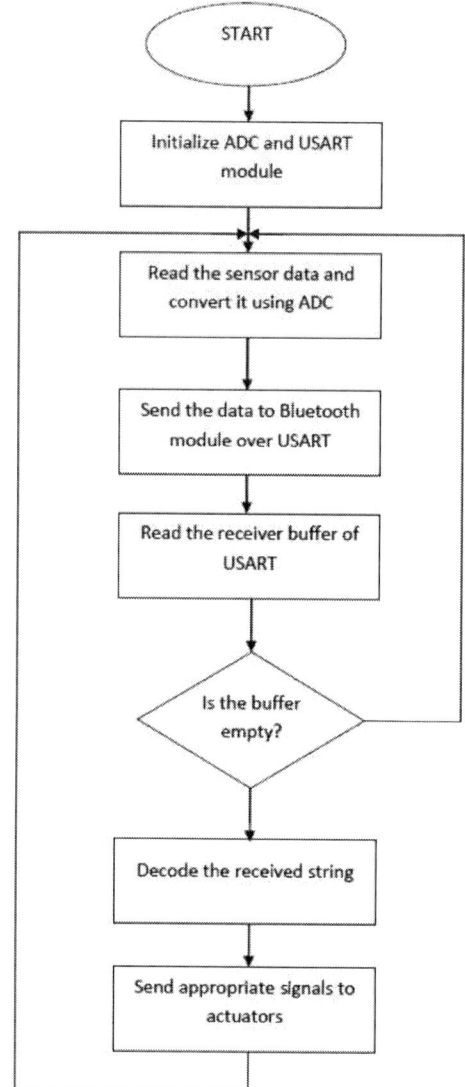

Figure 7. Flowchart of Microcontroller Operation

978-1-5386-1357-3/17 $31.00 © 2017 IEEE

V. TEST AND RESULTS

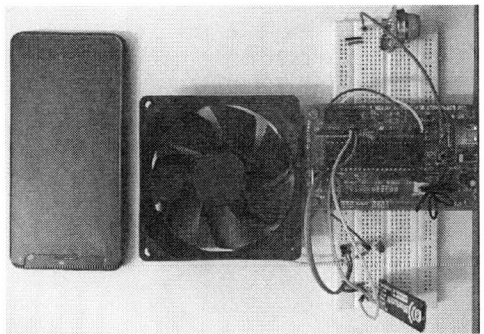

Figure 8. Final System

The final system can be seen in Figure 8. The sensor and the Bluetooth module are tested separately before assembling the final system. The MQ-6 sensor does not have a linear change in the conductivity. The change is logarithmic in nature. So appropriate conversion needs to be done before transmitting the data.

As explained in earlier sections, the sensor senses the LPG level and generates a corresponding voltage at its output. This voltage is read by the ADC which gives out an integer between 0-1023. This part of the operation is tested using a serial interface application called Putty. We can see the test results in the Figure 9.

Figure 9. Test results of the MQ-6 gas sensor showing different levels of LPG gas

Final testing is done with the android smartphone having android 4.3 system and Bluetooth v3.0. As we can see in the figure, the application provides the list of Bluetooth devices. Once the desired one is selected, it starts displaying the real time LPG levels. Also when the buttons are pressed, corresponding actuators are toggled accordingly. This is shown in Figure 10.

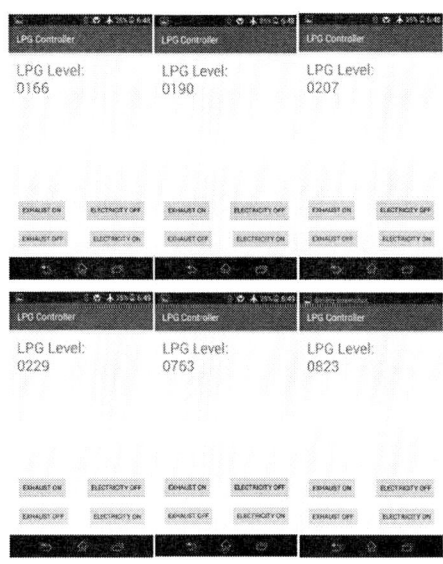

Figure 10. Final test results showing the LPG levels on android phone

VI. CONCLUSION

The implemented system is very efficient in detecting LPG gas leakage. It makes good use of the high sensitivity of the sensor so that the detection can occur as early as possible. The components used in the system are all low power devices. So the system can run on battery for a long time even in remote places. The designed android application provides a GUI to interact with multiple LPG detectors.

The LPG detection can be improved by using MEMS sensors which have a faster response or Poly-aniline/Magnesium Chromate composites which have a higher range. Also several data fusion algorithms can also be used to increase the accuracy of measurement of LPG leakage, especially in vehicles where higher sensitivity is desirable. The proposed system can be extended to detect any other gas or even have multiple sensors. It can be used to monitor multiple devices. This allows the system to be integrated into complex systems such as Wireless Sensor Networks, Advanced Driving Assistance Systems, Mobile data collectors, etc. Furthermore, the designed application can be configured to be used as a remote control to control various devices. This can particularly be useful in laboratories where the environment is not suitable for human presence. At such scenarios, this system can provide a simple yet convenient way to monitor and control various equipment.

VII. REFERENCES

[1] Praveen Kumar and K Krishna Naik, "Android based Wireless sensor network Application for Airborne Platforms," IEEE 2nd International Conference on Engineering and Technology, pp. 776-782, 2016

[2] Won-Jae Yi, Weidi Jia, and Jafar Saniie, "Mobile Sensor Data Collector using Android Smartphone," IEEE 55th International Midwest Symposium on Circuits and Systems (MWSCAS), pp 956-959, 2012

[3] A. N. Arvindan and Keerthika D., "Experimental Investigation of Remote Control via Android Smart Phone Of Arduino Based Automated Irrigation System Using Moisture Sensor," 3rd International Conference on Electrical Energy Systems, pp. 168-175, 2016.

[4] Santoshkumar, Udaykumar, R.Y., "Development of WSN for precision agriculture," International Conference of Innovations And Information, Embedded And Communications (ICIIECS), pp. 1-5, 2015

[5] Dmitry Duda, Pedro Martín-Mateos, Borja Jerez, Marta Ruiz Llata and Pablo Acedo, "Optical gas sensor based on an Android application for real-time, re-configurable spectroscopic data analysis," IEEE Sensors, pp 1054-1056, 2014

[6] S-L Chen, H-Y Lee, C-A Chen, H-Y Huang and C-H Luo "Wireless Body Sensor Network With Adaptive Low-Power Design for Biometrics and Healthcare Applications," Systems Journal IEEE, Vol.3, No.4, pp. 398-409, Dec. 2009

[7] Zeljko Jovanovic, Ranko Bacevic, Radoljub Markovic and Sinisa Randjic "Android Application for Obsrving Data Streams from Buit-in Sensors Using RxJAVA," 23rd Telecommunications Forum Telfor (TELFOR), pp 918-921, 201

[8] Stefano Rinaldi, Alessandro Depari, Alessandra Flammini, Angelo Vezzoli, "Integrating Remote Sensors in a Smartphone: the project - Sensors for ANDROID in Embedded systems," IEEE Sensors Applications Symposium (SAS), pp. 1-6, 2016

[9] Samer Hawayek, Claude Hargrove and Nabila A. BouSaba, "Real-Time Bluetooth Communication Between an FPGA Based Embedded System and an Android Phone," Proceedings of IEEE Southeastcon, pp. 1-4, 2013

[10] Anindya Nag, Asif Iqbal Zia, Xie Li, Subhas Chandra Mukhopadhyay and Jürgen Kosel, "Novel Sensing Approach for LPG Leakage Detection: Part.I — Operating Mechanism and Preliminary Results," IEEE Sensors, Vol. 16, No. 4, Feb. 2016

[11] Rasika S. Ransing and Manita Rajput, "Smart Home for Elderly Care, based on Wireless Sensor Network," International Conference on Nascent Technologies in the Engineering Field (ICNTE-2015), pp. 1-5, 2015

[12] ShariqSuhail Md, ViswanathaReddy G, Rambabu G, DharmaSavarni C. V. R and V. K. Mittal, "Multi-Functional Secured Smart Home" International Conference on Advances in Computing, Communications and Informatics (ICACCI), pp. 2629-2634, Sept. 2016

[13] Prashant A. Shinde, Mr.Y.B.Mane and Pandurang H. Tarange, "Real Time Vehicle Monitoring and Tracking System based on Embedded Linux Board and Android Application," International Conference on Circuit, Power and Computing Technologies [ICCPCT], pp. 1-7, 2015

[14] M. Prasasd, "Bluetooth Data Transfer Example" [Online] Available: http://manojprasaddevelopers.blogspot.in/2012/02/bluetooth-data-transfer-example.html

[15] T. Machappa, M. Sasikala, and M. V. N. Ambika Prasad, "Design of Gas Sensor Setup and Study of Gas(LPG) Sensing Behavior of Conducting Polyaniline/Magnesium Chromate (MgCrO) Composites," IEEE Sensors Journal Vol. 10, No. 4, April 2010

2017 IEEE International Symposium on Nanoelectronic and Information Systems

Reliability and Threat analysis of NBTI Stress on DSP cores

Anirban Sengupta
Computer Science and
Engineering
Indian Institute of Technology
Indore, India
asengupt@iiti.ac.in

Deepak Kachave
Computer Science and
Engineering
Indian Institute of Technology
Indore, India

Shuba Neema
Computer Science and
Engineering
Indian Institute of Technology
Indore, India

Sri Harsha Panugothu
Computer Science and
Engineering
Indian Institute of Technology
Indore, India

Abstract— **Device aging is a critical failure mechanism in nanoscale designs. Prolonged device degradation may result in failure. Delay degradation of a design depends on various factors such as threshold voltage, temperature, input vector pattern etc. An attacker who is aware of this phenomenon may exploit by accelerating the performance degradation mechanism. This paper proposes a novel reliability and threat analysis of negative bias temperature instability (NBTI) stress on digital signal processing (DSP) cores. The main contributions of this paper are as follows: (a) identifying input vectors that cause maximum degradation of DSP cores due to NBTI stress (b) analyzing impact of NBTI stress for varying stress time on DSP core in terms of delay degradation (c) analyzing performance comparison of stress vs. no-stress condition for various input vector samples.**

Keywords— DSP core, NBTI stress, input vector, delay

I. INTRODUCTION

Majority of the electronics products designed and manufactured are utilized in Consumer Electronics (CE) industry, thus making it as one of the central backbones of global economy. CE industry relies heavily on the economic scale for its profit. Additionally, due to surging competition, reduced time-to-market deadlines, and fast-changing consumer preferences, there is a huge risk involved with launch of every product. Thus, to reduce these risks CE industry relies heavily on third-party intellectual property (IP) facility during various stages of integrated circuit (IC) design flow. For example, design houses have to rely heavily on third party IP cores to meet time-to-market deadlines. This reliance on third-party facilities have increased security vulnerabilities. A malicious third party attacker may exploit the design to perform security attacks such as hardware trojan insertion, accelerated aging attacks, side channel attacks etc. Thus, a rigorous testing is required. One of the motivations of a rogue element could be to insert a malicious modification (in conjunction with the target hardware or through software program) to enable performance degradation such that a CE product fails within the warranty period [1]. Thus requiring replacement of device in full or in parts within warranty period. These attacks are termed as accelerated aging attacks. Aging occurs in transistor technology due to several physical phenomena such as Hot Carrier Injection (HCI), Time Dependent Dielectric Breakdown (TDDB), Negative Bias Temperature Instability (NBTI) etc [2]. Our proposed work considers accelerated aging due to NBTI stress. NBTI occurs when a negative bias is

applied between gate and source terminal of a PMOS transistor at higher temperature resulting in instability in transistor's parameters such as change in threshold voltage (Vth), transconductance (gm) etc [3] which manifests at gate level as degradation in delay (or performance) [4]. The attacker may attempt to exploit these parameter changes so as to accelerate aging process of device.

A large share of CE devices such as televisions, cameras, mobile phones, etc. requires Digital signal processing (DSP) component/core for applications such as audio processing, video compression, image processing, etc. DSP applications being a vital component of CE devices makes it one of the most ubiquitous and attractive target for malicious attacker(s). Further attacking a DSP application can cause huge impact on a CE product. Thus it becomes important to identify the impact of NBTI stress on DSP cores.

Novel contributions of the paper:

1. Identify the input vectors causing maximum degradation of DSP cores due to NBTI stress.
2. Analysis of the impact NBTI stress for varying stress time on DSP core in terms of delay performance degradation.
3. Identifying the presence threat posed due to NBTI stress. On identification, guiding the designer to remove the mechanism that enables threat for NBTI stress.

II. PROPOSED WORK

II. 1. Proposed Methodology
The section below discusses the proposed methodology.

A. Overview of proposed analysis

The proposed work investigates the impact of NBTI stress on DSP core. Fig. 1 summarizes the various steps involved in the proposed investigation. Investigation takes DSP core in the form of Control Data Flow Graph (CDFG) and perform High Level Synthesis (HLS) to obtain Register transfer level (RTL) datapath [5]. The modules of datapath thus obtained is converted into its equivalent gate level modules. In the next step critical path of gate level datapath is identified. Subsequently, we apply input vectors and evaluate its impact on change in threshold voltage as well as delay of gates in critical path with respect to time. These evaluations are performed for all possible combination of input test vectors. The results thus obtained is analyzed to identify set of input vectors that causes most degradation and thus must be checked for and avoided during standby mode of DSP applications.

This work is funded by Council of Scientific and Industrial Research (CSIR) under sanctioned grant no. 22/730/17/EMR-II

978-1-5386-1357-3/17 $31.00 © 2017 IEEE   11

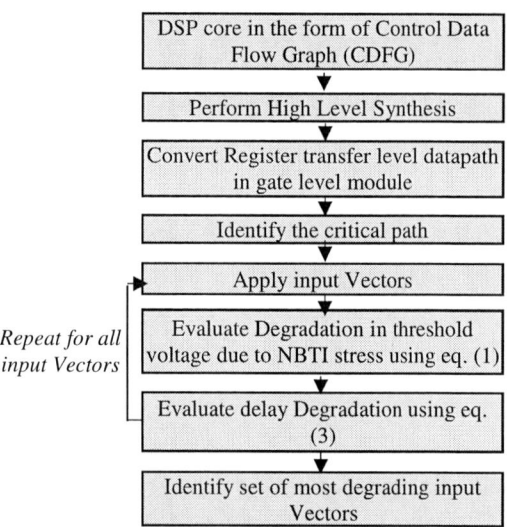

Fig. 1 Overview of proposed analysis

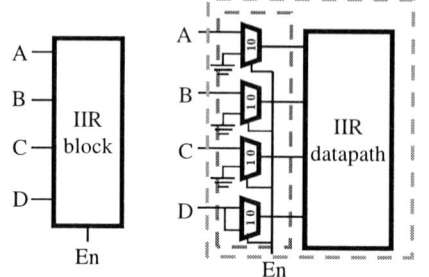

(a) IIR IP core block (b) Internal structure of IIR IP block
containing modified logic by the attacker

Fig. 2. Hardware modification model

B. Evaluating NBTI effect: Threshold Voltage and Performance

The aging of PMOS transistors is evaluated using a simplified model adopted from [1]. Equation 1 represents the degradation in threshold voltage.

$$\Delta V_{\text{th}} = b \cdot a^n \, t^n \tag{1}$$

Where, b = 3.9 x 10^{-3} V/s⁶, n=0.16, a = input signal probability, t = time in seconds. The new threshold voltage obtained for pmos under NBTI stress is given by eq. 2

$$V_{th}^{new} = V_{\text{th}} + \Delta V_{\text{th}} \tag{2}$$

The new threshold voltage of pmos thus obtained is utilized in eq. 3 [4] to obtain change in delay of gate.

$$T = K \frac{V}{(V - V_{th}^{new})^\alpha} \tag{3}$$

Where, T = change in delay of PMOS transistor, K is technology based proportionality constant, V = V_{DD}. For 65nm technology scale, V= 1.2V is adopted from [1], while α=1.4, K=155 x 10^{-6} assumed from [4].

C. CASE STUDY

In this sub-section, we will demonstrate investigation mechanism with the help of IIR benchmark. The initial step of proposed investigation involves performing high level synthesis to obtain register transfer level (RTL) datapath. HLS comprises of three sub-steps namely scheduling, allocation and binding. First the scheduling of the application is designed, followed by allocation and binding. Subsequently, the datapath through HLS process is generated at RTL. Then the modules of RTL datapath obtained are converted into their respective gate level structure as shown in fig. 3. NAND gates are utilized due to their universality and preference during fabrication in industrial applications. Further, the critical path (CP) is identified and all the gates of CP are numbered as shown in fig. 3. In the next step, we apply an input vector and identify its respective NBTI stress on each gate of the CP. The NBTI

stress on CMOS NAND gate occurs when a logic '0' is applied at its input. Logic '0' turns ON PMOS transistor by applying negative voltage at gate terminal.

II.2. Attack model

In this model, rouge element exploits the natural aging mechanism of PMOS transistor due to NBTI stress. The attacker accelerates aging process by continuously applying NBTI stress when a device is in Standby mode (i.e. outside of its active usage by the user). The NBTI stress can be applied through hardware as well as software modification as explained below.

a) Hardware based attack: The attacker modifies the DSP core application such that when the enable signal of DSP core is '1', the circuit will function as per the expectation. On the other hand, when the enable signal is '0' (a rare event), maximum NBTI stress causing input vectors are applied for accelerating the aging process. Fig. 2 shows the hardware based modification.

b) Software based attack: Another possible method of launching attack on the hardware is through reverse engineering (RE) or extracting information from a rogue insider of the design house. This may be accomplished through the following steps [1]:
(i) Obtaining netlist of the design through RE or rogue insider.
(ii) Determining the most threatful test pattern that can cause performance degradation (i.e. accelerate aging process).
(iii) Attacker creates a software program that can automatedly apply the determined test patterns on the target hardware in operating system mode.

II.3. Detection of NBTI Stress Based Attack

NBTI stress attack can be automatically launched either through hardware alteration to the design or software program such that input vector causing stress can be applied for a significant period of time. In case of hardware based attack, after the testing/validation phase, an attacker typically applies the input vector on the target design to generate its performance degradation effect. This is because in hardware based attack, the malicious signal (e.g. En signal in Fig.2) triggers (at En = '0') the threatful vectors only after validation phase is over so that during validation phase it goes un-detected. Thus to detect any possible NBTI stress based attack on the hardware, the system designer must re-evaluate the magnitude of delay of the datapath atleast after a significant time (say fifteen days) to verify whether the delay value is

978-1-5386-1357-3/17 $31.00 © 2017 IEEE   12

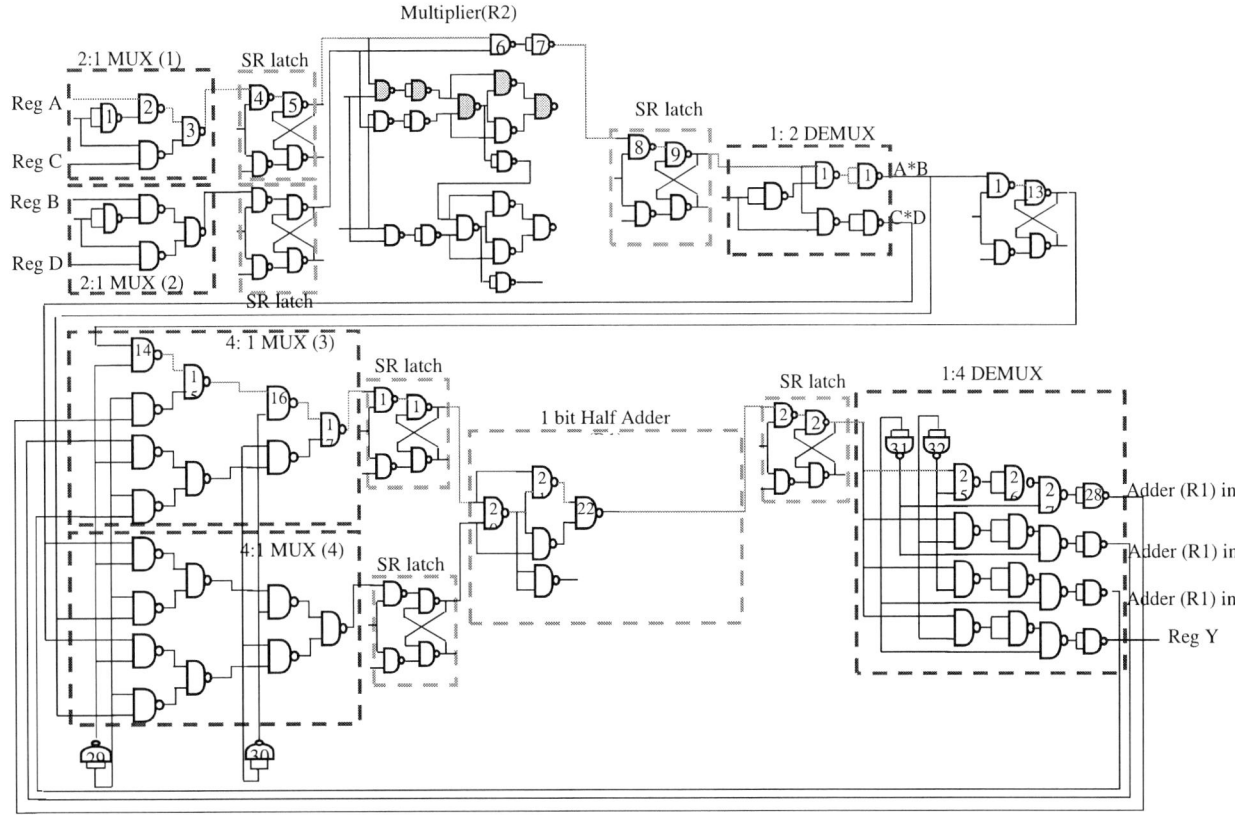

Fig. 3. Gate level Datapath of IIR Filter through HLS

same as it was determined at the end of validation phase earlier (i.e. fifteen days before). If the value does not match with the one determined earlier, then there is high chances that NBTI stress based attack has been launched internally without the knowledge of the designer. In such a scenario, the designer must check (and subsequently remove) any malicious alteration to the DSP RTL design or check for any unwanted software program operating in integration with the target hardware.

III. EXPERIMENTAL RESULTS

A. Change in Threshold Voltage Vs. Stress Time

NBTI stress affects several parameters of a device including threshold voltage, drain current, transconductance etc. In our experiments we have focused on the effect of NBTI stress on threshold voltage of the pmos. More the NBTI stress time, more is the increase in threshold voltage (as discussed in eqn. (1) & (2)). This has been shown by varying the stress time for evaluating the effect on threshold voltage. Fig. 4 shows the change in threshold voltage observed after applying NBTI stress for 1, 2 & 3 years respectively for distinct values of stress probability. Stress probability as defined in [6] is the fraction of the time the pmos transistor is under stress (it represents the workload of the device).

B. Delay Degradation Vs.Stress Time

Delay of the gate gets affected with change in threshold voltage (as shown in eqn. (3)). Thus when threshold voltage of the pmos increases due to NBTI stress, delay of the gate (corresponding to that pmos) also increases. This causes performance degradation of the entire datapath. However it also depends on the input vector applied at the gates. This is because not all input vectors are capable of tuning ON all (or majority of) the pmos in the critical path. Depending on the input applied, the number of pmos tuned ON in the critical path changes. Thus it is important to analyze the effect of each input vector on the critical path of the datapath, as critical path determines the delay of the circuit. Following process is performed to evaluate the delay of the gate level datapath for each input vector. First, for a specific test vector, the number of pmos in the critical path being turned ON is determined, followed by determination of ΔV_{Th} corresponding to a specific stress time (t). Once ΔV_{Th} is calculated, then the new threshold voltage (V_{Th}^{New}) corresponding to the pmos is calculated (using eqn. (2)). Subsequently, the V_{Th}^{New} is used to evaluate its gate delay (using eqn. (3)). In case a test vector is applied that does not turn a pmos of a gate ON, then the original threshold voltage corresponding to the nmos is used to evaluate delay of the gate. If a test vector affects both pmos and nmos of a gate, then the delay corresponding to the pmos

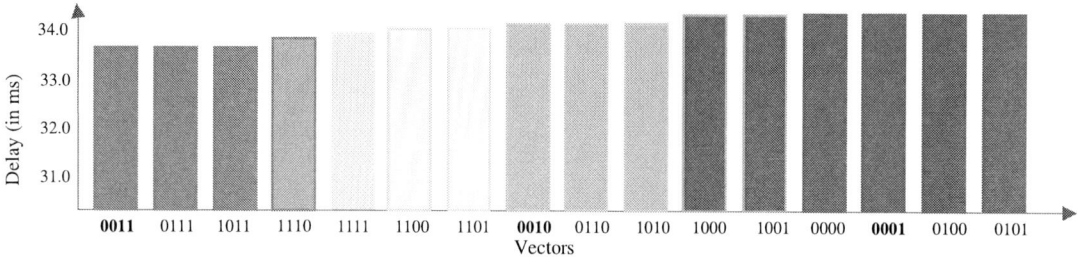

Fig.5. Delay of the datapath corresponding to each input vector applied

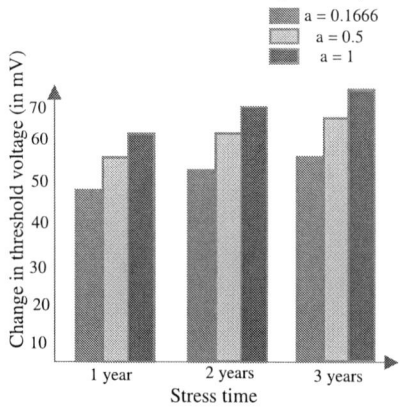

Fig.4. Change in threshold voltage with stress time

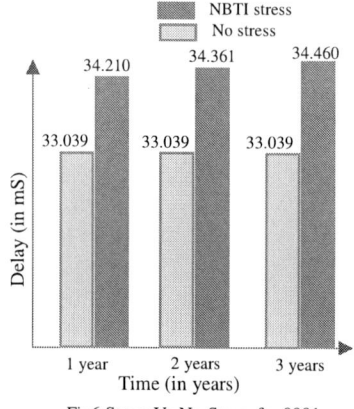

Fig6.Stress Vs No-Stress for 0001

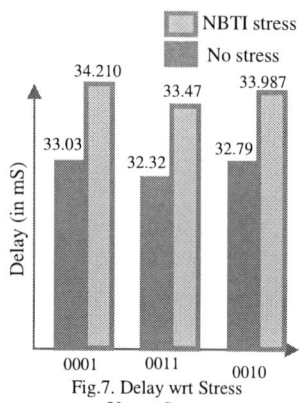

Fig.7. Delay wrt Stress Vs. no-Stress

is considered (as it is larger). Fig. 5 shows delay of the gate level datapath corresponding to each test (input) vector applied. As observed, the *red colored ones (0000, 0001, 0100, 0101)* are most threatful as they all incur same maximum performance degradation. The *green colored ones (0011, 0111, 1011)* produce least delay degradation.

C. Delay degradation due to NBTI Stress and No-Stress for most threatful input vector (for varying Stress time)

Fig. 6 shows the delay of the gate level datapath of IIR filter under NBTI stress and no-stress for most threatful input vector say *0001* (i.e. the one which causes maximum delay degradation as obtained in previous section). In other words, we analyze in this section how much degradation occurs when NBTI stress is applied due to specific input vector in contrast to when no-NBTI stress occurs. No-stress here indicates a theoretical condition when NBTI stress does not affect the pmos of the gate (i.e. its threshold voltage and corresponding delay). Three possible cases have been investigated for stress time (1 year, 2 year and 3 year) on datapath. As expected, with increase in stress time, the delay of the datapath has increased (due to increase in threshold voltage of corresponding pmos of the gate). However, there is no effect on delay when no NBTI stress is considered as threshold voltage remains same. This trend (in Fig.6) is likely to remain same as increase in stress time will always increase the threshold voltage.

D. Delay degradation due to NBTI Stress and No-Stress for different samples of input vector

In this section we investigate the effect of different samples of input vector on the delay of the datapath for both NBTI stress and no-NBTI stress condition. We have selected three samples

viz. *0001* (causing maximum delay degradation), *0011* (causing minimum delay degradation) and *0010* (causing median delay degradation) for this analysis. Fig. 7 shows the impact on delay of the datapath for the chosen sample vectors for NBTI stress and no-stress condition.

IV. CONCLUSION

This paper presented a novel reliability and threat analysis of NBTI stress on DSP cores. More explicitly, the paper presented the following: a) identification of input vectors causing maximum delay degradation b) analyzing the impact NBTI stress for varying stress time on DSP core in terms of delay performance degradation c) identifying the presence threat posed due to NBTI stress. After identification guiding the designer to remove the mechanism that enables threat for NBTI stress.

REFERENCES

[1] O. Sinanoglu et al., "Reconciling the IC test and security dichotomy," 2013 18th IEEE European Test Symposium (ETS), Avignon, 2013, pp. 1-6.

[2] W. Gos, "Hole Trapping and the Negative Bias Temperature Instability", PhD Dissertation, technische universität wien, Austria, December 2011.

[3] S. Mahapatra et al., "A Comparative Study of Different Physics-Based NBTI Models," in IEEE Transactions on Electron Devices, vol. 60, no. 3, pp. 901-916, March 2013.

[4] R. Gonzalez, et al., "Supply and threshold voltage scaling for low power CMOS," in IEEE Journal of Solid-State Circuits, vol. 32, no. 8, pp. 1210-1216, Aug 1997.

[5] Anirban Sengupta, et al., "A high level synthesis design flow with a novel approach for efficient design space exploration in case of multiparametric optimization objective", Microelectronics Reliability, Volume 50, Issue 3, March 2010, pp. 424-437.

[6] W. Wang, et al., "An efficient method to identify critical gates under circuit aging," 2007 IEEE/ACM International Conference on Computer-Aided Design, San Jose, CA, 2007, pp. 735-740.

A Firefly Algorithm Driven Approach for High Level Synthesis

Pallabi Sarkar[b], Anirban Sengupta[a], Santosh Rathlavat[a], Mrinal Kanti Naskar[b]

[a] Computer Science and Engineering, Indian Institute of Technology, Indore 452020

[b] Electronics and Telecommunication Engineering, Jadavpur University, Kolkata 700032

[a]asengupt@iiti.ac.in

Abstract: High Level Synthesis (HLS) for application-specific computer hardware design involves critical decision making steps such as design space exploration (DSE) of resources, scheduling, allocation and binding. Exploring an optimal design solution comprising of array of resource types is a complex problem and requires intelligent decision making in an automated manner. Exploration of an optimal design solution is considered a complex problem as it involves trade-off between contradictory parameters of hardware area and latency amongst numerous candidate design variants. This paper presents the following contributions: (a) novel FA driven DSE methodology during high level synthesis for application specific computing hardware based on area-latency trade-off (b) novel sensitivity analysis that provides optimal tuning of FA control parameters for performing DSE that leads to faster convergence.

Keywords: Firefly Algorithm, design space exploration, consumer electronics, high level synthesis

1. INTRODUCTION

High level synthesis popularly known as 'architectural synthesis' is a method of transforming an application specified through a control data flow graph (CDFG) to its register transfer level (RTL) counterpart. The method of HLS accepts set of user constraints for hardware area and latency, library and CDFG and produces an optimal design solution that implements the application behaviour. The vital foundation of HLS is the mechanism of design space exploration which is where critical decision making transpires. Design space exploration is an obligatory division of HLS that permits assessment of a number of design solutions through intelligent decision making and yields an optimal design solution that minimizes the design cost while meeting the constraints. Owing to this benefit over RTL synthesis, high-level synthesis has progressively gained approval to the CAD community for design of custom computing hardware. The process of DSE demanding intelligent pruning (decision making) can be accomplished through advanced metaheuristics such as population based algorithms [1] [3] [9], [10].

Though there are some population based algorithms (such as

genetic algorithm, hybrid genetic algorithm with particle swarm optimization, bacterial foraging, simulated annealing etc. [1], [2], [3], [4], [8], [9], [10]) that have been applied to solve similar problem, however they either did not always lead to global best solution, ending in pre-mature convergence or consumed large exploration time. Additionally, literature suggests that DSE approaches [1], [2], [3], [4] employing these population based algorithms have also incurred huge exploration time which is considered unacceptable from industry perspective.

2. PROPOSED WORK

2.1 Overview of proposed FA-DSE

The block diagram shown in Fig. 1 represents the overview of proposed multi-objective firefly algorithm driven design space exploration (FA-DSE). Based on the information provided in Fig. 1, the description of the algorithm is provided below:

The inputs to the proposed FA-DSE are data flow graph (DFG) which elaborates the behaviour of an application, user parametric constraints for hardware area and execution time, user weightage for design objectives and library. The library comprises following data as per 15nm open source NanGate library (NanGate Open CL, 2016 [5]): area occupied by each hardware resource in nanometre square (um²), delay (latency) expended by each hardware resource in pico-seconds (ps) and maximum resources. As stated earlier, for the proposed work, firefly algorithm has been embodied onto a DSE methodology of application specific computing system. **The mapping of FA is described below:**

Location of firefly → *Design solution (hardware resource configuration)*

Dimension of a firefly → *# of Hardware resource type*

Additionally, the fireflies are represented as set of original

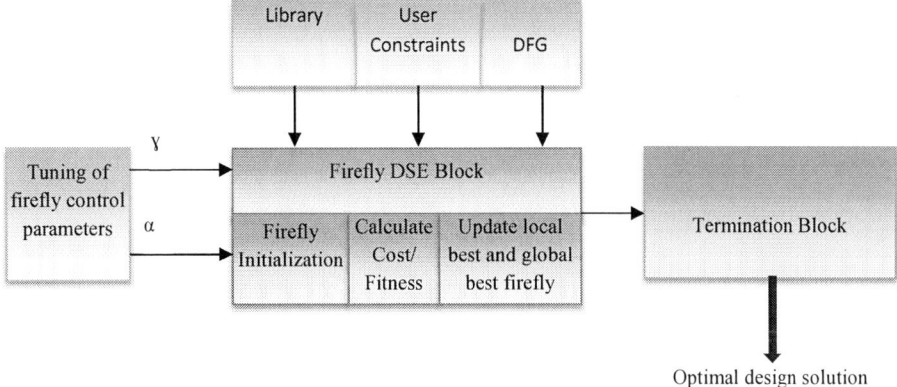

Fig.1. Block diagram of the Proposed FA-DSE

design solutions in a design space (that will be subjected to evolution in each iteration of the firefly algorithm). Additionally the firefly control parameters comprising of 'α', 'β' and *rand* are used as exploration drift factors to augment in exploration. In presented methodology, the initial firefly population may have numerous fireflies. The first firefly x_1: *(representing the first seed design solution)* is built by mapping the hardware resources with minimum value, the second firefly x_2 *(representing the second seed design solution)* is built by representing the hardware resources with maximum value, the third firefly x_3 is constructed by mapping the average value of minimum and maximum hardware resource. Once the fireflies are set, evaluation of fitness for each firefly is executed and the fittest firefly (which has minimum cost) is selected. The position of the fittest firefly assumes the global best and fitness of this firefly act as best fitness for the next iteration. After this step, iteration continues. Then in accordance to the proposed FA-DSE, in each iteration, a new firefly position (design solution indicating hardware resources) is determined. Then the calculation of the fitness of each new firefly position (new design solution) is performed, followed by updating the local best firefly position if the fitness of new firefly position is lower than previous firefly position. After the completion of the iterations, among all the local best firefly we choose the minimum cost value as global best firefly position. The above process is continued till we meet the conditions for terminating criterion (the terminating criterions are states in section 2.6). Therefore, after the stopping criterion the proposed algorithm yields the optimal design solution (hardware resource configuration).

2.2 Models for assessment of fireflies (design solutions) during FA-DSE
4.2.1 Proposed Area Model
$$A_T = \sum_{R_{i=1}}^{D}(A(R_i)) \quad (1)$$
$$= A(R_1) + A(R_2) + \cdots + A(R_n)$$
$$= n * a(R_1) + n * a(R_2) + \cdots + n * a(R_n)$$
Where $a(R_n)$ area of single instance of R_n , n belongs to hardware resource and $A(R_n)$ is total area of all instances of R_n.

4.2.2. Model for execution time
$$T_E = N * L \quad (2)$$
Where, the variables T_E, L and N are defined in Table 1.

4.2.3 Model for cost function
The cost function (C_{x_i}) created which incorporates execution time and hardware area shown in eq. (3).

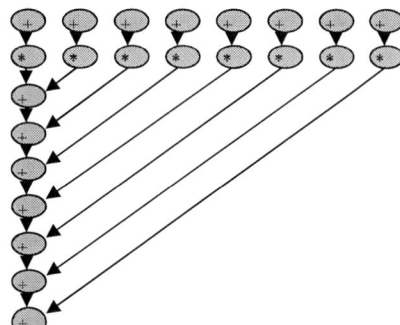

Fig.2. A FIR Data Flow Graph

$$C_{x_i} = \varphi_1 \frac{A_T - A_{cons}}{A_{max}} + \varphi_2 \frac{T_E - T_{cons}}{T_{max}} \quad (3)$$

Where C_{x_i} = Fitness of the firefly and the variables A_T, A_{cons}, A_{max}, T_E, T_{cons}, T_{max} are: area consumed by a design solution, Area constrained specified by user, Maximum area by resource configuration, Execution time taken by a design solution, Execution time constrained specified by user, Maximum execution time taken by a resource configuration.

φ_1, φ_2 = Weightage for area and execution time.

2.3 Initialization of fireflies
a) *Position*: Initialization is a key part of proposed FA-DSE for faster convergence. Firefly's positions are covered all over the design space uniformly. The presented algorithm initializes position by considering that an optimal solution to a multi-objective exploration problem lies amid the maximum parallel and serial implementation. Fig. 2 is used for demonstration where maximum adders and multipliers available are eight each. The general representation of firefly is given as follows:
$x_i = (R_1, R_2, R_3 \dots R_n)$

- First firefly $(x_1) = (R_1{}^{min}, R_2{}^{min}, \dots R_n{}^{min}) = (1, 1)$
- Second firefly $(f_2) = (R_1{}^{max}, R_2{}^{max}, \dots R_n{}^{max}) = (8, 8)$
- Third firefly $(x_3) = ((R_1{}^{min} + R_1{}^{max})/2 , (R_2{}^{min} + R_2{}^{max})/2, \dots (R_n{}^{min} + R_n{}^{max})/2) = (4, 4)$

In the above eqns, $R_n{}^{min}$ = minimum resource (hardware) and $R_n{}^{max}$ = maximum resources available.

2.4 Evaluation of local & global best firefly
After calculating the fitness of each firefly, the next step of the proposed FA-DSE is to determine the local best position (x_{lbi}) of each firefly and finally determine the global best firefly (x_{gb}). The x_{gb} of the population is determined using generic equation (4) as follows:
$$x_{gb} = x_j = [Min(C_{x_{lb1}}, C_{x_{lb2}} \dots C_{x_{lbn}})] \quad (4)$$
Based on the cost of the three fireflies (using eqn. 3) evaluated till iteration 1, the local best position of each firefly will be its current position i.e. $C_{x_{lb1}} = C_{x_1}$, $C_{x_{lb2}} = C_{x_2}$ and $C_{x_{lb3}} = C_{x_3}$. Thus the global best firefly position will be the minimum of the local best position of each firefly till that current iteration:
$$x_{gb} = x_j = [Min(C_{x_1}, C_{x_2,} C_{x_3})]$$
Therefore, substituting the values of fitness cost for each particle in eqn. (7) yields:
$$x_{gb} = x_j = [Min(-0.0645, 0.0229, -0.1771)]$$
$x_j = (4, 4)$ (i.e. the 3rd firefly is the most attractive so far)

2.5 Determination of new position of a firefly
Post-completion of the initialization stage in iteration1, the iteration procedure is restarted; the initial job of the iteration is evaluation of x_i^{t+1} (new resource configuration). The x_i^{t+1} is calculated using eqn. (5) inspired from [6] [7].
$$x_i^{t+1} = x_i^t + drift\ factor \quad (5)$$
The function of x_i^{t+1} consist of two parts: second part is drift factor and first part is previous position (x_i^t) of firefly. The drift factor is calculated using eqn. (6):
$$Drift\ factor = \beta(x_j - x_i) + \alpha(rand - 1/2) \quad (6)$$
In eqn. (6) two components contribute to determination of new firefly position:

978-1-5386-1357-3/17 $31.00 © 2017 IEEE

- First term in eq.6 $\beta(x_j - x_i)$ is called attraction factor. This component represents the attractiveness factor of a firefly and its relationship with the most attractive firefly.

Where the attractiveness (β) is given by:

$$\beta = \beta_0 e^{-\gamma r_{ij}^2} \qquad (7)$$

and distance between any two fireflies i and j at x_i and x_j, respectively is the Cartesian distance given by:

$$r_{ij} = |x_i - x_j| = \sqrt{\sum_{k=1}^{d}(x_{i,k} - x_{j,k})^2} \qquad (8)$$

- Where, $x_{i,k}$ is the kth component of the spatial coordinate x_i of ith firefly; β_0 is the attractiveness at zero distance and γ is absorption coefficient.
- $\alpha(rand - 1/2)$ is called step size controlling factor. 'α' for each dimension is initialized based on the maximum value of the resource type in that dimension

i) Firefly x_1 (new position x_1^{t+1}):

In '1st' dimension:

Using eqn. (6) drift factor is calculated as:

$$Drift\ factor = \beta * 3 + \alpha * \left(rand - \frac{1}{2}\right) = 8$$

Where $\beta = 1 * e^{-0.5*9} = 0.0111$, $\alpha = 8$ and $rand = 1.5$ (from [6] [7]). Using eqn. (5), the upgraded position for 1st dimension is:
$x_1^{t+1} = 1 + 8 = 9$.
However since this value is not within the range of the design space, thus *boundary outreach algorithm* is applied to bring the value within valid range. Thus $x_1^{t+1} = 9 - 4 = 5$

In '2nd' dimension:

Similarly,

$$Drift\ factor = \beta * 3 + \alpha * \left(rand - \frac{1}{2}\right) = 8$$

Where $\beta = 1 * e^{-0.5*9} = 0.0111$, $\alpha = 8$ and $rand = 1.5$ (assumed values from [6] [7]). Using eqn. (5), the upgraded position for 1st dimension is: $x_1^{t+1} = 1 + 8 = 9$.

By boundary outreach algorithm $x_1^{t+1} = 9 - 1 = 8$. Therefore 1st firefly position is $x_1 = (5, 8)$.

ii) Firefly x_2 (new position x_2^{t+1}):

Similarly, the value of x_2^{t+1} for f2 is found as:
$x_2 = (8, 8)$

iii) Firefly x_3 (new position x_3^{t+1}):

Similarly, the value of x_3^{t+1} for f3 is found as:
$x_3 = (4, 4)$

2.6 Termination criteria (Z)
Two conditions are as follows:
a) Stops when a maximum number of iteration is exceeded (Q = 100) or,
b) Stops if no improvement is spotted over a definite number of iterations. The stopping criteria is: no improvement is noticed in x_{gb} over 'w' number of iteration. (w=10)

3. RESULTS AND ANALYSIS
The presented FA-DSE is realized in java on Intel core i5-2450M processor with 3MB cache and 4GB DDR3 primary memory, at frequency of 2.5 GHz.

3.1 Effect on Proposed FA-DSE with variation in FA parameters
The impact of FA control parameters called 'absorption coefficient (γ)' and 'α' on the convergence and exploration time of proposed DSE methodology is discussed in this section. Further, this section also shows the impact of variation of firefly population sizes on the exploration time of the proposed DSE. The quality of solution (in terms of final cost) found and its assessment with existing DSE methodologies are discussed in the next sub-section. Additionally, sensitivity of analysis of the FA internal parameters also assists in pre-setting to an appropriate value while performing FA driven DSE during comparison with other approaches.

Table 1. Comparison of convergence time (milli-seconds) for FA parameter "γ"
(Note: p = 3)

Benchmark	Linearly Decreasing "γ"	"γ" constant (0.5)	"γ" Constant (0.01)
FIR	5741	10881	8458
IIR	5658	10671	7516
MPEG Motion Vectors	5651	7203	8861
MESA	5014	7100	10336
ARF	5776	6084	7530

Table2. Convergence time (milli-secs) w.r.t step size control parameter "α"
(Note: p = 3 and 'γ' linearly decreasing)

Benchmarks	Linearly Decreasing α	α=max	α=min
FIR	5741	8622	10652
IIR	5658	6599	5736
MPEG	5651	6368	7677
MESA	5014	9306	6159
ARF	5776	7337	7855

Table3. Exploration time (milli-secs) w.r.t firefly population size (p)
(Note: 'α' & 'γ' linearly decreasing)

Benchmarks	p=3	p=5	p=7
FIR	5744	5815	6859
IIR	5660	5891	8605
MPEG	5654	5686	12200
MESA	5019	7923	10902
ARF	6429	6603	7484

3.1.1 Absorption Coefficient (γ)

In the proposed FA-DSE approach, absorption coefficient is a major factor for controlling the exploration drift process of the firefly. As discussed in section 2.5 earlier, it is inversely proportional to attractiveness/brightness (β) (refer eqn. 6). As the FA-DSE process evolves, the attractiveness factor (or brightness) is expected to gradually increase due to smaller distance between the current firefly and brightest firefly (as the current firefly is slowly moving towards the brightest (global best) firefly). In order to mimic this phenomenon, absorption coefficient has to be gradually decreased with increase in DSE evolution. While experimentation, three variations of 'γ' is analysed and its influence on the performance of FA-DSE is presented:

a) Linearly decreasing 'γ' in each iteration starting from 0.5 (as suggested in [6, 7]) during the exploration process.

b) A constant value of 'γ' = 0.5 during the exploration process.

c) A constant value of 'γ' = 0.01 during the exploration process.

As apparent from Table 1, for all the tested benchmarks the convergence time to global optimal solution of proposed FA-DSE is lower for linearly decreasing value of 'γ', compared to constant 'γ'. This suggests proposed DSE converges to global optimal solution quicker when 'γ' is reduced linearly in each iteration.

3.1.2 Step Size Control Parameter (α)

The step size control parameter 'α' maintains the exploration-exploitation balance during the exploration process. Step size 'α' is supposed to be larger when the exploration starts with gradually decreasing as the exploration evolves. Therefore in order to mimic this phenomenon, step size 'α' is linearly decreased (from maximum value of each dimension) in each iteration as the DSE process evolves. As shown in Table 2, linearly decreasing 'α' converges to global optimal solution faster than constant values of 'α' = max and min respectively. This suggests proposed DSE converges to optimal solution quicker when 'α' is reduced linearly in each iteration.

3.1.3 Firefly Population Size (p)

A larger firefly population covers larger number of design solutions in the design space per iteration, but increases the computational complexity (time) per iteration. This results in higher exploration time with increase in value of 'p'. However, it should also be noted that using a large firefly population lower iteration of convergence (i.e. lower value of iteration count at which convergence to an optimal solution occurs). As shown in Table 3, three variations of firefly population sizes i.e. p = 3, p = 5 and p = 7 have been analysed. It is clear from the results obtained that the exploration time increases with increase in population size due to greater computational complexity per iteration. Further as shown in Table 4, the size of the firefly population does not have any impact on the quality of solution found. In other words, there is no improvement in the design solution found even if the population size increases. This is because, at p = 3, the global optimal design solution is found for all benchmarks. However as expected, the iteration of convergence for finding the global optimal solution decreases with increase in population size. Nevertheless since the computational complexity per iteration is large when the population size is greater, thus overall the

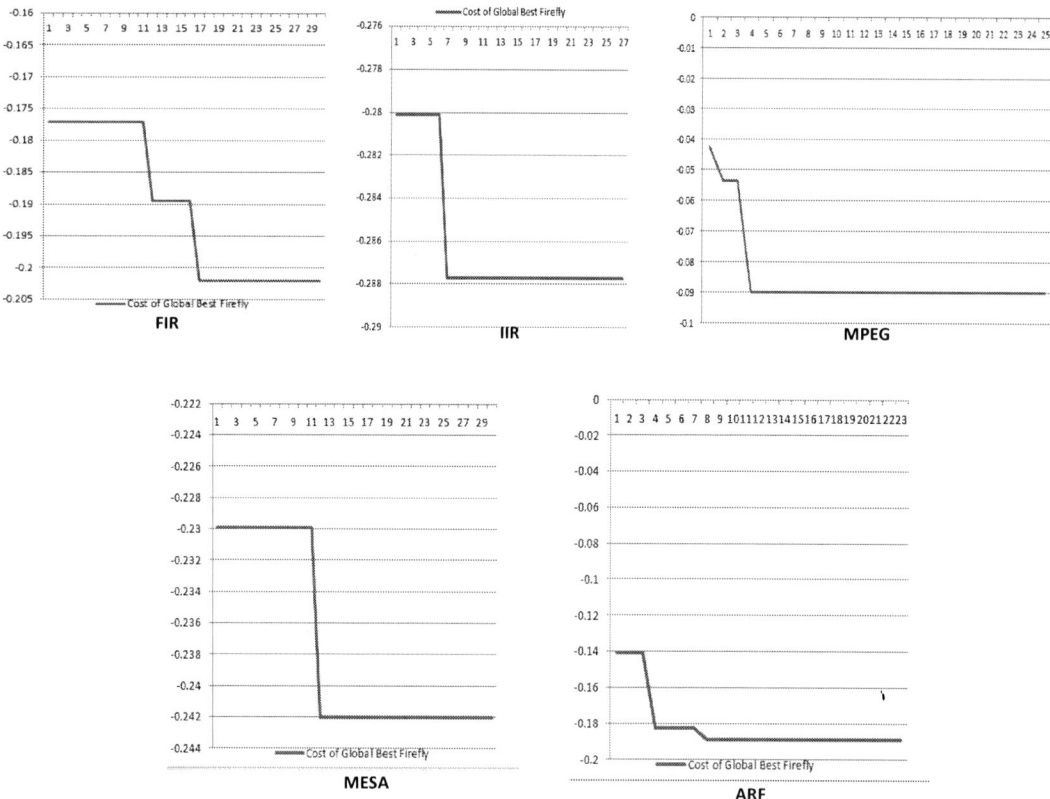

Fig.3 Trendline indicating reduction of global best cost of the firefly with evolution in proposed DSE (increase in iteration count)

Table 5. Results of proposed FA-DSE approach (Note: p = 3 and 'α' & 'γ' linearly decreasing)						
Benchmark	**Hardware Area**		**Execution time = N * L**		**Design Solution**	**Cost**
	Constraint (um sq.)	Proposed Area (um sq.)	Constraint (us)	Proposed execution time (us)	Proposed	Proposed
FIR	80	47.1859	55	39.7456	(2m, 2a)	-0.2021
IIR	80	42.4673	35	26.4971	(2m,1a)	-0.2877
MPEG	80	70.7788	56	39.7456	(3m,3a)	-0.0904
MESA	70	42.4673	50	39.7456	(2m,1a)	-0.2420
ARF	80	42.4673	70	64.5867	(2m,1a)	-0.1891

convergence time is more at p = 5 and 7 compared to p =3.

3.2 Results of proposed FA-DSE approach

This section first presents the results of proposed FA-DSE for respective benchmarks. As evident from Table 5, the hardware area and latency of the explored design solution always lies within the user constraints of hardware area and execution time respectively. For example in case of MPEG, based on user

Table 4. Population size Vs Iteration of convergence			
Benchmarks	**Population size (p)**	**Design solution (firefly position)**	**Iteration of convergence**
FIR	3	(2m, 1a)	20
	5	(2m, 1a)	15
	7	(2m, 1a)	10
IIR	3	(2m, 1a)	21
	5	(2m, 1a)	16
	7	(2m, 1a)	14
MPEG	3	(3m, 3a)	15
	5	(3m, 3a)	12
	7	(3m, 3a)	11
MESA	3	(2m, 1a)	20
	5	(2m, 1a)	15
	7	(2m, 1a)	10
ARF	3	(2m, 1a)	14
	5	(2m, 1a)	13
	7	(2m, 1a)	9

constraints of 80um^2 and 56us, the final solution (3m, 3a)

explored occupies an area of 70.77 um^2 and 39.74us respectively. This indicates that the final solution is always well within its upper budget specified. The associated final design cost is -0.0904 *(Note: negative indicates that area/execution time of explored solution is lower than its respective constraint)*. Similar trend has been observed for all other benchmarks. Fig. 3 shows the trendline of reduction of global best firefly cost of the proposed DSE (i.e. with the increase in iteration count, the global best cost of the firefly gradually decreases).

4. CONCLUSION

The proposed FA-DSE was capable of exploring optimal hardware area-execution time tradeoff while maintaining an acceptable implementation runtime.

REFERENCES

[1] D. Gajski, N.D. Dutt, A. Wu, and S. Lin, "High Level Synthesis: Introduction to Chip and System Design", *Kluwer Academic Publishers*, Norwell, USA 1992.

[2] Z Zeng, R Sedaghat, A Sengupta, "A Novel Framework of Optimizing Modular Computing Architecture for multi objective VLSI designs", *IEEE 21st International Conference on Microelectronics*, 2009 pp: 322-325.

[3] Anirban Sengupta, Saumya Bhadauria "Untrusted Third Party Digital IP cores: Power-Delay Trade-off Driven Exploration of Hardware Trojan Secured Datapath during High Level Synthesis", 25th IEEE/ACM Great Lake Symposium on VLSI (GLSVLSI), 2015, pp. 167 - 172.

[4] Vyas Krishnan, Srinivas Katkoori "A genetic algorithm for the design space exploration of datapaths during high-level synthesis", *IEEE Transactions on Evolutionary Computation*, 10 (3), 2006, pp. 213–229.

[5] NanGate Open CL, 2016, http://www.nangate.com/?page_id=2328.

[6] X.-S. Yang, "Firefly algorithms for multimodal optimization", Stochastic Algorithms: Foundations and Applications, SAGA 2009,Lecture Notes in Computer Sciences, Vol. 5792, pp. 169-178

[7] Xin-She Yang and Xingshi He, (2013). 'Firefly Algorithm: Recent Advances and Applications', Int. J. Swarm Intelligence, Vol. 1, No. 1, pp. 36–50. DOI: 10.1504/IJSI.2013.055801.

[8] Anirban Sengupta, Saumya Bhadauria, "'Automated Exploration of Datapath in High Level Synthesis using Temperature Dependent Bacterial Foraging Optimization Algorithm'", *Proceedings of 27th IEEE Canadian Conference on Electrical and Computer Engineering*, Toronto, May 2014, pp. 68- 73.

[9] Anirban Sengupta, Saumya Bhadauria, "Bacterial Foraging Driven Exploration of Multi Cycle Fault Tolerant Datapath based on Power-Performance Tradeoff in High Level Synthesis", *Elsevier Journal on Expert Systems With Applications*, Volume 42, Jan 2015, pp. 4719 – 4732.

[10] Summit Sehgal, Reza Sedaghat, Anirban Sengupta, Zhipeng Zeng, "Multi Parametric Optimized Architectural Synthesis of an Application Specific Processor", *Proceedings of 14th IEEE International CSI Computer Conference (CSICC)*, 2009, pp: 89-94

Mathematical Validation of HWT based Lossless Image Compression

Anirban Sengupta
Discipline of Computer Science and Engineering
Indian Institute of Technology Indore
Email: asengupt@iiti.ac.in

Dipanjan Roy
Discipline of Computer Science and Engineering
Indian Institute of Technology Indore

Abstract—**This paper proposes the mathematical framework and its validation for Haar Wavelet Transformation (HWT) based lossless image compression. The complete end to end transformation functions, starting from the input image to regenerate the original image without losing any data is proposed. The validation of each function is demonstrated with example. It proofs the capability of reconstructing the original image through the proposed framework. This function can be used to design the hardware for HWT-based lossless image compression.**

I. Introduction

Digital image compression plays a crucial role in applications like smart phone, camera, video-conference, scanner, medical imaging, satellite imaging, remote sensing. Due to the larger size of a digital image, it takes larger storage space, higher transmitting and file sharing time, and larger bandwidth to upload or download an image. An efficient image compression technique not only saves the storage space but also enables easy sharing of image files[1]. Image compression techniques can be classified into two categories: a)lossy compression[2] and b)lossless compression[3]. In lossy compression, the decompressed image may lose some data compare to the original image, whereas, in lossless compression, the exact original image can be obtained after decompression. Lossless image compression is very crucial for applications like medical imaging, satellite imaging, forensic imaging where small data loss can create a huge impact[4], [5]. There are several work have been proposed for lossy image compression based on discrete cosine transformation, discrete wavelet transformation, but there are no mathematical proof of concept is present for wavelet based end to end lossless image compression process. In this paper, we have proposed the proof of concept for Haar Wavelet Transformation (HWT) based lossless image compression for the first time.

II. Proposed Proof of Concept: HWT based Lossless Image Compression

A. Problem formulation

For a gray scale input image of size NxN shows the proof of concept for HWT-based lossless image compression for the input image.

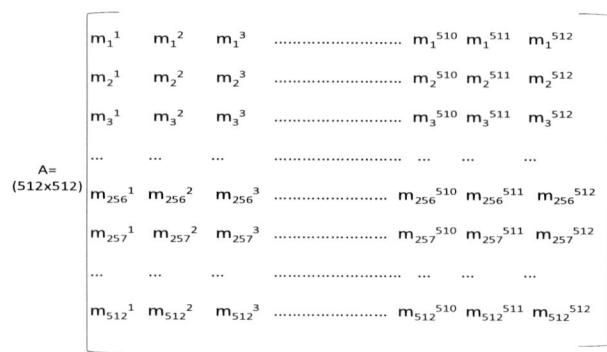

Fig. 1. Generic pixel value of an 512x512 image

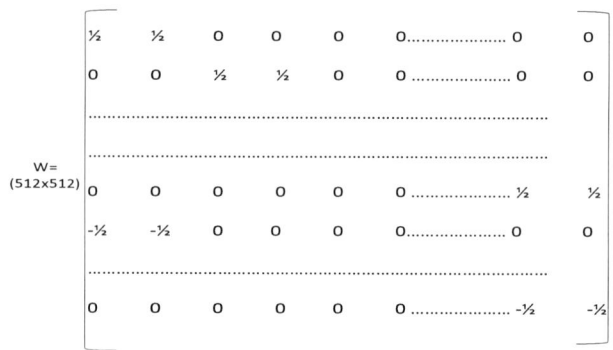

Fig. 2. Haar wavelet transformation matrix

B. Proposed Mathematical Framework for Pixel Computation

In this work, a novel proof of concept is proposed for Haar Wavelet Transformation (HWT) based lossless image compression. A generic 512x512 input image in the form of a matrix (A) is shown in Fig.1. The value of each element of the matrix indicates the pixel value or gray level of the image at that point. Each element is shown in the form of m_r^c, where r and c indicate the row number and column number of the element. For example, element m_{256}^1 represents the pixel value of 256th row and 1st column of matrix A. The Haar Wavelet based transformation of an input image is performed by a 512x512 HWT matrix shown in Fig.2. The

978-1-5386-1357-3/17 $31.00 © 2017 IEEE

Haar Wavelet transformation based compression on the input image is performed based on the following function:

$$B = WAW^T \qquad (1)$$

The input image matrix 'A' is first multiplied with HWT matrix 'W', this transforms the rows of matrix A. The resultant matrix is then multiplied with the transpose of HWT matrix to generate the matrix of compressed image (B).

The HWT based decompression is performed based on the following function:

$$A^{re} = W^{-1}B(W^{-1})^T \qquad (2)$$

The compressed image matrix 'B' is first multiplied with the inverse of HWT matrix 'W', then the resultant matrix is multiplied with the transpose of inverse HWT matrix to reconstruct the original input image (A^{re}) without any loss of information.

The end to end HWT-based lossless image compression can be expressed using following pixel equations:

$$A^{re} = (\phi_1 - \phi_2) - (\phi_3 - \phi_4)[for\ odd\ rows\ and\ odd\ columns] \qquad (3)$$

$$A^{re} = (\phi_1 - \phi_2) + (\phi_3 - \phi_4)[for\ odd\ rows\ and\ even\ columns] \qquad (4)$$

$$A^{re} = (\phi_1 + \phi_2) - (\phi_3 + \phi_4)[for\ even\ rows\ and\ odd\ columns] \qquad (5)$$

$$A^{re} = (\phi_1 + \phi_2) + (\phi_3 + \phi_4)[for\ even\ rows\ and\ even\ columns] \qquad (6)$$

where, $\phi_1, \phi_2, \phi_3,$ and ϕ_4 is shown in eqn.7, 8, 9 and 10 respectively.

$$\phi_1 = \left(\frac{m_{2n-1}^p + m_{2n}^p}{4}\right) + \left(\frac{m_{2n-1}^q + m_{2n}^q}{4}\right) \qquad (7)$$

$$\phi_2 = \left(\frac{m_{2n+i-j+1}^p - m_{2n+i-j}^p}{4}\right) + \left(\frac{m_{2n+i-j+1}^q - m_{2n+i-j}^q}{4}\right) \qquad (8)$$

$$\phi_3 = \left(\frac{m_{2n-1}^q + m_{2n}^q}{4}\right) - \left(\frac{m_{2n-1}^p + m_{2n}^p}{4}\right) \qquad (9)$$

$$\phi_4 = \left(\frac{m_{2n+i-j+1}^q - m_{2n+i-j}^q}{4}\right) - \left(\frac{m_{2n+i-j+1}^p - m_{2n+i-j}^p}{4}\right) \qquad (10)$$

where, N is the dimension of the square input image matrix; n is the variable ranging from 1 to N/2, increases in every alternate row; i is the variable ranging from 1 to N/2, increases in every alternate row; p is the odd variable ranging from 1 to N, increases in every alternate column; q is the even variable ranging from 2 to N, increases in every alternate column; j is the variable ranging from N/2 to 1, decreases as 'i' increases.

Fig. 3. Proof of generic equation 3 and 4

Fig. 4. Proof of generic equation 5 and 6

III. RESULTS AND VALIDATION OF FRAMEWORK

The validation of all the aforementioned generic pixel computation functions for specific examples are shown in Fig.3 and 4. Fig3 shows the validation of eqn.3 and 4 and Fig.4 shows the validation of eqn.5 and 6. As evident, using equations 3-6, exactly matching individual pixels of the original input image can be directly achieved. This demonstrates the capability of equations 3- 6 for lossless image compression.

IV. CONCLUSION

This paper presents a novel mathematical validation for HWT-based lossless image compression. All the generic equations proposed in this paper have been validated with examples which demonstrates the capability of the proposed framework to obtain original image.

REFERENCES

[1] S. Benchikh and M. Corinthios, "A hybrid image compression technique based on DWT and DCT transforms," International Conference on Advanced Infocom Technology 2011 (ICAIT 2011), Wuhan, China, 2011, pp. 1-8.

[2] A. M. G. Hnesh and H. Demirel, "DWT-DCT-SVD based hybrid lossy image compression technique," 2016 International Image Processing, Applications and Systems (IPAS), Hammamet, 2016, pp. 1-5.

[3] S. Li, H. Yin, X. Fang and H. Lu, "Lossless image compression algorithm and hardware architecture for bandwidth reduction of external memory," in IET Image Processing, vol. 11, no. 6, 6 2017, pp. 379-388.

[4] G. Scarmana and K. McDougall, "Exploring the application of some common raster scanning paths on lossless compression of elevation images," 2015 IEEE International Geoscience and Remote Sensing Symposium (IGARSS), Milan, 2015, pp. 4514-4517.

[5] A. Bilgin and M. W. Marcellin, "Applications of reversible integer wavelet transforms to lossless compression of medical image volumes," Proceedings. 1998 IEEE International Symposium on Information Theory (Cat. No.98CH36252), Cambridge, MA, 1998, pp. 411.

Lookup Table-Based Low-Power Implementation of Multi-Channel Filters for Software Defined Radio

Subhendu Kumar Sahoo
Department of Electrical and
Electronics Engineering
BITS-Pilani Hyderabad, India
Email: sahoo@hyderabad.bits-pilani.ac.in

P. K. Meher
Senior Member, IEEE
Nanyang Technological University
Singapore
Email: pkmeher@yahoo.com

Abstract—**In this work we present a low power multi-channel finite impulse response (FIR) filter using look-up table (LUT) approach. We have been able to reduce the LUT size by a factor of 8 over the conventional LUTs using Booth recoding, odd multiple storage and asymmetric product techniques. The proposed design provides 22%, 30% and 33% less duration of cycle period and 14%, 32% and 35% less energy per sample over the recently proposed design for multi-channel filter for tap size 4, 8 and 16 respectively. This design would therefore be highly useful for the implementation of multi-channel filters for certain applications like software defined radio systems.**

I. INTRODUCTION

The global sales of wireless consumer electronics products have been rapidly increasing over the years, where the sales of cell phones and portable tablets constitute a major part [1]. These wireless gadgets need to communicate with each other over a wireless network and should support different communication standards. Software defined radio (SDR), therefore, has gained tremendous popularity in the last two decades for its application to seamless wireless communication across the globe over several different standards involving different carrier frequencies for communication and different implementations of physical layer. To achieve such seamless communication over multiple standards, digital processing of communication systems is moved to the closest proximity of the receiving antenna, which demands a wideband receiver. The most computationally-exhaustive part of the wideband receiver of an SDR is the intermediate frequency (IF) processing block [2], [3], where finite impulse response (FIR) filtering is the main task. Since an SDR system can receive signals through multiple channels, multi-channel filter plays a very important role in SDR implementation [4]. A straight-forward implementation of multi-channel filter involves multiple FIR filters, to use a different filter for each of the different channels. To reduce the computational complexity of such a straight-forward implementation, a multi-channel filter is proposed in [5], where the input samples from different channels are time-multiplexed into the arithmetic unit of a single filter. Further, author in [6] have tried to implement a filter using only one multiply and accumulate(MAC) unit. So, a m-tap filter will need m-clock cycles to compute one filter output. For a n-channel multichannel filter, m*n clock

cycles will be required. This will be useful for extremely slow.

As the devices becoming smaller and smaller, semiconductor memory has become faster, cheaper, and more power-efficient. Furthermore, as to the forecasts of the international technology roadmap for semiconductors [ITRS], embedded memories will find dominating presence in the system-on-chips (SoCs), which may exceed 90% of the total SoC content. It has also been predicted that the transistor density of memory components will increase much faster than those of logic components. Apart from that, memory-based computing structures are more consistent than the multiply-accumulate structures and offer many other benefits, e.g., greater potential for high-throughput and low-latency operation and less dynamic power consumption. Memory-based computing is well suited for many digital signal processing (DSP) algorithms, which involve multiplication with a fixed set of coefficients.

In this paper we present a novel LUT-based design for low-power and high-speed implementation of multi-channel FIR filters. We have used Booth recoding along with odd multiple storage (OMS) and anti-symmetric product coding (APC) techniques to reduce the LUT size by a factor of 8 over the conventional implementation. The proposed multi-channel filter provides 22%, 30% and 33% less cycle period and 14%, 32% and 35% less energy per sample over the recently proposed design [5], of tap size 4, 8 and 16, respectively, for the multi-channel FIR filter.

The rest of the paper is organized as follows. In the next section we briefly discuss the concept of multi-channel filter. The proposed LUT-based design of multi-channel FIR filter using Booth recording, OMS and APC approach is discussed in section III. The synthesis results of the proposed design and the recently reported multi-channel filter along with their comparisons are presented in section IV. Conclusion is presented in section V.

II. MULTI CHANNEL FIR FILTER

A simplified functional block diagram of a multi-channel FIR filter is shown in Fig. 1. For simplicity of presentation we

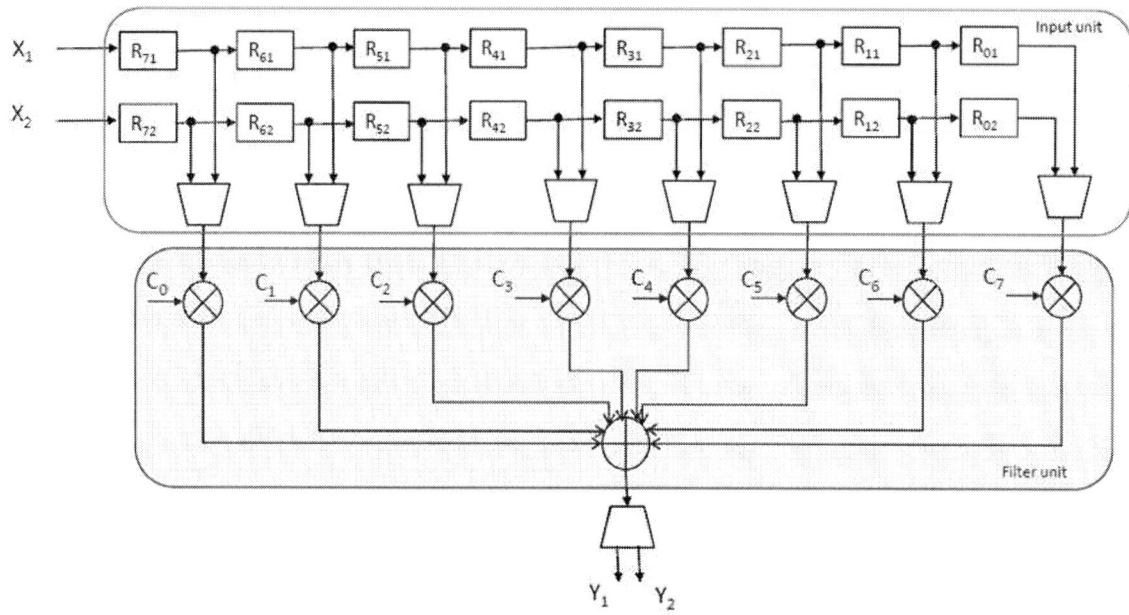

Fig. 1. Multichannel FIR filter structure.

have considered an 8-tap FIR filter, which can accept signals from a pair of channels, where the filter coefficients, in general, could be different for different channels. The input samples from the pair of channels are fed to a pair of input shift-registers. In each cycle, samples from one of the two channels are selected by the multiplexers and fed to the filter unit. Note that the sampling period for an n-channel filter is n times that of clock period such that in each sampling period one output of each of the channels is computed. The filter unit performs the necessary multiplications and additions to produce the output as shown in the figure. Finally the filter output is demultiplexed to produce the pair of channel outputs on two separate output lines. In this structure, a single filter unit is time-shared by two channels; and the throughput rate is nearly half of the actual throughput rate of a single channel FIR filter. For water acoustic digital signal processing applications, usually the sampling rate is less by many fold than the FIR filter operation. If the frequency of operation of FIR filter is f_1 and the sample generation rate is f_2 from different sources, then f_1/f_2 number of channels can be supported by this multichannel filter.

III. PROPOSED FIR FILTER ARCHITECTURE

As shown in Fig.1, in a multi-channel FIR filter, each coefficient is multiplied with the input selected by multiplexer using a multiplier. Multiplier is the most area-time and energy consuming component of FIR filter. In the proposed work, the multiplier is implemented using a novel LUT-based approach. To optimize the size of LUT three techniques are incorporated, namely, the Booth recoding [7], followed by OMS and APC [8]. Each of these techniques reduces the size of LUT by a factor of 2, so that the LUT size is reduced by a factor of 8. In

this section, we discuss these three techniques briefly followed by the implementation of the multiplier and multi-channel FIR filter.

A. Booth Encoding for Higher Radix Multiplier

In a non Booth conventional multiplication, the number of partial product rows is same as the number of bits in multiplier. Booth algorithm [7] decomposes an N-bit number X into $\lceil \frac{N}{k} \rceil$ groups, of k-bits each using radix-2^k encoding. Each group of $k+1$ overlapping operand bits are mapped to a singed digits D_i, where $i = 0, 1, \cdots, \lceil \frac{N}{k} \rceil$. The digit D_i can be obtained from $k+1$ bits, i.e. $D_i = X_{(i.k)+k-1} \cdots X_{(i.k)+1} X_{i.k} X_{(i.k)-1}$ where $X_{(i.k)-1}$ is the overlapping bit. This $X_{(i.k)-1}$ is the most significant bit of the previous digit D_{i-1} and is zero for $i = 0$. The decimal value of digit D_i is given by:

$$ D_i = X_{(i.k)+k-1}.2^{k-1} + \sum_{j=0}^{k-2} X_j.2^j + X_{(i.k)-1} $$

The digit D_i for a radix-2^k encoding can take one value from $(-2^{k-1}, -2^{k-1} + 1, ..., 1, 0, 1, ..., 2^{k-1} - 1, 2^{k-1})$. So the multiplication of a coefficient C with multiplicand X_i is given by:

$$ C.X_i = \sum_{i=0}^{\lceil \frac{N}{k} \rceil - 1} 2^{k.i}.D_i.C $$

The multiplication by $2^{k.i}$ can be realized just by hard-wired shifting without any logic complexity. One has to compute $D_i.C$ to perform the multiplication. For $k = 3$, the encoding is radix-8 Booth encoding. The number of partial product rows

978-1-5386-1357-3/17 $31.00 © 2017 IEEE   24

TABLE I
PARTIAL PRODUCTS REPRESENTATION IN TERMS OF 4C/-4C AND APC FOR RADIX-16 BOOTH ENCODING

$X_2X_1X_0X_{-1}$	Positive PP ($X_3 = 0$)	Negative PP ($X_3 = 1$)	Positive PP	Negative PP	Stored APC Word	Shift
0000	0	$-8C$	$4C - 4C$	$-4C - 4C$	C	2
0001	C	$-7C$	$4C - 3C$	$-4C - 3C$	$3C$	0
0010	C	$-7C$	$4C - 3C$	$-4C - 3C$	$3C$	0
0011	$2C$	$-6C$	$4C - 2C$	$-4C - 2C$	C	1
0100	$2C$	$-6C$	$4C - 2C$	$-4C - 2C$	C	1
0101	$3C$	$-5C$	$4C - C$	$-4C - C$	C	0
0110	$3C$	$-5C$	$4C - C$	$-4C - C$	C	0
0111	$4C$	$-4C$	$4C - 0$	$-4C - 0$	0	0
1000	$4C$	$-4C$	$4C - 0$	$-4C - 0$	0	0
1001	$5C$	$-3C$	$4C + C$	$-4C + C$	C	0
1010	$5C$	$-3C$	$4C + C$	$-4C + C$	C	0
1011	$6C$	$-2C$	$4C + 2C$	$-4C + 2C$	C	1
1100	$6C$	$-2C$	$4C + 2C$	$-4C + 2C$	C	1
1101	$7C$	$-C$	$4C + 3C$	$-4C + 3C$	$3C$	0
1110	$7C$	$-C$	$4C + 3C$	$-4C + 3C$	$3C$	0
1111	$8C$	0	$4C + 4C$	$-4C + 4C$	C	2

will be $\lceil \frac{16}{3} \rceil = 6$. So this need possible partial product values to be obtained are $-4.C, -3.C, .., -C, 0, C, .., 3.C, 4.C$. All these can be obtained by shifting from C and $3.C$. So the fundamental digit multiplied partial product(FDMPP) row need to be computed is $3C$.

For FIR filters, commonly used word-lengths are, 16, 24 and 32. Therefore, we use radix-4, radix-16, radix-256 Booth encoding to reduce the number of partial products by a factor of 2, 4 or 8 respectively. Use of radix-8, radix-32, radix-64 or radix-128 will not be efficient. Among Booth radix-4, radix-16, radix-256 encoding the mostly used one is radix-4 because of its simplicity and many researchers have given efficient circuits and methods for this. Again radix-256 can reduce the number of partial products significantly [9], but this needs a large number of FDMPP values which increases complexity. So, in the proposed work, we have tried to maintain a balance between complexity and number of partial products by choosing radix-16 Booth encoding [9]. The possible partial product values thus, required to be computed are $-8.C, -7.C, ..., -C, 0, C, ..., 7.C, 8.C$. All these can be obtained from the odd multiple values i.e. C and $3.C$.

B. APC for Radix-16 Booth Encoding

All possible product values for radix-16 Booth encoding for a group of four bits with an overlapping bit i.e. $X_3X_2X_1X_0X_{-1}$ are listed in Table I. Here X_{-1} is the overlapping bit. The first column lists all possible combination of $\{X_2X_1X_0X_{-1}\}$. In the 2nd and the 3rd columns, the partial product (PP) values corresponding to $X_3 = 0$ and $X_3 = 1$ for a specific combination of $\{X_2X_1X_0X_{-1}\}$ are listed. It is observed that, the PP values are positive for $X_3 = 0$ and are negative those for $X_3 = 1$. Further, the positive PP can be written as $4.C + APC$ word in the 4th column, where the APC word could be $-4.C, -3.C, -2.C, -C, 0, C, 2.C, 3.C$ or $4.C$. Similarly, the negative PP can be written as $-4.C + APC$ word in the 5th column [7]. It is also observed that the APC values listed in the 4th and 5th columns are the same for a specific bit values of $\{X_2X_1X_0X_{-1}\}$. These values are negative for $X_2 = 0$ and positive for $X_2 = 1$. Any APC word can be obtained only from C or $3.C$ as listed in the 6th column, and shifting by shift-value as indicated in the 7th column followed by negation when $X_2 = 0$. To compute any partial product we need only $C, 3.C$ and $4.C$ or $-4.C$. Therefore, we need to store only C and $3.C$ in the LUT to perform a multiplication using radix-16 Booth encoding.

C. Implementation of the Proposed Multiplier

The structure and function of the proposed optimized LUT-based multiplier for an operand of 4-bit width is shown in Fig. 2(a). It consists of an LUT, a barrel-shifter and a control unit. The LUT contains two words (the odd APC words) of size $w+2$ where w is the width of the filter coefficients. The control unit generates three control signals i.e. sel, $shift$ and $reset$ which depend on the values of X_1, X_0 and X_{-1}. The content of appropriate location of LUT is selected by sel signal. The LUT output is shifted by a barrel shifter through desired number of bit location(s) when the APC word is even. The number of bit-shifts is determined by the shift-control provided

978-1-5386-1357-3/17 $31.00 © 2017 IEEE

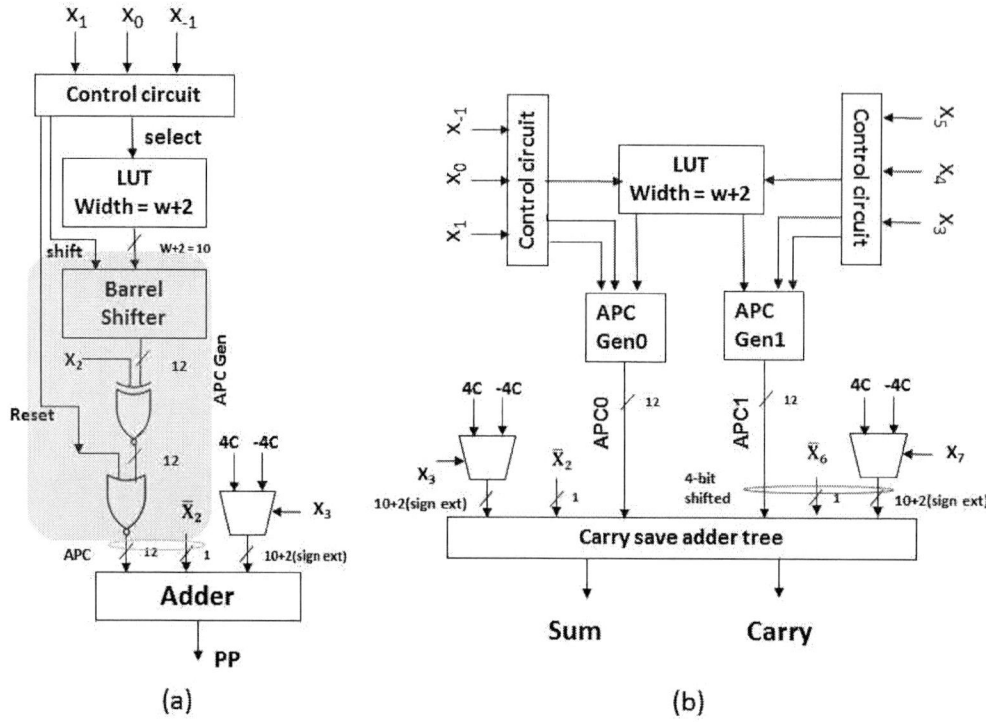

Fig. 2. Structure of the Proposed LUT-based multiplier for (a) 4-bits operand (b) 8-bit operand.

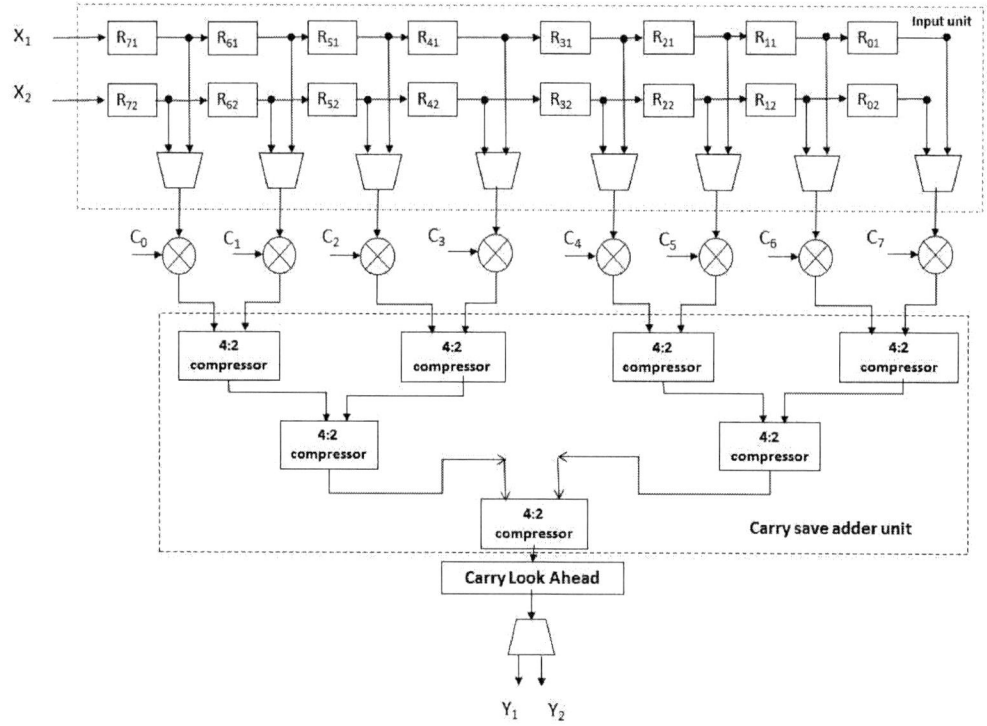

Fig. 3. Proposed multi-channel FIR filter structure.

978-1-5386-1357-3/17 $31.00 © 2017 IEEE

TABLE II
AREA, DELAY, POWER AND ENERGY COMPLEXITY FOR PROPOSED AND
EARLIER REPORTED MULTICHANNEL FILTER

Performance	Structure	Filter order		
		4	8	16
Area	[5]	13349	27404	55531
	Proposed	18529	36111.03	71224
Delay	[5]	14.64	18.07	21.57
	Proposed	11.38	12.48	14.33
	Advantage (in %)	22	30	33
Power	[5]	450.03	984	1941
	Proposed	492.39	968.86	1892.4
	more or less (in %)	-9	2	3
Energy	[5]	195429.4	495190.3	1197804
	Proposed	210860	450665.7	1020640
	Saving (in %)	14	32	35

to the barrel-shifter. The APC word thus obtained is negated when $X_2 = 0$ and is shifted through appropriate number of bit locations by the control word generated by the control circuit. The reset signal generated by the control circuit, resets the APC if X_1, X_0, X_{-1} all are 1. The final partial product is generated by adding $4.C$ or $-4.C$ depending on whether X_3 is 0 or 1. The structure of an 8-bit multiplier using one dual port LUT, two APC generation blocks and adder block is shown in the Fig. 2(b). Here the $APC0, X_2', 4.C/-4.C$ corresponding to X_3, X_2, X_1, X_0 and $APC1, X_6', 4.C/-4.C$ corresponding to X_7, X_6, X_5, X_4 are generated and fed to an adder tree to get the result in the form of a sum word and a carry word.

D. Proposed Structure of Multi-channel FIR Filter

The proposed structure for multi-channel FIR filter based on the proposed LUT-based multiplier and an adder tree is given in Fig 3. Here the LUT based multiplication units generates the results in Sum and Carry form. The results of these multiplier units are given to a carry save adder unit. The output of carry save adder structure is given as input to a carry-look-ahead (CLA) adder. The CLA output is demultiplexed to get the filtered output corresponding to two different channels.

IV. SIMULATION RESULTS AND DISCUSSION

The proposed multi-channel FIR filter is coded in Verilog and synthesized by Synopsys design Compiler using 180-nm library. The existing design [5] is also synthesized with the same technology library. We have used Wallace-tree and carry-look-ahead adder in the filtering part of the design of [5]. The delay, power and area complexity are estimated from the synthesis results and are listed in Table II. It is found that the proposed design has smaller critical path delay. This is because of two reasons. First, the multiplication is performed by using LUT of only two memory locations. Secondly, the final addition of carry and sum words resulting from the compression of partial products are transferred to the structural

adder. This avoids any carry propagation delay in the final addition stage of proposed LUT based multiplier. So, the proposed multi-channel filter provides 22, 30 and 33 % less cycle periods than [5]. There is marginal saving in power consumption for higher tap filters. However, this design need 14, 32 and 35 % less energy per sample over [5]. In terms of area, the proposed design requires higher area because of the use of control structures of the memory unit and generation of the desired partial products from the LUT output.

V. CONCLUSION

In this paper, we have shown the possibility of combining higher radix Booth encoding, OMS and APC and used it in LUT-based approach for multi channel FIR filter implementation for SDR system. The proposed multi-channel filter provides 22, 30 and 33 % less cycle period and 14, 32 and 35 % less energy per sample over the recently proposed design for multi-channel filter [5], for tap size 4, 8 and 16 respectively for the multi-channel FIR filter.

REFERENCES

[1] Eric Smith, Embedded Wireless (WWAN) and LTE CE Devices: Global Market Forecast. from Strategy Analytics, Jun 13 2013.

[2] Szlachetko, Bogusaw, and Andrzej Lewandowski. "A Multichannel Receiver of the Experimental FM Based Passive Radar Using Software Defined Radio Technology." International Journal of Electronics and Telecommunications 58, no. 4 (2012): 301-306.

[3] A.P.Vinod and E.M-K.Lai, Low power and high-speed implementation of FIR filters for software defined radio receivers, IEEE Transactions on Wireless Communications, vol. 5, no. 7, pp. 1669-1675, July 2006.

[4] Cleveland, John F., and Thomas G. Donich. "Multiple-channel software defined radios and systems using the same." U.S. Patent 8,279,796, issued October 2, 2012.

[5] Ming, Liu, and Yan Chao. "The Multiplexed Structure of Multi-channel FIR Filter and its Resources Evaluation." In Computer Distributed Control and Intelligent Environmental Monitoring (CDCIEM), 2012 International Conference on, pp. 764-768. IEEE, 2012.

[6] Pari, J. Britto, and D. Vaithiyanathan. "An Efficient Multichannel FIR Filter Architecture for FPGA and ASIC Realizations." International Journal of Applied Engineering Research 12, no. 10 (2017): 2209-2220.

[7] Booth, Andrew D. "A signed binary multiplication technique." The Quarterly Journal of Mechanics and Applied Mathematics 4, no. 2 (1951): 236-240.

[8] P. K. Meher. "LUT optimization for memory-based computation." Circuits and Systems II: Express Briefs, IEEE Transactions on 57, no. 4 (2010): 285-289.

[9] Sahoo, S. K., and K. Srinivasa Reddy. "A High Speed FIR Filter Architecture Based on Novel Higher Radix Algorithm." In VLSI Design (VLSID), 2012 25th International Conference on, pp. 68-73. IEEE, 2012.

Design of Practical Parity Generator and Parity Checker Circuits in QCA

Dharmendra Kumar, Chintoo Kumar, Shipra Gautam and Debasis Mitra
Department of Computer Science and Engineering
National Institute of Technology Durgapur, India
{dharmendrakrsp, chintook30, ships.gautam, debasis.mitra}@gmail.com

Abstract—**Quantum-dot Cellular Automata (QCA) has emerged as a possible alternative to CMOS in recent era of nanotechnology. Some attractive features of QCA include extremely low power consumption and dissipation, high device packing density, high speed (in order of THz). QCA based design of common digital modules have been studied extensively in recent past. Parity generator and parity checker circuits play important role in error detection and hence, act as essential components in communication circuits. However, very few efforts have been made for efficient design of QCA based parity generator and checker circuits so far. Moreover, these existing designs lack in practical realizability as they compromise a lot with commonly accepted design metrics such as area, delay, complexity, and cost of fabrication. This paper presents new designs of parity generator and parity checker circuits in QCA which outperform all the existing designs in terms of above mentioned metrics. The proposed designs can also be easily extended to handle large number of inputs with a linear increase in area and latency.**

Keywords-**Quantum-dot Cellular Automata; Parity generator; Parity checker; Exclusive-OR (XOR) gate.**

I. INTRODUCTION

Last six decades have seen tremendous growth in CMOS based integrated circuits. However, threatened by many physical constraints, further down-scaling of chip size seems to be reaching its limit. Consequently, the signs of deviation of chip production from the predicted course of Moore's Law have started to show [1]. Hence, the focus is shifting towards new emerging nanotechnologies which can make further down-scaling of integrated circuits possible. Quantum-dot cellular automata (QCA) is one of the promising nanotechnologies which has the potential to replace CMOS in upcoming nanotechnology era [2]. One of the most interesting feature of QCA is extremely low power dissipation and consumption. This is achieved by the fact that information flows in QCA devices without any flow of current [2]. Low power consumption and dissipation, high device packing density, high speed (in order of THz) enable realization of more dense circuits with fast switching speed, achieving room temperature operations [3]–[5] using QCA.

Design and simulation of common computing modules like adders, multipliers, multiplexers [6]–[8] have been studied enormously. However, lesser effort has been observed in the direction of designing communication circuits. Parity based method is one of the most widely used error detection techniques for the data transmission [9]. In digital systems, binary data being transmitted and processed, may be subjected to noise that may alter data bits from 0s to 1s and vice versa. A parity bit, that indicates whether the number of 1s present in the data word is even or odd, is added to the original data word during transmission from the transmitter. At the receiving end, parity bit of the received word is counted by counting the number of 1s in it and is compared with the transmitted one to detect the presence of an error in the data. A parity generator is a combinational logic circuit that generates the parity bit in the transmitter [9]. On the other hand, a circuit that checks the parity in the receiver is called parity checker [9]. A combined circuit or device consisting of parity generator and parity checker is commonly used in digital systems to detect the single bit errors in the transmitted data word.

A parity generator accepts an $(n-1)$-bit stream data and generates the additional parity bit that is to be transmitted with the bit stream. In even parity bit scheme, the parity bit is 0 (1) if there are even (odd) number of 1s in the data stream. In odd parity bit scheme, the parity bit is 1 (0) if there are even (odd) number of 1s in the data stream. A parity checker accepts an n-bit stream including $(n-1)$-bit data and the parity bit transmitted along with it and generates the parity bit for the data thus received. Parity checker at the receiver can be even or odd depending on the type of parity generator used at the transmitter end. For an even parity checker, an error is indicated by the output 1 (i.e., the number of 1s in its input is found to be odd instead of even). Similarly, for an odd parity checker, an error is indicated by the output 1 (i.e., the number of 1s in its input is found to be even instead of odd).

A few designs of parity generator and parity checker circuits in QCA have been presented in the literature [10]–[15]. However, existing designs lack in practical realizability as they compromise a lot with commonly accepted design metrics such as area, delay, complexity, and cost of fabrication. It may be noted that the basic principle involved in the implementation of parity circuits is that sum of odd number of 1s is always 1 and sum of even number of 1s is always zero. Hence, XOR function, that produces 0 (1) output when there are even (odd) number of 1s in the inputs, plays a pivotal role in implementing such circuits. For example, an $(n-1)$-bit parity generator can be realized by implementing an $(n-1)$-bit XOR function. Similarly, an n-bit parity checker, for checking the parity thus generated, can be realized by implementing an n-bit XOR function. Accordingly, overall efficiency of

978-1-5386-1357-3/17 $31.00 © 2017 IEEE

such circuits depends a lot on the implementation of XOR functions. A careful scrutiny of the existing designs of parity generator and checker circuits reveal that all these designs use cascaded 2-input XOR gates (without putting much effort in optimizing the individual XOR gates) for implementing the desired XOR function. In this paper, we have used a combination of 2-input and 3-input XOR gates to implement the desired XOR function for realizing parity generator and checker circuits in QCA. We have effectively utilized the fact that implementation of an n-bit XOR function in QCA can be optimized by using a combination of 2-input and 3-input XOR gates using ESOP based transformations [16] rather than using 2-input XOR gates only. It also helps in realizing larger parity generator and checker circuits using the smaller versions in a systematic manner. Simulation experiments performed to compare the proposed designs of QCA parity generator and checker circuits with the existing ones also demonstrate the expected benefit. The proposed ones are found to outperform all the existing designs in terms of commonly accepted design metrics.

The rest of the paper is organized as follows: Section II introduces the fundamentals of QCA technology. Section III reviews the related prior work. The proposed designs are presented in Section IV. Summary of comparative study between the proposed design with the existing ones is illustrated in Section V. Finally, Section VI draws the conclusion of this work.

II. BASICS OF QCA

The concept of QCA was first demonstrated by Metal-Island implementation [17]. Other possible implementation mechanism include semiconductor, molecular and magnetic [17]. In this work, we have considered semiconductor implementation of QCA. Basic operation of such QCA devices is based on the quantum mechanical effects and quantization of Coulombic charge [2]. The fundamental element, often referred to as QCA cell, is a square-shaped container-like structure to hold the charge. Each QCA cell has four potential wells (dots), one at each corner of the cell and two free electrons which are capable of tunnelling quantum mechanically, from one quantum dot to another. At equilibrium, the two electrons inside a cell always occupy the antipodal sites due to Coulombic repulsion. This gives way to two energetically equivalent arrangements. These two arrangements, as shown in the Fig. 1, are denoted as two different polarizations p = +1 and p = -1 which represent logic 1 and logic 0, respectively. Information flow in QCA is

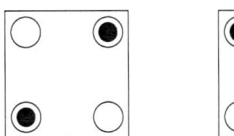

Fig. 1: Binary representation of QCA cells

achieved by Coulombic interactions between electrons present in neighboring cells without flow of electrons which leads to

extremely low power dissipation. In a series of QCA cell, every cell just rearranges their polarized state according to the adjacent cell to make the flow of information possible.

Majority gate or majority voter (M) and inverter gate (I) [6] are the two basic building blocks of any QCA circuits. Fig. 2 shows their design layout in QCA. Universal nature of the combination of these two gates facilitates implementation of any logic circuit using them.

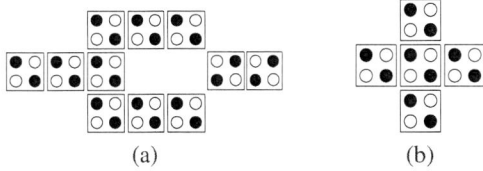

Fig. 2: Fundamental building blocks of QCA design layout (a) Inverter (b) Majority voter

QCA allows two wires to cross each other in the same layer without interfering each other. As shown in Fig. 3(a), such coplanar crossover [6] is realized using two different types of wires: a binary wire (consisting of a series of normal QCA cells) and an inverted chain (consisting of cells rotated $45°$ from their normal orientation). Wire crossings implemented using multiple layers (similar to metal wire crossovers in CMOS) are also possible in QCA (Fig. 3(b)). A quasi-adiabatic

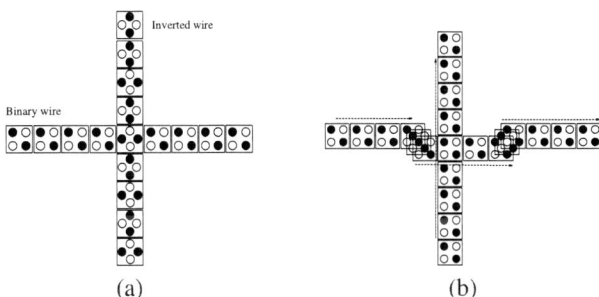

Fig. 3: (a) Coplanar and (b) Multilayer crossover in QCA

clocking mechanism is also used in QCA for synchronization of information and to meet the power requirement of the QCA device. A four-phase clock zone system, introduced by Lent et al. [2], with a $90°$ phase shift from one clock zone to the next is commonly used (Fig. 4). The four phases (zones) are named as switch, hold, release and relax, respectively. The orientations or state of electrons in a QCA cell is changed during switch or release phase only.

III. RELATED PRIOR WORK

A significant part of the research on QCA so far has focused on the design and simulation of basic logic gates [6] and various digital modules including adders [18], [19], multipliers [7], multiplexers [8]. However, as mentioned in Section I, very few such efforts can be found in the literature in designing communication circuits and their components such as parity generators and checkers. A 4-bit even parity checker

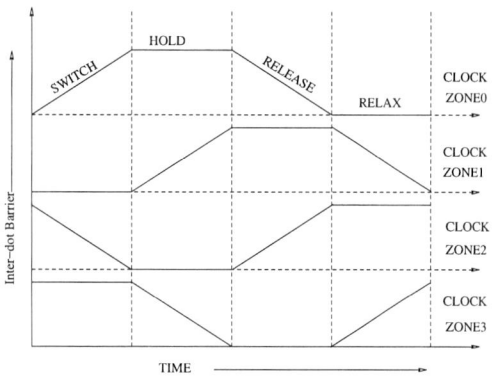

Fig. 4: QCA clocking

consisting of three 2-input XOR gates with both coplanar and multilayer crossovers was first proposed by Teja *et al.* [13]. Overall, the design consumes a large number of QCA cells (299 QCA cells) and incurs high latency (8 clock zones). With an intention of improving the design in terms of area, Mustafa and Beigh [15] proposed a 4-bit odd parity checker that consumes 145 QCA cells. However, this design is found to incur prohibitively large latency (12 clock zones). In [15], Mustafa and Beigh also presented a 3-bit even parity generator that consumes 99 QCA cells and incurs a latency of 8 clock zones. Later, Ahmad *et al.* [11] proposed designs of 4-input even and odd parity checkers both of which consume 94 QCA cells and incur a latency of 7 clock zones. Ahmad *et al.* [11] also proposed a 3-bit even parity generator and a 3-bit odd parity generator which consume 64 and 66 QCA cells, respectively, and incur a latency of 11 and 7 clock zones, respectively. An area efficient even parity generator and checker circuit was later proposed by Santra [14] where the 3-bit even parity generator consumes only 60 QCA cells and 4-bit even parity checker consumes 117 QCA cells. However, the latency incurred by them is 8 clock zones and 9 clock zones, respectively. A 3-bit odd parity generator and a 4-bit odd parity checker were proposed by Das and De in [10], using reversible logic for nano-communication. The design of parity generator consumes 72 QCA cells and incurs a latency of 7 clock zones whereas the design of the parity checker (that uses a 2×2 Feynman gate structure [20]) consumes 126 QCA cells and latency of 8 clock zones. It is apparent that all of these existing designs compromise a lot with one or more commonly accepted design metrics such as area, delay, complexity, and cost of fabrication. Moreover, most of these designs present smaller sized parity generators and checkers (3-bit/4-bit) and hardly provides any clue so that they can be extended to realize such circuits of bigger size (15-bit/16-bit or even bigger). The above mentioned drawbacks act as significant barrier against the practical realizability of these designs. Accordingly, efficient and practically realizable designs of parity generators and parity checkers have become very much essential.

IV. Proposed QCA Parity Generator and Parity Checker Circuits

As mentioned in Section I, XOR function, that produces 0 (1) output when there are even (odd) number of 1s in the inputs, plays a pivotal role in implementing parity generator and checker circuits. An n-input XOR function is usually implemented by combining several 2-input XOR gates. Moreover, use of existing designs of QCA 2-input XOR gates [15], [21], [22], which use a large number of majority gates, lead to significant compromise with the area as well as latency. For instance, implementation of a 4-bit XOR function using three 2-input XOR gates [15] takes at least 9 majority gates. However, we have observed that more efficient implementation of n-input XOR function is possible by using combinations of 2-input and 3-input XOR gates following ESOP based transformation [16]. Accordingly, we have used combination of 2-input and 3-input XOR gates to implement n-bit XOR function instead of relying solely on 2-input XOR gates which was the case in existing implementations. We have also used majority logic reduction [23]–[25] to further optimize the designs of individual XOR gates (both 2-input and 3-input).

The logical expression $(A\overline{B}+\overline{A}B)$, representing 2-input XOR function can be re-written equivalently as $M[M(A,B,1), \overline{M(A,B,0)}, 0]$ using majority logic reduction, where $M(X,Y,Z)$ represents a 3-input majority gate [6] with inputs X, Y, and Z. The above Boolean expression can be implemented using three 3-input majority gates and one inverter as shown in Fig. 5(a). Similarly, the logical expression $(\overline{A}\overline{B}C + \overline{A}B\overline{C} + A\overline{B}\overline{C} + ABC)$, representing a 3-input XOR function can be re-written as $M[M(\overline{A},\overline{B},\overline{C}),C,M(A,B,\overline{C})]$ using majority logic reduction, where $M(X,Y,Z)$ represents a 3-input majority gate with inputs X, Y, and Z. Fig. 5(b) shows the schematic diagram of the gate level implementation of the above expression. The logical expression for the

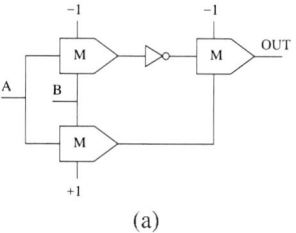

(a)

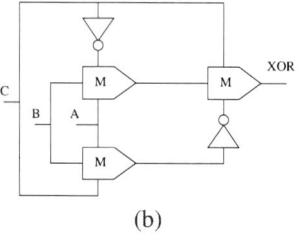

(b)

Fig. 5: Gate level implementation of (a) 2-input XOR and (b) 3-input XOR

978-1-5386-1357-3/17 $31.00 © 2017 IEEE

output of 3-bit even parity generator is $A \oplus B \oplus C$ and hence, it can be implemented simply by using the 3-input XOR gate of Fig. 5(b). The logical expression for the

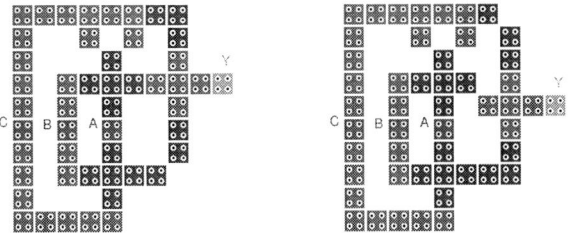

Fig. 6: Layout of the proposed 3-bit (a) even (b) odd parity generator

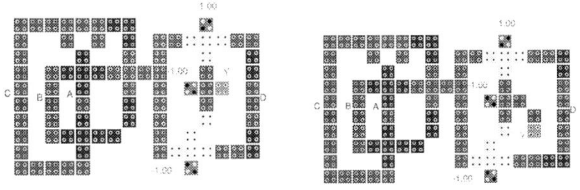

Fig. 7: Layout of the proposed 4-bit (a) even (b) odd parity checker

output of 3-bit odd parity generator (which is $A \oplus B \ominus C$) can be re-written as $(\bar{A}\bar{B}\bar{C} + \bar{A}BC + A\bar{B}C + AB\bar{C})$ i.e., $M[M(A,B,C), \bar{C}, M(\bar{A}, \bar{B}, C)]$ using majority logic reduction. The above expression indicates that 3-bit odd parity generator can also be implemented by using three majority gates. Fig. 6(a) and Fig. 6(b) show the layouts of the proposed 3-bit even and odd parity generators, respectively. As apparent from the figures, both the designs consist of 49 QCA cells without any crossover and incur 3 clock zones (0.75 clock cycle) latency. Assuming QCA cell size of $18nm \times 18nm$ with a gap of $2nm$ between two consecutive cells, each of the layouts consumes an area of $0.04\mu m^2$.

The logical expression for the output of 4-bit even parity checker (which is $A \oplus B \oplus C \oplus D$, where 'D' represents the transmitted parity bit) is same as that of a 4-bit XOR gate and hence, it can be implemented by combining a 3-input XOR gate and a 2-input XOR gate. Similarly, the logical expression for the output of 4-bit odd parity checker is $A \oplus B \ominus C \ominus D$ which can also be realized using a 3-input XOR gate and a 2-input XOR gate. Fig. 7(a) and Fig. 7(b) show the layouts of the proposed 4-bit even and odd parity checkers, respectively. As apparent from the figures, the proposed designs consist of 84 QCA cells and 88 cells, respectively. However, both of them consume same area ($0.08\mu m^2$) assuming QCA cell size of $18nm \times 18nm$ with a gap of $2nm$ between two consecutive cells. Moreover, both the designs incur latency of 5 clock zones (1.25 clock cycles) and have no crossover.

In order to verify the functional behavior of the proposed parity generator and checker circuits, we carried out simulations using the bistable simulation engine of QCADesigner [26] (version 2.0.3) with the following parameters: (i) QCA

cell dimension: $18nm \times 18nm$ with a gap of $2nm$ between two consecutive cells (ii) Radius of effect: $65nm$, (iii) Relative permittivity: 12.9, (iv) Convergence tolerance: 0.001000. It may be noted that the bistable simulation engine of QCADesigner uses intercellular Hartree approximation (ICHA) assuming a simple two-state system to represent each QCA cell. A little compromise in accuracy as compared to full-basis computation is often compensated by the significantly better scalability [27]. Simulation results are found to show significantly strong polarization (more than 0.954) at the output of all the circuits.

It may also be noted that the proposed designs can be systematically extended to handle any number of inputs (n). For example, Fig. 8 and Fig. 9 show the layouts of a 15-bit parity generator and a 16-bit parity checker circuit, respectively. In order to estimate the growth of various

TABLE I: Generalized expressions for various design metrics of proposed n-bit parity generator and parity checker

Circuits	Generalized expressions for		
	Area (μm^2)	Latency (clock cycles)	No. of Crossover
Parity Generator	$0.02(7n - 11) \times$ $0.02(1.5n + 11.5)$; $n \geq 5$	$n/4$	$1.5(n-3)$
Parity Checker	$0.02(7n - 9) \times$ $0.02(1.5n + 10)$; $n \geq 5$	$(n + 1)/4$	$1.5(n-4)$

design metrics as a function of the number of inputs (n), we have computed the general expressions for area, latency, and number of crossovers. Table I shows the expressions for n-bit parity generator and checker. Figs. 10-11 may be referred for the graphical representations of the growth in area and latency, respectively. It is apparent that both the parameters grow somewhat linearly with the increase in the value of n.

V. COMPARATIVE STUDY

In order to evaluate the effectiveness of the proposed designs of parity generators and checkers, we have compared each of them with existing ones in terms of commonly accepted design metrics such as area, latency, complexity, and the type and number of crossovers used. Note that the complexity of a QCA circuit can be expressed as $M+I+C$ [28], where M, I, and C refer to the number of majority gates, the number of inverters and the cost of crossovers used in the circuit, respectively. Table II and Table III show the summary of the comparative study made on 3-input parity generators and 4-input parity checkers, respectively. It is apparent from the tables that the proposed designs outperform all the existing designs in terms of all the design metrics.

As suggested by Liu *et al.* [28], instead of considering the individual metrics for comparison, cost functions combining multiple metrics may be more effective. For the sake of completeness, we have also included a case of comparison based on a cost function ($Cost = (M^2 + I + C^2) \times T$) specifically designed for QCA circuits [28]. Figs. 12-13 illustrate the comparison for 3-bit parity generators and 4-bit parity checkers, respectively. The proposed designs are found to be superior with respect to this cost function too.

978-1-5386-1357-3/17 $31.00 © 2017 IEEE

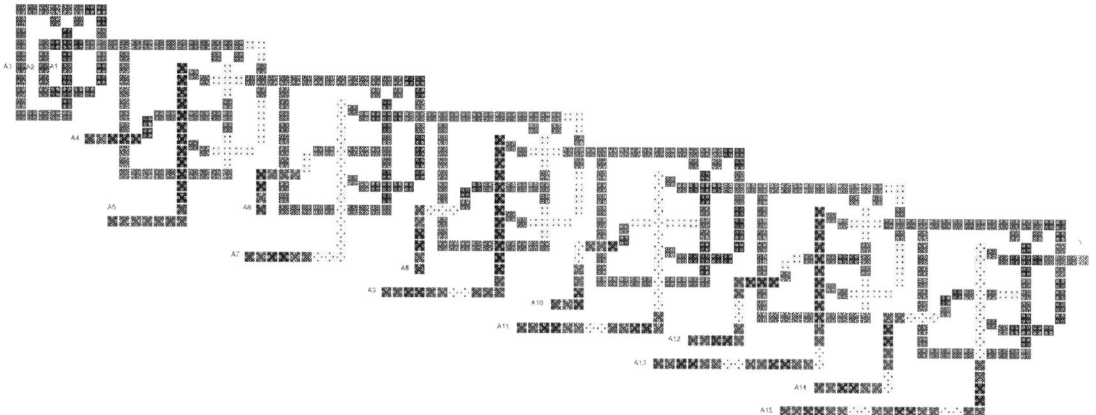

Fig. 8: Layout of the proposed 15-bit even parity generator

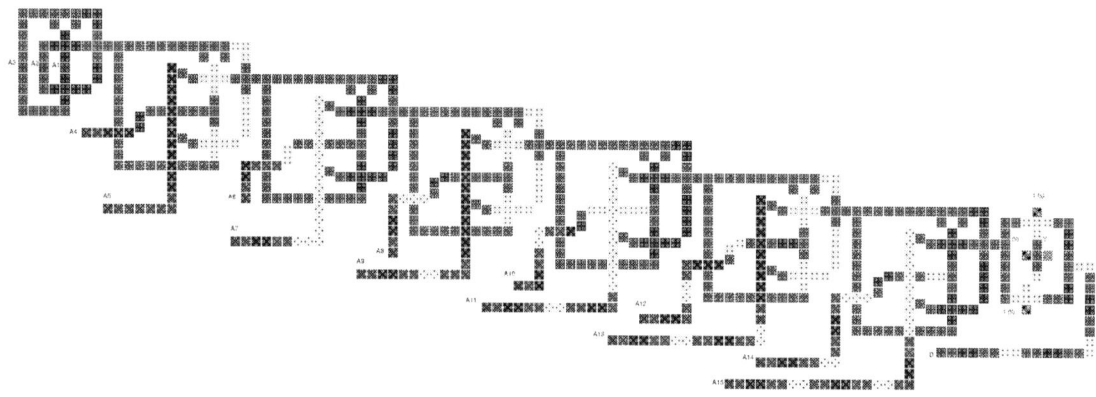

Fig. 9: Layout of the proposed 16-bit even parity checker

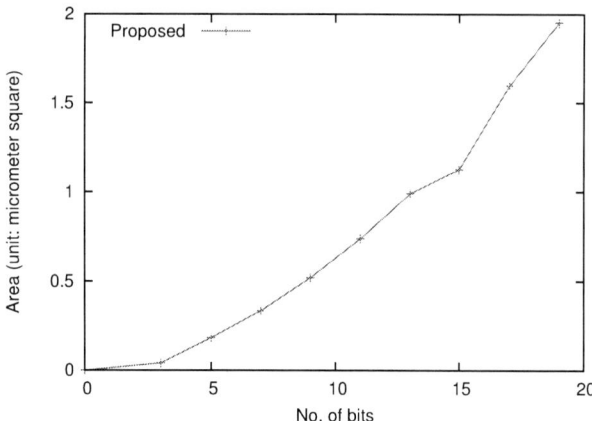

Fig. 10: Growth in area of the proposed n-bit parity generator with respect to increasing value of n

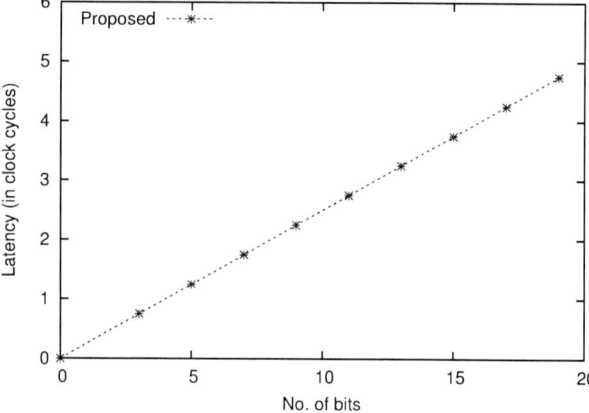

Fig. 11: Growth in latency of the proposed n-bit parity generator with respect to increasing value of n

VI. CONCLUSION

Efficient designs of 3-bit parity generator and 4-bit parity checker circuits in QCA have been presented. Both the designs are found to outperform all the existing designs in terms of common design metrics such as area, latency, and more importantly in-terms of cost function specially designed for QCA circuits. Moreover, both the designs can be extended easily for large number of inputs with linear increase in area and latency, thereby, making them suitable for practical realization.

REFERENCES

[1] C. C. Mann, "The end of Moore's law?" MIT Technology Review, May 2000, http://www.technologyreview.com/featuredstory/400710/the-end-of-moores-law/.

TABLE II: Comparisons of various 3-bit parity generators in terms of common design metrics

Parity Generator	Area (μm^2)	Latency (clock zones)	Type of crossover	Number of crossover	Complexity
[14], Even	0.06	8	None	–	8
[15], Even	0.17	8	None	–	8
[11], Even	0.09	11	Multilayer	2	14
[12], Even	0.09	12	Multilayer	2	14
Proposed (Even)	**0.04**	**3**	**None**	**–**	**5**
[10], Odd	0.08	7	None	–	11
[11], Odd	0.09	7	Multilayer	2	15
Proposed (Odd)	**0.04**	**3**	**None**	**–**	**6**

TABLE III: Comparisons of various 4-bit parity checkers in terms of common design metrics

Parity Checker	Area (μm^2)	Latency (clock zones)	Type of crossover	Number of crossover	Complexity
[14], Even	0.13	9	Coplanar	1	15
[15], Even	0.28	12	None	–	15
[11], Even	0.11	7	Multilayer	3	21
[12], Even	0.12	7	Multilayer	3	21
[13], Even	0.53	8	Coplanar & Multilayer	4 1	30
Proposed (Even)	**0.08**	**5**	**None**	**–**	**9**
[10], Odd	0.15	8	None	–	18
[11], Odd	0.13	7	Multilayer	3	22
Proposed (Odd)	**0.08**	**5**	**None**	**–**	**10**

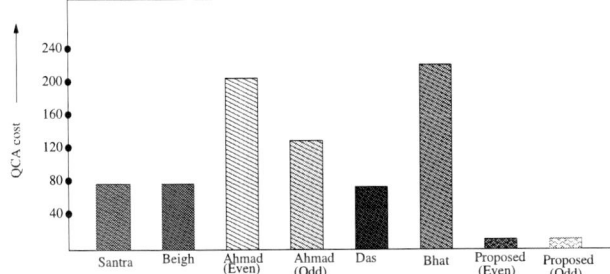

Existing and proposed 3−bit parity generators $\longrightarrow$

Fig. 12: Comparison of proposed 3-bit parity generator with the existing ones in terms of $Cost = (M^2 + I + C^2) \times T$

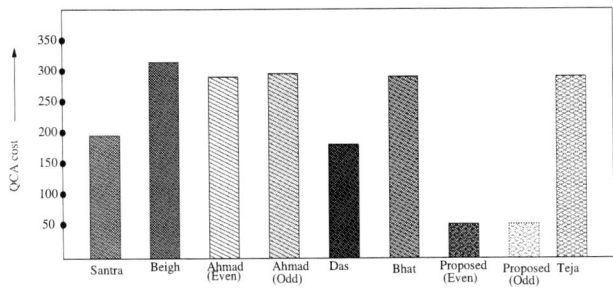

Existing and proposed 4−bit parity checkers $\longrightarrow$

Fig. 13: Comparison of proposed 4-bit parity checker with the existing ones in terms of $Cost = (M^2 + I + C^2) \times T$

[2] C. S. Lent et al., "Quantum-dot cellular automata," *Nanotechnology*, vol. 4, pp. 49–57, 1993.

[3] C. S. Lent and B. Isaksen, "Clocked molecular quantum-dot cellular automata," *IEEE Trans. Electron Devices*, vol. 50, pp. 1890–1896, 2003.

[4] R. P. Cowburn and M. E. Welland, "Room temperature magnetic quantum cellular automata," *Science*, vol. 287, no. 5457, pp. 1466–1468, 2000.

[5] Y. Wang and M. Lieberman, "Thermodynamic behavior of molecular-scale quantum-dot cellular automata QCA wires and logic devices," *IEEE Trans. on Nano*, vol. 3, no. 3, pp. 368–376, 2004.

[6] P. D. Tougaw and C. S. Lent, "Logical devices implemented using quantum cellular automata," *Journal of Applied Physics*, vol. 75, no. 3, pp. 1818–1825, 1994.

[7] H. Cho and E. E. Swartzlander, "Adder and multiplier design in quantum-dot cellular automata," *IEEE Trans. on Computers*, vol. 58, no. 6, pp. 721–727, 2009.

[8] V. A. Mardiris and I. G. Karafyllidis, "Design and simulation of modular 2^n to 1 quantum-dot cellular automata (QCA) multiplexers," *International Journal of Circuit theory and Applications*, vol. 38, pp. 771–785, 2010.

[9] M. M. Mano, Ed., *Digital Logic and Computer Design*. Prentice Hall of India Pvt. Ltd., 2004.

[10] J. C. Das and D. De, "Quantum-dot cellular automata based reversible low power parity generator and parity checker design for nanocommunication," *Frontiers of Information Technology and Electronic Engineering*, vol. 17, pp. 224–236, 2016.

[11] F. A. et al., "Design and analysis of odd and even parity generators and checkers using QCA," in *INDIACom*, 2015, pp. 187–193.

[12] F. Ahmad and G. M. Bhat, "Novel code converts based on QCA," *Intl. journal of science and research*, vol. 3, no. 5, 2014.

[13] V. C. Teja et al., "QCA based multiplexing of 16 arithmetic & logical subsystems-a paradigm for nano computing," in *Intl. Conf. on Nano/Micro Engineered and Molecular Systems*, 2008, pp. 758–763.

[14] S. Santra and U. Roy, "Design and optimization of parity generator and parity checker based on quantum-dot cellular automata," *IJCEACIE*, vol. 8, no. 3, 2014.

[15] M. Mustafa and M. R. Beigh, "Design and implementation of QCA based novel parity generator and checker circuit with minimum complexity and cell count," *Indian Journal of Pure and Applied Physics*, vol. 51, pp. 60–66, 2013.

[16] D. Y. Feinstein and M. A. Thornton, "ESOP transformation to majority gates for quantum-dot cellular automata logic synthesis," in *Reed-Muller Workshop*, 2007, pp. 43–50.

[17] M. O'Neill, A. Lau, and E. E. Swartzlander, *Design of Semiconductor QCA Systems*. Artech House Publishers, 2014.

[18] R. Zhang et al., "Performance comparison of quantum-dot cellular automata adders," in *ISCAS*, 2005, pp. 2522–2526.

[19] D. Kumar and D. Mitra, "Design of a practical fault-tolerant adder in QCA," *Microelectronics*, vol. 53, pp. 90–104, 2016.

[20] R. P. Feynman, "Quantum mechanical computer," *Optics News*, vol. 11, pp. 11–20, 1985.

[21] M. R. Beigh et al., "Performance evaluation of efficient XOR structures in quantum-dot cellular automata (QCA)," *Circuits and Systems*, vol. 4, pp. 147–156, 2013.

[22] W. S. Jahan et al., "Circuit nanotechnology: QCA adder gate layout designs," *IOSR Journal of Computer Engineering*, vol. 16, pp. 70–78, 2014.

[23] R. Z. et al., "A method of majority logic reduction for quantum cellular automata," *IEEE Trans. Nanotechnol*, vol. 3, no. 4, pp. 443–450, 2005.

[24] H. Mahmoud, "Systematic minimization technique for majority-majority digital combinational circuits," *Science alert*, vol. 11, no. 5, pp. 832–839, 2011.

[25] K. Kong et al., "An optimized majority logic synthesis methodology for quantum-dot cellular automata," *IEEE Trans. Nanotechnol*, vol. 9, no. 2, pp. 170–183, 2010.

[26] K. Walus et al., "QCADesigner: A rapid design and simulation tool for quantum-dot cellular automata," *IEEE Trans. Nanotechnol*, vol. 3, no. 1, pp. 26–31, 2004.

[27] M. LaRue et al., "Stray charge in quantum-dot cellular automata: A validation of the intercellular hartree approximation," *IEEE Transactions on Nanotechnology*, vol. 12, no. 2, pp. 225–233, 2013.

[28] W. Liu et al., "A first step toward cost functions for quantum-dot cellular automata designs," *IEEE Trans. Nanotechnol*, vol. 13, pp. 476–487, 2014.

978-1-5386-1357-3/17 $31.00 © 2017 IEEE

2017 IEEE International Symposium on Nanoelectronic and Information Systems

Design and Optimization of Single Electron Transistor based 4-Bit Arithmetic and Logic Unit at Room Temperature Operation

Rathin Joshi, Yash Agrawal, Rutu Parekh
DA-IICT, Gandhinagar, India

Abstract— Single electron transistor (SET) has been envisaged as a potential device to achieve high-end performance in deep sub-micron technologies. The paper innovatively presents a 4-bit ALU based on single electron transistor (SET). The proposed logic design model is encouragingly operational at room temperature. SET based ALU has been designed, simulated and optimized with incorporation of all the parameters in the feasible fabrication range. Design and optimization have been performed in hierarchical manner i.e. from basic cells (device or transistor level) to circuit level. Performance comparison between SET, MOS and hybrid SET-MOS based circuits has been evaluated. From the simulation results, it is investigated that ALU based on SET with optimized parameters is more efficient compared to its MOS counterpart. The percentage improvements in SET over CMOS based ALU design in power, delay and power-delay product are 65.4%, 79.7% and 92.9% respectively. These improvements in SET over hybrid SET-MOS based ALU are 0.6%, 33.7% and 33.8 % respectively. The analyses have been performed at 45nm technology node using Cadence EDA tool.

Keywords— *ALU, power-delay product (PDP), single electron transistor (SET), SET-CMOS hybridization, SET drivability.*

I. INTRODUCTION

Due to constant downscaling in feature size of the existing CMOS technology, performance degradation has been experienced. It is quite evident that further down scaling in technology node will not be maintained (mainly in sub-10 nm region) because of short channel effects [1]. Carbon nanotubes, quantum dots, single electron transistor (SET), and finFETs are the few candidates for replacement or co-existence with CMOS [1]. From recent research and literature studies, the use of a SET as a potential building block for future hybrid circuits is confirmed and justified. SET has wide range of applications, mainly categorized as SET logic and memory applications. In this paper, SET's application in 4 bit arithmetic and logic unit (ALU) is discussed.

SET's working and its formation is provided in literature [2, 3]. Coulomb Blockade and single electron tunneling are the two phenomenon's responsible for SET to work as a switching device. SET function as a p-switch and a n-switch can be obtained by adjusting back gate potential of a SET. This arrangement is ultimately known as nSET and pSET, which replicates nMOS and pMOS.

Much work has been done to address logic circuits comprising of only SETs [4]–[9] and hybrid SET-CMOS

circuits [10]-[12] with one or more constraints related to design and simulation. Hybrid SET- CMOS ALU has been addressed to achieve low power and better performance due to SET's drawback to low driving capability [13, 14]. But the design and simulation parameters are restricted to either, low temperature, unrealistic SET parameters and low output voltages. In case of hybrid SET-CMOS circuits, the fabrication is a challenge as it needs both the transistors with different fabrication process on the same IC. The main objective of the work discussed here is to design and simulate a 4 bit ALU with 12 functionalities for SET / SET-CMOS / CMOS implementation using Cadence virtuoso design environment at 45 nm technology node and compare their performances for same output voltages.

The SET parameters are within fabrication range using nanodamascene process [15], which is briefly discussed in [16]. Because of fabricating capacitors in sub attofarad (aF) range, SETs operating at temperatures exceeding 130 ℃ is obtained. The simulation is based on SETs Mahapatra–Ionescu–Banerjee model [17]. The SET logic has been simulated with realistic SET parameters based on previous work [16]. These parameters have been maintained throughout this paper. By proper design, simulation and optimization of the ALU circuit, it can be concluded that pure SET ALU is more efficient compared to SET – CMOS / CMOS ALU in terms of delay, power and power delay product (PDP) for the same CMOS comparable output voltage.

Paper formulation is as follows: Section 2 explains design of proposed 4-bit ALU in detail. Section 3 describes optimization and modifications made in design in order to enhance performance. Section 4 provides simulation results and finally section 5 is conclusion.

II. DESIGN OF A 4 BIT ALU

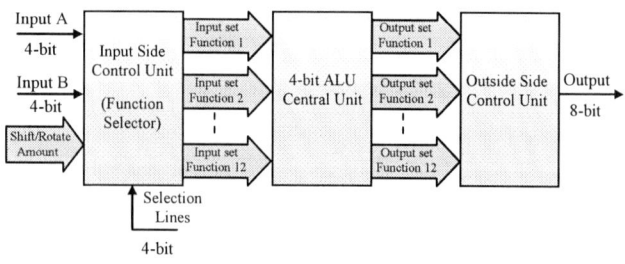

Fig. 1. Block Diagram of proposed ALU.

978-1-5386-1357-3/17 $31.00 © 2017 IEEE

Fig. 1 shows the block diagram of a proposed ALU. The basic building blocks are, (1) Central Unit (2) Input control unit and (3) Output control unit. The details are:

A. Central Unit

The basic logic operations like Inverter, NAND, NOR, XOR gate mimics CMOS [2]. NAND and NOR based SET circuit is presented in Fig. 2. The design parameters for all logic circuits are V_{dd} = input "high" = output "high" = 0.8 V, input "low" = output "low" = 0 V, gain = 1, T = 300 K, and R_t = 1 MΩ, C_j = 0.03 aF, C_g = 0.045 aF, and C_b = 0.05 aF [16] (all within the fabrication range).

The proposed ALU design has 12 functionalities as listed in Table I. As shown in Fig. 3, inputs A and B are 4 bit unsigned binary numbers, a 4 bit selection lines to select which operation is to be performed and 2 control bits for number of rotations / shifts to left / right position. Using NAND, NOR and XOR a full adder is constructed followed by other blocks. Following the central unit block diagram, the sub blocks architecture is explained.

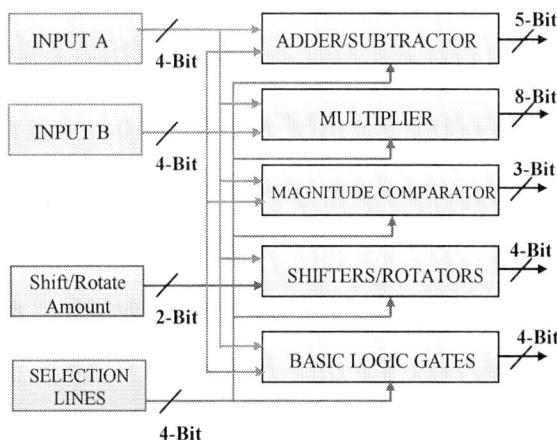

Fig. 3. Central unit block diagram.

1) Adder/Subtraction Block

Adder / subtraction block is implemented as shown in Fig. 4. In order to reduce hardware and make it power efficient, both the operations have been implemented using same adder by using XOR gates which differentiates between the addition and subtraction. For S = 1, it performs subtraction whereas for S = 0 it selects addition operation.

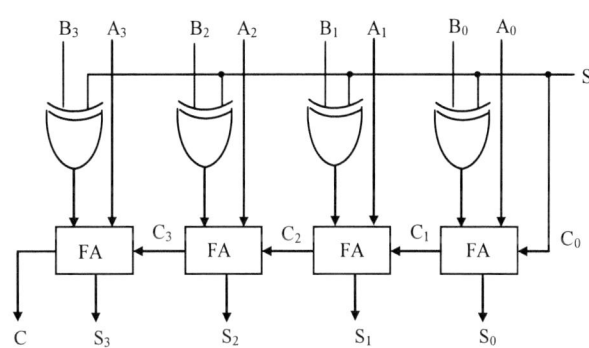

Fig. 4. 4-bit Adder/Subtractor block.

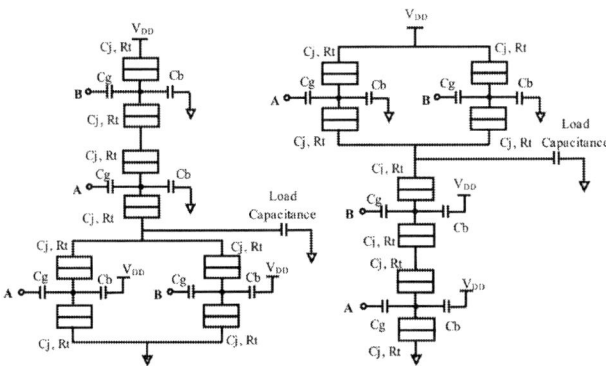

Fig. 2. SET logic circuit (a) NOR gate (b) NAND gate.

TABLE I. OPERATION TABLE FOR PROPOSED ALU

| Selection Lines | | | | Function | Output Width |
S3	S2	S1	S0		
0	0	0	0	Addition	5-bit
0	0	0	1	Subtraction	4-bit
0	0	1	0	Multiplier	8-bit
0	0	1	1	Magnitude Comparator	3-bit
0	1	0	0	Logical Left Shifter	4-bit
0	1	0	1	Logical Right Shifter	4-bit
0	1	1	0	Right Rotator	4-bit
0	1	1	1	Left Rotator	4-bit
1	0	0	0	Bitwise Inversion	4-bit
1	0	0	1	Bitwise AND	4-bit
1	0	1	0	Bitwise OR	4-bit
1	0	1	1	Bitwise XOR	4-bit

2) Multiplier

Two different configurations, which are array multiplier and Vedic Multiplier (which exist in ancient mathematics) have been tested and results shows that Vedic multiplier results in less hardware and also possess better speed [18]. In proposed design, multiplier is implemented using "Urdhva Tiryak-bhyam" formula mentioned in [18]. Critical path of proposed ALU is for multiplier block, which can be seen from delay comparison in simulation results. Structure for optimized multiplier is shown in Fig. 5.

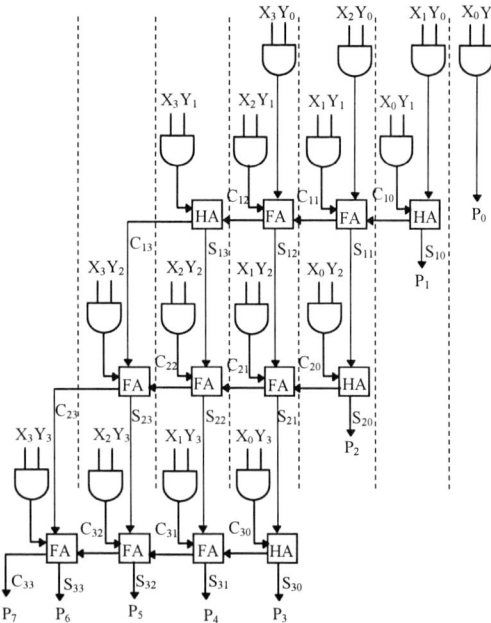

Fig. 5. 4-bit Vedic multiplier. X (x3x2x1x0) and Y (y3y2y1y0) are the two inputs whereas P7-P0 are the 8-bit final output.

3) Magnitude Comparator

Fig.6 (a) shows magnitude comparator that compares magnitude of a 4 bit unsigned binary numbers A and B. The bits in accumulator register reflect for A = B, A > B, and A < B as shown in Fig. 6 (b).

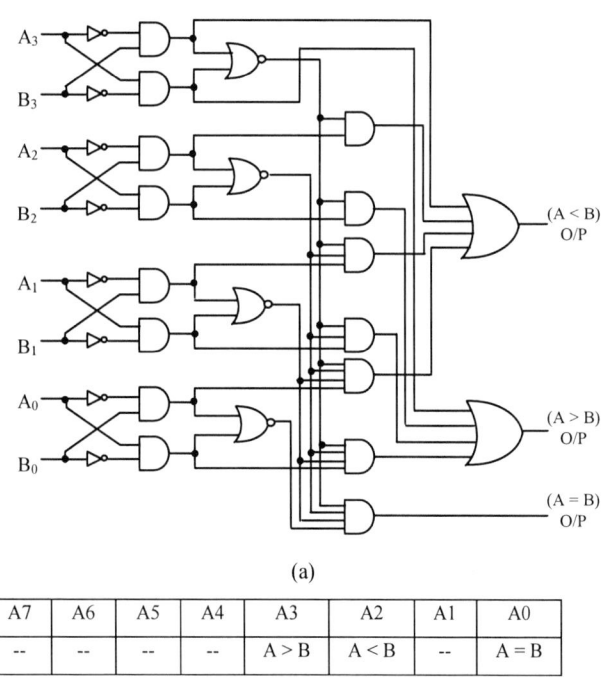

(a)

A7	A6	A5	A4	A3	A2	A1	A0
--	--	--	--	A > B	A < B	--	A = B

(b)

Fig. 6. (a) Magnitude Comparator Block Diagram.
(b) Accumulator bit allocation for comparator.

4) Logical Modules

Logical section mainly deals with bitwise logic gates, logical shifters and rotators, whose architectures are shown in Fig. 7. Shifters and rotators are implemented using 4 to 1 MUX. Since the input is 4-bit, numbers of shift/rotate input is 2-bit. These 2-bits are connected to selection line and depending on that, MUX selects any of four inputs and provides output. A 4-bit operand A (a3-a0) is input with L_Sh3 – L_Sh0 as output of left logical shifter, whereas R_Rt3 – R_Rt0 shows output of right rotator. Each Shifting or rotating operation requires four 4 to 1 MUXs.

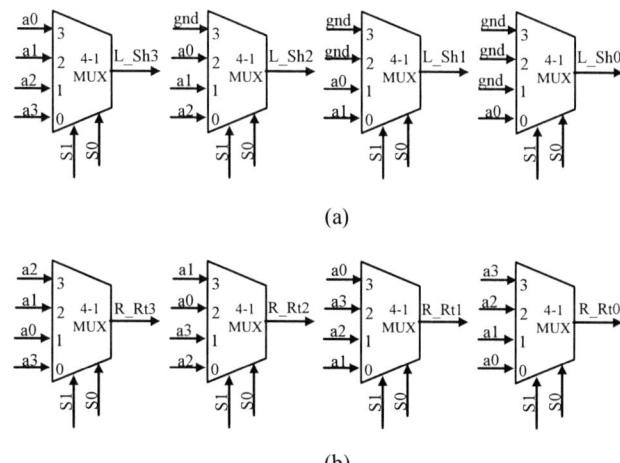

(a)

(b)

Fig. 7. (a) Left Shifter (b) Right Rotator.

B. Control Unit

Input side control unit turns on only one module at a time and makes design "power efficient" whereas output side control module merges outputs from each module and finally provide output into 8-bit accumulator. This will result in reduction of output bit width.

III. ALU OPTIMIZATION FOR ALU DESIGN

In order to improve the driving capability and hence the delay to be better than 45 nm CMOS and higher technology node, SETs can be added in parallel, which results in higher current, and hence less delay and better bandwidth [19]. Improved performance in terms of delay, bandwidth and output voltage swing can easily compensate small increment in power consumption in the range of nW because of adding SETs in parallel. Ultimately results show that PDP is significantly decreased while modifying design in this manner. In order to compare the performance parameters like delay, power, PDP of a SET / SET-CMOS ALU with CMOS ALU we shall simulate the circuits as under.

1. **SET based ALU:** Design is implemented with pure SET circuit which mimics CMOS.
2. **SET based ALU with optimized performance:** SET based ALU designed by increasing the driving capability of logic circuits by connecting SETs in

parallel [19] specially at the interface of the central unit and its input output control unit.

3. **SET-CMOS based ALU:** Arithmetic main unit is implemented in SET, whereas, control unit is implemented in 45nm CMOS. Input control unit is the driving unit whereas the output control unit acts as an interface between the central unit and the next stage.

4. **CMOS based ALU:** Design is implemented in 45 nm CMOS to compare the above circuit parameters with CMOS circuit.

From fabrication point of view SET-CMOS hybridization is feasible but with many challenges. The SETs considered in this work are metallic SETs that can be fabricated on a CMOS carrier with back-end-of-line (BEOL) compatible fabrication process. This has advantage of reduced footprint while minimizing interconnects complexity [20].

IV. SIMULATION RESULTS

Simulations have been done in Cadence Virtuoso Analog Design Environment (ADE). Supply voltage, V_{DD} = 0.8 V is maintained throughout this paper, which is compatible with 45 nm CMOS. Reason for choosing V_{DD} 0.8 V is to hybridize SET with CMOS and both can be driven by the same supply voltage. First, the gate level analysis is shown. After this, module wise analysis for power and delay in case of pure SET and CMOS logic with same architecture is presented. Further, functional verification and power delay analysis for ALU is shown. For 4-bit ALU, performance is analyzed using total power, delay and PDP.

A. Gate Level Simulation Results

Gate level analysis of basic building blocks like XOR, AND and inverter are shown in Table II. Mainly three cases are considered, these are (1) 45 nm CMOS logic (2) SET logic and (3) Optimized SET logic. Optimization and trade off of different parameters is done by adding SETs in parallel to enhance the driving capability of the logic.

TABLE II. GATE LEVEL PERFORMANCE IMPROVEMENT

Design	Logic	Power (nW)	Delay (ps)	PDP (nW*ps)
Inverter	45 nm CMOS	107.8	7.281	784.892
	SET (mimic CMOS)	1.842	6.977	12.8516
2 input XOR	45 nm CMOS	117.9	46.145	54128.1
	SET (mimic CMOS)	4.367	26.280	114.765
	SET Optimized	8.004	6	48.024
2 input AND	45 nm CMOS	134.5	34.210	4601.25
	SET (mimic CMOS)	2.428	23.062	55.9941
	SET Optimized	4.682	5.309	24.8657

B. Transient Analysis for Multiplier

Fig. 8 shows multiplier module transient analysis for both SET and CMOS logic. The multiplier has worst case delay. Waveforms for this module are shown for same set of inputs as in Fig. 8(a). Fig. 8(b) shows output transients for SET multiplier, which has better performance in terms of less spurious transitions, compared to 45 nm CMOS shown in Fig. 8(c). Red color box indicates that spurious transitions are less in SET logic compared to CMOS logic for same output voltage. These spurious transitions are responsible for overhead in power consumption. These transitions are unwanted and should be as minimum as possible.

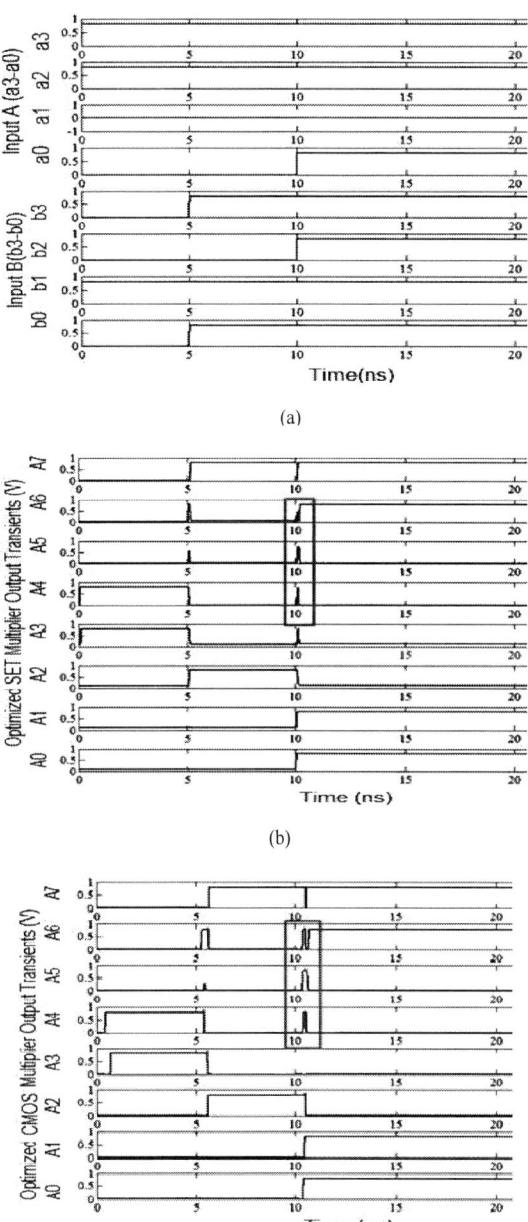

Fig. 8. (a) Common input transients for multiplier module. (b) Output transients for optimized SET Multiplier. (c) Output transients for optimized CMOS Multiplier. 8-bit product is generated in A7-A0.

C. Module Wise Delay and Power Analysis

Delay for each module was analyzed for a fixed set of inputs and based on that it can be analyzed that multiplier module has maximum delay. Module wise delay comparison is shown in Fig. 9 (a). Similar analysis has been performed for total power consumption and power for each module was analyzed for fixed set of inputs as shown in Fig. 9 (b).

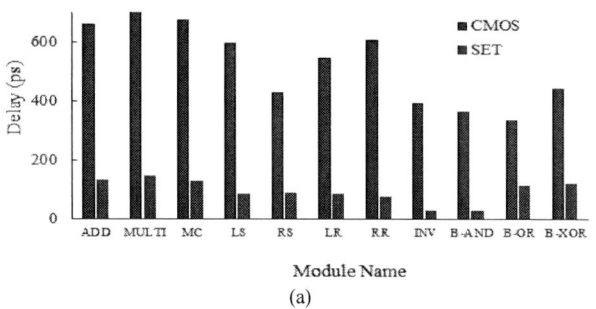

(a)

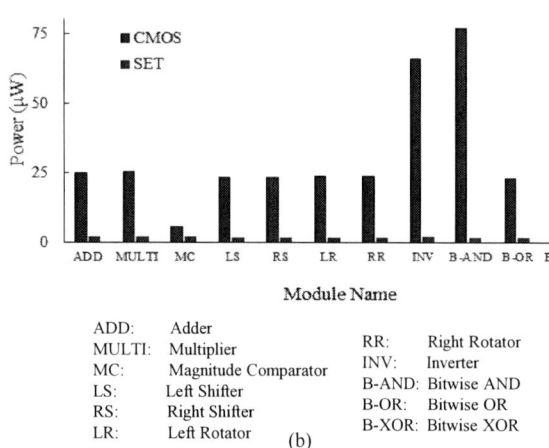

(b)

ADD:	Adder	RR:	Right Rotator
MULTI:	Multiplier	INV:	Inverter
MC:	Magnitude Comparator	B-AND:	Bitwise AND
LS:	Left Shifter	B-OR:	Bitwise OR
RS:	Right Shifter	B-XOR:	Bitwise XOR
LR:	Left Rotator		

Fig. 9. (a) Module wise delay Comparison. (b) Module wise Power comparison.

D. ALU Functional Verification

As shown in Fig. 10, in order to check the behaviour of SET ALU with optimized parameters, its 4-bit selection lines are set in such a way that it covers all the operations one after the other. Selection lines changes from 0000 to 1011 as shown as 4-bit bus. Considering function Table I, and for input data at that time, expected output has been calculated manually and for all operations its functionality has been verified and the accumulator provides expected output. Accumulator's transients are shown from A7 (MSB) to A0 (LSB). Input A and B are 4-bit input number, shown as analog values after converting digital transients into analog value. Similarly are shown 4-bit selection lines, 2-bit logical shift amount, and 2-bit rotation amount. Few example, for selection lines = 0000, it defines addition operation. Inputs are 12 and 2, hence output can be observed as 14. For accumulator transients, A7- A0 = 00001110. For selection lines = 0010, it defines multiplication. Inputs are 12 and 11, hence output can be observed as 132. For accumulator, transients A7-A0 = 10000100.

E. ALU Overall Performance Improvement

Four different versions of proposed ALU are implemented, and all of them have been analysed with same set of inputs. Performance comparison in terms of PDP is given below in table 3. Optimization is done for both SET and CMOS logic.

TABLE III.   ALU PERFORMANCE

Logic Family	Power (μW)	Delay (ps)	PDP (μW * ps)
CMOS	5.081	709.849	3606.7428
SET (mimic CMOS)	1.653	269.909	446.1595
SET (Optimized)	1.767	144.636	255.5718
SET-CMOS	1.770	218.19	386.196

From table 3, it is notable that optimized circuit of SET ALU is the most energy efficient compared to other implementations. Pure SET ALU (optimized) is more efficient compared to its CMOS counterpart with percentage difference in power, delay and PDP by 96.7 %, 132 % and 173 % respectively and when compared to hybrid SET-CMOS ALU, the percentage difference in PDP is 40 %.

I. CONCLUSION

It can be concluded that, an optimized SET based ALU design outperforms existing 45nm CMOS technology and its hybrid SET-CMOS counterpart. Due to SETs drawback of drivability it needs to be hybridized with CMOS. But the simulation results prove that pure SET based ALU operating at room temperature and CMOS comparable output voltage of 0.8 V is more efficient in terms of power, delay, and power delay product. The simulation results for power, delay and PDP for optimized SET ALU are 1.77 μW, 144.64 ps and 255.57 μW * ps when compared to 1.77 μW, 218.19 ps and 386.20 μW * ps for hybrid SET-CMOS ALU and 5.1 μW, 7.9.85 ps and 3606.74 μW * ps respectively for 45 nm CMOS. So rather than having hybrid circuit to improve the driving capability and hence the delay, pure SET circuit with optimized parameters is more efficient and less complex to fabricate when compared to hybrid circuits. Thus SET based logic circuits can be used for better power and delay performance. Hence low power SET logic can be stacked above a CMOS layer by BEOL compatible fabrication process. By such an approach, it is possible to pack more functionality into a small footprint in a 3D IC, which results in shorter critical path and faster operation.

978-1-5386-1357-3/17 $31.00 © 2017 IEEE

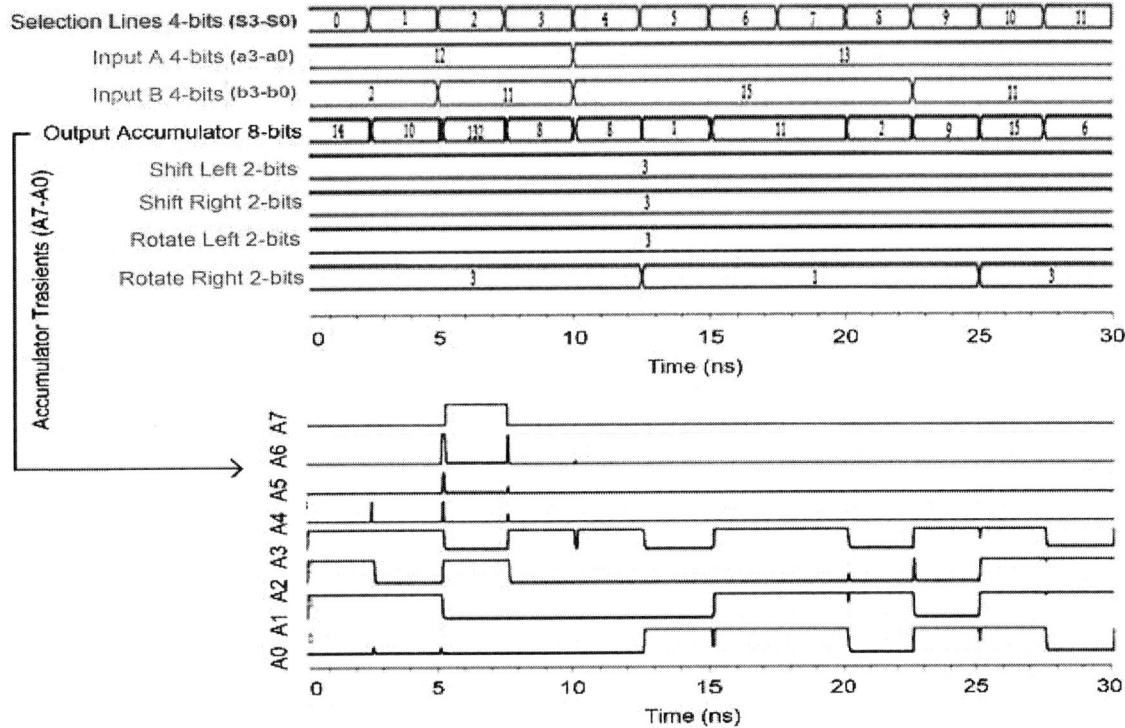

Fig. 10. 4-bit Vedic multiplier. Functional verification of proposed SET ALU. Both Input outputs and transients for accumulator output is shown, whereas inputs, selection lines and shift/rotate amount are shown in the form of bus.

REFERENCES

[1] ITRS 2013 Edition Report, Available at: http://www.itrs.net/reports.html

[2] S. Mahapatra and A. M. Ionencu, Hybrid CMOS Single electron transistor device and circuit design. Artech House, 2006.

[3] Zahid Ali Khan Durrani, Single-Electron Devices and Circuits in Silicon, Imperial College, UK 2010.

[4] V. Raut and P. K. Dakhole, "Design and implementation of single electron transistor N-BIT multiplier" International Conference on Circuit, Power and Computing Technologies (ICCPCT), 2014, pp. 1099 - 1104

[5] S. Mukherjee, T. S. Delwar, A. Jana, and S K Sarkar, "Hybrid single electron transistor based low power consuming 4-bit parallel adder/subtractor circuit in 65 nanometer technology" International Conference on Computer and Information Technology (ICCIT), Dec. 2014, pp. 136 – 140.

[6] M. A. Bounouar, A. Beaumont, K. E. Hajjam, F. Calmon, and D. Drouin "Room temperature double gate Single Electron Transistor based standard cell library" International Symposium on Nanoscale Architectures (NANOARCH), 2012 IEEE/ACM, pp. 146 - 151

[7] H. Zhong, Y. Chi, H. Sun, C. Zhang, and L. Fang, "Macromodeling of realistic single electron transistors for large scale circuit simulation," in Proc. INEC, 2010, pp. 193–194.

[8] M. Karimian, M. Dousti, M. Pouyan, and R. Faez, "An improved macro-model for simulation of single electron transistor (SET) using HSPICE," in Proc. IEEE TIC-STH, 2009, pp. 1000–1004.

[9] K. Zhou and H. Lu, "Simulation of single electronic device and robust circuit construction," in Proc. IEEE ICCA, 2007, pp. 211–213.

[10] D. Samanta and S. K. Sarkar, "A simple SET-MOS universal hybrid circuit for realization of all basic logic functions" 2012 International Conference on Advances in Engineering, Science and Management (ICAESM), March 2012, pp. 336 – 339.

[11] A. Venkataratnam and A. K. Goel, "Design and simulation of logic circuits with hybrid architectures of single-electron transistors and conventional MOS devices at room temperature," Microelectron. J., vol. 39, no. 12, pp. 1461–1468, Dec. 2008.

[12] S. Mahapatra, K. Banerjee, F. Pegeon, and A. M. Ionescu, "A CAD framework for co-design and analysis of CMOS-SET hybrid integrated circuits," in Proc. ICCAD, 2003, pp. 497–502.

[13] V. Raut and P. K. Dakhole, "Design and implementation of four bit arithmetic and logic unit using hybrid single electron transistor and MOSFET at 120nm technology", 2015 International Conference on Pervasive Computing (ICPC), pp. 1 – 6.

[14] B. Jana, A. Jana, S. Basak, J. K. Sing, and S. K. Sarkar, "Design and performance analysis of reversible logic based ALU using hybrid single electron transistor", 2014 Recent Advances in Engineering and Computational Sciences (RAECS), pp. 1 - 4

[15] C. Dubuc, J. Beauvais, and D. Drouin, "A Nanodamascene Process for Advanced Single-Electron Transistor Fabrication," Nanotechnology, IEEE Transactions on, vol.7, no.1, pp.68-73, 2008.

[16] R. Parekh, A. Beaumont, J. Beauvais, and D. Drouin, "Simulation and Design Methodology for Hybrid SET-CMOS Integrated Logic at 22-nm Room-Temperature Operation," Electron Devices, IEEE Transactions on, vol.59, no.4, pp.918-923, 2012

[17] S. Mahapatra and K. Banerjee, "Analytical modeling of single electron transistor for hybrid CMOS-SET analog IC design," IEEE Trans. Electron Devices, vol. 51, no. 11, pp. 1772–1782, Nov. 2004.

[18] S. Vaidya and D. Dandekar, "Delay-power performance Comparison of multipliers in VLSI circuit design," International Journal of Computer Networks & Communications (IJCNC), Vol.2, No.4, July 2010

[19] R. Parekh, J. Beauvais, and D. Drouin, "SET logic driving capability and its enhancement in 3-D integrated SET–CMOS circuit" Microelectronics Journal 45 (2014) pp. 11087–1092

[20] C. Dubuc, J. Beauvais, and D. Drouin, "Single-electron transistors with wide temperature operating range," Appl. Phys. Lett., vol. 90, no. 11, p. 113 104, Mar. 2007.

2017 IEEE International Symposium on Nanoelectronic and Information Systems

Realizing All Logic Operations using mRNA-Ribosome System as a Post Si Alternative

Pratima Chatterjee, Prasun Ghosal

Indian Institute of Engineering Science and Technology, Shibpur, Howrah 711103, WB, INDIA

Email: pratimachatterjee88@gmail.com, p_ghosal@it.iiests.ac.in

Abstract—Due to technological limitations in continuous scaling of transistors to cope up with the high performance requirements of today's Si based conventional computing researchers are in dying need of some non-conventional alternatives. Biomolecular computing is evolved as a promising candidate with powers of DNA and ribosome. Among these two ribosome has the power of automation, whereas DNA can provide inherent parallelism. The automated procedure of protein synthesis occurring within ribosome can be controlled properly using mutations. This has been used to implement logic gates and sequential logic in recent days. This paper goes another major step forward by realizing every kinds of shifters and comparators using this technique.

Index Terms—Ribosomal computing, mRNA-Ribosome system, Arithmetic and Logic operations.

I. Introduction

Since the birth of transistor, silicon-based conventional computer has been an integral part of human beings. Performance has also been increased following Moore's Law since then. In present scenario, size of transistor has come to such a reduced level, that it is facing some severe difficulties. Below *14 nm* scale electrons in transistors face quantum tunneling effect with large increase in leakage current [1]. Also, researchers have failed to find proper dielectric and insulating device interconnects for these scaled devices [2]. In simple words, conventional computing is losing its grip. Bio-molecular computing is evolved as a promising candidate with powers of DNA and ribosome. Among these two ribosome has the power of automation, whereas DNA can provide inherent parallelism. The automated procedure of protein synthesis occurring within ribosome can be controlled properly using mutations. This has been used to implement logic gates and sequential logic in recent days. This paper goes another major step forward by realizing every kinds of shifters and comparators using this technique.

II. Background

DNA (Deoxyribonucleic Acid) residing in nucleus acts as the carrier of genetic informations. Ribosome molecule resides within cytoplasm of a cell. Prior to protein synthesis, DNA is first transcribed into mRNA (messenger Ribonucleic Acid). This mRNA acts as a template in protein synthesis. On other hand, transfer RNA (tRNA) brings proper amino acid for protein synthesis. These amino acids are the basic building unit of protein chain. This section describes the protein synthesis process followed by structure of ribosome. It then concludes with an overview of mutations.

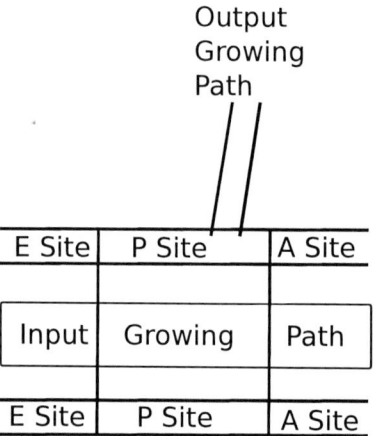

Fig. 1: Functional Block Diagram of a Ribosome.

A. Structure of Ribosome

Ribosome molecule [3] can be divided into two subunits *i.e.* larger and smaller. Each is composed of proteins, and ribosomal RNA(rRNA) or ribozyme. Ribosome has three tRNA binding sites *A, P* and *E* with one mRNA binding site in smaller subunit. *A, P,* and *E* sites are for tRNA entry, working, and tRNA exit sites respectively. PTC (Peptidyl transferase center) is also present in ribosome. An abstract figure of ribosome is also shown in figure 1.

B. Protein Synthesis

Protein synthesis is taken place in ribosome by following simple steps [4].

- First, mRNA binds to smaller ribosomal subunit by SD (Shine-Dalgarno) sequence [5]. Then upper subunit gets attached to this complex to build total mRNA-Ribosome system.
- Start codon is detected and corresponding aminoacyl tRNA comes directly to *P site* to get bound to mRNA. This initializes protein synthesis, with first amino acid being *Methionine*.
- Following steps are performed from the next codon of mRNA onwards upon reading the codon.
    - Proper aminoacyl tRNA according to mRNA codon gets bound to mRNA at *A site*.

978-1-5386-1357-3/17 $31.00 © 2017 IEEE

– Bound tRNA gets a push by the energy of GTP (Guanosine triphosphate) hydrolysis, to come to P *site*.
– Elongating chain of amino acid held at P *site* breaks its bond with the current tRNA of P *site* and gets hold by Peptidyl transferase center (PTC). Blank tRNA exits via E *site*.
– Pushed tRNA of A *site* comes to P *site* when held protein chain gets released from PTC and gets attached to the amino acid of new tRNA of P *site*.

• Last, stop codon is read and protein synthesis is stopped.

An overview of protein synthesis is shown in figure 2.

C. Mutation

rRNA is the controller of protein synthesis. The codon present in rRNA carries the specific *sense* for protein synthesis. If the codon sequence in rRNA is changed by altering, adding or deleting nucleotide bases then protein synthesis process gets effected. This change in nucleotide base is referred as mutation [6]. So, using mutation in rRNA protein synthesis process can be controlled.

III. RELATED WORK

Ribosome has been described as a molecular machine of a cell by [7]. As the world has been moving towards alternative computing platforms different biomolecules, such as, DNA, ribosome *etc.* have come to focus at present. But ribosome has been noticed since a long time due to its mechanical nature. Different mathematical approaches have been developed based on workings of ribosome [8], [9]. Not only a single ribosome, but use of more than one ribosome (polysome) [10] have also been found, and their mathematical modeling have also been developed.

Along with this, researchers have also been trying to develop models of protein synthesis process and ribosome structure [11], [12]. Artificial ribosome has been developed using synthetic biological process [13] and tethered sub-units [14]. But the concept using tethered sub-units has been proven more closer to natural ribosome in terms of protein synthesis in artificial ribosome.

This protein synthesis can be controlled using proper mutation [15]. Using protein synthesis and proper mutations, logic gates and sequential logic operations have already been performed in ribosome [16] and [17] respectively, that makes the ground of using ribosome for computing purposes.

IV. NOVELTY OF THE WORK

Ribosomal computing is rising as one of the promising biomolecular computing techniques due to following advantages.

1) It provides automation.
2) Components of this computing can be preserved for suitably long time.
3) A single ribosome acts as the main operating unit of a complete operation, and size of such complete operating unit is extremely minute [the diameter of a prokaryotic ribosome is of only *20 nm*].

V. REALIZATION OF COMPARATOR AND SHIFTERS USING MRNA-RIBOSOME SYSTEM

Amino acids are the basic units of a protein chain. In this work, two amino acids L and A with corresponding triplet codons $B_1B_2B_3$ and $B_4B_5B_6$ are being assumed to represent 0 and 1 respectively. Proposed mRNAs to perform comparison and shifts are quite similar to as proposed to implement logic gates in structure [17]. This mRNA can be constructed from DNA sticker model. Similar to logic gates name of this working unit is also mRNA-Ribosome system. How using proper mutation in this mRNA-Ribosome system, the realization of comparator and shifter logic is given below.

A. Comparator Realization

A comparator compares two values and indicates which one is larger. A ribosome can act as a comparator circuit with the help of proper mutations controlling translation process. In comparator, alongside A and L, another amino acid, say X is required having $B_7B_8B_9$ as associated triplet codon. Three kinds of mutations are required to design comparator in mRNA-Ribosome system as follows.

• Either a mutation reads current codon from mRNA and append appropriate amino acid to peptide chain, and skips next two codons.
• Or it skips current codon, reads next codon from mRNA and appends appropriate amino acid to peptide chain, and skips the following codon.
• Or it skips first two codons and reads third one to append appropriate amino acid to peptide chain.

Let us consider three mutations CM_1, CM_2, CM_3 to realize comparator. Description of these mutations is given in table I. Only one mutation among them can be deactivated and other two mutations can not be deactivated once activated.

The input $mRNA$ or reporter $mRNA$ is created from sticker model using following algorithm 1.

Where G_{SecM} and G_{SecA} are gene sequences of $SecM$ and $SecA$ respectively. $B_7B_8B_9$ codon has been selected to append amino acid X that represents equality of two values. $B_1B_2B_3$ triplet codon is chosen if 1st number is larger and append amino acid L. If 2nd number is larger $B_4B_5B_6$ triplet codon is selected, which attaches amino acid A to peptide chain. Input bits are taken in bit interleaved fashion starting from MSB and moving towards LSB. So, using algorithm1 corresponding $mRNA$ for input binary value 1100 and 1010 will be

$SD|\ AUG|\ G_{CM_1}|$
$\quad SecM B_4B_5B_6 B_4B_5B_6 SecA\ B_1B_2B_3\ B_4B_5B_6\ B_7B_8B_9$
$\quad SecM B_4B_5B_6 B_1B_2B_3 SecA\ B_1B_2B_3\ B_4B_5B_6\ B_7B_8B_9$
$\quad SecM B_1B_2B_3 B_4B_5B_6 SecA\ B_1B_2B_3\ B_4B_5B_6\ B_7B_8B_9$
$\quad SecM B_1B_2B_3 B_1B_2B_3 SecA\ B_1B_2B_3\ B_4B_5B_6\ B_7B_8B_9$
$\hspace{10cm}|UAA \quad (1)$

where G_{CM_1} represents activator for mutation CM_1.

So, using mRNA comparison between 1100 and 1010 is realized in ribosomal comparator as follows. The output protein chain may contain X amino acids followed by one

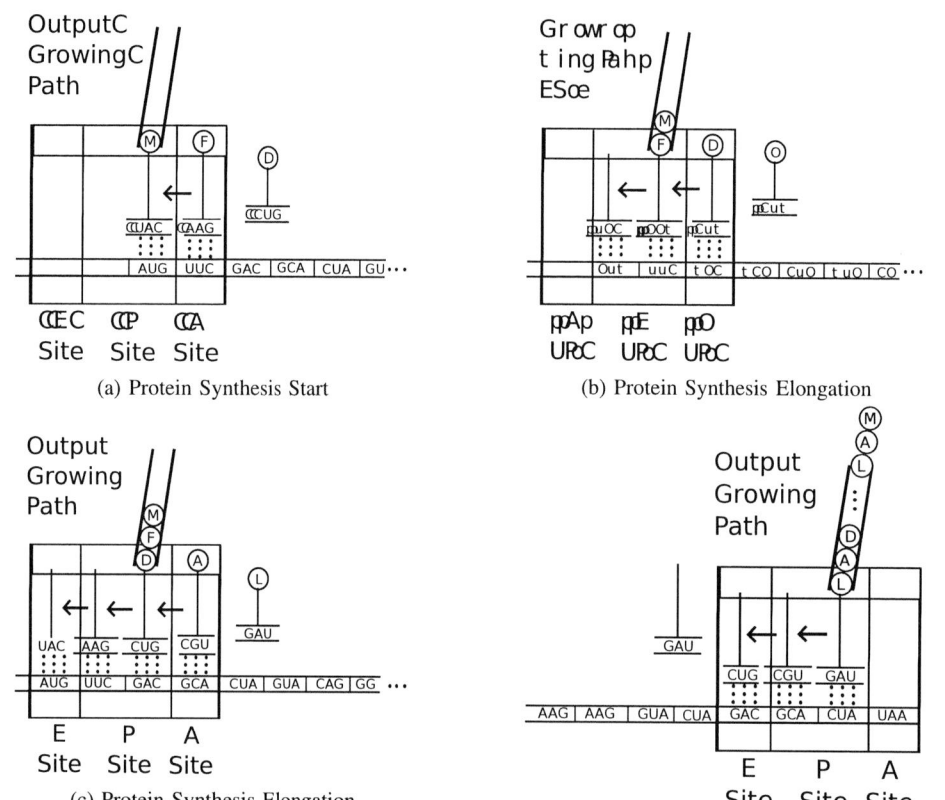

(a) Protein Synthesis Start

(b) Protein Synthesis Elongation

(c) Protein Synthesis Elongation

(d) Stop Sequence Detected and Protein synthesis Stop

Fig. 2: Steps of Protein Synthesis Procedure

TABLE I: Mutations for Comparator.

Name	Activator	De-activator	Performing operation	Initial State
CM_1	$B_1B_2B_3B_1B_2B_3 + CM_1$ $B_4B_5B_6B_4B_5B_6 + CM_1$	$B_4B_5B_6B_1B_2B_3,$ $B_1B_2B_3B_4B_5B_6$	Skip First codon Skip Second Codon Select Third Codon	Active
CM_2	$B_4B_5B_6B_1B_2B_3 + CM_1$		Select First Codon Skip Second Codon Skip Third Codon	Inactive
CM_3	$B_1B_2B_3B_4B_5B_6 + CM_1$		Skip First Codon Select Second Codon Skip Third Codon	Inactive

of A or L, but not both. If the peptide chain is built up of X only, then the two numbers are equal. Otherwise, if there is at least one L, the first number is larger and if there is at least one A, the second one is larger.

1) The Shine-Dalgarno(SD) sequence first choose proper mRNA-ribosome complex.
2) When the start codon AUG is read by $P\ site$ of a ribosome protein synthesis is started and output is created in the form of protein chain.
3) Activator mutation of CM_1 is then read, which activates CM_1.
4) Then $SecM$ reading stalls protein synthesis and input codon $B_4B_5B_6B_4B_5B_6$ is read and mutation CM_1 re-

mains active according to table I. $SecA$ reading releases the stall. Active CM_1 selects output bit $B_7B_8B_9$ and appends amino acid X to peptide chain.

5) Again $SecM$ stalls protein synthesis and input sequence $B_4B_5B_6B_1B_2B_3$ is read and mutation CM_2 is activated according to mutation table I. Mutation CM_1 gets deactivated at this instant. After releasing stall by $SecA$ $B_1B_2B_3$ output codon is selected, which appends amino acid L to growing peptide chain for active mutation.

6) $SecM$ again stalls protein synthesis. Then input is read but according to table I CM_2 remains activated and $B_1B_2B_3$ is selected again to attach amino acid L to protein chain after reading $SecA$.

7) In the similar manner $SecM$ stalls protein synthesis. Then input sequence read. $SecA$ releases stall to start

Algorithm 1: Algorithm to get required $mRNA$ for Comparator from user input

Input: Numbers $num1$ and $num2$ of n bits in DNA Sticker Model
Output: $mRNA$ for Comparator
Initialize $mRNA$ as blank;

Add Shine-Dalgarno(SD) sequence to mRNA;
Append AUG to $mRNA$;
Add activator of mutation CM_1 to $mRNA$;
for $i \leftarrow 1$ *to* n **do**
    Append G_{SecM};
    Separate($num1$, i, b_{on}, b_{off});
          // Separate $num1$ based on i^{th} bit
    if b_{on} *is not empty*   // i^{th} bit is on
    then
        | Append $B_4B_5B_6$;
    end
    else
        | Append $B_1B_2B_3$;
    end
    Empty b_{on} and b_{off};
    Separate($num2$, i, t_{on}, t_{off});
    if b_{on} *is not empty* **then**
        | Append $B_4B_5B_6$;
    end
    else
        | Append $B_1B_2B_3$;
    end
    Append $G_{SecA}B_1B_2B_3B_4B_5B_6B_7B_8B_9$;
end
Append UAA to $mRNA$;

Algorithm 2: Algorithm to get required mRNA for Left Shift from user input

Input: Number num of b bits in DNA Sticker Model, k as number of bits to shift
Output: $mRNA$ for left shift
Initialize $mRNA$ as blank;

Add Shine-Dalgarno(SD) sequence to mRNA;
Append AUG to $mRNA$;
for $i \leftarrow (k+1)$ *to* b **do**
    Separate(num, i, b_{on}, b_{off});
    if b_{on} *is not empty* **then**
        | Append($mRNA$, $B_4B_5B_6$);
    end
    else
        | Append($mRNA$, $B_1B_2B_3$);
    end
end
for $i \leftarrow 1$ *to* k **do**
    | Append($mRNA$, $B_1B_2B_3$);
end
Append UAA to $mRNA$;

This mRNA generate the protein chain in mRNA-Ribosome system is $LALL$ which means 0100. This is the correct result of logical left shift of 1010.

B. Logical Right Shift

Algorithm 3 converts the input number of b bits provided following DNA sticker model to desired mRNA molecule to be fed to a ribosome for logical right shift of k bits.

Algorithm 3: Algorithm to get required mRNA for Right Shift from user input

Input: Number num of b bits in DNA Sticker Model, k as number of bits to shift
Output: $mRNA$ for right shift
Initialize $mRNA$ as blank;

Add Shine-Dalgarno(SD) sequence to mRNA;
Append AUG to $mRNA$;
for $i \leftarrow 1$ *to* k **do**
    | Append($mRNA$, $B_1B_2B_3$);
end
for $i \leftarrow 1$ *to* $(b-k)$ **do**
    Separate(num, i, b_{on}, b_{off});
    if b_{on} *is not empty* **then**
        | Append($mRNA$, $B_4B_5B_6$);
    end
    else
        | Append($mRNA$, $B_1B_2B_3$);
    end
end
Append UAA to $mRNA$;

---

protein synthesis. But mutation CM_2 remains activated to append amino acid L to the growing chain.

8) Finally stop sequence UAA is read, which terminates the protein synthesis.
9) The resulted protein chain contains amino acid sequence XLLL, which represents that, $1st$ number is larger

VI. SHIFTER REALIZATION

Shifter moves a number some particular number of bits in a specified direction (left or right). Only logical left shift is designed as both arithmetic and logical are same. For realizing shifter logic only the mRNAs for different shifters are designed.

A. Logical Left Shift

Let, input number in DNA Sticker model be of b bits and logical left shift needs to be performed to shift k number of bits. The resultant $mRNA$ can be obtained by algorithm 2.

Following the algorithm 2 the mRNA for 1 bit left shifting logic of number 1010 is given below.

$$SD|AUG|B_1B_2B_3|B_4B_5B_6|B_1B_2B_3|B_1B_2B_3|UAA \quad (2)$$

Following the algorithm 3 the mRNA for left shifting logic of number 1010 is given below.

$$SD|AUG|B_1B_2B_3|B_4B_5B_6|B_1B_2B_3|B_4B_5B_6|UAA \quad (3)$$

The input mRNA VI-B generates $LALA$ amino acid sequence in protein synthesis which signifies 0101 number which is the correct result of logical right shift.

C. Arithmetic Right Shift

The only difference between logical right shift and arithmetic right shift is that first k bits are replaced by sign bit of the input number in place of 0s in arithmetic right shift of k bits. Algorithm 4 performs the arithmetic right shift of a b bit number by k bits.

Algorithm 4: Algorithm to get required mRNA for Arithmetic Right Shift from user input

Input: Number num of b bits in DNA Sticker Model, k as number of bits to shift
Output: $mRNA$ for arithmetic right shift
Initialize $mRNA$ as blank

Add Shine-Dalgarno(SD) sequence to $mRNA$;
Append AUG to $mRNA$;
Initialize $i \leftarrow 1$;
Separate(num, i , b_{on}, b_{off});
for $i \leftarrow 1$ *to* k **do**
   if b_{on} *is not empty* **then**
      Append($mRNA$, $B_4B_5B_6$);
   end
   else
      Append($mRNA$, $B_1B_2B_3$);
   end
end
for $i \leftarrow 1$ *to* $(b-k)$ **do**
   Separate(num, i , b_{on}, b_{off});
   if b_{on} *is not empty* **then**
      Append($mRNA$, $B_4B_5B_6$);
   end
   else
      Append($mRNA$, $B_1B_2B_3$);
   end
end
Append UAA to $mRNA$;

Following the algorithm 4 generates mRNA VI-C for arithmetic right shift of number 1010.

$$SD|AUG|B_4B_5B_6|B_4B_5B_6|B_1B_2B_3|B_4B_5B_6|UAA \quad (4)$$

So, mRNA VI-C generates protein chain containing amino acid $AALA$ which signifies number 1101 which is correct output result of arithmetic right shift of 1010.

D. Circular Left Shift

In circular left shift of single bit, first bit is appended at the end of number and the number is shifted left by 1 bit. This can be extended easily to k bits. Algorithm 5 shows how mRNA for circular left shift can be built from input number given in DNA sticker model.

Algorithm 5: Algorithm to get required mRNA for Circular Left Shift from user input

Input: Number num of b bits in DNA Sticker Model, k as number of bits to shift
Output: $mRNA$ for circular left shift
Initialize $mRNA$ as blank;

Add Shine-Dalgarno(SD) sequence to $mRNA$;
Append AUG to $mRNA$;
for $i \leftarrow (k+1)$ *to* b **do**
   Separate(num, i, b_{on}, b_{off});
   if b_{on} *is not empty* **then**
      Append($mRNA$, $B_4B_5B_6$);
   end
   else
      Append($mRNA$, $B_1B_2B_3$);
   end
end
for $i \leftarrow 1$ *to* k **do**
   Separate(num, i, b_{on}, b_{off});
   if b_{on} *is not empty* **then**
      Append($mRNA$, $B_4B_5B_6$);
   end
   else
      Append($mRNA$, $B_1B_2B_3$);
   end
end
Append UAA to $mRNA$;

Let us perform circular left shift on 1110 binary number. Using algorithm5 mRNA VI-D is developed for 1 bit circular shift.

$$SD|AUG|B_4B_5B_6|B_4B_5B_6|B_1B_2B_3|B_4B_5B_6|UAA \quad (5)$$

When this mRNA is sent through proper ribosome containing appropriate anti Shine-Dalgarno sequence, the output protein chain becomes $AALA$ that represents output 1101 i.e. result of 1 bit circular left shift of the input number.

E. Circular Right Shift

In circular right shift of single bit, last bit is appended at the beginning of number and the number is shifted right by 1 bit. This can be extended easily to k bits , where last k bits are appended at the beginning of input number one by one in the same sequence and then right shifting the number by k bits. Algorithm 6 shows how mRNA for circular right shift can be built from input number given in DNA sticker model.

978-1-5386-1357-3/17 $31.00 © 2017 IEEE

Algorithm 6: Algorithm to get required mRNA for Circular Right Shift from user input

---

Input: Number num of b bits in DNA Sticker Model, k as number of bits to shift

Output: $mRNA$ for circular right shift

Initialize $mRNA$ as blank;

Add Shine-Dalgarno(SD) sequence to $mRNA$;
Append AUG to $mRNA$;
for $i \leftarrow (b - k + 1)$ *to* b **do**
    Separate(num, i, b_{on}, b_{off});
    if b_{on} *is not empty* **then**
        Append($mRNA, B_4B_5B_6$);
    end
    else
        Append($mRNA, B_1B_2B_3$);
    end
end
for $i \leftarrow 1$ *to* $(b - k)$ **do**
    Separate(num, i, b_{on}, b_{off});
    if b_{on} *is not empty* **then**
        Append($mRNA, B_4B_5B_6$);
    end
    else
        Append($mRNA, B_1B_2B_3$);
    end
end
Append UAA to $mRNA$;

Similar to circular left shift the the mRNA for right circular shift will be mRNA VI-E for number 1110

$$SD|AUG|B_1B_2B_3|B_4B_5B_6|B_4B_5B_6|B_4B_5B_6|UAA \quad (6)$$

So, output protein chain for input mRNA is $LAAA$ amino acid sequence, which represents 0111, which is the result of 1 bit circular right shift of 1110.

VII. CONCLUSION

To make a new alternative computing technique widely usable, basic functionalities of a computing like *Arithmetic and Logic Unit (ALU)* needs to be realized first. Progress has already been made by implementing logic gates and basic sequential elements. This work takes another major forward step towards this goal by implementing two more important parts of an ALU, *viz.*, all shifters and comparators. So, this work is expected to be an important pathway towards making this promising ribosomal computing widely usable.

REFERENCES

[1] T.-c. Chen, "Challenges for silicon technology scaling in the nanoscale cra," in *ESSCIRC, 2009. ESSCIRC'09. Proceedings of.* IEEE, 2009, pp. 1–7.

[2] N. Z. Haron and S. Hamdioui, "Why is cmos scaling coming to an end?" in *Design and Test Workshop, 2008. IDT 2008. 3rd International.* IEEE, 2008, pp. 98–103.

[3] M. M. Yusupov, G. Z. Yusupova, A. Baucom, K. Lieberman, T. N. Earnest, J. Cate, and H. F. Noller, "Crystal structure of the ribosome at 5.5 å resolution," *science*, vol. 292, no. 5518, pp. 883–896, 2001.

[4] V. Ramakrishnan, "Ribosome structure and the mechanism of translation," *Cell*, vol. 108, no. 4, pp. 557 – 572, 2002.

[5] G.-W. Li, E. Oh, and J. S. Weissman, "The anti-shine-dalgarno sequence drives translational pausing and codon choice in bacteria," *Nature*, vol. 484, no. 7395, pp. 538–541, 2012.

[6] J. Hodgkin, "Genetic suppression," *WormBook: the online review of C. elegans biology*, vol. –, no. –, pp. 1–13, 2005. [Online]. Available: http://www.ncbi.nlm.nih.gov/books/NBK19667/

[7] A. S. Spirin, "Ribosome as a molecular machine," *FEBS letters*, vol. 514, no. 1, pp. 2–10, 2002.

[8] T. von der Haar, "Mathematical and computational modelling of ribosomal movement and protein synthesis: An overview," *Computational and Structural Biotechnology Journal*, vol. 1, no. 1, pp. 1 – 7, 2012.

[9] Y.-B. Zhao and J. Krishnan, "Probabilistic boolean network modelling and analysis framework for mrna translation," *IEEE/ACM Transactions on Computational Biology and Bioinformatics*, vol. 13, no. 4, pp. 754–766, 2016.

[10] E. W. Müllner and J. A. Garcia-Sanz, "Polysome gradients," *Manual of immunological methods*, vol. 1, pp. 457–462, 1997.

[11] M. A. Gilchrist and A. Wagner, "A model of protein translation including codon bias, nonsense errors, and ribosome recycling," *Journal of theoretical biology*, vol. 239, no. 4, pp. 417–434, 2006.

[12] A. Heyd and D. A. Drew, "A mathematical model for elongation of a peptide chain," *Bulletin of mathematical biology*, vol. 65, no. 6, pp. 1095–1109, 2003.

[13] B. Lewandowski, G. De Bo, J. W. Ward, M. Papmeyer, S. Kuschel, M. J. Aldegunde, P. M. E. Gramlich, D. Heckmann, S. M. Goldup, D. M. D'Souza, A. E. Fernandes, and D. A. Leigh, "Sequence-specific peptide synthesis by an artificial small-molecule machine," *Science*, vol. 339, no. 6116, pp. 189–193, 2013.

[14] C. Orelle, E. D. Carlson, T. Szal, T. Florin, M. C. Jewett, and A. S. Mankin, "Protein synthesis by ribosomes with tethered subunits," *Nature*, vol. 524, pp. 119–124, 2015.

[15] W. Sutton, W. Gerlach, D. Schwartz, and W. Peacock, "Molecular analysis of ds controlling element mutations at the adhl locus of maize," *Science*, vol. 223, pp. 1265–1269, 1984.

[16] P. Chatterjee, M. Sarkar, and P. Ghosal, "Computing in ribosomes: Performing boolean logic using mrna-ribosome system," in *VLSI (ISVLSI), 2016 IEEE Computer Society Annual Symposium on.* IEEE, 2016, pp. 260–265.

[17] ——, "Computing in ribosomes: Implementing sequential circuits using mrna-ribosome system," in *Nanoelectronic and Information Systems (iNIS), 2016 IEEE International Symposium on.* IEEE, 2016, pp. 230–235.

2017 IEEE International Symposium on Nanoelectronic and Information Systems

Design of an High Performance Carry Generation Circuit for Ternary Full Adder using CNTFET

Subhendu Kumar Sahoo, Gangishetty Akhilesh, Rasmita Sahoo*

Department of Electrical and Electronics Engineering

BITS-Pilani Hyderabad,

Nalla Narasimha Reddy School of Engeineering, Hyderabad*

Email: sahoo@hyderabad.bits-pilani.ac.in

Abstract—**The Full Adder is one of the most important and basic units of arithmetic logics which is used to design many complex circuits. Further, ternary logic is a promising alternative to the conventional binary logic in VLSI design as it provides the advantages of reduced interconnects, higher operating speeds and smaller chip area. The present work aims to design high performance ternary full adder (TFA) by using carbon nanotube field effect transistors (CNTFET). The proposed TFA designs were simulated using HSPICE to obtain critical delay, power delay product (PDP) and maximum operating frequency. Further the circuit performance also has been compared with recently reported TFA. The factors which are considered for comparison are transistor count, delay, PDP and maximum operating frequency. The comparison results infer that the proposed TFA gives outstanding performance as compared to the reported one. The proposed TFA has been demonstrated delay, PDP and maximum operating frequency improvement up to 29%, 20% and 39% respectively. The transistor count for the proposed circuit is also reduced by almost 15%.**

I. Introduction

Miniaturization of transistor is always a key to achieve better device performance and faster computation which is the root cause of scaling of transistors. Further to overcome the short channel effects associated with the scaling of traditional MOSFETS, now a days researchers are focusing on search for new devices. Among many proposed devices, CNTFET is found to be one of the best alternatives [1], [2], [3], [4]. So number of studies are going on in the field of digital circuits designed using CNTFETS. CNTFETs are also found to be very promising in designing multiple valued logics [5], [6], [7]. This is because the threshold voltage of a CNTFET varies with the diameter of the Carbon Nano Tube (CNT) used in the channel. Here we have concentrated our study on ternary logic which consists of three logic values.

Further adder circuit is the most basic circuit used for digital computation. Though some adder circuits, both half adder and full adder, are reported till date [5], [8], [9], still researchers are trying to design different adder circuits to obtain better performance in terms of transistor count, delay, power consumption, maximum operating frequency etc. One such recent design of a ternary full adder is presented in [7].

In the present work, the authors designed a carry generation circuit for ternary full adder which shows better performance

as compared to one of the recent reported full adder. The paper is organized as follow. A brief discussion on CNTFET and its use in ternary logic is discussed in section II. Ternary full adder and the previously best reported adders are discussed in section III. Our proposed carry generation circuit and its usefulness in ternary adder circuit is discussed in section IV. Simulation results as well as the comparisons with other reported circuits are analyzed in section V. The work done is finally concluded in section VI.

II. CNTFET and Ternary Logic

CNTFET is a field effect transistor in which CNTs are used in the channel region for carrier transport. The first CNTFET was demonstrated in 1998 [10]. The structure of the CNTFET considered for the present study is similar to that of a conventional MOSFET in which semiconducting single walled CNTs are used in the channel region. CNTFETs are of keen interest for the researchers working for multiple valued logics because the threshold voltage of a CNTFET can be adjusted by choosing proper diameter for the CNTs used in the particular CNTFET. This is because the threshold voltage for a particular CNTFET is inversely proportional to the diameter of the CNT [11] and is given as:

$$V_{th} = \frac{aV_\pi}{\sqrt{3}ed_{CNT}} \tag{1}$$

Here V_{th} is the threshold voltage, $a = 2.46A^0$, $V_\pi = 3.033$ eV is the carbon $\pi - \pi$ bond energy in the tight binding model, e is the unit electron charge and d_{CNT} is the diameter of the CNT used in the channel. Thus the diameter of the CNT and hence the threshold voltage can be adjusted to a required value by choosing proper chiral vector. This is the major advantage of CNTFET to be used for ternary logic.

In ternary logic we have three logic values and the symbols used are 0, 1 and 2. The voltage levels in terms of V_{DD} used to represent each symbol are 0, $V_{DD}/2$ and V_{DD} respectively. Some of the advantages of ternary logic over binary logic are reduced computational steps, requires lesser number of bits, memory and number of chip pins, simple and energy efficient, eliminate interconnect and pin-out problems etc. Because of these advantages now a days ternary logic is of keen attraction for many researches working in digital electronics. We have

978-1-5386-1357-3/17 $31.00 © 2017 IEEE

concentrated our study for exploring new efficient adder circuit which will be useful for computations using ternary logics.

III. TFA AND PREVIOUS WORK DONE ON TFA

Ternary full adder takes two ternary input signals to be added and the carry generated from previous steps and generates the output sum and carry. The truth table for a full adder circuit is given in table-1. Some adder circuits are proposed in last few years by different research group. The most recent one was reported by P. Keshavarzian and R. Sarikhani [7]. In this the authors have proposed two circuits one for sum and one for carry by using CNTFETs with CNTs of three different diameters. The circuits are designed in such a way that there are different flow paths for different combinations of input logics.

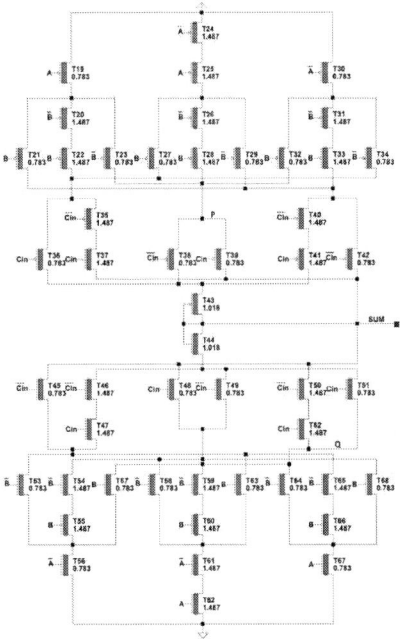

Fig. 1. SUM generator circuit for TFA [7].

TABLE I
TRUTH TABLE FOR TERNARY FULL ADDER

A	B	Cin = 0		Cin = 1		Cin = 2	
		Carry	Sum	Carry	Sum	Carry	Sum
0	0	0	0	0	1	0	2
0	1	0	1	0	2	1	0
0	2	0	2	1	0	1	1
1	0	0	1	0	2	1	0
1	1	0	2	1	0	1	1
1	2	1	0	1	1	1	2
2	0	0	2	1	0	1	1
2	1	1	0	1	1	1	2
2	2	1	1	1	2	2	0

IV. PROPOSED TFA

In this paper we have proposed a novel carry genetration circuit for a ternary full adder. This design consists of two circuits: sum generator and carry generator. The inputs given to these circuits are the two signals (A and B) to be added, the carry obtained from the previous step and their complements. The details of the two circuits are discussed below

A. Sum generator

The circuit used to obtain the sum is shown in Fig. 1 which was proposed by P. Keshavarzian and R. Sarikhani [7]. The inputs for this circuit are in ternary logics. The truth table for the sum circuit is given in Table-1. The circuit is designed in such a way that there will be different path for the logic flow depending on different combinations of the inputs. The control over the flow path is achieved by choosing three different CNTs with chiral vectors (19, 0), (13, 0) and (10, 0). For these three CNTs the diameters are obtained as 1.487 nm, 1.018nm and 0.783 nm. Further the threshold voltages for

the corresponding CNTFETs are found to be 0.293, 0.428and 0.557 respectively by using eq. (1). By selecting these three combinations of CNTs, proper switching of transistors are done according to different levels of the inputs and the required outputs are obtained. A typical example is discussed here to have a clear understanding.

For example, when both A and B have logic value 1 then the transistors T24, T25, T26, T28 in P-CNFET network are ON. So, there is a direct path between V_{DD} and node P. Hence logic 2 is obtained at node P. Further as T59, T60, T61, T62 are ON there is a direct path between node Q and ground which makes the voltage zero at node Q and the logic value at Q becomes 0. Now, if Cin is equal to 0 then T39 is ON and there is a direct path between node P and SUM because of which the SUM becomes 2. If Cin is equal to 1 T50 and T52 are ON and transmit 0 to SUM. For Cin equal to 2 T38 and T51 are ON and through voltage division between T43 and T44 we have logic 1 at the SUM. Similarly there are different allowed path for different combinations of inputs A, B and Cin and finally we obtain the desired outputs as per Table-1.

B. Carry generator

The circuit used to obtain the carry is shown in Fig. 2 which is proposed by us. The main advantage for the TFA comes from this circuit as this is designed with less number of transistors as compared to the reported one [7]. The inputs for this circuit are in ternary logics. The truth table for the carry circuit is given in Table-1. The circuit is also

978-1-5386-1357-3/17 $31.00 © 2017 IEEE   47

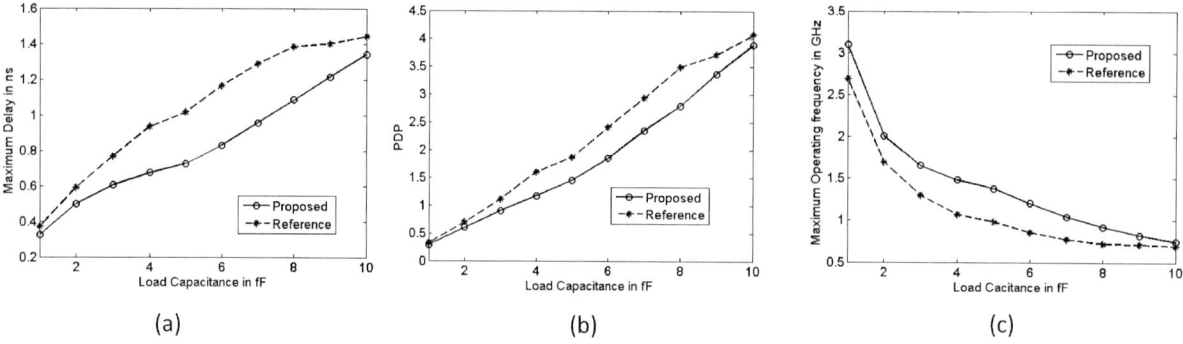

Fig. 3. Comparison of proposed circuit with [7] for varying load capacitance (a) critical delay (b) PDP (c) maximum operating frequency

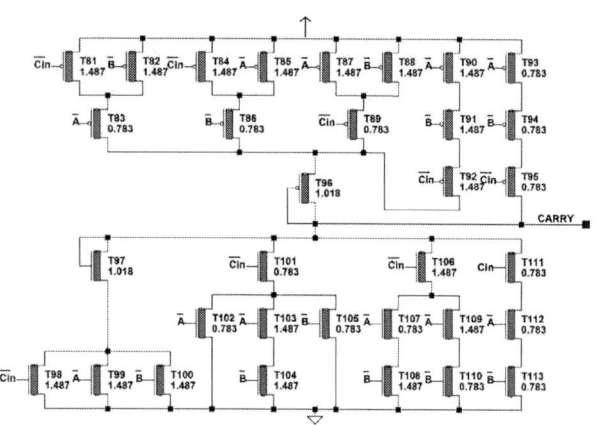

Fig. 2. Proposed Carry generator circuit for TFA

designed in such a way that there will be different path for the logic flow depending on different combinations of the inputs. One typical example is discussed here to have a clear understanding.

Considering the case A=1, B=1, and Cin=2 we have the transistors T99, T100, T90, T91, T92 are ON. Further T97 is grounded through T99, T100 and T96 is connected to VDD through T90, T91, T92. Now as T101, T106, and T112 are off , there is a voltage division between T96 and T97 giving the carry out as $V_{DD}/2$ which is nothing but logic 1. Similarly there are different allowed path for different combinations of inputs A,B and Cin and finally we obtain the desired outputs as per Table-1.

V. SIMULATION RESULTS AND COMPARISON

The proposed designs and the previous reported design [7] are implemented by HSPICE [12] using CNTFET model [13], [14], [15], [16]. All the simulations are done at room temperature with 0.9 V supply voltage. We have calculated the critical delay, average delay, power delay product and maximum operating frequency for both the circuits at different load capacitances. Finally the results are shown in graphical form in Fig. 3 to have a better comparison. From the graphs it is clear that our proposed circuit gives lower critical delay, lower PDP and higher operating frequency as compared to the reported one. Further the number of transistors used in the proposed case also reduced by 15% as compared to the reported one.

VI. CONCLUSION

This paper presents an improved carry generation circuit, for ternary full adder, using emerging futuristic CNTFET devices. Based on simulations with HSPICE, using the CNT-FET compactmodel of Standford, the circuit performances are compared with the best reported full adder circuit. From the simulation results it is clear that the proposed logic circuits achieve improvements in number of transistors, critical delay, PDP and operating frequency as compared to a most recently reported ternary full adder. Though the sum circuit is same in both the adder circuits, the proposed ternary adder outperforms because of the use of the proposed high performance novel carry generation circuit.

REFERENCES

[1] P. Avouris, J. Appenzeller, R. Martel, and S. J. Wind, "Carbon nanotube electronics," *Proceedings of the IEEE*, vol. 91, no. 11, pp. 1772–1784, 2003.

[2] J. Guo, S. Hasan, A. Javey, G. Bosman, and M. Lundstrom, "Assessment of high-frequency performance potential of carbon nanotube transistors," *IEEE transactions on Nanotechnology*, vol. 4, no. 6, pp. 715–721, 2005.

978-1-5386-1357-3/17 $31.00 © 2017 IEEE

[3] M.-H. Yang, K. B. Teo, L. Gangloff, W. I. Milne, D. G. Hasko, Y. Robert, and P. Legagneux, "Advantages of top-gate, high-k dielectric carbon nanotube field-effect transistors," *Applied Physics Letters*, vol. 88, no. 11, p. 113507, 2006.

[4] L. Wei, D. J. Frank, L. Chang, and H.-S. P. Wong, "Noniterative compact modeling for intrinsic carbon-nanotube fets: Quantum capacitance and ballistic transport," *IEEE Transactions on Electron Devices*, vol. 58, no. 8, pp. 2456–2465, 2011.

[5] S. Lin, Y.-B. Kim, and F. Lombardi, "Cntfet-based design of ternary logic gates and arithmetic circuits," *IEEE transactions on nanotechnology*, vol. 10, no. 2, pp. 217–225, 2011.

[6] M. H. Moaiyeri, R. F. Mirzaee, A. Doostaregan, K. Navi, and O. Hashemipour, "A universal method for designing low-power carbon nanotube fet-based multiple-valued logic circuits," *IET Computers & Digital Techniques*, vol. 7, no. 4, pp. 167–181, 2013.

[7] P. Keshavarzian and R. Sarikhani, "A novel cntfet-based ternary full adder," *Circuits, Systems, and Signal Processing*, vol. 33, no. 3, pp. 665–679, 2014.

[8] A. Dhande and V. Ingole, "Design and implementation of 2 bit ternary alu slice," in *Proc. Int. Conf. IEEE-Sci. Electron., Technol. Inf. Telecommun*, 2005, pp. 17–21.

[9] S. K. Sahoo, G. Akhilesh, R. Sahoo, and M. Muglikar, "High-performance ternary adder using cntfet," *IEEE Transactions on Nanotechnology*, vol. 16, no. 3, pp. 368–374, 2017.

[10] R. Martel, T. Schmidt, H. Shea, T. Hertel, and P. Avouris, "Single- and multi-wall carbon nanotube field-effect transistors," *Applied Physics Letters*, vol. 73, no. 17, pp. 2447–2449, 1998.

[11] R. Saito, G. Dresselhaus, and M. S. Dresselhaus, *Physical properties of carbon nanotubes*. World scientific, 1998.

[12] S. HSPICE, "Inc., dec. 2010," *Version E-2010.12*.

[13] J. Deng and H.-S. P. Wong, "Modeling and analysis of planar-gate electrostatic capacitance of 1-d fet with multiple cylindrical conducting channels," *IEEE Transactions on Electron Devices*, vol. 54, no. 9, pp. 2377–2385, 2007.

[14] ——, "A compact spice model for carbon-nanotube field-effect transistors including nonidealities and its applicationpart i: Model of the intrinsic channel region," *IEEE Transactions on Electron Devices*, vol. 54, no. 12, pp. 3186–3194, 2007.

[15] ——, "A compact spice model for carbon-nanotube field-effect transistors including nonidealities and its applicationpart ii: Full device model and circuit performance benchmarking," *IEEE Transactions on Electron Devices*, vol. 54, no. 12, pp. 3195–3205, 2007.

[16] ——, "A circuit-compatible spice model for enhancement mode carbon nanotube field effect transistors," in *Simulation of Semiconductor Processes and Devices, 2006 International Conference on*. IEEE, 2006, pp. 166–169.

Energy Efficient NoC Router for High Throughput Applications in Many-core GPUs

Shrestha Bansal, Hemanta Kumar Mondal, Sri Harsha Gade, Sujay Deb

Department of Electronics and Communication Engineering
Indraprastha Institute of Information Technology (IIIT), Delhi
E-Mail: {shrestha15113, hemantam, harshag, sdeb}@iiitd.ac.in

Abstract— With the development of many-core GPU clusters, researchers are forced to deal with an increased power consumption and a shortage of chip area. Numerous efforts have gone into creating a scalable and a power and area efficient Network-on-Chip solution. However most of these solutions either compromise on performance or are not scalable. In this work, we propose a scalable, minimally buffered router to minimize chip area and use the concept of Marching Memory to save power without having to compromise on the network performance. We achieve an average latency reduction of 27.9% and an average throughput increase of 30.46% as compared to conventional minimally buffered architectures.

Keywords—Network-on-chip, bufferless routing, marching memory through type (MMTH), scalable interconnects, many core GPU architectures.

I. INTRODUCTION

Many core architectures are becoming the de facto designs for implementing high performance computing systems desired by several applications like machine learning, IoT, etc. With applications becoming diverged and advancements being made in SIMD, SIMT and vector processors, architectures like GPUs and dedicated accelerators are gaining significance to achieve high performance. These architectures, in general, execute similar simple instructions on different copies of data and are highly suitable for operating on copious amounts of data. With several cores operating on different copies of data, they are required to communicate with each other and hence frequently inject packets onto the interconnection networks. Therefore, on-chip interconnection in such systems is desired to provide high throughput and plays a significant role in maintaining high performance. Network-on-Chip (NoC) infrastructures have emerged as a scalable and effective way to interconnect different cores in such many core architectures. NoC designs allow multiple core/memory to core/memory communications simultaneously and can handle very high network loads prevalent in GPU architectures.

However, with rapid growth in number of cores, NoCs are becoming large and NoC routers consume significant amount of chip resources, especially power and area. With several NoC switches active simultaneously for considerable lengths of time, NoC accounts for a significant portion of chip power in large systems. Studies have shown that NoCs consume up to 36% of total chip power [1] and it is only going to increase

with increasing core counts. Of all the router components, buffers consume the most of the router's power. Numerous efforts have been made to address the issue of high power consumption in NoC buffers. One of the approaches is to bypass the input buffers, when they are frequently empty to reduce the dynamic power consumption. This increases both design complexity and limits performance of the router. A disadvantage of such technique is that, it reduces only dynamic power, but does not impact static power of the buffers. Another prominent approach has been bufferless router designs that eliminate buffers completely [2]. It reduces buffer power consumption to a large extent, but results in serious performance degradation and cannot be used for high traffic injection rate. Though minimally buffered designs [3], [4] improve performance to some extent, they are still inferior as compared to conventional buffered routers. This is especially the case for GPUs that inject packets frequently and require high throughput network topologies. Hence, it is imperative to design buffer architectures that reduce power consumption while also maintaining high throughput performance desired by GPU systems.

In this work, we propose a novel bufferless design that reduces power and area consumption significantly and provides network throughput similar to that of conventional buffered routers. We achieve this by implementing minimally buffered deflection router that uses marching memory for buffers. Marching Memory Through Type (MMTH) [5] implementation reduces power consumption in buffers significantly as compared to conventional SRAM buffer designs with minimal impact on performance. In our proposed router architecture, we modify the side buffer to implement MMTH so as to obtain lower energy consumption as compared to conventional buffers. We also change the size of side buffer to increase the throughput and reduce unnecessary deflections. We thus propose a minimally buffered router architecture that can match the high requirements of GPU applications. We evaluate the proposed design using synthetic and application benchmarks for different system sizes using mesh network topology. We compare the achieved results with input buffered and minimally buffered router designs. Our implementation provides throughput in the range of input buffered router, while consuming significantly low power. It is area and power efficient and is easily scalable to any system size. The contributions of this work are summarized below:

978-1-5386-1357-3/17 $31.00 © 2017 IEEE

- Priority based minimally buffered router design that accepts high network load and provides optimum throughput for GPU applications.

- Marching memory implementation for side buffers to reduce power consumption with minimal performance impact.

The paper is organized as follows: Section II presents a brief overview of the related work done. Section III gives an insight into our motivation to pursue this work. Section IV details the proposed router architecture. In Section V, we present and discuss the performance evaluation of our design. We conclude in Section VI.

II. RELATED WORK

In this section, we present a brief review of the existing works in the field of Network-on-Chips for GPUs and CMPs. With the increased demand for data sharing between the GPUs and CMPs, there has been an increase in the contention between CPU and GPU cores on chip. Due to their fault-tolerance, congestion control mechanism and their capability to handle a high amount of load, NoCs have emerged as a preferred choice of interconnect networks for such resource sharing in the modern many-core GPU architectures.

Traditional NoC architecture consists of a crossbar, several input and output buffers, and some other logical blocks to perform the delivery of data packets from their source to destination. On a small CMP network, this NoC architecture poses no significant challenges to the design in terms of power and speed. However, as the number of cores increase, as in the case of GPUs, this design starts to consume a large amount of on-chip area and a significant part of total chip power. Studies have shown that buffers alone consume around 46% of the total router power [5].

Numerous researchers worldwide have identified this problem, and have come up with some remarkable design architectures. One such technique, bufferless routing, also known as hot-potato routing, was first introduced for distributed communication networks [3]. Bufferless routing, BLESS, eliminates the input and output buffers completely [3], thus helping save on chip area and make the network power efficient. This naturally comes at a cost of a reduced performance at higher network loads. Minimally Buffered deflection router (MinBD) was then proposed [3]. They introduced a small side buffer to store some of the flits. Using this design, although they increased chip area and power consumption (as compared to bufferless architecture), but enhanced the performance drastically. MinBSD reduced the clock delay to a single cycle [4], thus improving the performance even more. This design has been subsequently referred to as the baseline router. Recently, Marching Memory was proposed in [5]. This technique effectively utilized the existing buffer design by modifying the read and write pointer implementation. As a result of their design, they made the existing buffers more power efficient. Designs in [11], [13], [15] and [16] have also tried to minimize the power consumption in NoCs by using power gating technique.

Designs in [12] and [14], vary supply voltage to adapt to the network requirements, thus saving power consumption.

In the current work, we utilize minimally buffered deflection routing architecture to save on chip area and power, while using MMTH to save the power further. To compensate for performance loss, we modify the minimally buffered architecture as described in Section IV.

III. MOTIVATION

There have been numerous works proposing various minimally buffered router architectures. However, the problem with most of these designs is that they focus on optimizing only one particular parameter- power or performance. While many works use power optimization techniques to reduce power, others present new routing algorithms to optimize the performance. However, the common thing that accompanies these designs is the design complexity.

We have in this work addressed the above mentioned issues to a large extent. While we make our buffers more power efficient by incorporating the concept of MMTH, we also try to improve the performance of our design as compared to the previously proposed designs by modifying the router architecture. We thus try to save power consumption in the router while trying to improve the average throughput and reduce the average flit delay at the router. We also try to make our design scalable, which is a major challenge for most of the proposed designs.

IV. PROPOSED DESIGN

In this section we discuss the proposed router architecture with its design flow. We also discuss the deadlock-livelock free deflection routing strategy with its in-detail implementation.

A. Proposed Router Architecture

We propose the minimally buffered router architecture design as shown in Fig. 1. In the proposed design, we have used the following blocks:

Route Computation Unit: This unit comes into action when the header flit is encountered. It extracts the destination address from the flit and makes a decision on the output port of the flit. We use XY routing in our design.

Prioritization Unit: This unit assigns the priority to each flit based on its proximity to its destination. For this we calculate

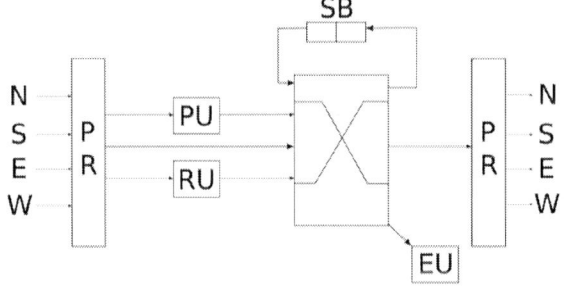

Fig. 1. Internal components of proposed router architecture

the hops-to-destination ($| D_X - C_X | + | D_Y - C_Y |$) for each flit, and based on its value assigns a priority ranging from 1-4. If multiple flits have a similar hops-to-destination value, we follow a round robin arbitration to allocate the priorities. Here (C_X, C_Y) are the co-ordinates of the current router and (D_X, D_Y) are the co-ordinates of the destination router.

Eject Unit (EU): This unit is used to route the flits to the local port of the router. In this architecture, the flits are divided into three categories- header, body and tail flits. When a header flit is received at the router, the prioritization unit (PU) is initiated. It assigns a priority to the flit. Simultaneously, the route computation unit (RU), based on its algorithm determines the output port of this flit. In one clock cycle, this action happens for all the header flits coming from all the four directions- North, South, East, and West as shown in Fig. 1.

Side Buffer (SB): This is the core block of our design. Unlike the baseline router, where we have a single depth side buffer, we modify the side buffer in our design. This helps improve the performance of our design. In our design, we are able to store two flits, thus preventing unnecessary deflections in case the desired output port is full. This helps ease network congestion as well as increases the overall throughput since most of the flits are routed to their exact locations. This double depth of the side buffer facilitates us to use MMTH here instead of conventional buffers.

A minimally buffered router usually routes one flit, stores one flit and deflects all the other flits demanding the same output port. So in any given scenario there could be a maximum of three flits demanding the same output port. On including a side buffer, there could be a maximum of four (One from side buffer, three from input ports) flits that could demand the same output port. There could be two approaches to handle these flits- to keep a single side buffer with a double depth, or to keep two side buffers. The latter approach has been explored in HiPAD [6] and it gives some minor improvements in the average router delay as compared to the baseline router and MinBD. We explore the first option. While the area required for both the cases may be same, we try to exploit the larger depth side buffer to save power as well, by using MMTH in its place.

For our evaluation purposes, we have kept the depth of side buffer as two. We increase the depth of this side buffer to accommodate two flits in it. This helps reduce congestion and avoid unnecessary flit deflection. This value of side buffer depth was also varied from one to three. And upon various observations of area occupation and power consumption with respect to the performance enhancements, it was decided that a depth of two would be optimal for our design.

Our design reserves the output port for a particular packet when the header flit arrives at any router. This reservation is cancelled once the entire packet has traversed that particular output link. Then upon encountering the tail flit, this reservation is cancelled. This allows one packet to pass completely, so reduces the need for re-ordering of the flits, thus saving design complexity.

B. Communication Protocol

In the proposed design, we eliminate the Permutation Deflection Network (PDN) used in the baseline router and MinBD and instead use a conventional crossbar in its place. This is done because of two reasons- one, our routing algorithm is robust enough to avoid deadlocks and livelocks, two, it eases the design complexity.

We adapt the routing flow as described in the chart in Fig. 2. In our proposed design, we make channel reservations based on the flit's destination. When a flit arrives at a router (Router X), it is checked if this flit is a header flit or body or tail flit. If it is a header flit, its destination address is extracted and according to the flow in Fig. 2 it is assigned an output port. If the desired destination port is not available, we check for the availability of the side buffer. If it is free, then the flit is pushed into the side buffer and reservation is made for all the flits up to the tail flit from the incoming direction. Whenever the desired output port becomes free, this flit is assigned to its desired destination port on the highest priority. This avoids any chances of the flit to get stuck into the side buffer. To prevent a situation where only flits from a particular port are stored in the side buffer, we use round-robin arbitration in our design, so that every input port gets to write into the side buffer whenever the need be.

In our architecture, we can store a header flit and one extra, body flit in the side buffer. This helps reducing the time for which the channel in the source router is blocked due to unavailability of the desired output port. We even ensure that the flits that get deflected, do not reach their source point, thus avoiding any formation of cycles in the network. We do this by using XY routing strategy. So, even though we deflect our flits,

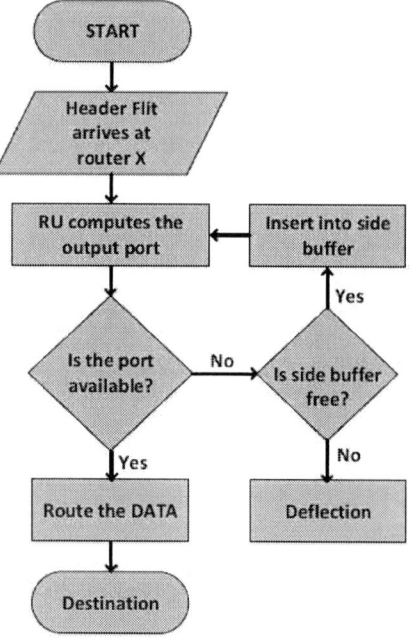

Fig. 2. Data Routing Strategy

978-1-5386-1357-3/17 $31.00 © 2017 IEEE

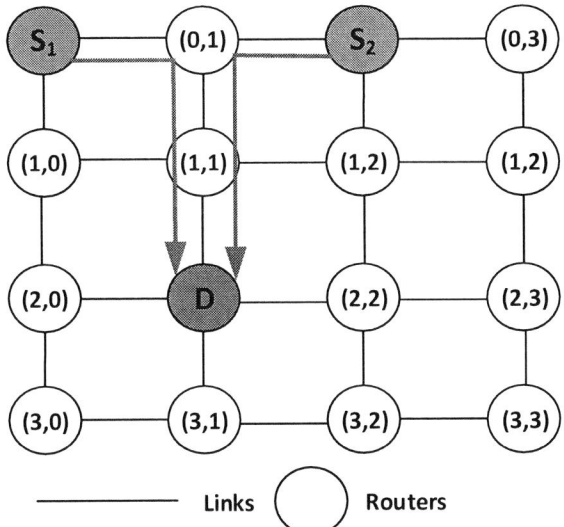

Fig. 3. Walkthrough example for proposed architecture

TABLE I
Area and Power of the building blocks of the proposed design

Block	Area (μm^2)	Power (μW)
Routing Unit	0.176435	0.065854
Side Buffer	0.117624	0.032927

TABLE II
Area and Power overheads of MMTH implementation

Block	Area (μm^2)	Power (μW)
Read Pointer	23.007802	0.23062
Write Pointer	98.159000	8.0092

the deflection is in a way that is deterministic. We actually deflect the flits towards their destination rather than just deflecting them anywhere.

C. Walkthrough Example

For illustration purposes, assume a system size of 4X4. Suppose a flit originating from router R_{00} (S_1), flit1, wants to go to router R_{21} (D). At the same time a flit from router R_{02} (S_2), flit2, also wants to go to the same destination. These flits reach router R_{01} from West and East input ports respectively. Now only one flit can go to router R_{11}. Assuming initial conditions, the round-robin arbiter's pointer is set to East port. So, the RCU will allocate the South port to flit2 and insert flit1 into the side buffer of R_{11}. It will also reserve this output port for flits from East port until the tail flit is encountered from that port. Now, another packet is generated from router R_{11} which also wants to go to South. Considering all the packets are of size 3, after 3 clock cycles, flit1 will be routed to the South port and two other flits from same source will also be routed to this destination via the side buffer. Meanwhile, flit3 will be routed

to the next router depending on its y-coordinates. In case this flit also wants to go to R_{21}, it cannot do so until the side buffer is free. So this will be deflected to either East or West depending on which port is free. If both East and West ports are reserved, then it will not move out of the pipelined registers. In this way, we avoid livelocks and deadlocks. Having round-robin arbitration ensures that next time a flit wants South, first, this flit3 is given that port. Fig. 3 shows the diagrammatic illustration of the above problem.

V. PERFORMANCE EVALUATION

In this section, we discuss the simulation setup and the observations we made based on our simulations for the proposed design.

A. Simulation Setup

We compare our work with traditional buffered router architecture as well as with the baseline router. We developed the Verilog models for the baseline router architecture, our proposed architecture as well as for the marching memory implementation. We extracted the area and power values for both, our proposed architecture and for MMTH implementation by synthesizing and analyzing it using Synopsys Design Compiler at 65nm technology node. Table 1 lists the area and power consumption of the various designed blocks, while

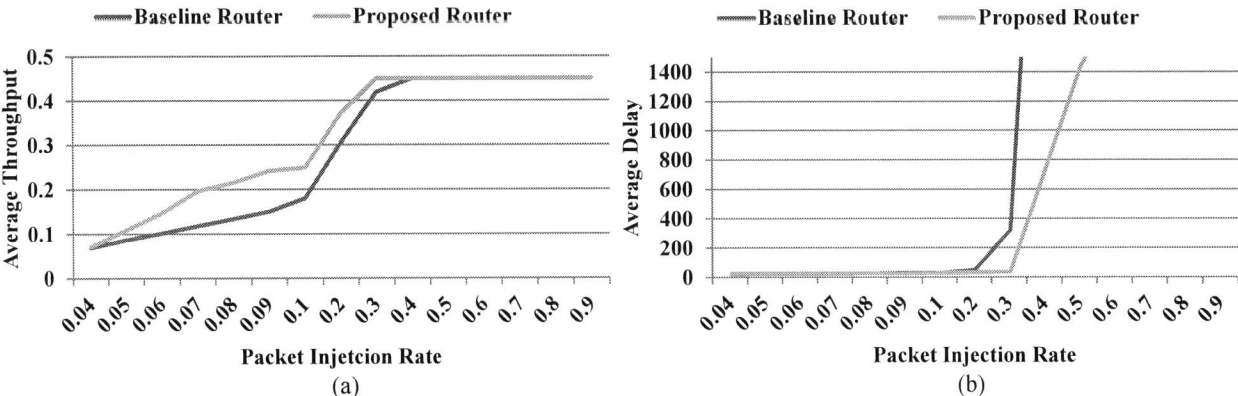

Fig. 4. Performance comparison of 8X8 proposed architecture with MinBSD using Synthetic traffic. Fig. 4(a) illustrates average throughput and Fig. 4(b) illustrates average delay.

978-1-5386-1357-3/17 $31.00 © 2017 IEEE   53

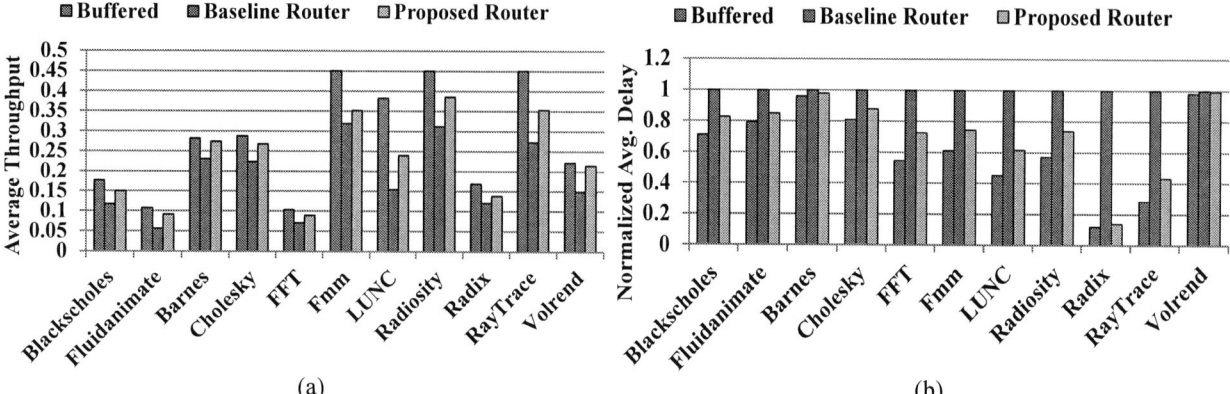

(a)   (b)

Fig. 5. Performance comparison of 8X8 proposed architecture with MinBSD and traditional Buffered design using Real traffic generated from GPGPU-SIM simulator. Fig 5(a) shows average throughput and 5(b) shows average delay.

Table 2 reports the area and power overheads associated with the MMTH implementation.

We have tried to make our design scalable, so we simulate and observe our design on uniform synthetic traffic patterns on 8X8, 16X16 and 16X32 mesh networks. We used Noxim Network Simulator [7] to model the proposed router design and compare it with the baseline architecture and extracted the performance parameters for all the three system sizes mentioned above. We also ran the real time workloads, which we obtained by collecting the application level traffic from GPGPU-SIM [8] full system simulator using PARSEC [9] and SPLASH-2 [10] benchmarks on both, the baseline router architecture and on the proposed architecture for a system size of 8X8, and obtained the performance parameters. We ran our system for a large enough number of clock cycles (11000 cycles), so that we could observe the proper functioning of our side buffer and so as to bombard the network with fairly high traffic load to attain saturation and explicitly induce congestion.

B. Throughput and Latency Analysis

In this section we discuss about the average throughput and delay for a 64-core system size under uniform traffic pattern.

Throughput: Our evaluations show an average 11.42% increase in average throughput as compared to the baseline router, while the saturation point is achieved at a 21.7% higher injection load. This happens because lesser number of flits are deflected, thus resulting in reduced congestion and a higher rate of load delivery to its destination, due to prevention of unnecessary livelocks and deadlocks. Fig. 4(a) shows the average throughput comparison between proposed and the baseline router architecture.

Latency: We observe that the average flit latency reduces by around 55.4% on an average over the baseline router architecture for the synthetic traffic patterns we evaluated for. Fig. 4(b) presents the average flit delay comparison between proposed and the baseline router architecture.

C. Performance Evaluation with Real Traffic

We also analyzed our system with the SPLASH2 benchmarks. On analysis, it is observed that the proposed design performed better than the baseline router in all the traffic patterns analyzed. The proposed design gave a 27.9% reduction in latency and an improvement of 30.46% in the total average throughput over the baseline router design.

Our simulations show a performance improvement over the

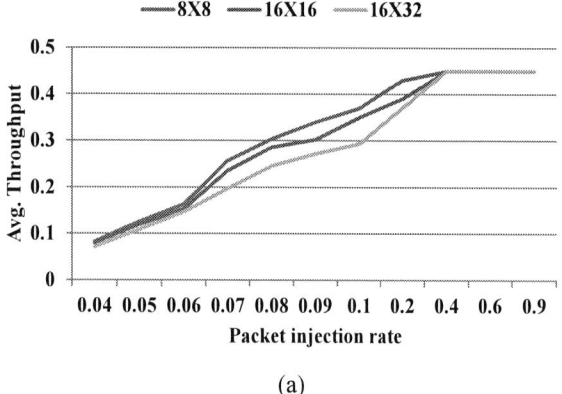

(a)

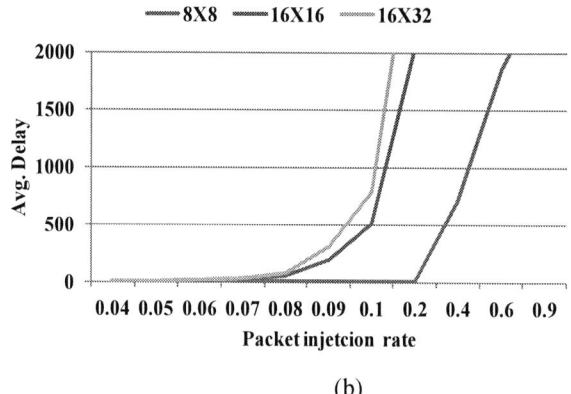

(b)

Fig. 6. Performance comparison of proposed architecture at varying system sizes. Fig. 6(a) illustrates average throughput and Fig. 6(b) shows average delay.

978-1-5386-1357-3/17 $31.00 © 2017 IEEE

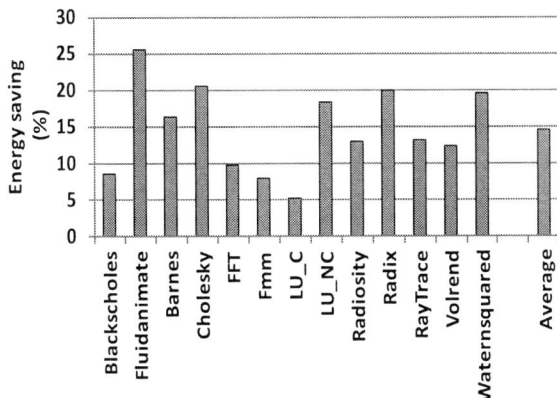

Fig. 7. Comparison of energy savings using proposed architecture for various PARSEC and SPLASH2 benchmarks for a 8X8 system

conventional minimally buffered router architecture. Even for large system sizes, the proposed design provides a better average throughput and a better average flit latency as compared to the baseline router. Fig. 5(a) illustrates the average throughput comparison between proposed and the baseline router architecture using real traffic loads, and Fig. 5(b) presents the normalized average delay comparison between proposed and the baseline router architecture.

D. Scalability

Introduction of a larger side buffer ensured storage of more flits that would have been otherwise deflected. This helps our design to perform optimally when the system scales. As seen from Fig. 6(a.) and Fig. 6(b.), which show a comparison of average throughput and average delay at varying system sizes, our design performs well at even larger system sizes. This is because, when a large number of flits are injected into the system, it is able to store them, rather than deflecting them, thereby reducing the interconnect load and improving average throughput.

E. Energy saving

We increased the size of the side buffer but eliminated the core buffer in the minimally buffered router design. As a result, the total number of buffers in the system remain same. We exploited this and saved on an average 14% of power over the baseline router architecture as a result of using MMTH in our design. This came even as MMTH had some overheads as mentioned in Table 2. Fig. 7 presents the comparison of energy savings at various real traffic loads, with a maximum power saving of 26%.

VI. CONCLUSION

In this work, we proposed a novel minimally buffered deflection router, which is area and power efficient, reduces average flit latency, increases the average throughput and is scalable. We analyzed multiple works pertaining to minimally buffered deflection architecture and observed their limitations, and in our design rectified these limitations. Using a larger

depth side buffer in our design helped us achieve a 30.46% higher throughput and a 27.9% lower average flit latency for real applications. Use of MMTH further enabled us to save power consumption in the router by 14%.

REFERENCES

[1] Y. Hoskote, S. Vangal, A. Singh, N. Borkar and S. Borkar, "A 5-GHz Mesh Interconnect for a Teraflops Processor," in IEEE Micro, vol. 27, no. 5, pp. 51-61, Sept.-Oct. 2007.

[2] Thomas Moscibroda and Onur Mutlu. 2009. A case for bufferless routing in on-chip networks. SIGARCH Comput. Archit. News 37, 3 (June 2009), 196-207.

[3] C. Fallin, G. Nazario, X. Yu, K. Chang, R. Ausavarungnirun and O. Mutlu, "MinBD: Minimally-Buffered Deflection Routing for Energy-Efficient Interconnect," 2012 IEEE/ACM Sixth International Symposium on Networks-on-Chip, Copenhagen, 2012, pp. 1-10.

[4] G. R. Jonna, J. Jose, R. Radhakrishnan and M. Mutyam, "Minimally buffered single-cycle deflection router," 2014 Design, Automation & Test in Europe Conference & Exhibition (DATE), Dresden, 2014, pp. 1-4.

[5] R. Yasudo et al., "Design of a low power NoC router using Marching Memory Through type," 2014 Eighth IEEE/ACM International Symposium on Networks-on-Chip (NoCS), Ferrara, 2014, pp. 111-118.

[6] S. Z. Sleeba, J. Jose and M. M. G., "HiPAD: High Performance Adaptive Deflection Router for On-Chip Mesh Networks," 2015 Fifth International Conference on Advances in Computing and Communications (ICACC), Kochi, 2015, pp. 16-19.

[7] V. Catania, A. Mineo, S. Monteleone, M. Palesi, D. Patti. Noxim: An Open, Extensible and Cycle-accurate Network on Chip Simulator. IEEE International Conference on Application-specific Systems, Architectures and Processors 2015, July 27-29, 2015, Toronto, Canada.

[8] A. Bakhoda, G. L. Yuan, W. W. L. Fung, H. Wong and T. M. Aamodt, "Analyzing CUDA workloads using a detailed GPU simulator," 2009 IEEE International Symposium on Performance Analysis of Systems and Software, Boston, MA, 2009, pp. 163-174.

[9] Bienia, C., & Li, K. (2011). Benchmarking modern multiprocessors. New York: Princeton University.

[10] S. C. Woo, M. Ohara, E. Torrie, J. P. Singh and A. Gupta, "The SPLASH-2 programs: characterization and methodological considerations," Proceedings 22nd Annual International Symposium on Computer Architecture, Santa Margherita Ligure, Italy, 1995, pp. 24-36.

[11] H. K. Mondal, S. H. Gade, R. Kishore and S. Deb, "P2NoC: Power-and Performance-aware NoC Architectures for Sustainable Computing." Sustainable Computing: Informatics and Systems (2017).

[12] H. K. Mondal, S. H. Gade, S. Kaushik and S. Deb, "Adaptive Multi-Voltage Scaling with Utilization Prediction for Energy-efficient Wireless NoC," in IEEE Transactions on Sustainable Computing, vol. PP, no. 99, pp. 1-1.

[13] H. K. Mondal, S. H. Gade, R. Kishore, S. Kaushik and S. Deb, "Power efficient router architecture for wireless Network-on-Chip," 2016 17th International Symposium on Quality Electronic Design (ISQED), Santa Clara, CA, 2016, pp. 227-233.

[14] H. K. Mondal, S. H. Gade, R. Kishore and S. Deb, "Adaptive multi-voltage scaling in wireless NoC for high performance low power applications," 2016 Design, Automation & Test in Europe Conference & Exhibition (DATE), Dresden, 2016, pp. 1315-1320.

[15] H. K. Mondal and S. Deb, "An energy efficient wireless Network-on-Chip using power-gated transceivers," 2014 27th IEEE International System-on-Chip Conference (SOCC), Las Vegas, NV, 2014, pp. 243-248.

[16] H. K. Mondal and S. Deb, "Energy efficient on-chip wireless interconnects with sleepy transceivers," 2013 8th IEEE Design and Test Symposium, Marrakesh, 2013, pp. 1-6.

2017 IEEE International Symposium on Nanoelectronic and Information Systems

Routing Algorithm for Application-Specific Network-on-Chip with Irregular Core Sizes

Grandhi Sai Anirudh, Soumya J.
Department of Electrical and Electronics Engineering
Birla Institute of Technology and Science-Pilani, Hyderabad Campus, Hyderabad, India
gsaianirudh96@gmail.com, soumyatkgp@gmail.com

Abstract— Routers, physical links and routing algorithms together perform the task of routing a message between different cores on a Network-on-Chip (NoC) architecture, they are among the most important parts of its structure because of their huge power consumption compared to other components of a NoC. In order to minimize the power consumed by these components it is essential to design an efficient routing algorithm. In this paper we propose an application specific routing algorithm for mesh based topologies with irregular core sizes. While there are algorithms already present for this purpose the aim of this study is to develop a more efficient algorithm by reducing the length of the path between the two communicating cores. The algorithm proposed in this paper follows wormhole switching and uses XY routing algorithm as a basis and improves on it. In this algorithm routing is performed in two steps, Mapping in X-direction followed by Y-direction. Efficiency of the proposed algorithm has been tested on different mesh networks with varying sizes and also with different locations of oversized cores in the network. Experimental results show that the proposed algorithm performs better compared to existing algorithms in the literature.

Keywords: Network-on-Chip, Mesh based topologies, Routing Algorithm.

I. INTRODUCTION

Moore's law [1] states that number of transistors per square inch on integrated circuits has doubled every year since their invention. In recent years, with the advancement of Multi-Processor System-on-Chip (MPSoC) [2], intra chip communication has become more and more important in determining the performance of the entire system [3]. NoCs [4] have gradually gained acceptance as the dominant communication platform for emerging Chip Multi-Processing systems (CMP) with multiple cores on a single chip. Due to their high performance, scalability, modularity and predictability they have replaced traditional bus based communication architectures on CMP systems. NoCs have also shown to provide better communication bandwidth when compared to bus based architectures [5].

While performance plays a key role in determining the efficiency of a communication system, power consumption is the most important factor influencing it. It is shown that NoCs consume a significant part of the total system power, so relevant steps are to be taken to reduce the power consumption of NoC. Main components of a NoC are physical links, Intellectual Property (IP) cores, Network Interface (NI) and routers. IP cores are the components in the digital systems

which have a particular task to perform ex: CPU, DSP, memory and I/O units. NI ensures that the heterogeneous IP cores with different protocols communicate transparently. NI is also responsible for packetization and de- packetization of data. Routers connected by the physical links ensure that efficient data transfer takes place between any two IP cores in the system [6]. Power consumed by NoC depends on several factors such as the IP cores (the processing load of each core influences the power consumption), network interface, routing algorithm, switch placement. The routing algorithm (along with the physical links and routers) determines the path taken by the message in a NoC, consumes a significant part of the total chip power (as high as 40% in [7]) and hence the path followed by the message will have a key role in determining the power consumption of a NoC.

There are numerous possible topologies available for NoC, however, routing methodologies discussed in this paper are limited to mesh topologies. They are frequently used in NoC because of their network scalability and employment of a simple routing algorithm [8]. There are two types of mesh topologies, the first one where all the cores are of uniform size and focus is on mapping cores on to mesh network. The second one - where application specific NoCs having cores of irregular sizes and possess diverse communication flow requirements [9]. In practical systems we rarely find a NoC with uniform core sizes, modern CMP systems have heterogeneous cores such as Application specific Integrated Circuits (ASICs), programmable processors and accelerators that are non-uniform in size.

Rest of the paper is organized as follows. Brief overview of similar algorithms is discussed in section II. The proposed algorithm is discussed in section III. In section IV experimental results are shown and section V concludes the paper.

II. RELATED WORK

This section describes different routing algorithms available in the literature. In [6], XY routing algorithm and different variations of it are discussed. In this algorithm wormhole routing is followed, header flit has the destination node address. Routing is done such that the message first travels horizontally mapping its X-axis and then travels vertically mapping its Y-axis. This algorithm is frequently used in mesh topologies and it is only applicable for mesh network with uniformly sized cores and fails for a mesh with irregular core sizes. In [3], an irregular routing algorithm is discussed in which message is injected into the network as a row message. If there are no oversized nodes, it follows XY routing and its status is set to normal. When it comes across an oversized

---

This work is partially supported by the research project No. ECR/2016/001389 dt. 06/03/2017, sponsored by the SERB, Govt. of India

978-1-5386-1357-3/17 $31.00 © 2017 IEEE   56

node it gets assigned a predefined status and depending on the location of the destination certain rules are followed to determine the direction the message has to follow. In [3] another irregular routing algorithm is discussed which is an improvement to the previous one, it routes the message along the oversized node orientation and it is allowed to use all the four virtual channels unlike the previous algorithm. 3-D ICs are mainly affected by temperature hotspots. [14] Proposes an algorithm to overcome such temperature issues resulting from increased power density in some parts of the chip. This algorithm considers thermal and communication factors while mapping and placement of application tasks in a 3-D NoC. [11] Presents an adaptive routing algorithm to overcome thermal issues in a 3-D IC. Routers whose temperature increases beyond a specific limit are blocked and in order to overcome traffic congestion around these routers packets travel to the bottom most layer and then come back to the destination layer. In this way overheated routers are quickly cooled.

In [10] a thermal aware application specific routing algorithm is discussed which makes use of the traffic information of the entire network to derive multiple paths for routing to avoid deadlock in the network. Given a set of possible routing paths, the problem of distributing traffic among the paths is done such that each path possess optimal traffic and is tackled as a mathematical programming problem. A given application is specified as a task flow graph which carries information about the required bandwidths and communication dependencies. A tile based mesh topology is used in this algorithm. Depending on the task flow graph and target topology processor mapping and task allocation is done before the routing phase. [13] Discusses about increasing temperature stress in design of digital circuits. This stress is due to technology scaling factors such as increased transistor density. It presents a hardware infrastructure which provides thermal control of MPSoC architectures and uses the NoC interconnections of the base system as an active component in the coordination and communication between temperature sensors located around the chip. [15] Concentrates on algorithmic design approaches, reviews the thermal management systems on 3-D NoC systems. It also analyzes some packet routing algorithms which are currently being used.

III. OUR APPROACH

In this section, the proposed routing algorithm for mesh networks is discussed which considers non-uniform sized cores.

Proposed Algorithm

According to the proposed algorithm, X-axis is first mapped (co-ordinates of source and destination made equal) then Y-axis is mapped, which implies that the message first travels horizontally and then vertically. Let the co-ordinates of the source and destination be (A_x, A_y) and (D_x, D_y) respectively. Analysis of the algorithm in three cases which can arise in any given mesh network is discussed as follows.

Case-1: When the message encounters an oversized node before mapping its X-axis, it cannot go any further in the X direction because of which it changes its direction and travels

in the Y direction to equal its Y co-ordinate ($A_y=D_y$). Then it again travels in X direction equaling its X co-ordinate ($A_x=D_x$) and reaching the destination i.e., ($A_x=D_x$ and $A_y=D_y$).

Case-2: Similar to the previous case, message first encounters an oversized core before mapping its X co-ordinate and starts travelling in the Y direction, but here before equaling its Y co-ordinate it comes across another oversized core. In this situation, the algorithm restarts i.e., it again looks to map its X co-ordinate after which it maps its Y co-ordinate.

Case-3: When there are no oversized cores between the source and destination it follows standard XY routing algorithm [6].

It may be noted that the in all the three cases message travels from west to east when $D_x>A_x$ and from east to west when $D_x<A_x$. Similarly when $D_y<A_y$ it travels from south to north and from north to south when $D_y>A_y$. The proposed algorithm is explained in the following by taking an example given in Fig. 1.

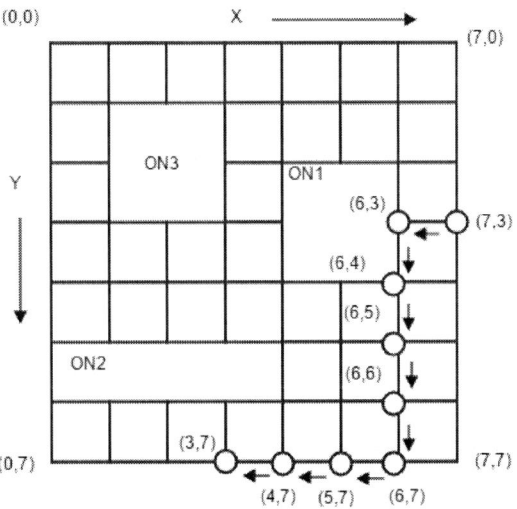

Fig 1: Proposed Algorithm (case 1) Source (7, 3) Destination (3, 7)

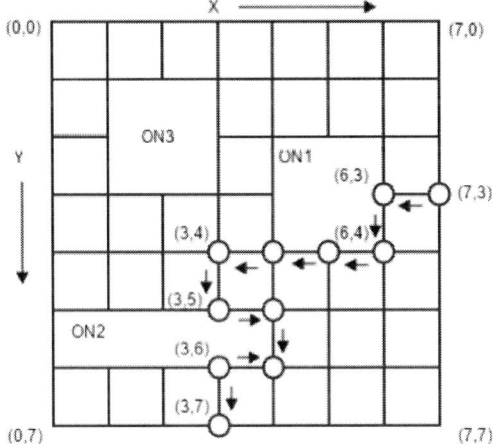

Fig 2: Algorithm [3] (case 1) Source (7, 3) Destination (3, 7)

Suppose a message has to travel from source (7, 3) to destination (3, 7). According to our algorithm (Fig. 1) it has to map its X-axis first by travelling horizontally to (3, 3) and

978-1-5386-1357-3/17 $31.00 © 2017 IEEE   57

then vertically to (3, 7) mapping its Y-axis. Due to the presence of oversized node ON1, it cannot travel horizontally from (6,3) to (5,3), hence it should look to map its Y-axis instead and it does so by travelling to (6,7) and then it travels horizontally to (3,7) reaching its destination in a total of 8 hops. The same network when routed using the algorithm [3] takes 10 hops to reach the destination as shown in Fig. 2. In this case, the message encounters two oversized cores in its path from source to destination.

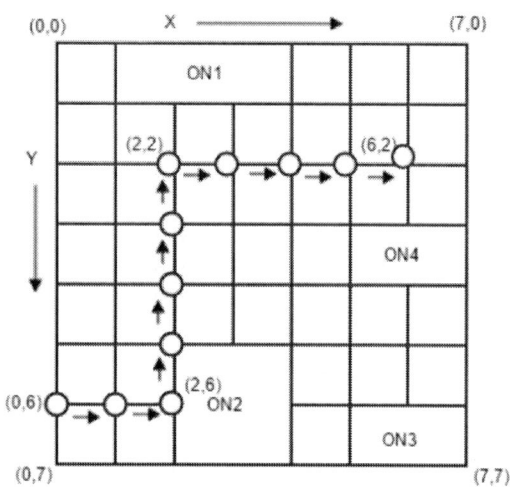

Fig 3: Source-(0, 6), Destination (6, 2) - Proposed Algorithm-10 hops

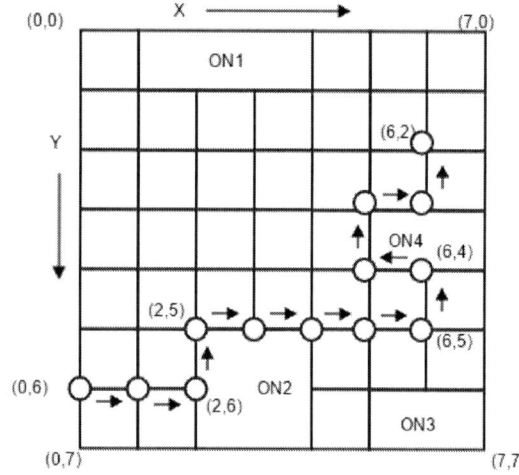

Fig 4: Source-(0, 6), Destination (6, 2) – Algorithm [3]-12 hops

Another example is taken to demonstrate the efficiency of our algorithm, where the number of hops taken by the message to reach the destination is lesser using proposed algorithm compared to the algorithm in [3] (Illustrated in Fig. 3 and 4). In this case, message first comes across an oversized core at (2, 6). According to the proposed algorithm (Fig. 3) it now looks to map its Y-axis and travels to (2, 2). Since its Y-axis is mapped it looks to map its X-axis now. It travels from (2, 2) to (6, 2) reaching its destination in a total of 10 hops. The same network when routed using algorithm given in [3] (Fig. 4), after getting blocked at (2, 6), it travels along the oversized

core from (2, 5) to (6, 5). But it again gets blocked at (6, 4) and travels along the oversized core and reaches its destination in 12 hops. However, in some cases, message cannot reach the destination using the algorithm proposed in [3] due to the positions of the oversized nodes, whereas in the proposed algorithm, message could reach the destination. This case is shown in Fig. 5. Using the proposed algorithm two paths (because of direction of de-map (discussed next) are possible from source to destination, one of which is shown in the Figure (longer one).

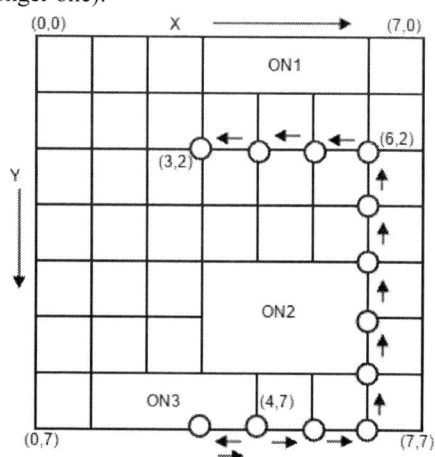

Fig. 5: Source-(4, 7), Destination (3, 2)

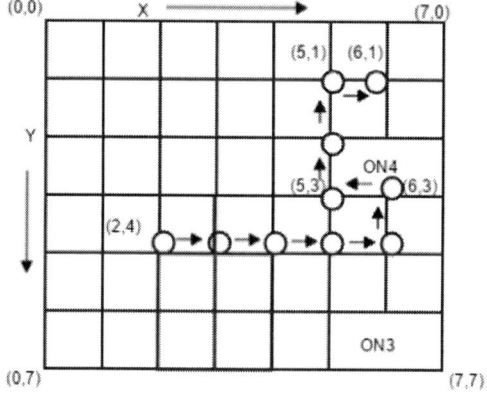

Fig. 6: Source-(2, 4), Destination (6, 1)

Even though the proposed algorithm takes less number of hops in routing a message from source to destination, in some cases it fails to do so. One such case is shown in Fig. 6, where X-axis is mapped and it is not possible to map Y axis. In order to overcome this problem, when a message comes across such an oversized node it de-maps its already mapped axis, travels in one direction (it can choose any direction) till it is possible to map the previously blocked axis. As shown in Fig. 6, the message gets blocked at (6, 3) then it de-maps its X-axis and moves to (5, 3), here it has a chance to map its Y- axis (previously blocked axis), so now it moves to (5, 1) and finally to (6, 1) reaching its destination.

There might be a network where the message might still be blocked at (5, 3). If such a case arises then the message

978-1-5386-1357-3/17 $31.00 © 2017 IEEE   58

continues to move in the same direction (to the west) until it gets a chance to map its Y-axis. Continuing this way if the message reaches end of the network (i.e., it does not get a chance to map its Y-axis), it then reverses its direction and comes back in the same way and crosses the point where it was blocked and continues to travel in this direction. It stops when it is capable of mapping its previously blocked axis (such a case arises only when more than 50% of the mesh is occupied by oversized nodes). While de-mapping its originally mapped axis it chooses to de-map depending on the direction. The message gets two possible paths, both being equally likely. This de-mapping is the limiting factor of the proposed algorithm, however using our algorithm message could reach the destination but using the algorithm given in [3], it could not. Below figure summarizes the proposed algorithm.

Input: Mesh network with positions of oversized cores, Source (Ax, Ay) and Destination (Dx, Dy)
Output: Possible routing path from source to destination
Begin
    If Ax not equal to Dx, then
        Travel east or west to achieve Ax = Dx
        Travel north or south to achieve Ay = Dy
    If blocked before Ax = Dx, then
        Travel north or south to achieve Ay = Dy
        Travel east or west to achieve Ax = Dx
    If blocked twice before Ax = Dx, and before Ay = Dy then
        Restart
End

IV. EXPERIMENTAL RESULTS

In this section we present the simulation results obtained and compare the results of the proposed algorithm and algorithm in [3] in routing messages from source to destination. A code on java platform was developed to generate networks with varying sizes, number of oversized cores and size of the network. In order to study the impact of size of mesh on routing, different networks with sizes varying from 2 to 12 (size m indicates a mxm network) are taken with the percentage of oversized cores in the network remaining constant at 20 and routing is done using both the algorithms (proposed algorithm and algorithm mentioned in [3]), results of which are shown in Fig. 7. It is evident from the Figure that using the proposed algorithm, greater number of messages reaches the destination. For example, in a mesh of size 8 using proposed algorithm 58.54% of messages could reach the destination where using the algorithm in [3] only 37.41% could reach the destination. For mesh networks of small sizes the difference between the proposed algorithm and algorithm in [3] is small (Fig. 7) but as the size of the mesh increases it is evident that proposed algorithm has a better performance. Hence, the performance of our algorithm increases as the size of the mesh increases, which clearly shows the applicability of

our approach to multi-core architectures with non-uniform sizes.

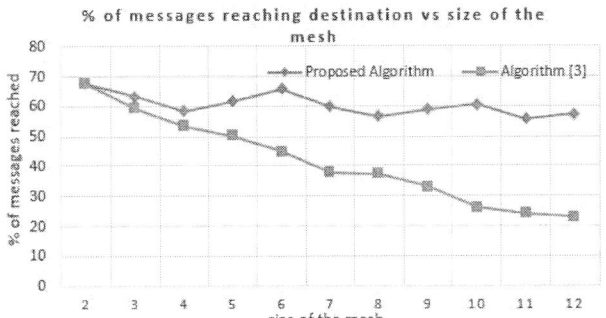

Fig. 7: No. of messages reaching destination vs Size of mesh

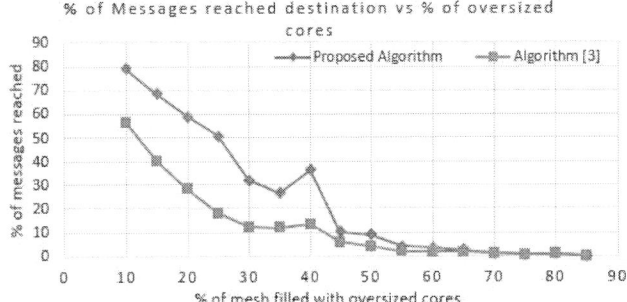

Fig. 8: No. of messages reaching destination vs No. of oversized cores

Performance of the proposed algorithm is also tested with the increase in number of oversized cores in the network. Keeping the size of the network constant at 10, percentage of oversized cores is increased from 10 to 90. Results are shown in the Fig. 8 where the Y-axis contains the percentage of messages reaching the destination and X-axis contains the percentage of mesh occupied by oversized cores. It can be observed that the proposed algorithm has superior performance over the algorithm in [3]. According to the results, in a mesh which is 40% filled with oversized cores proposed algorithms delivers 36.8% of messages while the algorithm [3] delivers only 13.26% of the messages. Also, if the number of oversized cores is less than 50%, our algorithm performs better but as it increases beyond this point, both the algorithms perform almost the same. It may be because as there are more number of oversized cores, there may be few available paths in the network. However, in other cases, proposed algorithm outperforms the algorithms proposed in the literature. This clearly shows the advantage of the proposed algorithm with the increasing number of oversized nodes, making it more suitable for ASNoCs.

978-1-5386-1357-3/17 $31.00 © 2017 IEEE

% of network occupied by oversized cores(size 10)	% of messages reaching destination using		Percentage improvement
	Proposed Algorithm	Algorithm in [3]	
10	79.15	56.44	40.23
20	58.94	28.21	108.9
30	31.92	12.19	161.8
40	36.3	13.26	173.75
50	8.98	4.04	122.2
60	3.4	1.7	100
80	0.7	0.7	0

Table 1: Comparison of No. of messages reaching destination in the proposed algorithm and [3]

Table-1 summarizes the simulation results of our algorithm in comparison with the algorithm proposed in [3]. In some cases number of messages reaching the destination using the proposed algorithm is high compared to the algorithm given in [3]. For example, during the simulations we tested for a network of size 10 and 25% of the mesh was occupied with oversized cores. In this network proposed algorithm routed 50% of messages to the destination whereas the algorithm in [3] routed only 18% of messages to their respective destinations.

V. CONCLUSION

To address the routing problems faced by mesh networks with oversized cores, in this paper we propose an application specific routing algorithm which is based on XY routing algorithm. Proposed algorithm routes a message from source to destination by mapping the X and Y axis one after the other, it behaves as XY routing algorithm in the absence of oversized cores. Simulation results show that in some cases, using the proposed algorithm almost 80% of the generated messages reach the destination and in some cases there was an improvement of 170% over previously existing routing algorithms.

REFERENCES

[1]"Moore's Law", http://www.mooreslaw.org/

[2]"MPSoC", https://en.wikipedia.org/wiki/MPSoC

[3]Ladan Momeni, Arshin Rezazadeh, "Throughput Evaluation of Irregular Routing Algorithm for 2-Dimensional Mesh Network-on-Chip", pp.331-336, 2015.

[4] Ankur Agarwal, Cyril Iskander and Ravi Shankar, "Survey of Network on Chip (NoC) Architectures & Contributions", Engineering, Computing and Architecture, vol. 3, no. 1, 2009.

[5] S. Kwon, S. Pasricha and J. Cho, "POSEIDON: A framework for application-specific Network-on-Chip synthesis for heterogeneous chip multiprocessors," 2011 12th International Symposium on Quality Electronic Design, Santa Clara, CA, 2011, pp. 1-7.

[6] Shubhangi D Chawade, Mahendra A Gaikwad and Rajendra M Patrikar. Article: Review of XY Routing Algorithm for Network-on-Chip Architecture. International Journal of Computer Applications 43(21/973-93-80867-69-8):20-23, April 2012.

[7] Vangal, Sriram & Singh, Arvind & Howard, Jason & Dighe, Saurabh & Borkar, Nitin & Alvandpour, Atila. (2007). A 5.1GHz 0.34mm2 Router for Network-on-Chip Applications. 42 - 43. 10.1109/VLSIC.2007.4342758.

[8] Vaishali V Ingle and Mahendra A.gaikwad. Article: Review of Mesh Topology of NoC Architecture using Source Routing Algorithms. IJCA Special Issue on Recent Trends in Engineering Technology RETRET:30-34, March 2013.

[9] Bei Yu, Sheqin Dong, Song Chen and S. Goto, "Floorplanning and topology generation for application-specific Network-on-Chip," 2010 15th Asia and South Pacific Design Automation Conference (ASP-DAC), Taipei, 2010, pp. 535-540.

[10] Z. Qian and C. Y. Tsui, "A thermal-aware application specific routing algorithm for Network-on-Chip design," 16th Asia and South Pacific Design Automation Conference (ASP-DAC 2011), Yokohama, 2011, pp. 449-454.

[11] S. Y. Lin, T. C. Yin, H. Y. Wang and A. Y. Wu, "Traffic-and thermal-aware routing for throttled three-dimensional Network-on-Chip systems," Proceedings of 2011 International Symposium on VLSI Design, Automation and Test, Hsinchu, 2011, pp. 1-4.

[12] S. Pasricha, "A Framework for TSV Serialization-aware Synthesis of Application Specific 3D Networks-on-Chip," 2012 25th International Conference on VLSI Design, Hyderabad, 2012, pp. 268-273.

[13] D. Atienza and E. Martinez, "Inducing Thermal-Awareness in Multicore Systems Using Networks-on-Chip," 2009 IEEE Computer Society Annual Symposium on VLSI, Tampa, FL, 2009, pp. 187-192.

[14] Addo-Quaye, Charles. (2005). Thermal-aware mapping and placement for 3-D NoC designs. 25 - 28. 10.1109/SOCC.2005.1554447.

[15] K. C. J. Chen, C. H. Chao and A. Y. A. Wu, "Thermal-Aware 3D Network-On-Chip (3D NoC) Designs: Routing Algorithms and Thermal Managements," in IEEE Circuits and Systems Magazine, vol. 15, no. 4, pp. 45-69, Fourthquarter 2015.

[16] T. Mak, R. Al-Dujaily, K. Zhou, K.-P. Lam, Y. Meng, A. Yakovlev, C.-S. Poon, "Dynamic programming networks for large-scale 3D chip integration", IEEE Circuits Syst. Mag., vol. 11, no. 3, pp. 51-62, Aug. 2011.

[17] Y. Jin, E. J. Kim, T. M. Pinkston, "Communication-aware globally-coordinated on-chip networks", IEEE Trans. Parallel Distrib. Syst., vol. 23, no. 2, pp. 242-254, Feb. 2012.

[18 Y. Hoskote, S. Vangal, A. Singh, N. Borkar and S. Borkar, "A 5-GHz Mesh Interconnect for a Teraflops Processor," in IEEE Micro, vol. 27, no. 5, pp. 51-61, Sept.-Oct. 2007.

[19] J.-J. Lecler, G. Baillieu, "Application driven network-on-chip architecture exploration & refinement for a complex SoC", Des. Automat. Embedded Syst. (DAES), vol. 15, no. 2, pp. 133-158, June 2011.

[20] B. S. Feero, P. O. Pande, "Networks-on-chip in a three dimensional environment: a performance evaluation", IEEE Trans. Comput., vol. 58, no. 1, pp. 32-45, Jan. 2009.

[21] B. Black, M. Annavaram, N. Brekelbaum, J. DeVale, L. Jiang, G. H. Loh, D. McCauley, P. Morrow, D. W. Nelson, D. Pantuso, P. Reed, J. Rupley, S. Shankar, J. Shen, C. Webb, "Die stacking (3D) microarchitecture", Proc. IEEE/ACM Int. Symp. Microarchitecture (Micro), pp. 469-479, 2006-Dec.

[22] R. Holsmark and S. Kumar, "Design issues and performance evaluation of mesh NoC with regions," 2005 NORCHIP, 2005, pp. 40-43.

2017 IEEE International Symposium on Nanoelectronic and Information Systems

STT-MRAM for Low Power Access for Read-Intensive Parallel Deep-Learning Architectures

Saranyu Chattopadhyay[1], Kaustav Brahma[2], Arkaprova Ray[3], Mrigank Sharad
Department of Electronics and Electrical Communication Engineering, Indian Institute of Technology, Kharagpur,
West Bengal, INDIA-721302

Abstract—**Spin Transfer Torque-Magnetic Random Access Memory (STT-MRAM) is one of the most promising alternatives to Static Random Access Memory (SRAM) for integrated memory, owing to high density, low leakage and possibility of high-speed read-write operation. In this work, we explore the advantages of STT-MRAM for a parallel processing application specific integrated circuit (ASIC) architecture of fully connected deep learning network. Low power architectures for deep learning networks have gained significant attention in recent years, owing to their potential for edge computing. For a deep-net, memory read operations for fetching network weights constitute major fraction of total power dissipation. In this work, we show that read-optimized STT-MRAM bit-cells can achieve 80% low power read access power as compared to SRAM, through voltage-mode read operation. Such a read scheme can facilitate low voltage operation, which significantly adds to the energy benefits. Hence, apart from the benefits of area and leakage, low power STT-MRAM can also help in achieving low power read access for memory intensive deep-learning architectures.**

Keywords-memory; low power; MRAM; circuit; deep-learning architecture

I. INTRODUCTION

In the current years, machine learning has gained utmost importance mainly because of the huge amount of data we have and because it can solve a wide variety of problems simply on the basis of the statistics and not going deep into intricacies and specificity of the problem. Of the known algorithms of machine learning, deep learning and its implementation through neural networks (NN) is by far the most popular [1]. In fully connected deep neural networks, there are several layers, each consisting of a large number of nodes interconnected via weighing networks. For high performance applications, like real time image processing, the deep-net architecture must offer high degree of parallelism [17]. Recent implementations employ array of processing elements with co-located memory for storing network weights. Each processing element (PE) in such an architecture may implement a number of nodes. For every input cycle, all the weights associated with each node need to be read, thereby leading to significant power dissipation [2,3].

Development of alternative memory technologies for on-chip integration has taken center-stage in last few years. Achieving higher density and lower leakage have been the prime targets of such developments, while maintaining performance comparable to standard CMOS [15,16]. However, in case of a deep-net architecture, evidently, the read-access energy can dominate the power budget. Hence, it may be crucial to assess the chosen technology in terms of access energy.

STT-MRAM has been projected as a promising candidate for future on-chip memory, owing to zero leakage, excellent scalability and acceptable read-write speed [2-10].

In this paper, we show that read optimized STT-MRAM can also facilitate low-voltage, fast read operations that can be advantageous for highly parallel deep learning architectures with an array of PEs and co-located memories.

Rest of the paper is organized as follows. Section-II presents a high-level overview of a parallel architecture for deep-net. Section-III describes the basic operation of STT-MRAM and the design choices for read optimization. Low power, low voltage, pre-charge based voltage mode read operation for read-optimized STT-MRAM is presented in Section-IV, along with comparison with SRAM. Section-V concludes the paper.

II. DEEP LEARNING ARCHITECTURE

The basic building blocks of deep neural networks are multiply and accumulate units (MACs), which implement the functionality of neurons [1]. Several neurons form a single layer of the neural network (Fig. 1). For fully connected deep-nets, each neuron is connected with all the neurons in the next layer via weighted networks [1]. During training phase, when an input pattern is applied at the input layer, it propagates forward to every neuron in the next layer, scaled by the weight of the connecting network. The weighted signals from the previous layer are summed and processed through a nonlinear transformation, like sigmoid or ReLU [1]. The result obtained from the output layer is compared with the training example. Thereafter, the weights are modified via back propagation algorithm [11] to reduce the deviation of the model output. The trained network weights are written into the network memory. In general, the training phase is less frequent and hence write operation for the weight memory [1]. During evaluation phase, NNs are fed with unknown input patterns to generate plausible inferences.

978-1-5386-1357-3/17 $31.00 © 2017 IEEE

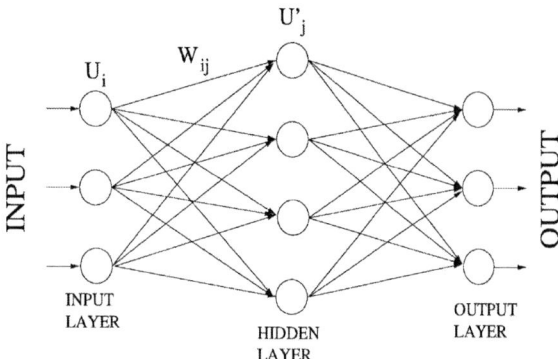

Figure 1: Neural network

In the evaluation phase, from hardware point of view, it would take a huge time to access so many weights sequentially, thus mandating great degree of parallelism. A preferred solution would be a multi-core architecture where each core or PE implements a certain number of neurons, along with a memory bank for storing the associated weights (Fig. 2) [18]. A standard SRAM implementation generally mandates appropriate partitioning and optimal block-size to arrive at optimal read-access energy, by balancing the power dissipated in bit-line charging and that associated with decoding [3]. Such partitioning further degrades the area density of SRAM. Note that for a parallel deep-net architecture with thousands of neurons in each layer, the memory size for a PE implementing few 10's of neurons can approach 100kB.

In the following sections, we explore STT-MRAM as an alternative to SRAM for distributed memory in a parallel deep-net architecture.

III. STT-MRAM: STRUCTURE AND OPERATION

STT-MRAM has been identified as a candidate for future on-chip memory due the advantages of low leakage,

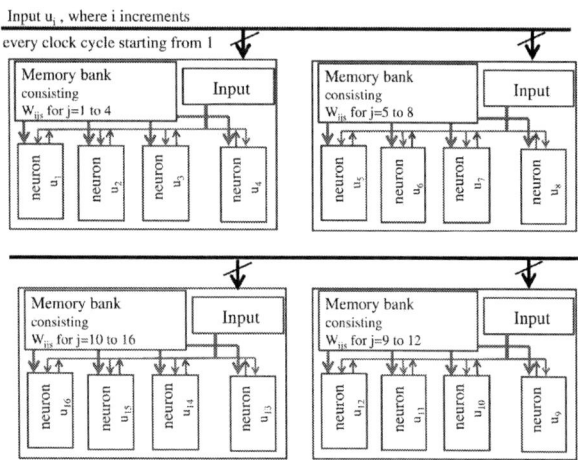

Figure 2: Multi-core architecture

higher area density and comparable read-write speed with respect to SRAM [18]. STT-MRAM is a magnetoresistive memory. Each cell contains one free layer (FL) and one pinned layer (PL). The pinned magnetic layer acts as a spin polarizer. Electrons flowing from PL to FL make the magnetic moment of the FL parallel to the PL. Whereas, electrons flowing from FL to PL makes it antiparallel by spin transfer torque [18], thereby changing the resistance of the two layers (higher for antiparallel state). This change can then be detected by a current or voltage-mode sensing circuit [18]. The simplest MRAM bit cell constitutes of just one transistor for read-write access (1T-1R). Write current happens to be significantly higher than the sensing current used for read operation. In recent literature, several alternate device structures and bit-cells have been proposed for STT-MRAM that can achieve decoupled read-write operation and lower write power [18]. The write energy is higher for standard 1T-1R cell. However, deep-neural network being a read intensive architecture (due to large number of network weights), the work emphasizes the read-energy benefits of STT-MRAM with simplest and most common configuration, optimized for low power read [3, 6].

IV. ENERGY EFFICIENT VOLTAGE MODE READ SCHEME FOR STT-MRAM

The STT-MRAM bit cell (Fig. 3) consists of a bitline (BL), a wordline (WL) and a source line (SL). The bit cell comprises one transistor and one MTJ. [12].

The MTJs are modelled as resistors in the simulation environment [6]. Tunnel Magnetoresistance (TMR)s of up to 500% have been reported in experiments [8]. We have used physics based device models calibrated with experimental data for the MTJ device parameters [8]. In this work, we employ 300% TMR.

The read operation can be implemented through voltage-mode or current-mode sensing [18].

In voltage-mode sensing (Fig. 3b), the source-line capacitance (C_{SL}) is precharged to a reference voltage, after which the word line (WL) is enabled, which provides a discharge path for the charge stored in the source-line capacitor. Depending on the value of the MTJ resistance, the discharge path where R_{ap} (R_p) value is larger (smaller) has a larger (smaller) RC time constant and hence, the charge decays slowly (swiftly). The sense-amplifier (Fig. 4) is switched *ON* when the voltage difference between the V_{ref} and V_{MTJ} is approximately 100mV (evaluation phase of Fig. 5) and this enables the sense amplifier to read the bit stored in the STT-MRAM bit cell. Higher oxide thickness and low voltage read operation is conducive to higher TMR. Hence, voltage mode read operation proposed in this work has the inherent advantage of high TMR.

In the current-mode sensing, a current source is supplied continually to the source-line capacitance and the sink path consisting of MOS transistor and MTJs is always enabled.

978-1-5386-1357-3/17 $31.00 © 2017 IEEE

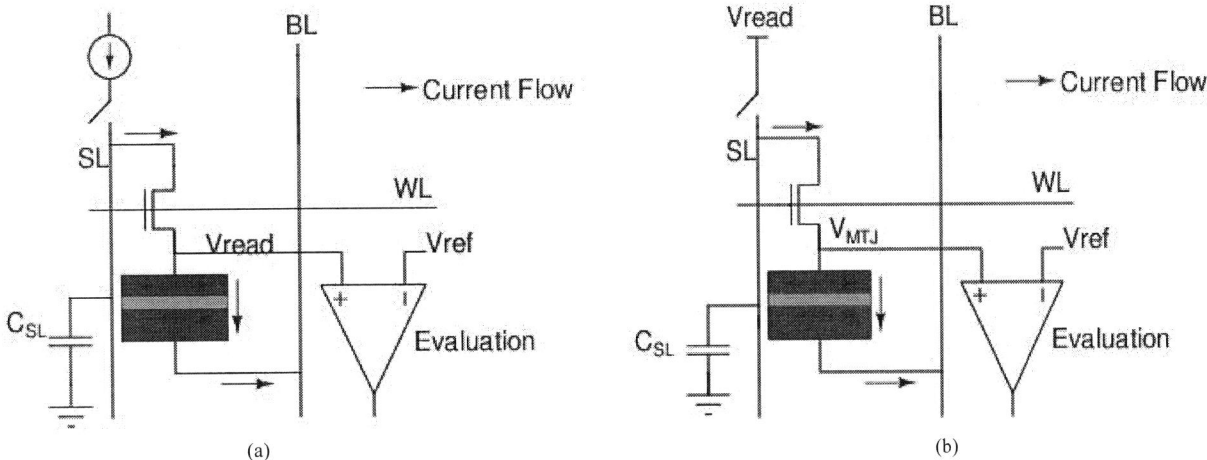

(a)

(b)

Figure 3: 1T-1R STT-MRAM bit cell for (a) Current-mode sensing (b) Voltage-mode sensing

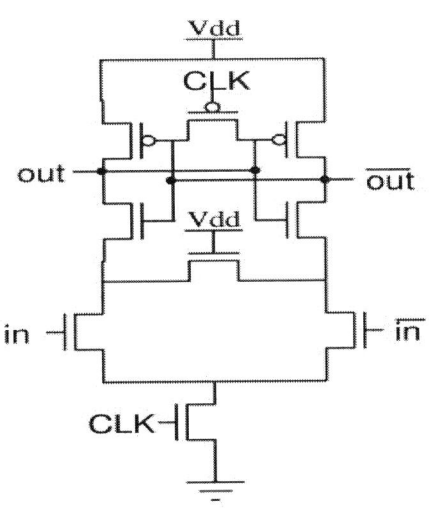

Figure 4: CRSABL(Charge recycle sense amplifier based logic). The charge used during read operation is recycled, leading to greater scaling of the supply voltage and hence, lesser power dissipation [12]

The capacitance(C_{SL}) takes finite time to charge depending on the RC time constant. The final voltage developed across the MTJ is compared with that of a reference cell to determine its state.

The current-mode sensing is inefficient as compared to voltage-mode sensing as there is a constant current sink in the read circuitry. Also, voltage-scalability is limited by the operating constraints of the current sources.

In the voltage-mode read-scheme on the other hand, the precharge voltage applied to the bit-line capacitances can be aggressively scaled down to 300mV. Additionally, there is no static current path in the read circuitry.

For the 1000×1000 STT-MRAM memory bit cell, the peak current through the MTJ in current sensing read operation is 17.17uA (Fig. 6), whereas, the peak current supplied by the precharge voltage source (of 300mV) to the

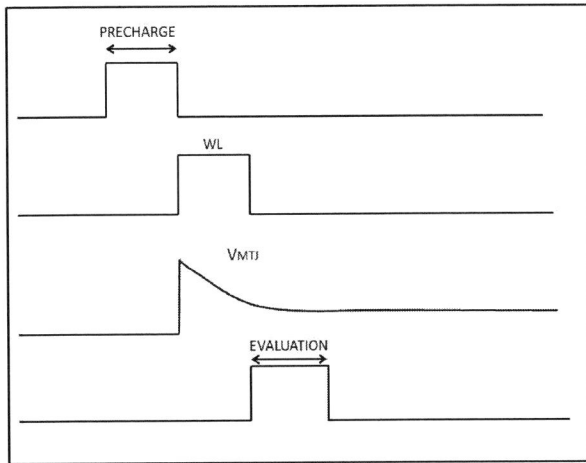

Figure 5: Voltage waveforms for one read cycle

MTJ is 2.89uA (Fig. 7). The total energy consumption of the current mode sensing for one period of 4.5ns is 159.71fJ, and for one period of 5ns the voltage mode sensing of STT-MRAM consumes a total of 23.9fJ (Fig. 8). Since the threshold of switching current through MTJ for a switching time of 4.5 ns is 50-100uA[3] (much greater than 17.17uA and 2.89uA), therefore the possibility of read-disturbance during voltage mode read operation is highly suppressed.

When the precharge voltage is reduced to a lower value, for the same RC time constant of the bit cell, the read latency of the STT-MRAM cell increases, but the power consumption of the bit cell decreases.

The bit cell was modelled as shown in Fig. 3. In order to model the read-write access, appropriate values of parasitic capacitances and leakage currents have been added to the BL, SL and WL to emulate the presence of other bit-cells and metal interconnects. In this work we chose an array size of 1000×1000 for 1T-1R STT-MRAM. Note that, partitioning the memory into smaller banks would lead to area overhead due to read-write interface circuits. SRAM,

978-1-5386-1357-3/17 $31.00 © 2017 IEEE   63

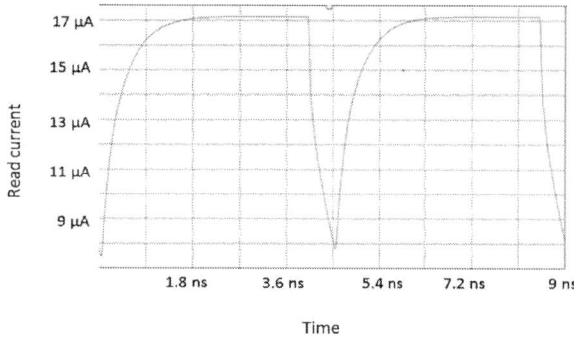

Figure 6: Current through MTJ in current-mode sensing

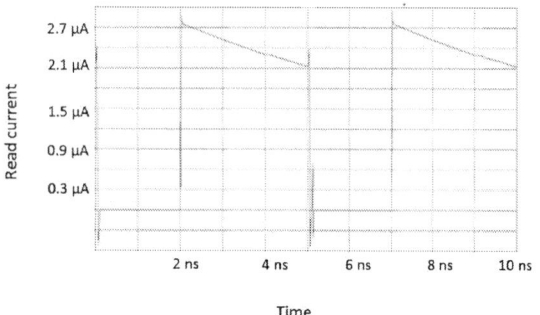

Figure 7: Current through MTJ in voltage-mode sensing

owing to larger area, would mandate partitioning, else, read-write power due to longer metal interconnects would add to power overhead. MRAM, being area efficient, achieves low power for read-access, even with large array size. Moreover, voltage-mode read access proposed in this work facilitates low voltage operation and hence lower read-access energy.

For the SRAM, the read latency is 3.2ns (Fig. 9) as compared to 5ns read latency of STT-MRAM (Fig. 8). But the SRAM read energy is 118.02fJ whereas the STT-MRAM read energy is 26.52fJ for the proposed voltage mode read operation.

Another major disadvantage of SRAM as compared to STT-MRAM is the leakage power . Since in STT-MRAM, the electron spin is used to store information, there is no leakage. But for one bit cell the leakage power of SRAM is 0.429nW. However, leakage power of SRAM can be reduced to some extent through dynamic voltage scaling during idle mode. In this work we mainly emphasise the advantage of low energy read operation of MRAM for read-intensive deep-neural network computing.

V. ARCHITECTURE SIMULATION

In the memory architecture proposed, we assume that each layer of the fully connected deep neural network has 1000 nodes [17] . And each node of the previous layer is connected to each node of the

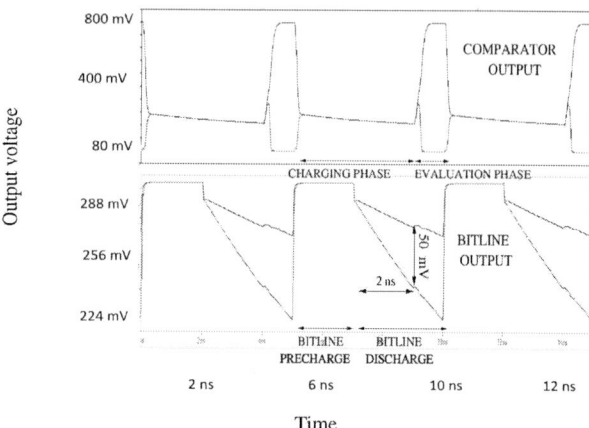

Figure 8: Bit line and sense amplifier output for 1000×1000 STT-MRAM

next layer via weights. To implement this architecture, we assume a multi-core architecture. Each core consists of an 8×1000 byte memory unit, 8 MAC units (to calculate the weighted sum corresponding to the node of the next layer) and a 64 bit data bus to carry the weights from the memory unit to the MAC unit [19].

Thus, to map 1000 input neurons to 1000 neurons in the next stage, 125 cores are needed.

For simulating 8×1000 byte memory units, we modelled the bit-line capacitance as 0.08pF and 0.16pF for STT-MRAM and SRAM respectively as the effective bit-line length hence the capacitance will be reduced to 1/125th of the previous length (1000×1000 memory units) of the bitline.

On reducing the size of the memory units, it was observed that power consumption of the bit cells reduce further, but the leakage of the SRAM remains somewhat same and also the advantage of STT-MRAM over SRAM in terms of read operation decreases.

The 1000×8 STT-MRAM has a read latency of 1.1ns and consumes 4.36fJ of read energy, whereas the 1000×8 SRAM has a read latency of 0.83ns and consumes 10.3fJ of read energy.

It is observed that the order of advantage of STT-MRAM over SRAM was 5 times (80% low power) in case of 1000×1000, but is 2 times in case of 1000×8. As mentioned earlier, partitioning the memory into smaller banks would lead to area overhead due to read-write interface circuits. SRAM, owing to larger area, would mandate partitioning, else, read-write power due to longer metal interconnects would add to power overhead. MRAM, being area efficient, achieves low power for read-access, even with large array size. Moreover, voltage-mode read access proposed in this work facilitates low voltage operation and hence lower read-access energy.

978-1-5386-1357-3/17 $31.00 © 2017 IEEE   64

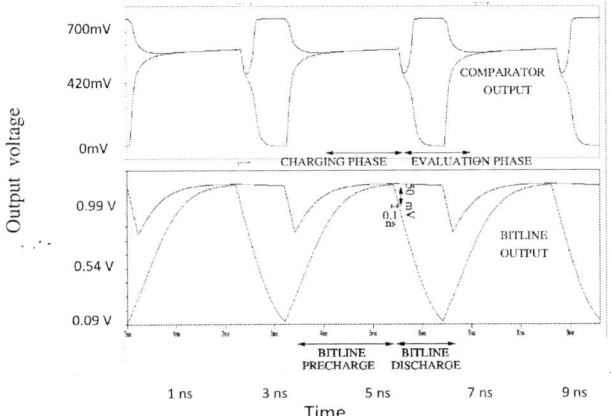

Figure 9: Bit line and sense amplifier output for 1000×1000 SRAM

VI. CONCLUSION

We explored energy benefits of STT-MRAM for read intensive deep-neural network architecture. It was shown that voltage mode read operation can facilitate low power read operation for MRAM. MRAM also facilitates larger memory array owing to higher area density, thereby reducing the area overhead due to read-write interfaces in highly partitioned memory. Owing to higher scalability, negligible leakage and lower read power, STT-MRAM can be suitable for parallel multi-core architecture with distributed memory banks.

(*123 authors contributed equally*)

REFERENCES

[1] Juergen Schmidhuber "Deep Learning in Neural Networks: An Overview"

[2] C.W. Smullen et al. Relaxing non-volatility for fast and energy-efficient stt-ram caches. In High Performance Computer Architecture (HPCA), 2011 IEEE 17th International Symposium on, pages 50 –61, feb. 2011.

[3] X. Guo et al. "Resistive computation: Avoiding the power wall with low-leakage STT-MRAM based computing" Proceedings of the 37th annual International Symposium on Computer Architecture pp. 371-382 2010.

[4] Q. Guo et al. "AC-DIMM: Associative Computing with STT-MRAM," in Proceedings of ISCA, 2013

[5] M.-T. Chang et al. Technology comparison for large last-level caches (l3cs): Low-leakage sram, low write-energy stt-ram, and refresh-optimized edram. In 2013 IEEE 19th International Symposium on High Performance Computer Architecture (HPCA2013), pages 143–154, Feb 2013.

[6] X. Dong et al. "Circuit and microarchitecture evaluation of 3-D stacking magnetic RAM (STT-MRAM) as a universal memory replacement," in Proc. 45th ACM/IEEE Design Autom. Conf., Jun. 2008, pp. 554–559

[7] H. Noguchi et al., "Highly reliable and low-power nonvolatile cache memory with advanced perpendicular STT-MRAM for

high-performance CPU," in Proc. 2014 Symp. IEEE VLSI Circuits Dig. Tech. Papers, 2014, pp. 1–2

[8] Sharad, Mrigank, et al. "Multi-level magnetic RAM using domain wall shift for energy-efficient, high-density caches." Low Power Electronics and Design (ISLPED), 2013 IEEE International Symposium on. IEEE, 2013.

[9] X. Fong et al. "Bit-cell level optimization for non-volatile memories using magnetic tunnel junctions and spin transfer torque switching," IEEE Trans. Nanotechnol., vol. 11, no. 1, pp. 172–181, Jan. 2012.

[10] Chenyun Pan and Azad Naeemi, "Nonvolatile Spintronic Memory Array Performance Benchmarking Based on Three-Terminal Memory Cell", IEEE Journal on Exploratory Solid-State Computational Devices and Circuits, 2017

[11] Tanyawat Sanguanchue, Kietikul Jearanaitanakij, "Hybrid algorithm for training feed-forward neural networks using PSO-information gain with back propagation algorithm", 2012 9th International Conference on Electrical Engineering/Electronics, Computer, Telecommunications and Information Technology.

[12] Kris Tiril and Ingrid Verbauwhede, "Charge Recycling Sense Amplifier Based Logic: Securing Low Power Security IC's against DPA," Solid-State Circuits Conference, 2004. ESSCIRC 2004. Proceeding of the 30th European

[13] W. Zhang et al. "A write-back-free 2T1D embedded DRAM with local voltage sensing and a dual-row-access low power mode," in Proc. Custom Integr. Circuits Conf., 2012, pp. 1–4

[14] Gamiz et al., "A 20nm low-power triple-gate multibody 1T-DRAM cell." VLSI Technology, Systems and Applications (VLSI-TSA), 2012 International Symposium on. IEEE, 2012, pp.

[15] I. G. Baek et al. "Highly scalable nonvolatile resistive memory using simple binary oxide driven by asymmetric unipolar voltage pulses," IEEE Elec. Dev. Meet., 2004, 587-590

[16] H. Y. Lee et al. "Low power and high speed bipolar switching with a thin reactive Ti buffer layer in robust HfO2 based RRAM," in Proc. IEEE Int. Electron Devices Meeting, Dec. 2008, pp. 297–300

[17] Pai-Yu Chen, Shimeng Yu, "Partition SRAM and RRAM based Synaptic Arrays for Neuro-inspired Computing", 2016 IEEE International Symposium on Circuits and Systems (ISCAS)

[18] Augustine, Charles, et al. "Spin-transfer torque MRAMs for low power memories: Perspective and prospective." IEEE Sensors Journal 12.4 (2012): 756-766.

[19] Ackland, Bryan, et al. "A single-chip, 1.6-billion, 16-b MAC/s multiprocessor DSP." *IEEE Journal of Solid-state circuits* 35.3 (2000): 412-424.

2017 IEEE International Symposium on Nanoelectronic and Information Systems

Comprehensive Operation Chaining Based Schedule Delay Estimation during High Level Synthesis

Vipul Kumar Mishra[1], Anirban Sengupta [2]

[1]Computer Science and Engineering
[1]Bennett University, India
[2]Computer Science and Engineering
[2]Indian Institute of Technology Indore, India
[2]asengupt@iiti.ac.in

Abstract- Design space exploration (DSE) during high level synthesis (HLS) involves a major step called scheduling which is responsible for estimating the delay of a control data flow graph (CDFG). However, a DSE process which concurrently estimates schedule delay by considering functional unit (FU), switching devices (such as mux, demux) and storage elements (such as latches), much before creation of its controller timing sequence, is an unsolved problem in the literature. Current DSE approaches either consider only FU during scheduling, or generate the complete controller timing sequence for delay evaluation of a CDFG based on provided resource constraint. The prior case, though fast but is not realistic in delay estimation. The latter case, though very slow, but provides realistic delay estimation. This paper solves the aforesaid problem by proposing a balanced DSE methodology that includes comprehensive delay estimation by considering combined delay of FU, switching devices and storage elements directly from scheduling. Results indicate improvement in achieving more realistic delay estimation process than previous approaches.

I. INTRODUCTION

Realistic delay estimation in high level synthesis considering delay of switching devices and storage elements, besides FU is the need of the hour due to increasing complexity resulting from heavy sharing of resources involved during scheduling. Ignoring delay of switching devices (multiplexer and demultiplexer) and storage elements (latches) during schedule results in inaccurate delay evaluation of a design solution. This becomes particularly misleading in the context of design space exploration process where design solution must meet the user delay constraint [1,5].

None of the previous work so far aimed at incorporating switching device and storage element delay from scheduling during delay estimation. Some authors have considered interconnect delay after describing the timing sequence of the controller for its equivalent register transfer level (RTL) circuit however not from scheduling (and thereby not during DSE) [2,3,4]. Particle swarm optimization (PSO) and bacterial foraging optimization algorithm (BFOA) based DSE methodology were presented in [2, 4] which did not consider mux, demux and latch delay during delay evaluation from scheduling (only considered functional unit from schedule for delay estimation). Further, genetic algorithm (GA) based DSE in [1], also only considered functional unit during scheduling for delay estimation. This paper resolves the aforesaid problem by encompassing the delays of switching devices, storage elements and FU through an delay estimation module incorporated in proposed DSE framework.

II. PROBLEM FORMULATION

Given a CDFG, find a minimal cost solution which satisfies the user constraints. The formulation is as follows:
Find: Optimal $(X_i) = (R_x,)$

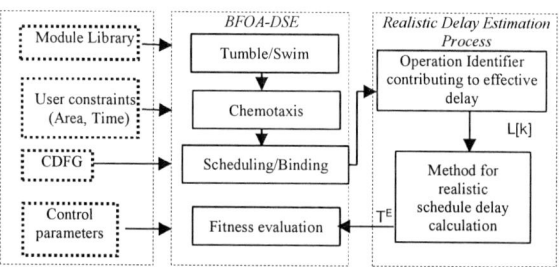

Fig 1. Block diagram of proposed methodology

Position of a Bacterium (X_i) → Resource combination
Tumble of a Bacterium in d^{th} dimension → Change in Direction
Dimension (D) → Number of resource types

Fig 2. Mapping Process

With minimum hybrid Cost (A_T, T_E)
Subjected to: $A_T <= A_{cons}, T_E <= T_{cons}$
where, 'X_i' is a set comprising of datapath resources combination as:
$X_i = (N(R_1), (N(R_2),..(N(R_d)..(N(R_D))$

where, '$N(R_d)$' is the number of instances of resource type 'R_d'; 'D' is the total number of resource types; 'A^R' and 'T_E' are the area used by a
and execution delay a solution respectively; 'A_{cons}' and 'T_{cons}' is area and execution delay constraint specified by the user.

Area Model: The fitness function is defined as [4]:

$$A_T = \left[\sum_{i=1}^{v} (N_{R_i} \cdot K_{R_i}) + N_{MUX/DMUX} \cdot K_{MUX/DMUX} \right] \quad (1)$$

where, 'N_{Ri}' represents the number of instance of resource Ri; 'K_{Ri}' represents the area occupied by resource R_i (# of transistors), 'v' is the number of resource types, '$N_{MUX/DMUX}$' is number of the multiplexer or demultiplexer, and '$K_{MUX/DMUX}$' is area occupied by the multiplexer or demultiplexer *(Note:- area is calculated in au; 1 au = 1 transistor).*

Cost Model: The fitness function is defined as [4]:

$$C_f(x_i) = \varphi_1 \frac{A_T - A_{cons}}{A_{max}} + \varphi_2 \frac{T_E - T_{cons}}{A_{max}} \quad (2)$$

$C_f(x_i)=$ Cost of particle denoted by X_i; T_{max} = Maximum execution time of a solution calculated using minimum resources; A_{max} = Maximum area of a solution calculated using maximum resources.

III. PROPOSED FRAMEWORK

The block diagram of the proposed methodology is given in Fig 1. The input module comprises of blocks such as module library consisting of various switching devices (mux &

978-1-5386-1357-3/17 $31.00 © 2017 IEEE

demux) information, FU information and storage element information; user constraints such as for area and delay (time), CDFG and control parameters of DSE engine such as stochastic variables (*Note: the system proposed in Fig.1 is adaptable to any DSE engine. However for the sake of demonstration, BFOA-DSE engine has been used*). The corresponding mapping of current DSE problem with BFOA is given in Fig 2 which includes the following major steps [4]:

Step 1. Bacterium initialization (X_i)
Step 2. Generate Tumble vector
Step 3. Perform Chemotaxis (responsible for bacterium movement)
Step 4. Fitness evaluation Process (based on eqn. 2)
Step 5. Accept the current solution if found fitter and then swim, Else Tumble and jump to step 3
Step 6. Update global best solution
Step 7. Repeat step 3-6 until stopping criteria does not meet

The control parameters used in BFOA-DSE are population size, terminating crieria and tumble vector (used for chemtactic movement) which are applied experimentally for performing sensitivity analysis. For the sake of brevity, details have been ommitted. The BFOA-DSE engine performs scheduling (using as soon as possible algorithm) and binding which is fed to the proposed '*realistic delay estimation*' module. The schedule delay estimation block operates in two dependent phases viz. (a) determination of operations that contribute to effective delay and (b) comprehensive dealay calculation.

IV. PROPOSED REALISTIC DELAY ESTIMATION MODULE

The proposed comprehensive delay estimation module is only valid for operation chaining/multi-cycling based scheduling technique. Schedule delay estimation through the proposed module is divided into two phases where the second phase (processing block) is dependent on the output of the first phase (or processing block) as shown in Fig 1. Identification of the operations which result in effective delay estimation is performed through the first block. The effective contribution to delay is only made by those operations that do not lie within the lifetime of the other larger delay operations i.e. operations that lie (or overlap) within the lifetime of the other larger delay operations are discarded (since they do not effectively contribute in the

comprehensive delay estimation). Fig 3 highlights the major steps of the algorithm [5]. As evident, the output of the first phase produces a list L[k] containing the operations that effectively contribute to the delay. The second phase calls the list L[k] operation wise to comprehensively estimate schedule delay with consideration of interconnect and storage elements delays of those corresponding operations as show in Fig 4.

V. EXPERIMENTAL RESULT

The proposed approach as well as [2], [4] all have been implemented in Java and run on Intel core i5-2450M processor, 2.5 GHz frequency with 3MB L3 cache memory and 4GB DDR3 RAM. The comparisons between proposed approach and [4], [2] have been shown in Fig 5 & 6 respectively. For efficiency (accuracy) evaluation of the proposed approach, the actual delay was found by generating the controller timing sequence corresponding to the final solution for each benchmark. This actual controller delay was compared with the delay estimated by proposed and previous approaches [2], [4]. The lesser the difference of estimated delay with actual controller delay (*indicated by: Δ*), more is the accuracy. As evident in Fig 5 & 6, the proposed approach produces lower Δ when compared to the Δ obtained by [2], [4]. This is because in [2] and [4], comprehensive schedule delay estimation was not made, by ignoring the delays of switching devices and storage elements from schedule during DSE. By only considering the delay of FU in scheduling, the estimation was not realistic enough. On other hand in proposed DSE methodology, more realistic estimation was made, which enabled to achieve closer to actual controller delay (i.e. higher accuracy indicated by lower Δ).

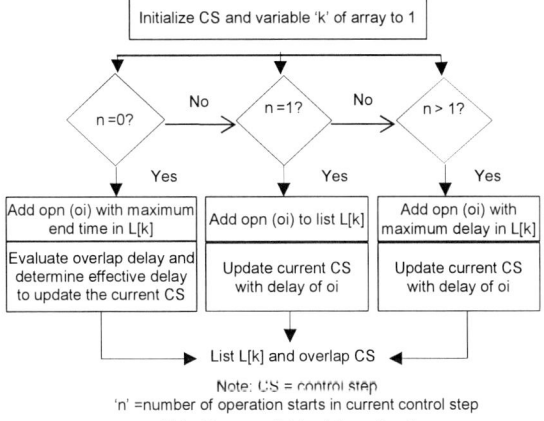

Note: CS = control step
'n' =number of operation starts in current control step
L[k] – Resource list for delay estimation
Fig 3 Operation Identifier algorithm

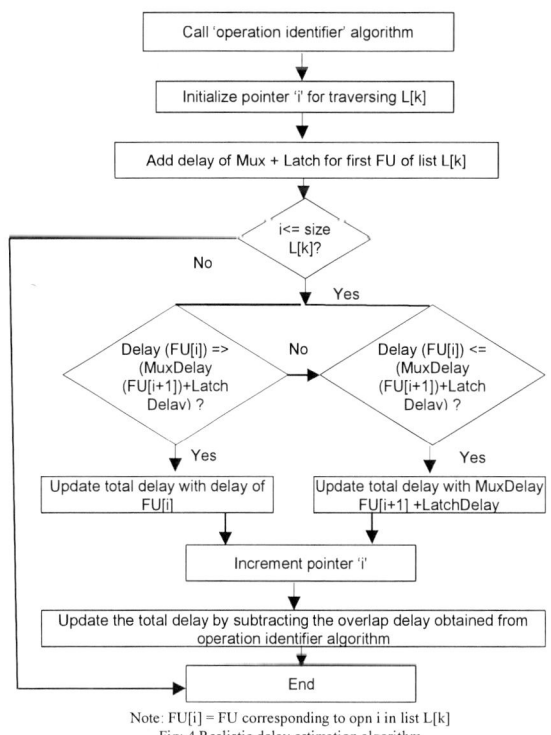

Note: FU[i] = FU corresponding to opn i in list L[k]
Fig: 4 Realistic delay estimation algorithm

978-1-5386-1357-3/17 $31.00 © 2017 IEEE   67

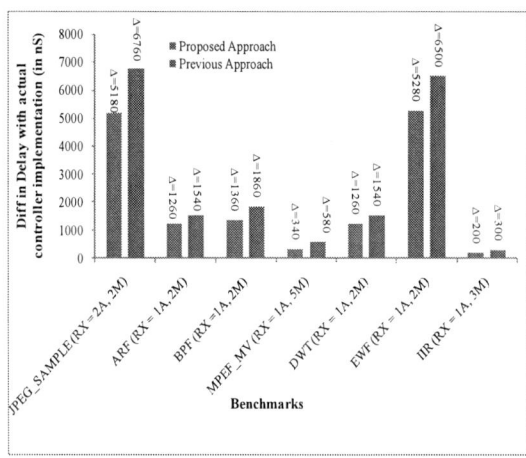

Fig: 5 Indicates improvements in obtaining realistic schedule delay through proposed approach compared to previous approach [4]

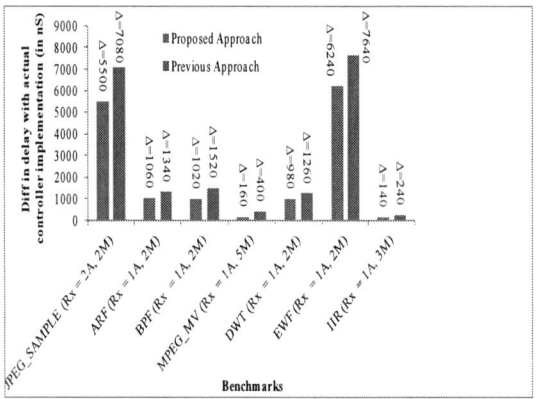

Fig: 6 Indicates improvements in obtaining realistic schedule delay through proposed approach compared to previous approach [2]

VI. CONCLUSION

This paper presented a novel design space exploration approach with realistic schedule delay estimation process that considers delays of switching devices and storage elements, besides FU delay, directly from a schedule for delay estimation. Results indicated that more realistic (comprehensive) delay estimation is possible compared to previous approaches [2], [4] that only provide a sketchy delay estimate from scheduling.

ACKNOWLEDGMENT

This work is financially supported by Council of Scientific and Industrial Research (CSIR) under sanctioned grant no. 22/730/17/EMR-II.

References

[1] V. Krishnan, and S. Katkoori., A Genetic Algorithm for the Design Space Exploration of Datapaths During High-Level Synthesis, *IEEE Transactions on Evolutionary Computation*, Vol.10 (3), 2006, pp. 213-229.

[2] Harish R. D. S., Bhuvaneswari M. C., Prabhu S. S "A Novel Framework for Applying Multiobjective GA and PSO Based Approaches for Simultaneous Area, Delay, and Power Optimization in High Level Synthesis of Datapaths", *VLSI Design Hindaw*i, Article ID 273276, 12 page, 2012.

[3] V. M.J.M.Heijlingers, L.J.M.Cluitmans, J.A.G.Jess, "High-level synthesis scheduling and allocation using genetic algorithms", *Proceedings of ACM Asia South Pacific Design Automation Conference*, 1995, pp.61–66

[4] A. Sengupta, S. Bhadauria, "'Automated Exploration of Datapath in High Level Synthesis using Temperature Dependent Bacterial Foraging Optimization Algorithm'", *Proceedings of 27th IEEE Canadian Conference on Electrical and Computer Engineering*, 2014, pp. 69 – 73.

[5] A. Sengupta, V.K Mishra, "A Methodology for Comprehensive Schedule Delay Estimation during Design space Exploration in Architectural Synthesis", *IEEE VLSI Circuits & Systems Paper*, Vol.1 (1), 2015, pp. – 2-8

Cost Aware Majority Logic Synthesis for Emerging Technologies

Vipul Kumar Mishra

School of Engineering

Bennett University, Greater Noida, India

Email: vipul.mishra@bennett.edu.in

Abstract—Continuous growth of semiconductor industry based n More's law is in danger due to the physical limitation of CMOS technology. In order to sustain more's law, researchers are aggressively discovering other alternatives such as nano magnetic logic (NML) and quantum dot cellular automata (QCA). Majority gate is the base element for these technologies for synthesis of logic circuit. This paper presents a novel majority logic synthesis (MLS) which optimize Area delay using a novel Cost of Circuit (CoC) parameter during the synthesis process which was faced by currently available synthesis algorithms which are mainly focused on area optimization. In addition, an updated library is presented for majority logic synthesis based on 3-input(M3) and 5-input(m5) majority gates for area-Delay optimization during MLS. Experiments on microelectronics center of north Carolina (MCNC) benchmarks indicate that, the proposed approach has achieved an average reduction of 39% in delay, an average reduction of 14% in cost of circuit with the 2% overhead in the circuit area.

Index Terms—Majority logic synthesis, Cost of Circuit, Heuristic, Emerging technology.

I. INTRODUCTION

The success of More's law was based on continues scaling down of transistors, they get less costly to produce, consume less energy and get faster. However, further scaling generate new challenges such as increase gate leakage current, electromigration failures, and become more difficult in lithography [2], due to the physical limits of CMOS technology. Therefore, Emerging technologies such as QCA [15], NML [4], and SET [5], are considered as possible alternatives of CMOS technology to sustaining Moore's law. Majority gate acts as foundation logic device in these technologies. Therefore, synthesis of the boolean network into majority network using majority gate and inverter is known as majority logic synthesis [?]. Moreover, a number of methodologies are available for logic synthesis of CMOS technology but, these methodologies are not suitable for majority logic synthesis. Therefore, fresh research is required for majority logic synthesis to produce optimal results.

This paper presents a novel majority logic synthesis approach which optimize the result by using a novel parameter named as *cost of circut(CoC)* during majority logic synthesis which will maintain tradeoff between delay and area of the circuit. Moreover, we include 5-input majority gates in the synthesis process as primitives which helps to reduce circuit delay. The contributions of the paper are as follows:

1) Proposed a novel methodology for majority logic synthesis which maintain the tradeoff between area and delay.

2) A novel parameter named as cost of circuit is introduced in the paper to maintain tradeoff between area and delay of the circuit.

3) Proposed an extended list of primitives using 5-input and 3-input majority gates.

4) A novel post synthesis optimization technique has been proposed in the paper.

5) A novel comparative result have been presented in the paper based on CoC.

II. RELATED WORK

Researchers proposed a majority synthesis technique in [8] was the first automated method for majority logic synthesis, it failed to produce optimal results in terms of majority gate count and number of majority levels. Further improvement on this approach was proposed in [11] and [9]. In [11], authors used disjoint concepts for further improvement, whereas in [9] the author proposed a cost metric to evaluate a majority network which helps on improving the majority circuit. Moreover, the authors in [10] presented a methodology which can handle 4 variable networks by using 3-input majority gates. The drawback of [9], [10] is that they are exhaustive in nature. In another work [12] the author presented a binary decision diagram (BDD) based majority logic synthesis. The methodologies presented in [8]–[11] used a SIS tool [13] for decomposition of the multi variable Boolean network into max 3-input Boolean network. Most of these approaches used K-maps for majority logic synthesis, but the K-map method become too complicated to handle 5-input and 7-input majority gates during majority logic synthesis, because the handling of the K-map for more than four variable is an intricate task [1]. Therefore, these approaches suffer from scalability problems.

Therefore, majority logic synthesis requires a method which is not exhaustive in nature and able to produce good quality results. Moreover, the inclusion of 5 input majority gates in majority logic synthesis can improve the circuit in terms of majority level and number of majority gates which is directly proportional to circuit delay and circuit area respectively. Furthermore, utilizing advanced tools for decomposition such as ABC [14] can further improve the majority logic synthesis.

III. BACKGROUND

A. QCA cell

A basic QCA cell is comprised of 4 quantum dots in a square shape coupled by tunnel fences. Electrons cant leave the QCA cell, but can move between these dots. when two

978-1-5386-1357-3/17 $31.00 © 2017 IEEE

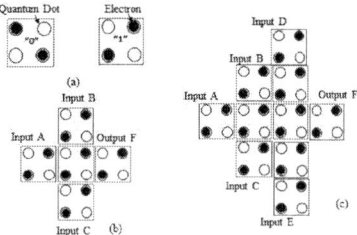

Fig. 1: Basis QCA primitives (a) Basic quantum cell, (b) 3-input majority gate, (c) 5-input majority gate

extra electrons are placed in the cell, Coulomb repulsion will force the electrons into dots on opposite corners. There are thus two actively equivalent ground state polarizations which can be labeled logic 0 and 1 as shown in Fig. 1(a) [3].

B. QCA Majority Gate

The fundamental QCA logic device is a 3-input majority gate [3] as shown in Figure 1(b). The functionality of the majority gate is presented in eqn. 1 [3].

A three input majority gate can be programmed to act as a 2 input AND or a 2 input OR gate by fixing one of the majority gate inputs to LOW (0) or HIGH (1) respectively. The AND and OR implementation of a majority gate is given in eqn. 2 and 3, where input "C" is fixed with value "0" and "1" to implement AND and OR respectively [3].

$$M(X,Y,Z) = XY + XZ + YZ \qquad (1)$$

$$M(X,Y,0) = XY \qquad (2)$$

$$M(X,Y,1) = X + Y \qquad (3)$$

Moreover, a 5-input majority [7] gate is also presented in the literature as shown in the Figure 1(c). The functionality of a 5-input majority gate can be expressed using eqn. 4.

$$M(A,B,C,D,E) = ABC + ABD + ABE + ACD + \\ ACE + ADE + BCD + BCE + BDE + CDE \qquad (4)$$

IV. PROPOSED METHODOLOGY

The proposed methodology begins with the input of a minimized multi output combinational network N. After taking input of network N, the second phase is pre-processing of the given Boolean network. During pre-processing the network is decomposed in such a way so that, every function in the network have maximum fan-in equal to three. (Because one can transform any boolean function with 3 fanin to majority function with maximum 4 majority gate and 2 levels). After decomposition, the next task is to transform boolean network N into majority network N_M using proposed methodology (explain in subsequent sections). Once we have synthesized majority network, then for further improvement in majority circuit has been done using post synthesis optimization.

A. Preprocessing

Simplification and factorization of the Boolean network is the first step of proposed methodology. This provides an algebraic factored multiple output form of the input Boolean network. After factorization, decomposition of Boolean network into the algebraic factored in such a manner so that every function have maximum three fan-in. Because, any three variable Boolean function can be realized by four or less three input majority gates with maximum two level. Therefore if we have such a Boolean network with K Boolean function then this network can be synthesized maximum 4*K and minimum K majority gates. Therefore, pre-processing try to minimize K. SIS [13] and ABC [14] tools are used to perform pre-processing of the circuit. The script for pre-processing using SIS is adapted from [9]. On The other hand, to the best of authors' knowledge, ABC tool for the pre-processing and decomposition during majority logic synthesis is used first time in proposed methodology. The proposed script for pre-processing are shown in Figure 3.

B. Proposed Primitives

Primitives act as a library (basic units) for the majority logic synthesis process. These Boolean functions required only one majority gate to transform into majority function. In this paper we proposed an extendedlibrary which includes 5-input majority gate along with the 3-input majority gate as primitives presented in the Table I and II respectivelly. Using these premitives one can convert any boolean function into majority function.

TABLE I: 3-input majority gates primitives

S. no.	Primitive function	majority expression	S. no.	Primitive function	majority expression
1	0	0	21	B'+C'	M(B,0,C)'
2	1	1	22	A'+B	M(A',1,B)
3	A	A	23	A'+C	M(A',1,C)
4	B	B	24	B'+C	M(B',1,C)
5	C	C	25	A'B	M(A,1,B)'
6	A'	A'	26	A'C	M(A,1,C)'
7	B'	B'	27	B'C'	M(B,1,C)'
8	C'	C'	28	A'B	M(A',0,B)
9	AB	M(A,0,B)	29	A+B'	M(A,1,B')
10	AC	M(A,0,C)	30	A'C	M(A',0,C)
11	BC	M(B,0,C)	31	A+C'	M(A,1,C')
12	A+B	M(A,1,B)	32	B'C	M(B',0,C)
13	A+C	M(A,1,C)	33	B+C'	M(B,1,C')
14	B+C	M(B,1,C)	34	A'B+A'C+B'C'	M(A,B,C')
15	AB+AC+BC	M(A,B,C)	35	A'B+A'C+BC	M(A',B,C)
16	AB'	M(A,0,B')	36	AB'+AC+B'C'	M(A,B',C)
17	AC'	M(A,0,C')	37	AB'+AC'+B'C'	M(A,B',C')
18	BC'	M(B,0,C')	38	A'B+A'C'+BC'	M(A',B,C')
19	A'+B'	M(A,0,B)'	39	AB+AC'+BC'	M(A,B,C')
20	A'+C'	M(A,0,C)'	40	A'B+A'C+B'C	M(A',B',C)

TABLE II: 5-Input majority gates primitives

S. no.	Primitive function	majority expression	S. no.	Primitive function	majority expression
1	A+B+C	M(A,B,C,1,1)	17	A+(B*C)	M(A,A,1,B,C)
2	A+B+C'	M(A,B,C',1,1)	18	A+(B*C')	M(A,A,1,B,C')
3	A+B'+C	M(A,B',C,1,1)	19	A+(B'*C)	M(A,A,1,B',C)
4	A+B'+C'	M(A,B',C',1,1)	20	A+(B'*C')	M(A,A,1,B',C')
5	A'+B+C	M(A',B,C,1,1)	21	A'+(B*C)	M(A',A',1,B,C)
6	A'+B+C'	M(A',B,C',1,1)	22	A'+(B*C')	M(A,A,0,B,C')'
7	A'+B'+C	M(A',B',C,1,1)	23	A'+(B'*C)	M(A,A,0,B,C)'
8	A+B+C	M(A,B,C,0,0)'	24	A'+(B'*C')	M(A,A,0,B,C)
9	A*B*C	M(A,B,C,0,0)	25	A*(B+C)	M(A,A,0,B,C)
10	A*B*C'	M(A,B,C',0,0)	26	A*(B+C')	M(A,A,0,B,C')
11	A*B'*C	M(A,B',C,0,0)	27	A*(B'+C)	M(A,A,0,B',C)
12	A*B'*C'	M(A,B',C',0,0)	28	A*(B'+C')	M(A,A,0,B',C')
13	A'*B*C	M(A',B,C,0,0)	29	A'*(B+C)	M(A',A',0,B,C)
14	A'*B*C'	M(A',B,C',0,0)	20	A'*(B+C')	M(A,A,1,B',C')'
15	A'*B'*C	M(A',B',C,0,0)	21	A'*(B'+C)	M(A,A,1,B,C')'
16	A'*B'*C'	M(A,B,C,1,1)'	22	A'*(B'+C')	M(A,A,1,B,C')'

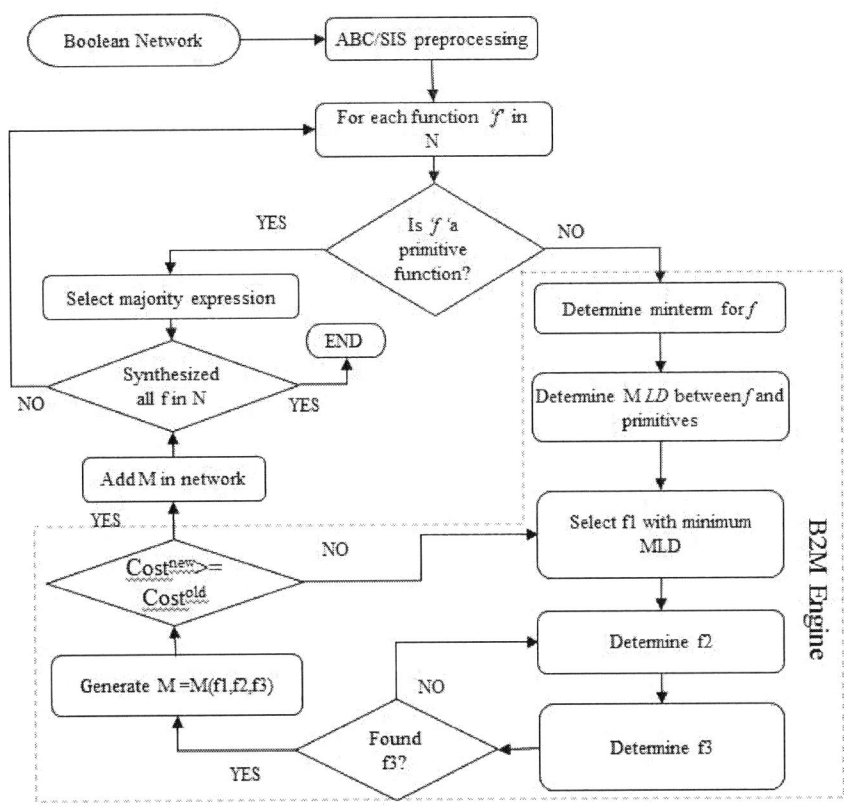

Fig. 2: Proposed Majority Logic Synthesis

C. Proposed Majority logic Synthesis

The flow diagram of proposed Majority logic synthesis is shown in Figure 2. The proposed procedure starts with taking input of a Boolean Network. Then preprocessing is performed using the SIS or ABC tool in such a way so that preprocessed network should not have more than three fan-in. After preprocessing process boolean network will have 'N' boolean functions. The next step is to convert all 'N' boolean function into majority function one by one using B2M engin shown in Figure 2. For each function f present in preprocessed network, first selected function f is checked that f is a primitive function or non primitive function. If f is a primitive function then add it to the majority network from Table I or Table II. If f is not a primitive function then we determine an equivalent majority expression M using proposed *B2M engine* (shown in Figure ??) and add majority function M into the majority network. This process repeats until all of the functions in the N' will not transform into equivalent majority functions.

The detail description of proposed B2M engin to transform a Boolean function into a majority function is as follows:

The first step of proposed B2M engine is determine the minterm of the boolean function f. Then determine Modified Levenshtein distance (MLD) between f and the primitives p_i(given in Table I) using eqn. 5

(*Note- In case of composit majority function 5-input*

majority premitives results in area overhead without any improvement in delay. Therefore primitives present in Table II have not taken into account for B2M engin.).

$$mld\,(I,J) = max \begin{cases} |I| - |I \cap J| \\ |J| - |I \cap J| \end{cases} \quad (5)$$

Where I and J are the minterm set, mld(I,J) is the MLD between I and J, $|I|$ length of J.

1) Determine f_1: In order to determine f1, First, we select a function f_x which has minimum MLD from the primitives. If there is tie among functions then select the function f_x with minimum cost. (The procedure for cost calculation for a function is given in Agorithm 1.)

2) Determine f_2: After selecting f_2, the next task is to determine f_2. select a function f_i from Table I such that f_i should satisfy rule describe in eqn6.

$$f_2 = \begin{cases} (f_i \neq f_1) \\ ((f_1^{mt} - f^{mt}) \cap f_i^{mt} = \phi) \\ ((f^{mt} - f_1^{mt}) - f_i^{mt} = \phi) \end{cases} \quad (6)$$

3) Determine f_3: After selecting f_3, the next task is to determine f_3. select a function f_i from Table I such that f_i should satisfy rule describe in eqn7. Once we determine

```
Method 1:                    Preprocessing
xl_split -n 3
                             collapse
Method 2:                    sweep
xl_imp -n 3                  eliminate 5
                             simplify -m nocomp -d
Method 3:                    resub -a -d
xl_part_coll -n 3 -m -g 2    gkx -a -b -t 30
xl_coll_ck -n 3              resub -a -d
xl_partition -n 3 -m         sweep
full_simplify                gcx -b -t 30
xl_imp -n 3                  resub -a -d
xl_partition -n 3 -t         sweep
xl_cover -n 3 -e 30 -u 200   gkx -a -b -t 10
xl_coll_ck -n 3 -k           resub -a -d
                             sweep
Method 4:                    gcx -b -t 10
xl_part_coll -n 3 -m -g 2    resub -a -d
xl_coll_ck -n 3              sweep
xl_partition -n 3 -m         gkx -a -b
sweep                        resub -a -d
eliminate -1                 sweep
simplify -m nocomp           gcx -b
eliminate -1                 resub -a -d
sweep                        sweep
eliminate 5                  eliminate 0
simplify -m nocomp
resub -a
fx                       (b) Simplification using SIS
resub -a
sweep                    read <input file>
eliminate -1             if -K 3
sweep                    write <output file>
full_simplify -m nocomp
xl_imp -n 3
xl_partition -n 3 -t      (c) Preprocessing using ABC
xl_cover -n 3 -e 30 -u 200
xl_coll_ck -n 3 -k
```

(a) Decomposition using SIS

Fig. 3: Preprocessing

f_1, f_2, f_3 then generate majority function as shown in step 20. and add $M(f_1, f_2, f_3)$ in the majority network.

$$f_3 = \begin{cases} (f_i \neq f_2 \neq f_1) \\ ((f_1^{mt} - f^{mt}) \cup (f_1^{mt} - f^{mt}) \cap f_i^{mt} = \phi) \\ ((f^{mt} - f_1^{mt}) \cup (f^{mt} - f_1^{mt}) - f_i^{mt} = \phi) \end{cases} \quad (7)$$

This whole process repeats untill all the functions in the boolean network are not converted into the majority function. Finally we have an equivalent majority network for a given boolean network.

Algorithm 1 Majority function Cost

1: **procedure** COST($Mexp$)
2: $\quad M \leftarrow$ NUMBER OF MAJORITY GATES IN ($Mexp$)
3: $\quad I \leftarrow$ NUMBER OF INVERTERS IN ($Mexp$)
4: $\quad Cost \leftarrow M * 10 + I$
5: $\quad$ **return** $Cost$
6: **end procedure**

D. Post Synthesis Optimization

After synthesis it was observed that the transformed network is not optimize and can further optimize in terms of majority

gate count, inverter count and majority level using majority Algebra presented in [16] as shown in eqn 8.

$$\Omega = \begin{cases} Commutativity: \\ M_3(A, B, C) = M_3(B, A, C) = M_3(C, B, A) \\ Majority: \\ \begin{cases} if (A = B): M_3(A, B, C) = A = B \\ if (A = B'): M_3(A, B, C) = C \end{cases} \\ Assocativity: \\ M_3(A, D, M_3(B, D, C)) = M_3(C, D, M_3(B, D, A)) \\ Distributivity: \\ M_3(A, B, M_3(D, E, C)) = \\ M_3(M_3(A, B, D), M_3(A, B, E), C) \\ Inverter\,Propogation: \\ M_3(A, B, C)' = M_3(A', B', C') \end{cases}$$
$$(8)$$

It is clearly seen from example shown in Figure 4. After post synthesis optimization we are able to reduce from three majority gate, three inverter and two levels to only one majority gate, one inverter and one level.

Example of Post synthesis
 Consider the majority function
$M\left(M(a, b, c)', M(a, b, 1)', c'\right)$

$M\left(M(a, b, c)', M(a, b, 1)', c'\right)$ applying inverter prorogation
$\Rightarrow M\left(M(a, b, c), M(a, b, 1), c\right)'$ applying Distributivity
$\Rightarrow M\left(a, b, M(c, 1, c)\right)'$ applying Commutativity and Majority
$\Rightarrow M(a, b, c)'$ Final optimized majority function

Fig. 4: Example of Post synthesis

V. RESULT AND ANALYSIS

A comparison between the proposed approach and previous approaches [8]–[10] has been presented in this section. The proposed approach has been implemented in Java and run on an Intel core i5-6200U processor with 2.3 GHz 8GB DDR3 RAM. To compare proposed approach with previous approches by experiment with microelectronics center of north Carolina (MCNC) benchmarks.

TABLE III: Area and Delay [17], [18]

	Area (μm^2)	Delay (ns)
Majority3	$4.0 * 10^{-3}$	$1.4 * 10^{-2}$
Majority5	$7.64 * 10^{-3}$	$1.4 * 10^{-2}$
Inverter	$1.2 * 10^{-3}$	$4 * 10^{-3}$

A. Comparison in terms of area, delay and CoC

This section present a comparative study of proposed approach and previous works [8]–[10] in terms of area, delay and CoC of the majority network. Area and delay value of majority gate is given in Table III. The calculation of area, delay and CoC is performed using eqn 9, eqn 10 and 11 respectively. The comparitive results are shown in Figure 5. It is clearly

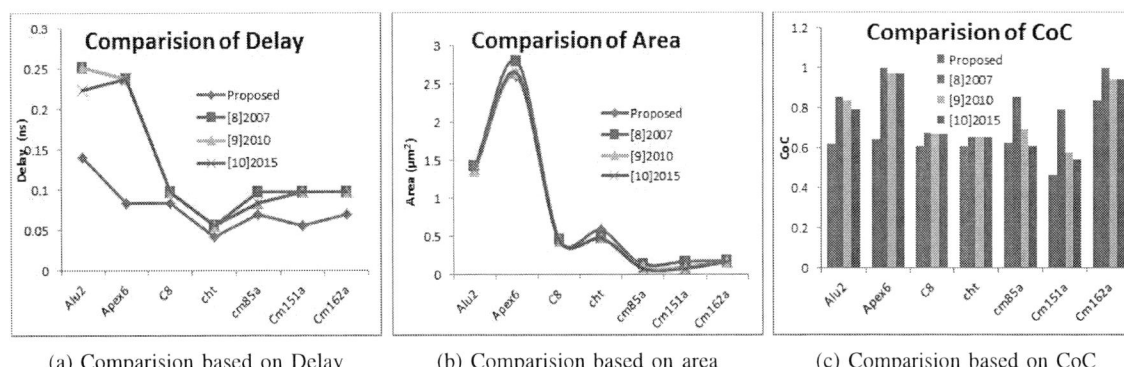

(a) Comparision based on Delay   (b) Comparision based on area   (c) Comparision based on CoC

Fig. 5: Comparision of proposed approach and [8], [9], [10] in terms of Area, Delay and CoC

shown in the Figure 5a that proposed approach able to achive a significant improvement (39 % in delay) and also shown in Figure 5c that proposed approach able to achive 14% in Cost of cicuit with the panality of 2% in area as shown in Figure 5b. During calculation of CoC we give equal priority to delay and area of the circuit. Thus, w1 = 0.5 and w2 = 0.5.

$$Area = (\sharp M3 * A(M3)) + (\sharp M5 * A(M5)) + (\sharp inv * A(inv)) \quad (9)$$

$$Delay = Level * D(M) \quad (10)$$

$$CoC = w_1 \left(\frac{Delay}{maxDelay} \right) + w_1 \left(\frac{Area}{maxArea} \right) \quad (11)$$

where $\sharp M3$ denotes the number of 3-input majority gates, A(M3) denotes the area of a 3-input majority gate, $\sharp M5$ denote number of 5-input majority gates, and A(M5) denotes the area of a 5-input majority gate, D(M) denotes delay of majority gate. *For the sake of simplicity we did not consider interconnection area into circuit area.*

VI. CONCLUSION

This paper presented a novel cost aware majority logic synthesis methodology for emerging technologies. We introduced a novel parameter CoC to determine cost of cicuit which helps to maintain tradeoff between area and delay. Furthermore, we presented an extended library of majority networks which includes 5-input majority gates along with 3-input majority gates. We proposed a post synthesis optimization process for further reduction in area and delay if the majority circuit. We performed experiments on MCNC benchmarks that show the proposed methodology achieved an average reduction indelay by 39% and overall cost of the circuit is reduced by 14% with and penality of 2 % in area of the circuit when compared with the most recent works presented in the literature.

REFERENCES

[1] Brown and Stephen Fundamentals of digital logic design with VHDL, Tata McGraw Hill, 2010.

[2] Semiconductor industries association roadmap. Available: http://public.itrs.net.

[3] P Douglas Tougaw and Craig S Lent. Logical devices implemented using quantum cellular automata. Journal of Applied physics, 75(3):1818–1825, 1994.

[4] Stuart A Wolf, Jiwei Lu, Mircea R Stan, Eugene Chen, and Daryl M Treger. The promise of nanomagnetics and spintronics for future logic and universal memory. Proceedings of the IEEE, 98(12):2155–2168, 2010.

[5] Takahide Oya, Tetsuya Asai, Takashi Fukui, and Yoshihito Amemiya. A majority-logic device using an irreversible single-electron box. Nanotechnology, IEEE Transactions on, 2(1):15–22, 2003.

[6] Stephan Breitkreutz, Josef Kiermaier, Irina Eichwald, Xueming Ju, Gyorgy Csaba, Doris Schmitt-Landsiedel, and Markus Becherer. Majority gate for nanomagnetic logic with perpendicular magnetic anisotropy. Magnetics, IEEE Transactions on, 48(11):4336–4339, 2012.

[7] Keivan Navi, Samira Sayedsalehi, Razieh Farazkish, and Mostafa Rahimi Azghadi. Five-input majority gate, a new device for quantum-dot cellular automata. Journal of Computational and Theoretical Nanoscience, 7(8):1546–1553, 2010.

[8] Rui Zhang, Pallav Gupta, and Niraj K Jha. Majority and minority network synthesis with application to qca-, set-, and tpl-based nanotechnologies. Computer-Aided Design of Integrated Circuits and Systems, IEEE Transactions on, 26(7):1233–1245, 2007.

[9] Kun Kong, Yun Shang, and Ruqian Lu. An optimized majority logic synthesis methodology for quantum-dot cellular automata. Nanotechnology, IEEE Transactions on, 9(2):170–183, 2010.

[10] Peng Wang, Mohammed Niamat, Srinivasa Vemuru, Mansoor Alam, and Taylor Killian. Synthesis of majority/minority logic networks. 2015.

[11] Suresh Rai. Majority gate based design for combinational quantum cellular automata (qca) circuits. In System Theory, 2008. SSST 2008. 40th Southeastern Symposium on, pages 222–224. IEEE, 2008.

[12] Luca Amarú, Pierre-Emmanuel Gaillardon, and Giovanni De Micheli. Bds-maj: A bdd-based logic synthesis tool exploiting majority logic decomposition. In Proceedings of the 50th Annual Design Automation Conference, page 47. ACM, 2013.

[13] Ellen M Sentovich, Kanwar Jit Singh, Luciano Lavagno, Cho Moon, Rajeev Murgai, Alexander Saldanha, Hamid Savoj, Paul R Stephan, Robert K Brayton, and Alberto Sangiovanni-Vincentelli. Sis: A system for sequential circuit synthesis. 1992.

[14] Alan Mishchenko et al. Abc: A system for sequential synthesis and verification (2007). URL http://www. eecs. berkeley. edu/alanmi/abc, 2010.

[15] Jaberipur, Ghassem and Parhami, Behrooz and Abedi, Dariush. A Formulation of Fast Carry Chains Suitable for Efficient Implementation with Majority Elements. IEEE 23nd Symposium on Computer Arithmetic (ARITH), 8–15, 2016.

[16] Amaru L, Gaillardon P. E., Chattopadhyay A., De Micheli G. A Sound and Complete Axiomatization of Majority-n Logic. IEEE Transactions on Computers, 65(9), 2889–2895.

[17] Lent, C. S., Tougaw, P. D. (1997). A device architecture for computing with quantum dots. Proceedings of the IEEE, 85(4), 541-557.

[18] Navi, K., Farazkish, R., Sayedsalehi, S., Azghadi, M. R. (2010). A new quantum-dot cellular automata full-adder. Microelectronics Journal, 41(12), 820-826.

2017 IEEE International Symposium on Nanoelectronic and Information Systems

Implementation of a 6 GHz MEMS Switch

Saurabh Chaturvedi[1,2], *Senior Member, IEEE*, Mladen Božanić[1,2], *Senior Member, IEEE*,
and Saurabh Sinha[2], *Senior Member, IEEE*

[1]Department of Electrical and Electronic Engineering Science, [2]Faculty of Engineering and the Built Environment
University of Johannesburg, Auckland Park Kingsway Campus
Johannesburg, South Africa
E-mail: chaturvedi.s.in@ieee.org, mbozanic@ieee.org, ssinha@uj.ac.za

Abstract - The design and simulation of a microelectromechanical systems (MEMS) switch at 6 GHz are presented in this paper. A MEMS shunt capacitive switch is implemented and simulated using the Keysight Technologies Advanced Design System. With the Momentum electromagnetic simulator, the scattering parameter plots for the up and down states of the MEMS switch are obtained to compare the switch characteristics.

Keywords - Microelectromechanical systems (MEMS), MEMS switch, *S*-parameters, coplanar waveguide (CPW), membrane.

I. INTRODUCTION

Microelectromechanical systems (MEMS) switches use mechanical movement to result in an open circuit or a short circuit in the transmission line (TL). Radio frequency (RF) MEMS switches are operated at RF to millimeter-wave frequency range for high-performance applications. The primary advantages of RF MEMS switches over conventional solid-state switches, such as diodes and transistors, are low power dissipation, low insertion loss (IL), high isolation, low intermodulation products, and low manufacturing cost. The main shortcomings of MEMS switches are low speed, low power handling, low reliability, high-voltage drive, and packaging issues. RF MEMS switches are used mainly in radar systems, wireless and satellite communication systems, instrumentation systems, and reconfigurable integrated circuits [1]-[5].

RF switches can be classified into two categories: series switches and shunt switches. When a direct current (DC) bias voltage is not applied, a series switch achieves an open circuit in the TL. When a bias voltage is applied, it results in a short circuit in the TL. A shunt switch is positioned in shunt between the TL and the ground. Depending on the application of a bias

voltage, a shunt switch either leaves the TL unconnected or connects it to the ground [1], [6].

The rest of the paper is organized as follows: Section II explains a complete design procedure for an RF MEMS switch using the Keysight Technologies Advanced Design System (ADS) 2016.01. The physical parameters of the coplanar waveguide (CPW) TL are calculated, and its layout is shown. The layout of the MEMS switch and substrate diagram are depicted. The electromagnetic (EM) simulation results for the up and down positions of the MEMS switch are presented in Section III. Finally, the paper is concluded in Section IV.

II. DESIGN PROCEDURE FOR A MEMS SWITCH

An RF MEMS shunt capacitive switch comprises a thin metallic membrane bridge that is electrostatically actuated using a DC bias voltage. Depending on the applied voltage, two states of a MEMS switch are possible, namely up and down. The membrane bridge is suspended over the signal line of a CPW TL, and both ends of the bridge are riveted to the ground planes of the CPW, as shown in Fig. 1. This arrangement creates a parallel plate capacitor system, where the suspended membrane is a vertically movable plate and the signal line of the CPW is a fixed plate. On applying an electric field, the induced electrostatic force on the movable membrane drives the membrane down towards the fixed signal line.

After reaching the threshold bias voltage, the membrane collapses on the fixed plate. The voltage that is responsible for pulling the switch completely down to the down position is called pull-down voltage (V_p) [2], [6]. The pull-down voltage is defined by (1):

$$V_p \cong \sqrt{\frac{8kg_0^3}{27\varepsilon W w}},\qquad(1)$$

978-1-5386-1357-3/17 $31.00 © 2017 IEEE

where k, g_0, and w are the spring constant, initial height, and width of the MEMS membrane, respectively, and W is the width of the CPW signal line.

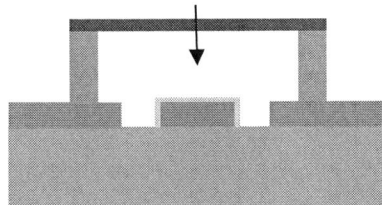

Fig. 1. Cross-section of a MEMS shunt capacitive switch.

A. Calculation of Physical Parameters of CPW TL

The physical parameters of the CPW TL used in the MEMS switch are synthesized with the ADS LineCalc [7] tool for the parameter specifications listed in Table I.

TABLE I. PARAMETER SPECIFICATIONS

Parameter	Value
Substrate thickness	600 μm
Substrate dielectric constant	11.9
Conductor thickness	1 μm
Metal conductivity	5.8×10^7 S/m
Electrical length	$\lambda/4$
Characteristic impedance	50 Ω
Frequency	6 GHz

The physical parameters of the CPW TL synthesized by LineCalc are displayed in Fig. 2 and given below:

Width of signal line = 2.24 mm
Gap between signal and ground lines = 0.61 mm
Length of signal and ground lines = 6.17 mm

B. Layout of MEMS Switch

The dimensions of the MEMS switch are given below:

Width = 0.2 mm
Length = 4.5 mm
Thickness = 2 μm

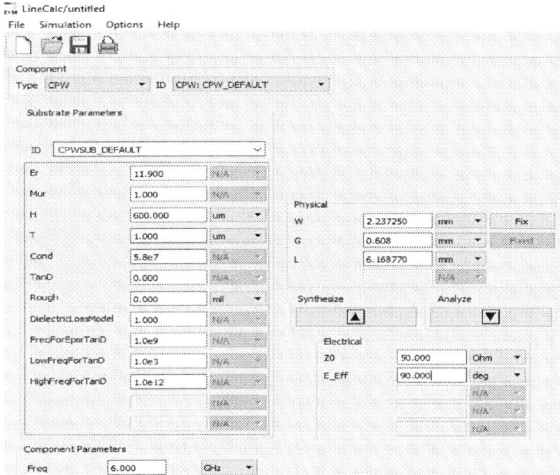

Fig. 2. Physical parameters of CPW line.

The layout of the CPW TL with the aforementioned specifications is drawn in ADS, as shown in Fig. 3. After assigning pins to the signal line and ground planes of the CPW, posts on the ground planes are drawn. These posts provide support to the membrane. Subsequently, the MEMS membrane is drawn over the posts, as illustrated in Fig. 4.

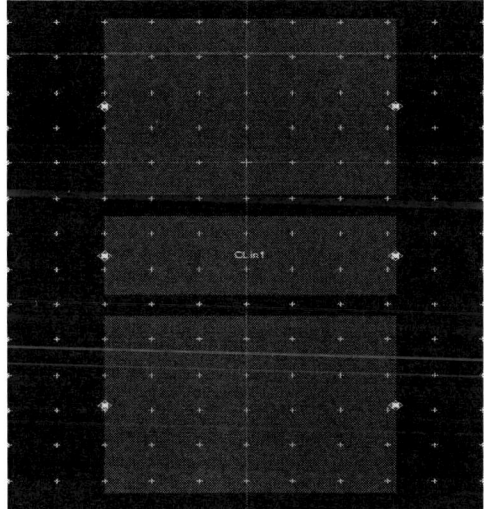

Fig. 3. Layout of CPW line.

The substrate structure can be defined in ADS from the Substrate Editor option. Fig. 5 shows the substrate structure for EM simulation of the MEMS switch in the up state. The layers `cond2` and `cond` represent the membrane and CPW signal line, respectively. The `hole` via layer is used for the post. The initial gap between the membrane and signal line is 20 μm.

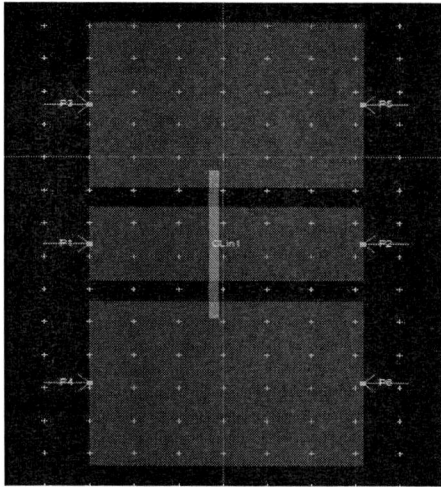

Fig. 4. Layout of MEMS switch.

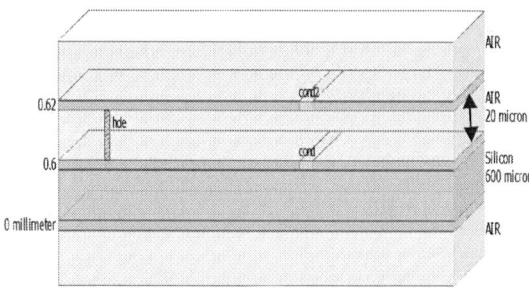

Fig. 5. Substrate structure for the up state.

The *S*-parameter ports can be observed from the Ports option of the ADS EM Setup window, as demonstrated in Fig. 6.

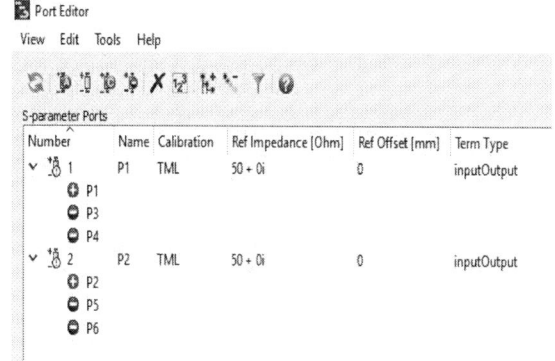

Fig. 6. Ports list.

In the ADS Momentum simulation, the layout pins can be combined to form the *S*-parameter ports. Fig. 7 demonstrates the rearrangement of ports in the Port Editor window, where one positive pin (P1) and two negative pins (P3 and P4) are combined into a single port (Port1) to create a ground-signal-ground

configuration for the CPW TL. Similarly, the pins P2, P5, and P6 are combined for Port2.

Fig. 7. Ports configuration.

On applying a DC bias voltage, the membrane starts moving down towards the CPW signal line. After reaching the threshold bias voltage, the membrane snaps down to the CPW line. Ideally, the gap between the membrane and the fixed plate becomes 0 μm for a capacitive switch in the down state.

III. SIMULATION RESULTS

The MEMS switch is simulated for the up and down states using the Momentum Microwave simulator in ADS. Fig. 8(a) presents the scattering parameter (*S*-parameter) plots for the up state of the switch. The simulated *S*-parameter plots for the down state are given in Fig. 8(b). The simulated S_{11} value is -10.01 dB and the S_{21} is -0.48 dB at 6 GHz in the up state of the switch. The S_{11} and S_{21} values for the switch in the down state are -0.36 dB and -12.10 dB, respectively. Therefore, this MEMS switch shows an IL of 0.48 dB and an isolation of 12.10 dB at 6 GHz.

The EM simulation results of an RF MEMS switch [8] demonstrate the values of S_{11} as -17.63 dB and S_{21} as -0.09 dB at 4 GHz in the up state. In the down state of the switch, the S_{11} is -0.04 dB and S_{21} is -20.62 dB at the same frequency. An RF MEMS switch presented in [9] has an IL of 0.18 dB and an isolation of 32 dB up to 3 GHz.

IV. CONCLUSION

This paper discusses the design and simulation of a shunt capacitive MEMS switch at 6 GHz using ADS. The *S*-parameter-based EM simulation results are obtained for the up and down states of the switch using the Momentum Microwave simulator. The simulated IL and isolation of the switch are observed as 0.48 dB and 12.10 dB, respectively, at 6 GHz.

978-1-5386-1357-3/17 $31.00 © 2017 IEEE

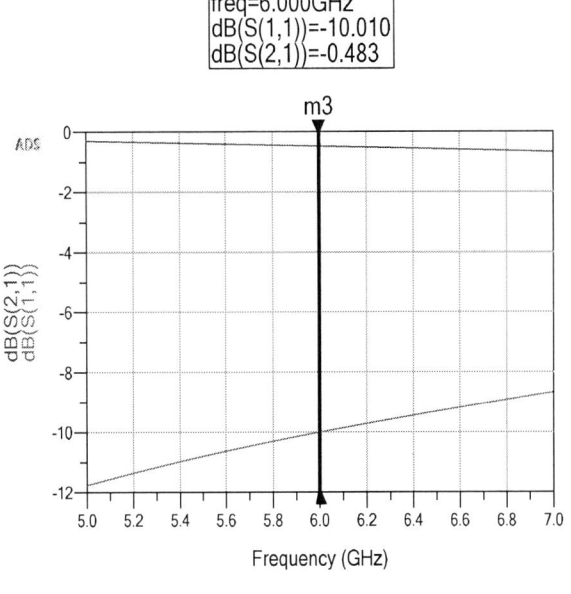

(a)

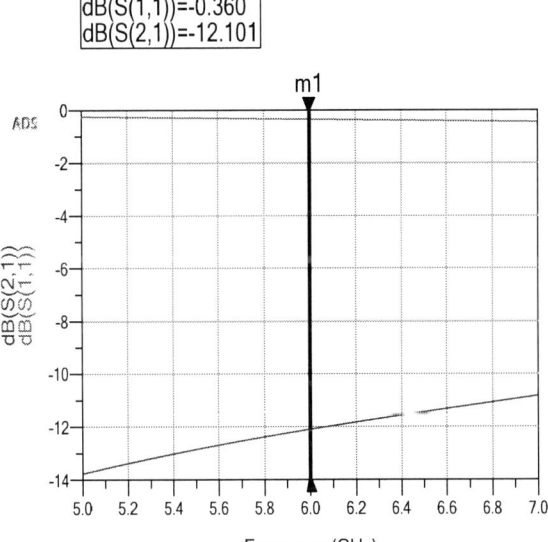

(b)

Fig. 8. Switch characteristics: (a) up state, (b) down state.

REFERENCES

[1] G. M. Rebeiz, *RF MEMS: Theory, Design, and Technology*, 2nd ed. New Jersey: John Wiley & Sons, 2003.

[2] G. M. Rebeiz and J. B. Muldavin, "RF MEMS switches and switch circuits," *IEEE Microw. Mag.*, vol. 2, no. 4, pp. 59-71, Dec. 2001.

[3] Z. Jiang, Z. Gong, and Z. Liu, "Copper-based multimetal-contact RF MEMS switch," in *Proc. IEEE 2016 17th Int. Conf. Electron. Packag. Technol. (ICEPT)*, 2016, pp. 546-550.

[4] M. Daneshmand and R. R. Mansour, "RF MEMS satellite switch matrices," *IEEE Microw. Mag.*, vol. 12, no. 5, pp. 92-109, Aug. 2011.

[5] E. R. Brown, "RF-MEMS switches for reconfigurable integrated circuits," *IEEE Trans. Microw. Theory Tech.*, vol. 46, no. 11, pp. 1868-1880, Nov. 1998.

[6] Keysight Technologies, *Keysight EEsof EDA Advanced Design System: Circuit Design Cookbook 2.0*, 2012.

[7] Keysight Technologies. [Online]. Available: http://literature.cdn.keysight.com/litweb/pdf/ads2002c/pdf/linecalc.pdf

[8] M. Spasos, N. Charalampidis, N. Mallios, D. Kampitaki, K. Tsiakmakis, P. T. Soel, and R. Nilavalan, "On the design of an ohmic RF MEMS switch for reconfigurable microstrip antenna applications," *WSEAS Trans. Commun.*, vol. 8, no. 1, pp. 153-161, Jan. 2009.

[9] T. Fujiwara, T. Seki, F. Sato, and M. Oba, "Development of RF-MEMS ohmic contact switch for mobile handsets applications," in *Proc. 42nd European Microw. Conf.*, 2012, pp. 180-183.

2017 IEEE International Symposium on Nanoelectronic and Information Systems

Towards the Approximation of Cell Wise Switching Time in Quantum-dot Cellular Automata

Soudip Sinha Roy
Department of Nanotechnology
Amity Institute of Nanotechnology
Noida, UP, India
soudipsinharoy@gmail.com

Abstract— **In purpose of preserving the congenial and emergent flow in nanoscience it is the requisite stage to accomplish an eminent aspect of nano-molecular devices to attain fastest operative speed with highest dynamic accuracy. Quantum dot cellular automata prescribes the high operative speed with lower fault occurrence possibility to the nanoelectronic circuits. In this contribution one new binary to gray code converter layout has proposed which presents 29.17% area optimization in comparison to the best reported designs in this paper. The proposed design involves the multilayer concept of circuit design whereas more than one layers have undertaken to pass the signals error-freely to avoid the crossovers complications. Moreover, an elegant method for cell to cell switching time approximation has proposed theoretically which puts a major aspect on signal propagation delay and the device operating delay computation. The switching time has expressed by a sophisticated mathematical equation which enables the theoretical computation of device latency with better accuracy. All the simulations in this paper have performed with the open source software QCADesigner 2.0.3.**

Keywords—gray code converter; switching time; latency; quantum computing; QCA.

I.   INTRODUCTION

In competition of creating the novel nanoscale devices the quantum-dot cellular automata have been a vastly accepted technology which is transistor-less, conduction channel less quantum scale computational approach to attain the lightning speed of the device. This quantum device is operated only by two electrons which are bounded into a square quantum cell through columbic repulsion. The electrostatic repulsive force between the electrons keeps them apart at the maximum distance according to the coulomb's law of electronic force. The sensitivity of electrons to the positive or negative charges is the main fundamental concept of quantum dot cellular automata. The quantum cells are filled with four quantum dots where only two dots can contain two electrons consistently under a specific polarization effect either it is P+ or P-, as fig. 1 (a) [1]. The long chain arrangement of a few quantum cells produce a binary QCA wire which is used to convey the quantum information from one cell to another cell simultaneously, fig. 1 (b) [2]. This is a 90^0 binary wire. Fig. 1 (c) illustrates the operating principle of a 3 input majority gate which can be configured either as OR gate or AND gate with the alternation of the control polarization/fixed polarization C [3]. In fig. 1(d) the binary 45^0 wire is shown. In case of the inverter the cells are placed at cornerwise to invert the electron orientation due to electrostatic repulsion force, fig. 1 (e) [4] – [7]. The orientation of the

electrons are the foremost factors for defining quantum binary logic states. The right handed orientation of electrons is defined as the binary logic high state '1' and for left handed orientation stands for the binary lower logic state '0'. In QCA each clock lags 90^0 in phase compared to the previous clock pulse. When the clock pulse seems low the cells get latched to its next stage and for higher clock pulse the cells get depolarized [6]. In this practice the information (signal) flows through the Q-cells [8]–[11].

This paper introduces the multilayer working principle of QCA devices and proposes optimized 2 bit, 4 bit and 16 bit binary to gray code converter layouts. The proposed 2 bit code converter circuit has achieved 29.17% optimization in area occupancy compared to the best QCA binary to gray code converters have proposed ever.

In this paper a new approach of switching time computation has proposed which mathematically illustrates the approximated value of cell to cell switching time during circuit operation. Moreover, the tunneling rate has also calculated in terms of cell count and switching time which is useful in many equations those are obviously found in QCA device physics articles [3], [4].

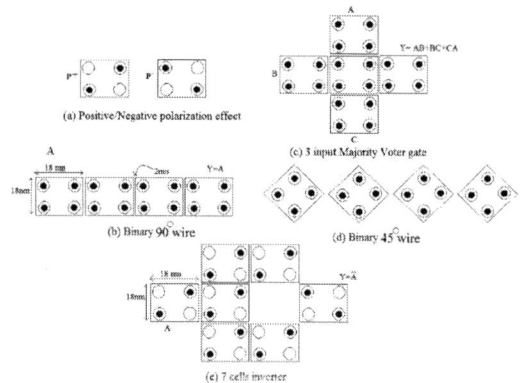

Fig. 1: (a) Positive/Negative polarization effect, (b) 90 degree binary wire, (c) 3 input Majority gate, (d) binary 45 degree wire, (e) 7 cells inverter

II.   TRY-LAYER CONCEPT OF QCA SIGNALING

In certain phenomena of QCA circuit planning due to heavy conjunction of wires in main cell layers (layer 1) it becomes obligatory to take a crossover between the binary wires, fig. 2, 3. The crossover between the wires increases signal propagation delay and therefore that reduces the circuit efficiency [5]. The switching time between two cells

978-1-5386-1357-3/17 $31.00 © 2017 IEEE

fundamentally depends upon the displacement between the cells in nanometer. Moreover, the semiconductor property, tunneling junction properties, number of used cells etc. are also some influencing factors of the switching time. When the several crossovers are taken in the circuit the switching time increases [6]−[9].

In this purpose multilayer configuration has undertaken to simplify the dynamic inaccuracy of signal propagation through the cells. Below figures explain the multilayer arrangement of QCA cells in three dimensionally, fig. 2, 3.

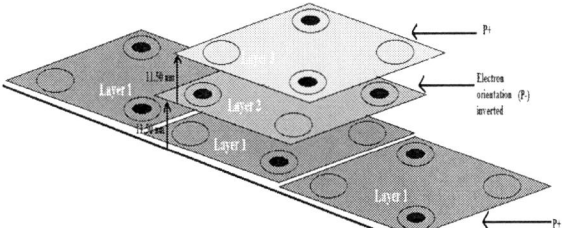

Fig. 2: Try layer positive polarization influence to the main cell layer

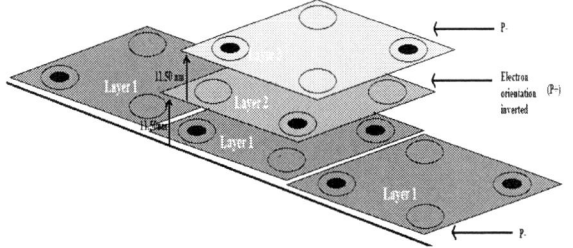

Fig. 3: Try layer negative polarization influence to the main cell layer

III. PROPOSED BNARY TO GRAY CODE CONVERETERS

Application of code converters in digital circuits have been an essential part of design. It is the emergent key sense of any digital circuits which can perceive the features for instance data encryption, data compression etc. [9]−[14], [16]. In the same purpose the optimization of these code converters becomes most exigent. Binary to gray code converters are those circuit which completes the task of doing Ex-OR operation between the input bit streams to provide the output bits [15], [17]. This XOR process continues up to the required number of bits whichever is needed [15]−[20].

In this paraphrase three layouts are proposed, those are 2 bit, 4 bit and 16 bit binary to gray code converters fig. 5, 7, and 8 respectively. All of these code converter circuits are proposed in multilayer designing methodology [8], [12].

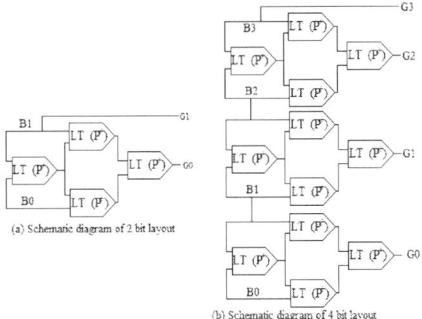

(a) Schematic diagram of 2 bit layout

(b) Schematic diagram of 4 bit layout

Fig. 4: (a) Schematic diagram of 2 bit layout, (b) Schematic diagram of 4 bit layout

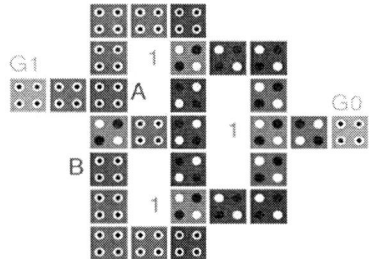

Fig. 5: Two bit binary to gray code converter

Fig. 4, explicates the schematic design of 2 bit and 4 bit code converters, fig. 4 (a,b). Above fig. 5 illustrates the 2 bit binary to gray code converter layout which has minimum area occupancy of 25,200 nm^2 with minimum total clock latency of 0.75. It is shaped only with 28 cells. The proposed layout is the best design which is highly optimized in area, cell count, clock latency and leads by more other factors as table III.

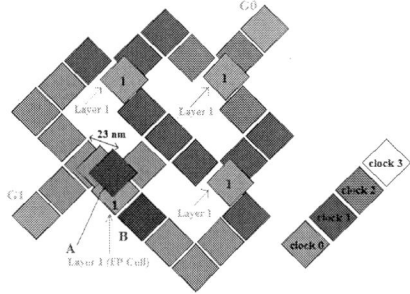

Fig. 6: Nanoscale level scaling of proposed 2 bit code converter

In above fig. 6, the cell and layer wise segmentations have performed for the utmost visualization of the proposed code converter circuit. This 2 bit binary to gray code converter module is operating through two different layer of cells separated by 11.50 nm each. In quantum dot technology if two layers are used just upon the main cell layer then the intermediate cell always inverts the signal and the whole layered wire of cells works as a buffer. In this practice the same working principle is appreciated to avoid the crossover for inclusion of the input signal. In this layout two different cell layers are involved, in the first layer (layer 1: 11.50 nm above from the main cell layer) the fixed polarization cells are implemented and in the second layer (layer 2: 11.50 nm above from the layer 1) the input itself is situated, fig. 6. The fixed polarizations cells are implemented upon the layer 1 and the input 'A' is provided from the layer 2. In this layout only the input 'A' is in layer 2. Contrariwise input 'A' can also be provided from the main cell layer then the output waveform also becomes so exact. But there will be a cell allocation problem which will cause the wire crossovers that results the raise of design complexity.

978-1-5386-1357-3/17 $31.00 © 2017 IEEE   79

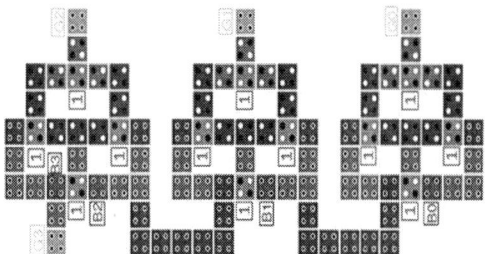

Fig. 7: 4 bit binary to gray code converter

The fig. 7, illustrates the layout of 4 bit binary to gray code converter circuit. In this paradigm the proposed two bit gray code converter is used as the fundamental primitive block of the 4 bit code converter circuit. The resultant expressions for the 4 bit code converter are,

$G_0 = B_0 \oplus B_1$, $G_1 = B_1 \oplus B_2$, $G_2 = B_2 \oplus B_3$, $G_3 = B_3$

This layout occupies only 82800 nm^2 area upon the cell bed. In terms of cell count this design is highly optimized. This layout involves only 95 cells with 0.75 clock latency and performs its task successfully.

Fig. 8: 16 bit binary to gray code converter

Fig. 8, gives the layout of 16 bit binary to gray code converter. This design involves 490 cells which occupies 508341.14 nm^2 area upon the cell bed. The circuit total clock latency is 0.75, which is significantly acceptable in terms of device performance and accuracy.

IV. OUTPUT POLARIZATION STRENGTH ANALYSIS AND SWITCHING TIME CALCULATION

TABLE I. TESTING OF OUTPUT POLARIZATIONS DEPENDING UPON SEVERAL SEMICONDUCTORS

Sr. No	Selected semiconductors for fabrication	Relative permittivity (F/m) or ($m^{-3}kg^{-1}s^4A^2$)	Output polarization strength analysis							
			G0		G1		G2		G3	
			Max.	Min.	Max.	Min.	Max.	Min.	Max.	Min.
1	Indium Antimonide (InSb)	16.8	9.25	-9.14	9.25	-9.14	9.18	-9.14	9.24	-9.26
2	Germanium (Ge)	16.2	9.30	-9.20	9.30	-9.20	9.24	-9.20	9.29	-9.31
3	Gallium Antimonide (GaSb)	15.7	9.34	-9.25	9.34	-9.25	9.28	-9.25	9.33	-9.34
4	Indium Arsenide (InAs)	15.15	9.38	-9.30	9.38	-9.30	9.33	-9.30	9.37	-9.39
5	Gallium Arcenide (GaAs)	12.9	9.51	-9.49	9.52	-9.49	9.51	-9.49	9.54	-9.55
6	Indium Phosphide (InP)	12.5	9.54	-9.53	9.54	-9.53	9.54	-9.53	9.57	-9.57
7	Silicon (Si)	11.7	9.60	-9.58	9.60	-9.58	9.60	-9.58	9.62	-9.63
8	Gallium Phosphide (GaP)	11.1	9.64	-9.63	9.64	-9.63	9.64	-9.63	9.66	-9.66
9	Aluminum Nitride(AlN)	9.14	9.76	-9.75	9.76	-9.75	9.76	-9.75	9.76	-9.77
10	Gallium Nitride (GaN)	8.9	9.77	-9.76	9.77	-9.76	9.77	-9.76	9.78	-9.78

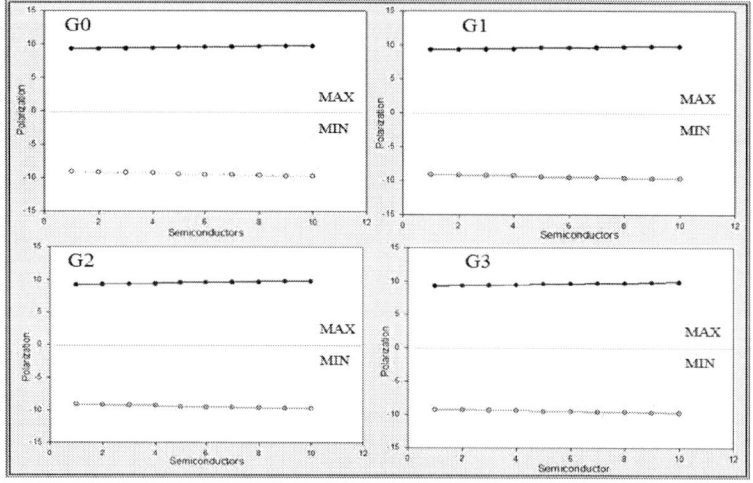

Fig. 9: Result of output polarizations with respect to different semiconductors from table I

TABLE II.   COHERENCE VECTOR SIMULATION OPTIONS IN QCADESIGNER 2.0.3

Temperature	1.000000 K
Relaxation Time	1.000000e-015 sec
Time step	1.000000e-016 sec
Total simulation time	7.000000e-011 sec
Clock high	9.800000e-022 J
Clock low	3.800000e-023 J
Clock shift	0.000000e+000
Clock amplitude factor	2.000000
Radius of effect	80.000000 nm
Relative permittivity	*12.900000 F/m*
Layer separation	11.500000 nm
Eular Method	Yes
Runge Kutta	No
Randomize simulation order	Yes
Animate	No

This output polarization strength variation analysis have performed for 4 bit gray code converter circuit. From above graphical analysis it is observed that the output polarizations vary with when the semiconductor for device fabrication is changed. Different semiconductors have different relative permittivity, which is a built in property of any semiconductor that influences the quantum (electron) tunneling. From this observation it is found that the tunneling rate is inversely proportional to the relative permittivity. In mathematics it is expressed as [4], [5],

$$\Gamma = \frac{1}{\varepsilon_r \varepsilon_0 e R_T (1 - e^{\frac{E_{ij}}{KT}})} \quad (1)$$

$$\Gamma = \frac{1}{\varepsilon_r \varepsilon_0 e R_T} \cdot \frac{1}{1 + \left(-e^{\frac{E_{ij}}{KT}}\right)}$$

$$\Gamma = \left| \frac{1}{\varepsilon_r \varepsilon_0 e R_T} \cdot f_{(E,T)} \right|$$

From this above calculation it is proposed that the tunneling rate of the electron is verily dependent upon the two major factors. Number one is the relative permittivity ε_r of the substrate semiconductor and the number two is energy probability distribution function $f_{(E,T)}$.

For ununiformed energy distribution in the tunneling barrier, the tunneling rate could have different values in different zone of energy. As the energy density changes, the tunneling rate through the zone also gets influenced. Therefore, the probability P_e of the occupying another quantum dot becomes a dependent function of the tunneling rate. $P_e = f(\Gamma)$

In this following calculation the rate of change of Γ is traced by the help of a simple derivation with respect to energy.

$$\frac{d\Gamma}{dE} = -\frac{1}{\varepsilon_r \varepsilon_0 e KT R_T} \cdot \frac{1}{e^{\frac{E_{ij}}{KT}} (1 - e^{\frac{-E_{ij}}{KT}})^2}$$

Considering only the MOD value, the eq. 2 is obtained,

$$\hbar v = \frac{d\Gamma}{dE_B} = \left| \frac{1}{\varepsilon_r \varepsilon_0 e KT R_T} \cdot \frac{1}{e^{\frac{E_{ij}}{KT}} (1 - e^{\frac{-E_{ij}}{KT}})^2} \right| \quad (2)$$

In this equation $\hbar v$ is named as "*S-instantaneous tunneling density*" (SITD) where the unite is given by $J^{-1} sec^{-1}$. E_{ij} is the energy difference between the initial and final state of electron. Therefore, the instantaneous energy of the electron during tunneling is defined as,

$$E_{ie} = \left(\frac{d\Gamma}{dE_B} \right)^{-1} v = \hbar v^{-1} v = m_e^* c^2 \quad (3)$$

Where, v is the frequency of electron, m_e^* indicates the effective mass of the electron and c is the speed of light.

CASE 1: If the energy difference between initial state and final state is $E_{ij} = 0$; then the maximum electron tunneling probability is observed with maximum tunneling rate.

$$\hbar v|_{max} = \frac{d\Gamma}{dE}\Big|_{max} = \left| \frac{1}{\varepsilon_r \varepsilon_0 e KT R_T} \right|$$

CASE 2: If the energy difference is equals to thermal energy KT,

$$\frac{d\Gamma}{dE}\Big|_{E_{ij}=KT} = \left| \frac{0.9206}{\varepsilon_r \varepsilon_0 e KT R_T} \right|$$

CASE 3: If the energy difference E_{ij}, is greater than KT, then there will be a gradual decay of tunneling probability. The electrons will get more repulsive forces from the tunneling energy barrier that causes a falling edge of tunneling energy. In the practical context this energy difference E_{ij} is denoting directly the barrier energy of the tunneling junction, which is indicated as E_B.

In comprehended it is mathematically realized as $E_{ij} = E_B$, which explicates that the tunneling energy barrier is nothing but the replica of the dissipated energy through the tunneling junction as a result of quantum tunneling.

The switching time for a binary wire containing C number cells is expressed as,

$$t_s = |(1 - C)\Gamma^{-1}| \quad (4)$$

$$= \left| C\varepsilon_r \varepsilon_0 e R_T (1 - e^{\frac{E_B}{KT}}) - \varepsilon_r \varepsilon_0 e R_T (1 - e^{\frac{E_B}{KT}}) \right|$$

$$= \left| \varepsilon_r \varepsilon_0 e R_T (1 - C) \left(1 - e^{\frac{E_B}{KT}}\right) \right| \quad (5)$$

Above equation 5 has been redefined by the help of eq. 4 in paper [4]. That equation is defining the speed of tunneling is given by,

$$v = \Gamma d \left(1 - e^{\frac{E_B}{KT}}\right)$$

Therefore, in eq. 5, $\left(1 - e^{\frac{E_B}{KT}}\right)$ is replaced by $\frac{v}{\Gamma d}$. The quantum tunneling occurs at speed of light hence, v is written as c in the below expression. Accordingly, the switching time is manipulated as,

$$t_s = \left| \varepsilon_r \varepsilon_0 e R_T (1 - C) \frac{c}{\Gamma d} \right|$$

In above equation d defines the effective tunneling distance which is covered by the electron while it penetrates the tunnel barrier.

The tunneling rate is modified as, $\Gamma = |(1 - C)t_s^{-1}|$ (6)

In the paper [4], a wave nature model of electron tunneling was proposed where the tunneling power, tunneling rate and dissipated power were calculated in wave nature of electrons. The proposed expressions have required the exact value of the tunneling rate in particle nature of electrons to make a mathematical conversion between the particle nature and wave nature of the electronic behavior. Therefore, QCA circuit power dissipation is calculated as,

$$P_{diss} = \frac{\hbar}{2} \Gamma \frac{d}{dt} \vec{\lambda}$$

$$P_{diss} = \frac{\hbar}{2} (1 - C) t_s^{-1} \left(\frac{d}{dt} \vec{\lambda}\right) \quad (7)$$

Where, $\vec{\lambda}= \mathrm{Tr}\{\hat{\rho}^{ss}\hat{\sigma}\} = \dfrac{\tanh\Delta}{\Omega}\begin{bmatrix} -2\gamma \\ 0 \\ G \end{bmatrix}$

$\Delta = \Omega / KT$ this term is known as thermal ratio where, $\Omega = \sqrt{4\gamma^2 + G^2}$.

Rest of the equations in [4] can also be simplified by using the tunneling rate equation, eq. 6.

The eq. 6 provides the exact value of tunneling rate if the propagation delay and the number of used cells are the two known parameters. Below fig. 10 is the graphical exposure of the eq. 5 which explicates the switching time for several semiconductors. Following are the set of conditions which have assumed for switching time calculation purpose,

1. The number of cells in a binary wire is, C=10. The cell count could be changed if the wire is having a different number of cells.
2. The spacing between all the cells are identical.
3. All the cells should be in a single layer on cell bed.
4. Tunneling resistance is unite, R_T=1.
5. Energy barrier is equals to the thermal energy. E_B=KT.
6. Electronic charge is 1.6×10^{-19}C.
7. Permittivity in vacuum $\varepsilon_0 = 8.854\times 10^{-12}$ F/m.
8. Relative permittivity ε_r varies for different semiconductors, as table I.

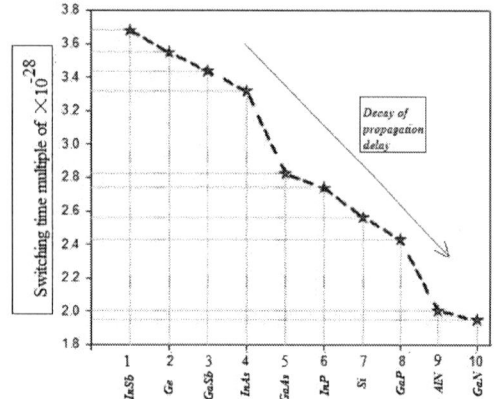

Fig. 10: Switching time variation curve in application of different semiconductors

This same approach is applicable for the purpose of switching time computation for Q-cells in various conditions. The equation 5 is a sophisticated mathematical way which involves the crucial and fundamental parameters in purpose of switching time (signal propagation delay) approximation. In other cases the variability of the right hand side parameters of the eq. 5 influences the value of t_s. In the same way it becomes accurate to approximate the device latency (signal propagation delay). The following subsection gives some possible conditions which can be used as the variable parameters in case of QCA circuit switching delay calculation.

The tunneling junction energies have been classified into three different values whereas it is observed that E_B is directly proportional to the device kink energy. The assumed values are, $E_B=0.5E_k$, $E_B=1E_k$ & $E_B=1.5E_k$

E_k is the kink energy for QCA cells which typically varies from 0.18meV to 0.75meV. As a variable parameter the surrounding temperature is also a crucial factor. Usually for QCA devices the standard operating temperature lies within 1K to 15 K, in some cases it is extended up to 17K. Because more than 15 K temperature can eject the electron from the quantum dot for that the device starts to misbehave. Temperature should be varied from 1K to 15K for several approximations on switching time. Therefore, it is possible to find out the value of t_s from the following combinations notably, but not limited to. The conditions can be altered depending on the designing and fabrication process.

T=1K: Energies $(0.5E_k, 1E_k, 1.5E_k)$

T=2K: Energies $(0.5E_k, 1E_k, 1.5E_k)$

………...

T=15K: Energies $(0.5E_k, 1E_k, 1.5E_k)$

V. COMPARISON WITH EARLIER DESIGNS IN TERMS OF AREA, O-COST, CLOCK LATENCY, LAYERS, CROSSOVERS

In this section proposed binary to gray code converter has compared to other erstwhile designs in terms of a few major parameters of QCA design alike area occupancy of the circuit, number of used cells, clock latency, used layers etc. From the below analysis it is concluded that proposed code converter layout has achieved 29.17% optimization in area occupancy, 28.94% cell reduction compared to the earlier best design as reported in table III.

TABLE III. COMPARISON TABLE FOR 2 BIT CODE CONVERTER

Sr. No.	Designs	Area (nm^2)	O-cost	Latency	Layers	Crossover types
1	[8]	63,000	389	8	No	coplanar
2	[9]	50,000	41	0.75	No	No
3	[10]	40,000	40	0.75	No	No
4	[11]	35,582	38	1	No	No
5	[12]	25,600	29	0.75	No	No
6	Proposed	25,200	28	0.75	2	No

TABLE IV. COMPARISON TABLE FOR 4 BIT CODE CONVERTER

Sr. No.	Designs	Area (nm^2)	O-cost	Latency	Layers	Crossover types
1	[9]	1,80,000	131	0.75	No	No
2	[10]	1,50,000	126	0.75	No	No
3	[13]	1,40,000	133	0.75	No	coplanar
4	[12]	158400	137	0.75	2	coplanar & multilayer
5	[12]	128000	127	1.00	2	coplanar & multilayer
6	Proposed	82,800	95	0.75	2	No

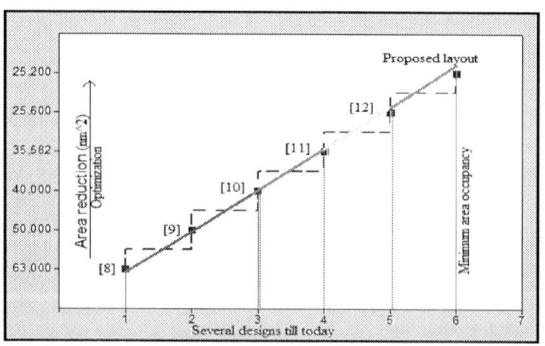

Fig. 11: Area optimization graph for proposed 2 bit code converter (data from table III)

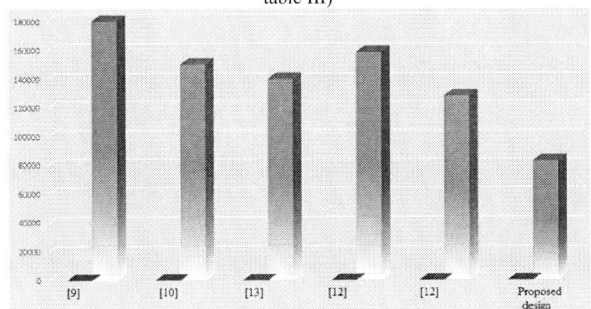

Fig. 12: Area comparison for 4 bit binary to gray code converter (data from table IV)

VI. CONCLUSION

In present quantum technology there are numerous instances alike carbon nanotubes, electron spin devices etc. which has resolved abundant complexities in COMS designs for example elevated outlay of lithography, impurity inconsistency in semiconductors etc. Since the last 25 years Quantum-dot Cellular Automata have been playing a major role in quantum binary logic device designing field. It has been widely popular as the research field of a huge number of researchers.

In this letter one novel 2 bit binary to gray code converter has proposed which layout has been extended up to 16 bit binary to gray code converter circuit implementation. Proposed 2 bit code converter circuit has acquired 29.17% optimization in overall area utilization. The circuit clock latency has also been reached at a very high acceptable value which is 0.75. In the second most highlighted contribution of this paper is the switching time calculation which is represented by t_s. The proposed mathematical model of this switching time explains that the nature of signal propagation time along the Q-cells. This propagation time is approximated by several complex calculations in section IV.

VII. FUTURE WORK

This concept will be extended much beyond of the existing theoretical models on quantum tunneling. In future proposals this theory will become very apparent and generalized for detecting the quantum tunneling effects accurately.

REFERENCES

[1] ITRS: International Technology Roadmap for Semiconductors (ITRS), website (2011). http://www.itrs.net/

[2] C. S. Lent, P. D. Tougaw, W. Porod, G. H. Bernstein, "Quantum cellular automata," Nanotechnology, vol. 4, pp. 49–57, 1993

[3] Gino A. Dilabio, Edmonton (CA); Robert A. WolkoW, Edmonton (CA); Jason L. Pitters, Edmonton (CA); Paul G. Piva, Edmonton (CA), "Atomistic Quantum Dots", Pub. No. US2015/0060771 A1, 5th May 2015.

[4] Soudip Sinha Roy, "Simplification of master power expression and effective power detection of QCA device (Wave nature tunneling of electron in QCA device)," *2016 IEEE Students' Technology Symposium*

(TechSym), Kharagpur, pp. 272-277, 2016, doi: 10.1109/TechSym.2016.7872695.

[5] Soudip Sinha Roy, "An intelligent mathematical QCA power analysis technique in wave nature of electrons," *2016 6th International Conference - Cloud System and Big Data Engineering (Confluence)*, Noida, pp. 680-684, 2016, doi: 10.1109/CONFLUENCE.2016.7508204.

[6] Chiradeep Mukherjee, Soudip Sinha Roy, Dr. Saradindu Panda, Dr. Bansibadan Majhi, "T- Gate: concept of partial polarization in quantum dot cellular automata", proc. of IEEE 20th international conf. on VLSI Design and Test (VDAT), in press, India, 2016.

[7] Yuhui Lu, Mo Liu, and Craig Lent, "Molecular quantum-dot cellular automata: From molecular structure to circuit dynamics", JOURNAL OF APPLIED PHYSICS, 102, 034311, 2007

[8] J. Iqbal, F. A. Khanday, N. A. Shah, "Efficient Quantum Dot Cellular Automata (QCA) Implementation of Code Converters", Communications in Information Science and Management Engineering, Vol. 3 Iss. 10, PP. 504-515, Oct. 2013.

[9] SHIFATUL ISLAM, MOHAMMAD ABDULLAH-AL-SHAFI AND ALI NEWAZ BAHAR, "Implementation of Binary to Gray Code Converters in Quantum Dot Cellular Automata", DOI: 10.15415/jotitt.2015.32010

[10] Md. Abdullah-Al-Shafi, Ali Newaz Bahar, "Novel binary to gray cde converter with power dissipation analysis", International journal of multimedia and ubiquitous engineering, vol. 11, no. 8, pp. 379-396, 2016.

[11] Young-Won You, Jun-Cheol Jeon, "Low Complexity QCA Binary to Gray Code Converter", Advanced Science and Technology LettersVol.144 (UBWCN 2017), pp.46-50 DOI: 10.14257/astl.2017.144.06.

[12] Nandini G. Raoa, P.C. Srikantha, Preeta Sharanb, "A novel quantum dot cellular automata for 4-bit code converters", Optik, Elsevier, DOI: 10.1016/j.ijleo.2015.12.119, 2015.

[13] M. G. Waje and P. K. Dakhole, "Design and simulation of new XOR gate and code converters using Quantum Dot Cellular Automata with reduced number of wire crossings," 2014 International Conference on Circuits, Power and Computing Technologies [ICCPCT-2014], Nagercoil, 2014, pp. 1245-1250. doi: 10.1109/ICCPCT.2014.7054942

[14] Mohammad Thaerifard, Mahmmod Fathy, "Improving logic function synthesis through wire crossing reduction in quantum-dot cellular automata layout", IEEE trans. on Devices and Syatems, IET ciruits, vol. 9, no. 4, pp. 265-274, July 2015.

[15] V. K. Mishra and H. Thapliyal, "Heuristic Based Majority/Minority Logic Synthesis for Emerging Technologies," 2017 30th International Conference on VLSI Design and 2017 16th International Conference on Embedded Systems (VLSID), Hyderabad, 2017, pp. 295-300. doi: 10.1109/VLSID.2017.27.

[16] Mo Liu and C. S. Lent, "Power dissipation in clocked quantum-dot cellular automata circuits," 63rd Device Research Conference Digest, 2005. DRC '05., Santa Barbara, CA, 2005, pp. 123-124. doi: 10.1109/DRC.2005.1553086.

[17] Soudip Sinha Roy"pGate: An Introduction to A Novel Universal Gate and Power Drop Calculation of QCA circuits" *International Journal Of Engineering And Computer Science (IJECS)*, https://www.ijecs.in, Volume 6 Issue 4, 20967-20972, April 2017, DOI: 10.18535/ijecs/v6i4.31.

[18] Manisha G. Waje, P.K. Dakhole, "Design and simulation of new XOR gate and code converters using Quantum Dot Cellular Automata with reduced number of wire crossings", proc. of IEEE on International Conference on Circuit, Power and Computing Technologies (ICCPCT), DOI: 10.1109/ICCPCT.2014.7054942, 2014.

[19] Davoud Bahrepour, "A Novel Full Comparator Design Based on Quantum-Dot Cellular Automata", International Journal of Information and Electronics Engineering, Vol. 5, No. 6, pp. 406-410, November 2015, DOI:10.7763/IJIEE.2015.V5.568.

[20] Qcadesigner online. Avaliable at http://www.qcadesigner.ca/

Fault Tolerance and Temperature Stability: The Dynamic Error Estimation in Quantum-dot Cellular Automata

Soudip Sinha Roy
Department of Nanotechnology
Amity Institute of Nanotechnology
Noida, UP, India
soudipsinharoy@gmail.com

Abstract— **Quantum-dot cellular automata is a transistorless quantum-scale computational approach which perceives the digital logic operations by fastest quantum tunneling. This paper contributes a proficient comparator layout and a decoder segment design without involving any wire crossing. The proposed 32 cells comparator has realized by Layered T gate approach of QCA, which has 19.20% less effective area as reported in paper. The fault tolerance analysis of proposed comparator has graphically analyzed to exhibit the robustness of the proposed comparator. Furthermore, one 2 to 4 line decoder has proposed which exhibits 37.82 % optimization in area occupancy compared to earlier decoder design. Temperature stability factor is proposed for recognizing the circuit dynamic faults. Dynamic temperature stability of the 2 to 4 line proposed decoder has been analyzed which estimates the circuit stability and specifies the suitable operating temperature zone. This contribution proposes a novel technique for estimating the dynamic errors in quantum-dot cellular automata circuits. QCADesigner 2.0.3 an open source software for QCA circuit simulation have used in purpose of all the circuit simulations in this paper.**

Keywords— *Quantum binary logic; robustness analysis; Temperature Stability Factor; quantum dots.*

I. INTRODUCTION

Friedrich Hund was the first observer of the electron tunneling, that in spite of high energy barrier presence the electrons have the ability to penetrate and cross the energy barrier through revealing its wave nature. Quantum dot cellular automata is the modified and up-graded application of quantum tunneling effect. It allows only a pair of electron in a single quantum cell for conveying a particular binary message from one flank to another flank of the cell [1], [2]. The sequential fabrication of quantum cells produces a binary wire, which conveys the binary information under the suitable clock pulse.

This paper presents one efficient 1 bit comparator layout which comprises only thirty two quantum cells to acquire the highest position among all the mentioned previously proposed comparators. In table I, the comparison is propounded by relating the crucial and fundamental factors for QCA circuit design. The robustness analysis of the comparator makes the section IV sophisticated, which reveals the information about the fault tolerance for the comparator during fabrication.

Moreover, the temperature stability factor for QCA circuit dynamic fault remedy has proposed which exhibits the dynamic

operating state change of the output polarizations. It is reported that, the TSF is a superior computational practice which is useful to observe theoretically about the temperature stability of quantum-dot cellular circuits.

II. FUNDAMENTAL OF QCA DESIGN

Since the last ten years of 20th century, quantum dot technology has been an utmost interest in the field of logic circuit design. Some prime designing schemes of logic gates like binary wire, inverter, Majority Voter (MV) have been exemplified in this section, fig. 1 (a,b,c). In fig.1, six cells binary wire, eight cells inverter and five cells MV gate have figured out [3]–[5]. Fig. 1 (c) assures, the multi-operation of MV gate as AND, OR gate due to the alternation of control polarization.

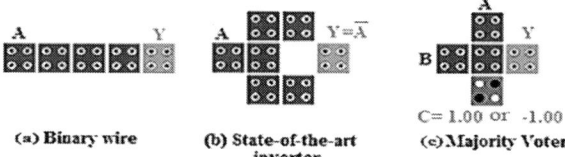

(a) Binary wire   (b) State-of-the-art inverter   (c) Majority Voter

Fig. 1 (a) Binary Wire, (b) State-of-the-art inverter, (c) Majority voter gate

In the digital logic family, Layered T gate is one of the convenient designing technique for the universal logic gates. Principally LT gate has been evolved by placing cells in different layers upon the cell bed [4]. Fundamentally this gate uses five cells, but proceeds its work through two different layers simultaneously. The fixed polarization cell of this gate is permanently established in a new layer which is 11.50 nm above of the main cell layer, fig. 2. The variable polarization of FP cell originates a partial polarization effect onto the existing cells of the main cell layer [5]–[7].

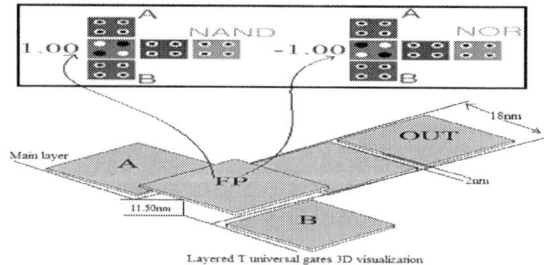

Fig. 2: Layered T gate layout

978-1-5386-1357-3/17 $31.00 © 2017 IEEE

8	In [13]	100	>32400nm^2	1.25	0	7
9	In [14]	117	171084 nm^2	0.50	0	28
10	In [15]	221	>71604nm^2	1.75	0	357
11	In [16]	262	150000 µm^2	1	0	28

A circuit becomes efficient if it is implanted inside a very small area efficaciously. Area utilization factor is a ratio between the total occupied area and the area is consumed by the used cells, which displays the proficiency of design in the sense of area occupancy [4]–[8].

III. PROPOSED LAYERED COMPARATOR DESIGN

A binary comparator is a combinational circuit which accomplishes the task of calculation of two or more binary numbers either those are equals to, greater or lesser than. Comparator is a vital part of any processor which operates in digital logic system [9]–[14]. This section covers an efficient comparator design, which is proposed with its performance analysis in terms of cell utilization, occupied area, and circuit delay. Presented comparator has shaped by 32 cells which exhibits a sufficient strength in account of cell reduction.

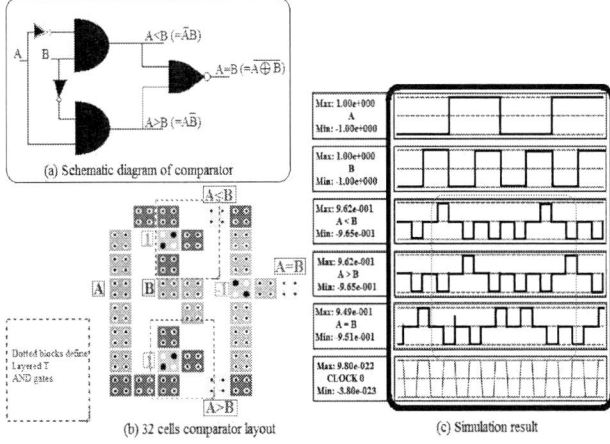

Fig. 3 (a) Schematic diagram of proposed comparator, (b) proposed comparator layout, (c) Simulation result

The following output waveform justifies the designing accuracy of the comparator. In fact, it resembles that, the occurred output is so viable at the point of their polarization strengths, which is technically efficient to be a 1 bit comparator primitive block, for developing further large scale comparators, fig. 3 (c). It is found that the polarization strength of A<B is {max: 9.62 e-001 and min: -9.65 e-001}. Similarly for A>B and A=B the polarization strengths are {max: 9.62 e-001, min: -9.65 e-001} and {max: 9.49 e-001, min: -9.49 e-001} respectively.

TABLE I.   COMPARATIVE ANALYSIS WITH PREVIOUSLY PROPOSED COMPARATORS

Sl. No.	Proposed Designs	O-Cost	Occupied Effective Area	Delay	Layer used	Cost$_1$
1	Proposed comparator	32	25600nm^2	0.75	1	3
2	In [7]	35	31684nm^2	–	0	9
3	In [8]	43	71724 nm^2	1.25	0	–
4	In [9]	48	56604nm^2	0.25	0	–
5	In [10]	65	80444nm^2	0.25	0	5.5
6	In [11]	73	64964nm^2	1	1	16
7	In [12]	81	>50000 nm^2	0.75	0	39.75

IV. ROBUSTNESS AND FAULT TOLERANCE ANALYSIS

As the final step for the production of any integrated circuit the fabrication is the ultimate process. In the present technology no satisfactory instances are present those are completely errorless. During fabrication the placement of the cells severally not happen prominently, but causes a bit displacement [15], [16]. The segmented visualized layout in the fig. 4, has been analyzed by QCADesigner appropriately by translating the entire columns individually with the maximum possible displacements, until the output waveform loses its consistency [17]–[20].

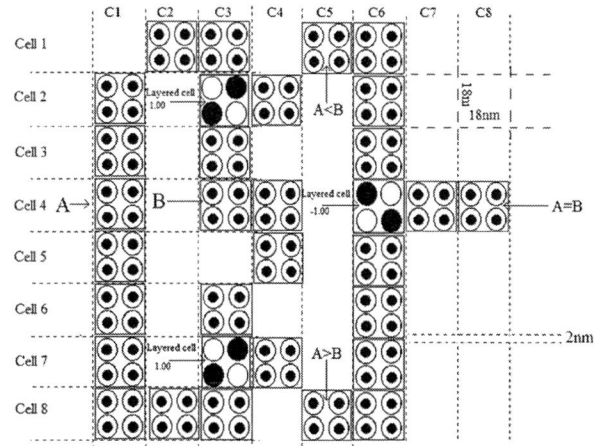

Fig. 4: Cell-wise segmented view of proposed comparator

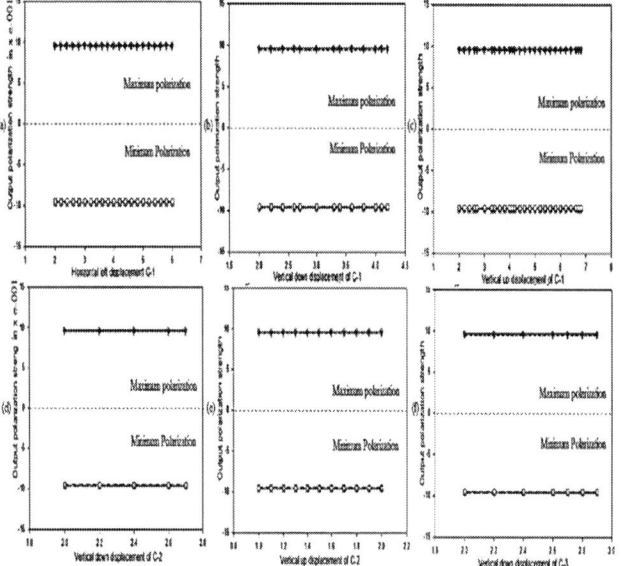

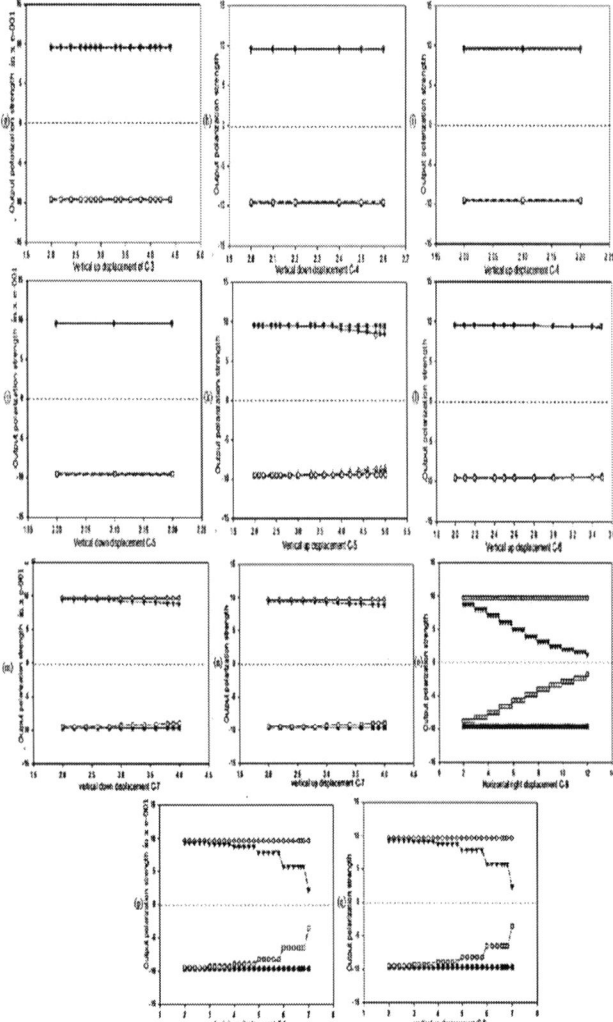

Column 4 is displaced vertically down, (i) Column 4 is displaced vertically up, (j) Column 5 is displaced vertically down, (k) Column 5 is displaced vertically up, (l) Column 6 is displaced vertically up, (m) Column 7 is displaced vertically down. (n) Column 7 is displaced vertically up, (o) Column 8 is displaced horizontally right, (p) Column 8 is displaced vertically down and (q) Column 8 is displaced vertically up.

After examining the faults of the circuit by displacing all the columns (C1 to C8) horizontally and vertically whichever is appropriate, the following graphs have been accomplished. The rigorous cell by cell investigation reveals the strength of the circuit to inhibit the distortion in output, if either displacement occurs notably. The obtained information from the fault analysis have been graphically plotted for vivid understanding of the robust characteristics of proposed comparator.

From the above fault analysis it is observed that the proposed layout has a very tiny fluctuation in the output when the column of cells are displaced within C1 to C7. In this observation, only the displacement of column 8 (C8) results a rapid fluctuation in the output. Due to the high stability of the comparator circuit the output polarization fluctuation is not so high. Therefore, the intermediate gaps in between the output polarization curves are not so apparent fig. 5. This is apparent only when the polarization fluctuates rapidly in high scale, as for the column 8, fig. 5 (o – q).

V. 2 TO 4 DECODER SEGMENT DESIGN

Decoder is a binary logic circuit which decodes two binary bits and generates 4 bit decoded output. In several cases for instance ROM (Read Only Memory), PLA, PAL design etc. the decoder circuits have been most prevalent segment of design. Often, decoders are also used as universal gates which is compatible enough to generate further high fan-out logic circuits. Using layered T gate one 2×4 decoder segment has been propounded which is designed in very less area.

In molecular level design of the binary logic circuits some fundamental metrics alike cell uses, area occupancy, circuit latency etc. are most crucial to be inspected. In table II, those are presented and compared with several earlier decoder designs.

Fig. 5: (a) Column 1 is displaced horizontally left, (b) Column 1 is displaced vertically down, (c) Column 1 is displaced vertically up, (d) Column 2 is displaced vertically down, (e) Column 2 is displaced vertically up, (f) Column 3 is displaced vertically down, (g) Column 3 is displaced vertically up, (h)

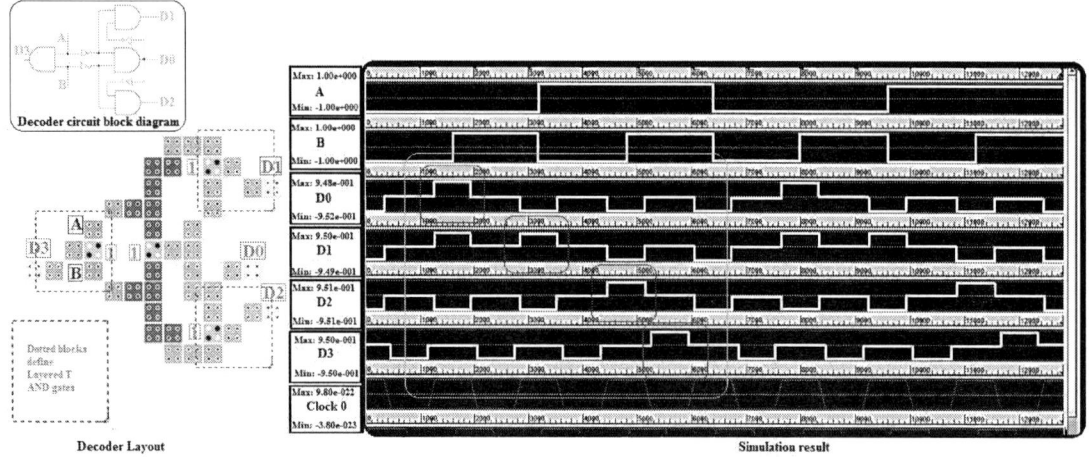

Fig. 6: (a) Decoder schematic design, (b) proposed decoder layout, (c) output result

TABLE II. AREA DELAY COST COMPARISON

Sr. No	Designs	Area nm^2	Latency	O-Cost
1	Proposed	57200	0.75	46
2	In [26]	264000	0.75	216
3	In [27]	129624	0.75	110
4	In [28]	528000	1	364
5	In [29]	92000	1	93

After the proficient comparison with the erstwhile designs, it is reported that the proposed decoder is optimized on area occupancy and O-cost sufficiently.

A. Temperature attack: the dynamic error estimation

Temperature is an inevitable parameter which leads the quantum scale circuits towards the unstable state of operation due to the dynamic energy state change of quantum tunneling and the extra energy of thermalized quantum cells. On this significant viewpoint the large range temperature stability provides a better dynamic fault tolerance to the circuits. In this section it is reported that, the proposed decoder circuit has a very high temperature stability within the range of 0 K to 17 K. The four output lines of the decoder D0, D1, D2 & D3 have identical effects of temperature on their polarizations, fig. 7.

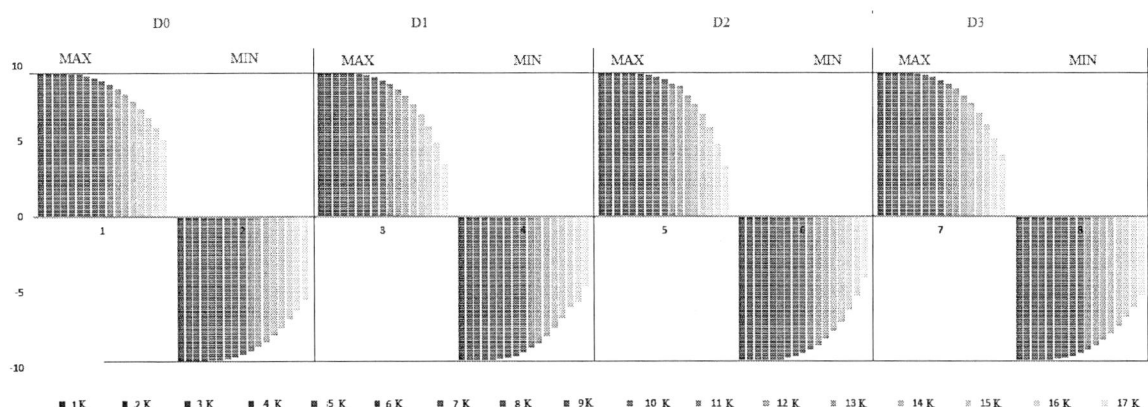

Fig. 7: Existing symmetry in output strengths under temperature attack

B. Proposed Temperature Stability Factor (TSF) calculation for QCA circuits dyanamic error remedy

Temperature stability factor defines that, how much stable one circuit is under the temperature attack. The QCA circuits can operate in a particular range of temperature but the operating temperature varies when the designing strategy changes. The variation of temperature casus a certain extra energy restoration into the quantum cells due to thermal energy kT ('k' stands for Boltzmann constant, 'T' is temperature in Kelvin unit). This energy attacks the QCA circuit and enforces its output waveforms to get decayed and gradually distorted.

Temperature stability is a dynamic type of stability of the circuits which should be highly achieved during the circuit operation for obtaining the maximum stability and the computing accuracy from the circuit. It is the requisite matter to accomplish the same level temperature stability factor for all the output lines of a circuit. TSF value of any circuit indicates that how strongly the circuit is stable under the specified range of temperature. In case when the logic operations are being performed through the QCA circuits, there has possibility of surrounding temperature variation. For high fan out circuits, if the TSF value of the output lines are not closer to '1', then it becomes extremely critical to predict the effect of temperature on the entire circuit. The temperature variation could result the

gradual decaying of the output polarization strengths even towards the rapid fluctuation. Still today there has no any instances of any ideal insulator. Therefore, the temperature variation is realized as a common effect which can cause a significant error. The entire circuit becomes unstable if the temperature variation occurs during the operation of the circuit. The algorithm for temperature stability factor calculation has illustrated below.

Algorithm for TSF calculation

Begin

1. *Use coherence vector simulation options in QCAD.*
2. *Extract the exact output polarization strengths (max and min) by varying the temperature from 1 K.*
3. *Apply for all output lines.*
4. *If (output polarization strength remains as desired)*
   {
   Continue the same process of simulation by varying temperature gradually upwards.
   }
5. *Else {*
   Stop the simulation immediately at that particular temperature value, where any

> *one of the output of the circuit gets distorted.*
> *}*

6.   *Store the entire data matrix.*
7.   *Find out the ratio of MAX polarizations between two output lines according to the temperature value extracted by simulation.*
8.   *Continue this process for all possible combinations of output lines.*
9.   *Perform the same operation for MIN polarizations between two output lines according to the temperature value.*
10. *Store the data matrix as TSF data.*
11. *Plot the TSF matrix individually for MAX and MIN polarizations obtained from step 2.*

End

C. Mathematical approach of Temperature Stability Factor (TSF)

In case of dynamic operation of QCA circuits the operating temperature is a crucial factor which influences the polarization variation. For each and every change in operating temperature the polarization strength fluctuation is noticed.

For T_1 K temperature if the output polarization strengths are P_{i_1} and P_{j_1}, then TSF value is expressed as $TSF_1 = \frac{P_i}{P_{j_1}}\Big|_{T1}$. For T_2 K temperature TSF is $TSF_2 = \frac{P_{i_2}}{P_{j_2}}\Big|_{T2}$. Therefore $TSF_n = \frac{P_{i_n}}{P_{j_n}}\Big|_{T_n}$, where n is either integer or decimal depending upon exact temperature value. Through this approach all the TSF values could make the graphical exposure of the temperature stability for any QCA circuit.

D. TSF calculation for proposed Layered 2 to 4 decoder

Through this rigorous investigation on the 2 to 4 line decoder segment it is reported that, the output lines MAX-{D0, D1, D2, D3} are in same level of TSF approximately within the temperature range of $1\,K - 9.8\,K$, fig. 8. This reveals that the TSF value lies between 1 and 1.001 within the mentioned temperature range, which exhibits the outputs are very stable if the operating temperature is within the specified range.

Furthermore, the output lines MIN-{D0, D1, D2, D3} are in same level of TSF approximately within the temperature range of $1\,K - 8\,K$, fig. 9. TSF value lies between 1 and 1.001 within the reported temperature zone. Therefore, effectively the entire circuit is fully stable under $1\,K - 8\,K$ temperature range. Because the MAX polarizations are stable within $1\,K - 9.8\,K$ and the MIN polarizations are stable within $1\,K - 8\,K$, so the effective stability region for the entire 2 to 4 decoder output lines is $1\,K - 8\,K$ where both of the polarization strengths (MAX & MIN) are in a same level.

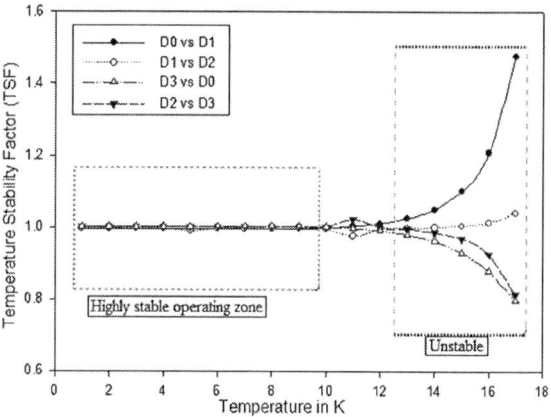

Fig. 8: TSF (Temperature Stability Factor) calculation of MAX polarizations for proposed decoder

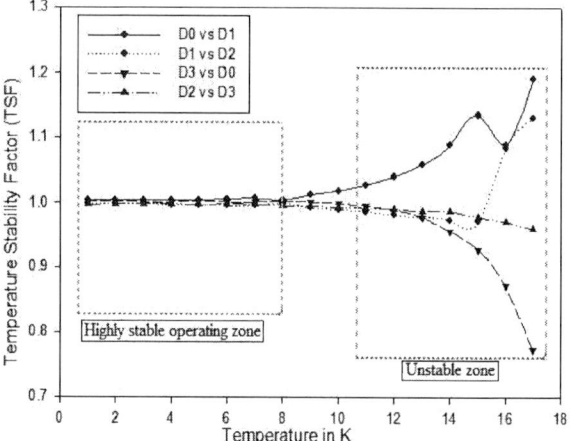

Fig. 9: TSF calculation of MIN polarizations for proposed decoder

VI. CONCLUSION

In certain previews of quantum dot cellular designs, the proposers have contributed previously numerous comparator designs most challengingly. In this research contribution one novel comparator layout is propounded keeping the point of attraction on highly cell reduction and area optimization. In table I, it is revealed that, the proposed comparator layout has $25600\,nm^2$ area occupancy which is reliably acceptable to prove the effectiveness of the layout in comparison of other erstwhile designs by all the mentioned authors. It is the requisite matter for any integrated circuit being to have a high fault tolerance strength during fabrication. Therefore, the robustness analysis of the proposed comparator layout is performed to demonstrate the displacement dependency of the cells and the preventing proficiency in output waveform from fluctuation.

Moreover, the proposed temperature stability factor judges on the temperature sensitivity of the proposed 2 to 4 line decoder, which approves that this decoder is highly stable under the temperature attack during dynamic operation. It is reported that the proposed 2 to 4 decoder circuit is highly stable within $1\,K - 8\,K$ temperature zone.

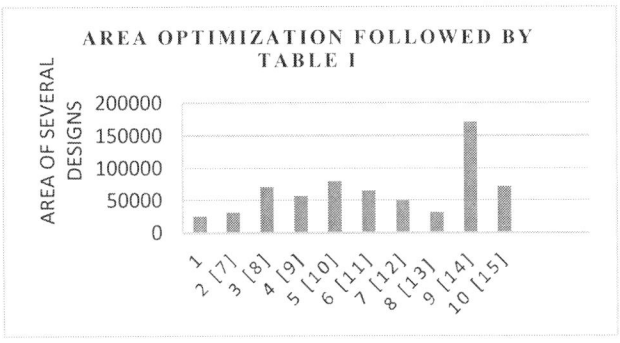

AREA OPTIMIZATION FOLLOWED BY TABLE I

REFERENCES

[1] ITRS: International Technology Roadmap for Semiconductors (ITRS), website (2011). http://www.itrs.net/

[2] S. Perri, P. Corsonello, G. Cocorullo, "Design of Efficient Binary Comparators in Quantum-Dot Cellular Automata", IEEE transactions on nanotechnology, vol. 13, no. 2, March 2014.

[3] J. R. Janulis, P. D. Tougaw, S. C. Henderson, E. W. Johnson, "Serial Bit-Stream Analysis Using Quantum-Dot Cellular Automata", IEEE transactions on nanotechnology, vol. 3, no. 1, pp. 158-164, March 2004.

[4] C. Mukherjee, D. Das, "Layered T full adder using Quantum-dot Cellular Automata" in Proceedings of IEEE International Conference on Electronics, Computing and Communication Technologies (CONECCT-2015), pp. 1-6, July 2015, DOI:10.1109/CONECCT.2015.7383867

[5] C. Mukherjee, Soudip. S. Roy, S. Panda, B. Maji, "T-Gate: Concept of partial polarization in Quantum dot Cellular Automata", presented at the 20th International Symposium on VLSI Design and Test (VDAT), Guwahati, India, May 24-27, 2016, Paper 47.

[6] Yuhui Lu, Mo Liu, and Craig Lent, "Molecular quantum-dot cellular automata: From molecular structure to circuit dynamics", *JOURNAL OF APPLIED PHYSICS*, 102, 034311, 2007.

[7] Mika Helsingius, Pauli Kuosmanen, and Jaakko Astola,"Quantum-dot Cells and their Suitability for Nonlinear Signal Processing", Proceedings of the IEEE-EURASIP Workshop on Nonlinear Signal and Image Processing (NSIP'99),pp. 659-663, June 1999, ISBN 975518-133-4.

[8] Davoud Bahrepour, "A Novel Full Comparator Design Based on Quantum-Dot Cellular Automata", International Journal of Information and Electronics Engineering, Vol. 5, No. 6, pp. 406-410, November 2015, DOI:10.7763/IJIEE.2015.V5.568.

[9] Anuradha.S.S, Ravi.B.D, Vishal.M.Pasar, "Design of five input majority gate Full Comparator using Quantum-Dot Cellular Automata", International Journal of Ethics in Engineering & Management Education, Vol. 1, Issue 4, April 2014, ISSN: 23484748.

[10] Jason R. Janulis, P. Douglas Tougaw, Steven C. Henderson, and Eric W. Johnson, Serial Bit-Stream Analysis Using Quantum-Dot Cellular Automata, IEEE Transaction on Nanotechnology, vol.3, no. 1, pp. 158-164, March 2004, DOI:10.1109/TNANO.2004.824014

[11] Bahniman Ghosh, Shoubhik Gupta, and Smriti Kumari, "Quantum Dot Cellular Automata Magnitude Comparators", Proceedings of 2012 IEEE International Conference on Electron Devices and Solid State Circuit (EDSSC), Dec.2012, DOI: 10.1109/EDSSC.2012.6482766.

[12] D.Ajitha, K.Venkata Ramanaiah, and V.Sumalatha, "A Novel Design of Cascading Serial Bit-Stream Magnitude Comparator Using QCA", Proceedings of 2014 International Conference on Advances in Electronics, Computers and Communications (ICAECC), October 2014, DOI:10.1109/ICAECC.2014.7002449.

[13] Y. Xia, K. Qiu, "Comparator design based on quantum-dot cellular automata," Journal of Electronics & Information Technology, vol 31, no. 6, pp. 1517–1520, 2009, DOI: 10.3724/SP.J.1146.2008.00838.

[14] Md. Abdullah-Al-Shafi, Ali Newaz Bahar,"Optimized design and performance analysis of novel comparator and full adder in nanoscale", Cogent Engineering, pp. 1-14, September 2016, DOI: http://dx.doi.org/10.1080/23311916.2016.1237864.

[15] Blaz Lampreht, Luka Stepancic, Igor Vizec, Bostjan Zankar, "Quantum-Dot Cellular Automata Serial Comparator", Proceedings of 11th

EUROMICRO Conference on Digital System Design Architectures, Methods and Tools, pp. 447-452, November 2008, DOI: 10.1109/DSD.2008.49.

[16] G. Rajesh Kumar, D. Venkatarami Reddy, Sk. Subhan, "A Novel Approach for Efficient Binary Comparators in Quantum-dot Cellular Automata", Vol.3, Issue 5, pp. 0721-0728, July 2015, ISSN: 23220929.

[17] P. K. Rahi, S. Dewangan, S. Mirania, Md Muzaheru, "Low Power and Area Efficient Design of 1-BitCMOS Comparator Using Different Foundry", International journal for research in emerging science and technology, vol. 2, no. 5, May 2015.

[18] A. Sharma, R. Singh, P. Kajla, "Area Efficient 1-Bit Comparator Design by using Hybridized Full Adder Module based on PTL and GDI Logic", International Journal of Computer Applications (0975 – 8887), Vol. 82, No.10, Nov. 2013.

[19] Mehmood ul Hassan, Rajesh Mehra, "Design Analysis of 1-bit CMOS comparator", International Journal of Scientific Research Engineering & Technology (IJSRET) ISSN: 2278–0882, EATHD-2015 Conference Proceeding, 14-15 March, 2015.

[20] Jien-Chung Lo, "A Novel Area-Time Efficient Static CMOS Totally Self-checking Comparator", proc. of IEEE on journal of solid-state circuits, vol. 28, no. 2. Feb. 1993.

[21] Satyabrata Nanda, Avipsa S. Panda, G.L.K. Moganti, "Design of a high speed and low area latch-based comparator in 90-nm CMOS technology having low offset voltage", 2015 International Conference on Energy Systems and Applications (ICESA 2015).

[22] A. Sharma, P. Sharma, "Area and Power Efficient 4 – Bit Comparator Design by Using 1- Bit Full Adder Module", proc. of IEEE on International Conference on Parallel, Distributed and Grid Computing, 2014.

[23] V. Shekhawat, T. Sharma, K. G. Sharma, "2-Bit Magnitude Comparator using GDI Technique", IEEE International Conference on Recent Advances and Innovations in Engineering (ICRAIE-2014), May 09-11, 2014.

[24] S. S. Roy, "Simplification of Master Power Expression and Effective Power Detection of QCA Device", Proceedings of IEEE Students' Technology Symposium, pp. 272-277, 2016, DOI: 10.1109/TechSym.2016.7872695.

[25] S. S. Roy, "An intelligent mathematical QCA power analysis technique in wave nature of electrons", Proceedings of the IEEE conf. on Cloud system and Bigdata Engineering (Confluence), pp. 680-684, 2016, DOI: 10.1109/CONFLUENCE.2016.7508204.

[26] Jun-Cheol Jeon, "Extendable Quantum-Dot Cellular Automata Decoding Architecture Using 5-Input Majority Gate", International Journal of Control and Automation Vol.8, No.12 (2015), pp.107-11

[27] Kondwani Makanda, Jun-Cheol Jeon, "Improvement of Quantum-Dot Cellular Automata Decoder Using Inverter Chain", Advanced Science and Technology Letters Vol.29 (CA 2013), pp.227-229

[28] Moein Kianpour, Reza Sabbaghi-Nadooshan, "A Novel Modular Decoder Implementation In Quantum-dot Cellular Automata (QCA)", IEEE proc.

[29] Debashis De, Tamoghna Purkayastha, Tanay Chattopadhyay, "Design of QCA based Programmable Logic Arrayusing decoder", Microelectronics Journal.

[30] QCADesigner. Available online at http://www.qcadesigner.ca/

Tunneling Field Effect Transistors for Energy Efficient Logic, Sensor Interface and 3D IC Circuits for IoT Platforms

Japa Aditya, T. Nagateja, and Ramesh Vaddi

Nano Scale Integrated Circuits and Systems for Self-Powered IoT Laboratory, Electronics and Communication Engineering
DSPM InternationalInstitute of Information Technology, Naya Raipur,Chhattishgarh-493661,India
Contact e-mails:{aditya,tnagateja,ramesh}@iiitnr.edu.in

Abstract—Tunneling Field-Effect Transistors (TFET) have emerged as a leading future transistor option for energy efficient and next generation VLSI. In this paper, we propose novel logic design exploiting TFET device unique asymmetrical characteristics and performance benchmarked with conventional TFET transmission gate (TG) based designs for energy efficiency. Proposed logic gates are ~7-10x energy efficient than the conventional TFET TG logic designs along with small on-chip area. This work further explores the scope of TFETs for dual mode transceiver design with resistive sensing circuit for 3D IC and time-to-digital converter (TDC) used in sensor interface circuits for IoT. TFET's steep subthreshold slope characteristics enable designing high throughput and energy efficient transceiver circuits for 3D IC and sensor interface TDC circuits with highly precision and linearity.

INTRODUCTION

Tunneling Field-Effect Transistor as an alternative low power device candidate offers promising solution to build future energy efficient ICs [1-2].TFET based digital/analog/mixed signal and RF circuit designs were demonstrated with high energy efficiency in the recent times [3-5]. TFETs possess unique device characteristics such as enhanced on-channel Miller capacitance effect, unidirectional current conduction and ambipolarity [6-8] and pose significant challenges to circuit designers for exploring alternative circuit and architectural techniques to get full benefit of TFETs for future IC design. Not many have explored designing alternative logic designs for TFETs along with TFET's scope for 3D IC and IoT sensor interface circuits. CMOS transceiver designs in subthreshold region either fall in the low energy or high throughput region, but achieving satisfactory values both is very challenging [9-14]. Highly precise TDCs function as the fundamental circuits of numerous accurate instrumentation systems. CMOS TDC in sub-threshold/sub-micron region does not meet the requirements like high precision and linearity [15-16]. TFET with its steep subthreshold slope, can able to build highly precise and linear TDC at supply voltage of 0.3V. Exploiting the unidirectional current conduction (asymmetry) property of TFETs, energy efficient and area efficient logic designs are presented in this paper. This work further explores the scope of Tunnel FETs for high throughput and energy efficient transceivers and time-to-digital converters (TDC) used in sensor interface circuits.

TFET DEVICE CHARECTERISTICS FOR ENERGY EFFICIENT CIRCUIT DESIGNS

In this work, LUT based Gasb-InAshetero junction TFET (HTFET) models, Si FinFET models [17] and a universal analytical double-gate (DG) InAs TFET Verilog-A models [18] with channel length L=20nm are explored for circuit designs in *Cadence Virtuoso* environment. Fig. 1(a-b) present the TFET and FinFET I_D-V_G characteristics (log scale) achieving high on-current (7x improvement at V_{DD}=0.3V) and lower subthreshold swing (30mV/dec) compared to the base line FinFET along with the presence of significant ambipolarity in InAs TFETs.

TFET BASED ENERGY EFFICIENT AND ALTERNATIVE LOGIC DESIGN USING TFET UNIDIRECTIONAL CURRENT CONDUCTION CHARCATERISTICS

This section presents a circuit technique exploiting the unidirectional current conduction of TFETs for building novel and energy efficient logic gates compared to the conventional TFET based transmission gate logic designs. Fig. 2 presents novel TFET based two input AND, OR and XOR logic gate designs that have been built using the unique unidirectional current conduction characteristics of TFETs. To illustrate the functionality of the proposed designs, TFET AND logic design has been considered. In this circuit (Fig.2 (a)), since TFETs are unidirectional devices, circuit current flows only from drain to source in case of NTFET and from source to drain in PTFET. Its functionality is further explained in Fig. 3. For example, when the input A is at logic '0' and B is at logic '1', transistor T_1 switches OFF, T_2 switches ON and the output of the AND gate discharges through T2 and becomes logic '0'. Simultaneously, Transistor T3 enters into the ambipolar region but no current flows due to unidirectional current conduction property of TFET. As a result, power consumption of proposed logic gates ~7-10x lower when compared to the conventional designs as shown in Fig.4. Fig. 5 shows the transient characteristics of proposed TFET logic gates at supply voltage of 0.5V.

TUNNEL FET BASED DUAL MODE TRANSCEIVER DESIGN WITH RESISTIVE SENSING CIRCUIT FOR 3D IC

Fig. 6 presents a proposed TFET based dual mode transceiver with resistive sensing circuit for 3D IC applications. In this work, plates of the capacitor Cc are considered as transmitter and receiver pads. A resistance sensor is embedded in the proposed transceiver, observe the interface condition at the initial stage and independently reconfigures the RX to work at either the ohmic contact mode (enable =1) when two pads are intact or the capacitive coupling mode (enable =0) when pads are loosely connected. In the capacitive coupling mode, the pass gate (consist of bidirectional switch (BS)) is connected to the RX, working principle was explained in [14]. To make transceiver work in ohmic contact mode, the pass gate in Fig. 6 should be disabled. In this work, a new resistance sensor structure which consists of sense enable transistors (M1, M2), write transistors (W1, W2), set transistor (M3) for transceiver is proposed, which enables or disables the pass gate according to pads' interference distance. Fig. 7 summarize and give the comparison of throughput (Gbps) and energy consumption (pJ/bit) of TFET based dual mode transceiver with resistive sensing circuit for 3D IC [9-14] with recently published CMOS 3D IC designs. TFET based transceiver design achieves high energy efficiency and throughput compared to reported CMOS transceivers.

TUNNEL FET BASED TIME TO DIGITAL CONVERTER

Fig. 8 presents a TFET based ring oscillator TDC circuit. When time signal Φ_{err} is at logic '1', the capacitors c1 and c2 are charged to logic '0' and '1' respectively as control and control bar. At this condition, ring oscillator starts oscillating and convert the time signal into digital code until Φ_{err} becomes logic'0'. Fig. 9 presents the transient characteristics of the TFET TDC showing the output oscillations for a given input pulse width at supply voltage of 0.3V. Fig. 10 shows the digital code vs input pulse width of TFET TDC design and comparison with FinFET TDC at supply voltage of 0.3V. From this, it can be observed that TFET TDC can achieve highly linear characteristics than that of FinFET due to the steep slope characteristics. TFET TDC achieves highly linear characteristics and the TFET TDC has ~ 4x better precision value than the equivalent FinFET based TDC at supply voltage of 0.3 V.

CONCLUSION

This work presents TFET based alternative logic gates for energy efficiency by exploiting TFET's unique unidirectional current conduction property of tunnel FETs energy and area efficient logic gates. The proposed TFET based novel logic gates are ~7-10x energy efficient over TFET based conventional transmission logic gate designs. Taking the specific TFET device characteristics into consideration, the proposed TFET based dual mode transceiver with resistive sensing circuit for 3D IC demonstrate enhanced performance in terms of throughput and energy efficiency not achievable with CMOS designs at low V_{DD}. TFET based TDC achieves high precision and linearity compared to FinFET based design at supply voltage of 0.3V and TFETs suit for energy efficient, precise, area efficient circuits for next generation VLSI systems/IoT .

REFERENCES

[1] Pandey Rahul, Sauarbh Mookerjea, and Suman Datta, "Opportunities and Challenges of Tunnel FETs," *IEEE Trans. Circuits and Systems I*, vol.63, no. 12, pp.2128-2138, Dec. 2016.

[2] Cristolovean Sorin, Jing Wan, and Alexander Zaslavsky, "A review of sharp-switching devices for ultra low power applications," *IEEE Journalof the Electron Devices Society,* vol.4, no. 5, pp. 215-226, Sep. 2016.

[3] V. Harshita, *et. al,* "Designing Energy Efficient Logic Gates with Hetero Junction Tunnel FETs at 20nm" IEEE Trans. Electron Devices sponsored Device, Circuit, System Conference, India, March 2014.

[4] Sedighi Behnam, Xiaobo Sharon Hu, Huichu Liu, Joseph J. Nahas, and Michael Niemier, "Analog circuit design using tunnel-FETs," *IEEE Trans. Circuits and Systems I: Regular Papers,* vol. 62, no. 1, pp. 39-48, Jan. 2015.

[5] Huichu Liu, Xueqing Li, Ramesh Vaddi, Kaisheng Ma, Suman Datta, Vijaykrishnan Narayanan " Tunnel FET RF Rectifier Design for Energy Harvesting Applications", *IEEE Journal on Emerging and Selected Topics in Circuits and Systems (JETCAS),* Vol.4, Iss.4, pp. 400-411, Dec,2014.

[6] Japa Aditya, Harshita Vallabhaneni, and Ramesh Vaddi, "Reliability enhancement of a steep slope tunnel transistor based ring oscillator designs with circuit interaction," *IET Circuits, Devices & Systems,* vol. 10, no. 6, pp. 522-527, Nov. 2016.

[7] Daniel H. Morris, Uygar E. Avci, Rafael Rios, and Ian A. Young, "Design of low voltage tunneling-FET logic circuits considering asymmetric conduction characteristics," *IEEE J. Emerging and Selected Topics in Circuits and Systems*, vol. 4, no. 4, pp. 380-8, Dec. 2014.

[8] Sebastiano Strangio, Pierpaolo Palestri, David Esseni, Luca Selmi, Felice Crupi, Simon Richter, Qing-Tai Zhao, and Siegfried Mantl, "Impact of TFET unidirectionality and ambipolarity on the performance of 6T SRAM cells," *IEEE J. of the Electron Devices Society*, vol. 3, issue. 3, pp.223-232, May 2015.

[9] Mari Inoue , Noriyuki Miura , Kiichi Niitsu , Yoshihiro Nakagawa , Masamoto Tago , Muneo Fukaishi , Takayasu Sakurai and Tadahiro Kuroda, "Daisy Chain for Power Reduction in Inductive-Coupling CMOS Link," in Proc. *IEEE Symposium on VLSI Circuits Digest of Technical Papers,* 2006, pp. 65–66.

[10] Noriyuki Miura, Hiroki Ishikuro, Kiichi Niitsu, Takayasu Sakurai, and Tadahiro Kuroda, "A 0.14 pJ/b Inductive-Coupling Transceiver With Digitally-Controlled Precise Pulse Shaping," in Proc. *IEEE Int. Solid-State Circuits Conf. Dig. Tech. Papers*, 2007, pp. 358–608.

[11] Qun Gu, Zhiwei Xu, Jenwei Ko, Mau-Chung Frank Chang, "Two 10 Gb/s/pin low-power interconnect methods for 3D ICs," in Proc. *IEEE Int. Solid-State Circuits Conf. Dig. Tech. Papers*, 2007, pp. 448–614.

[12] Myat Thu Linn Aung, Eric Lim, Takefumi Yoshikawa, and Tony Tae-Hyoung Kim, "A 3-Gb/s/ch Simultaneous Bidirectional Capacitive Coupling Transceiver for 3DICs," *IEEE Trans. on VLSI*, vol. 61, no. 9, pp. 706-710, Sept. 2014.

[13] Alberto Fazzi, Luca Magagni, Mauro Mirandola, Barbara Charlet, Léa Di Cioccio, Erik Jung, Roberto Canegallo, and Roberto Guerrieri, "3-D Capacitive Interconnections for Wafer-Level and Die-Level Assembly," *IEEE J. of Solid-State Circuits*, vol. 42, no. 10, pp. 2270-2282 Oct. 2007.

[14] Myat-Thu-Linn Aung, Taketumi Yoshikawa, Chuan-Seng Tan and Tony Tae-Hyoung Kim, "Yield Enhancement of Face-to-Face Cu–Cu Bonding With Dual-Mode Transceivers in 3DICs," *IEEE Trans. on VLSI*, vol. 25, no. 3, pp. 1023-1031, Mar. 2017.

[15] Jussi-Pekka Jansson, Vesa Koskinen, Antti Mäntyniemi, and Juha Kostamovaara, "A Multichannel High-Precision CMOS Time-to-Digital Converter for Laser-Scanner-Based Perception Systems," *IEEE Trans. on Instrumentation and Measurement*, vol. 61, no. 9, pp. 2581-2590, Sept. 2012.

[16] Chun-Chi Chen, Shih-Hao Lin, and Chorng-Sii Hwang, "An Area-Efficient CMOS Time-to-Digital Converter Based on a Pulse-Shrinking Scheme," *IEEE Trans. on Circuits and Systems*—II, vol. 61, no. 3, pp. 163-167, Mar. 2014.

[17] Penn State University Verilog-A Models for III-V TunnelFETs.[Online].http://www.ndcl.ee.psu.edu/downloads.asp.

[18] Verilog-A Models for universal Tunnel FETs [online]:https://nanohub.org/publications/31/serve/1/138?el=6&download=1.

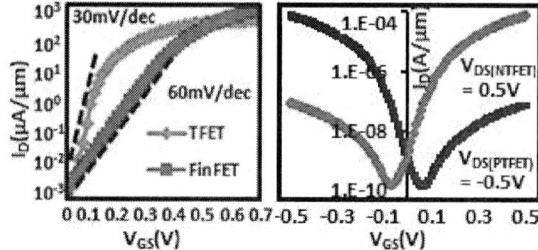

Fig. 1.(a) Comparison of I_{DS}-V_{GS} characteristics of Double gate Hetero-junction TFET with Double gate Si FinFET (log scale) (b) I_{DS}-V_{GS} characteristics of InAsHomo-junction PTFET and NTFETs showing ambipolarity.

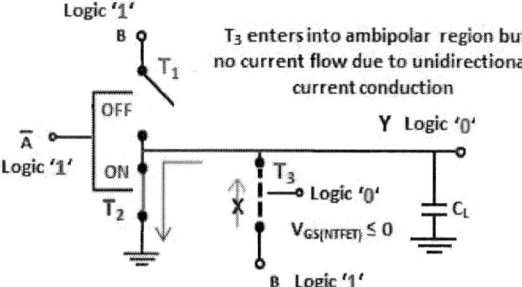

Fig. 2 Proposed Tunnel FET based alternative Logic gates exploiting unidirectional current conduction characteristics (a) 2 I/p AND (b) 2 I/p OR (c) 2 I/p XOR.

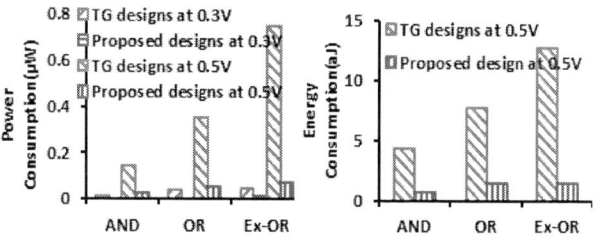

Fig.4.Power and energy consumption comparison of proposed tunnel FET based logic gates with TFET transmission gate designs.

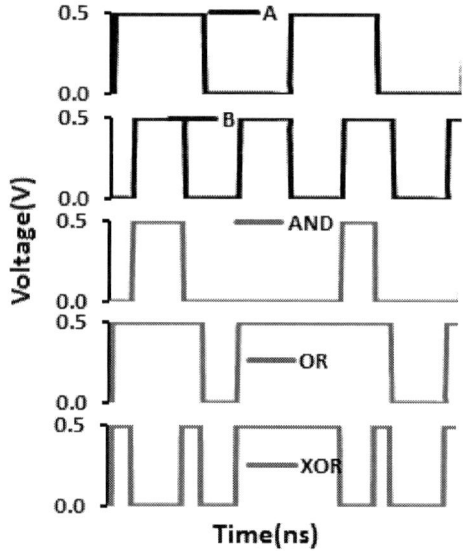

Fig. 3 Functional Illustration of proposed TFET based AND logic for given inputs (A = logic '0', B= logic '1').

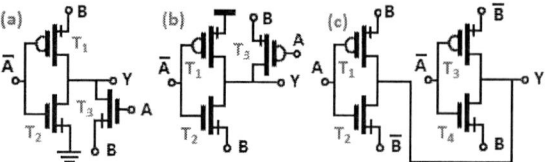

Fig. 6TFET based Dual mode transceiver design with resistive sensing circuit for 3D IC.

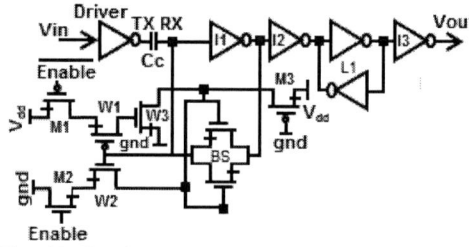

Fig. 7 Energy effiiciency vs throughput benchmarking of TFET based dual mode transceiver design with resistive sensing circuit for 3D IC.

Fig. 5 Transient response of proposed Tunnel FET based alternative Logic gates at 0.5V supply voltage.

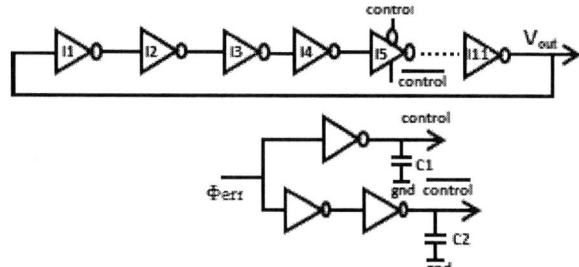

Fig. 8 TFET ring oscillator based Time-to-Digital Converter (TDC).

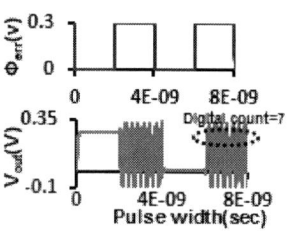

Fig.9 Transient characteristics of TFET TDC for 2ns input pulse at a supply voltage of 0.3V.

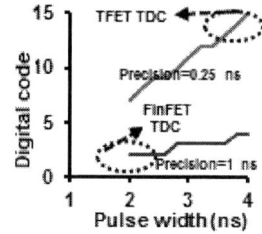

Fig.10 Linearity (precision) comparison of TFET TDC with FinFET at a supply voltage of 0.3V.

978-1-5386-1357-3/17 $31.00 © 2017 IEEE

SiGe source charge plasma TFET for biosensing applications

Nawaz Shafi[1], Chitrakant Sahu[2], *C. Periasamy*[3] and Jawar Singh[4]

Malaviya National Institute of Technology Jaipur[1, 2, 3]

PDPM IIITDM Jabalpur[4]

Contact e-mail: chitrakant.ece@mnit.ac.in

Abstract—**In this paper, a new TFET biosensor architecture has been proposed that is based on charge plasma technique in conjugation with a low energy bandgap source material i.e. SiGe. The proposed device sensitivity is compared with a conventional doped DG-TFET for biosensing applications. Charge density and dielectric constant of biosamples are used as the intrinsic properties that modulate the device characteristics. The charge plasma TFET shows excellent I_{ON}/I_{OFF} ratio and a subthreshold swing steeper than the DG-TFET. In biosamples with higher dielectric constants, the sensitivity in CP-TFET is about 62% more than DG-TFET biosensor.**

Keywords—charge plasma, double-gate, TFET, ambipolar current, dopingless, dielectric modulated.

I. Introduction

Human life and its related issues have always been a priority for the researchers but early detection and diagnosis of disease causing biological pathogens at an early stage of their development has been ever since a challenging task. Hence, development of ultra-sensitive, robust and reliable devices is the current need of the hour. Label free detection of bio-samples technique has evolved as a resolve measure by inculcating the knowledge of material science and operation of nano devices. Nano-volume devices turn out to be effective devices in early diagnosis and detection of diseases. FET's use the basic intrinsic properties like dielectric constant and charge density for free label biosensing. The underlying idea is to modulate the electrostatic action of gate terminal on the FET based biosensors in inculcation with the bio-sample generally present in cavity using its elementary properties such as dielectric constant and charge density [1]. Various types of FET biosensors based on ion sensitive FETs, tunnel FETs have been proposed time to time with good improving results. However ISFETs work on the principle of the charge present due to ions in the ionic biosample solution under the dielectric gate. This limits the application of ISFETs as biosensors as uncharged molecules of any biosample cannot be detected [2]-[3]. Tunnel field effect transistor has proved to be a potential device that can be used as a biosensor and can effectively detect the charged as well as uncharged molecules as the device characteristics depend on charge as well as dielectric constant of the biosample solution, such devices at a general level have been named as dielectric modulated FET biosensors as a whole [4]-[7]. With the inception of other type FET biosensors and their derivatives, limitations also arise. Various performance metrics such as I_{on}/I_{off}, g_m/I_D, threshold voltage shift, and capacitance can be employed to analyze the type of biosamples. Higher the sensitivity of the device for a biosample

better is the device performance [6]. As scalability and level of integration has approached the lower realm of nano scale regime, various performance factors such as off current, threshold voltage, power dissipation are compromised for the level of integration achieved. The DM-FET biosensor incorporates a nano-cavity in which the biosample is lodged. The gate capacitance variation depends on the dielectric strength (K) and charge density (ρ) of the biosample solution. Since, the drain current in a TFET is governed by the process of BTBT, hence, a steeper subthreshold slope (SS) is obtained less than the conventional Boltzman's thermal limit op 60mV/dec. By combining all the advantageous features of DM biosensors, TFETs and charge plasma concept, a dielectric modulated CP-TFET biosensor has been developed and its performance has been compared to a simple DM-TFET biosensor. TFET has been labeled as 'green transistor' due to its distinguished feature of attaining lower subthreshold swing [8], which implies a very low voltage change can cause a decade shift in the drain current of the device, thus when the gate action is controlled by the intrinsic properties of the biosample in the nanogap cavity, it modulates the property of the device to be used as a biosensor. Charge plasma devices do not suffer from random dopant fluctuations (RDF) due to the dopingless mechanism of creating the source and drain regions [9]. Due to the dopingless mechanism of creating source and drain regions in the device, the potential variations caused due to variability in temperature are drastically reduced, hence, the sensitivity of such devices is more or less constant irrespective of process variations [10-12].

II. Device Structure and Simulation Parameters

The cross-sectional views of SiGe source doped double gate TFET biosensor and charge plasma concept based TFET biosensor are shown in Fig. 1 (a & b). The physical dimensions of both the devices are maintained same in order to compare their performance as biosensors. The doping parameters for the conventional DG-TFET biosensor used in this simulation are: channel region doping concentration $N_D = 10^{15}$ cm^{-3}, P$^+$ source doping $N_A = 10^{19}$ cm^{-3}, N$^+$ drain doping $N_D = 10^{19}$ cm^{-3}. The physical dimensions for both the biosensors are: gate length (L_g) = 53 nm, the length of the cavity for bio-sample (L_{cavity}) = 23 nm, height of the biosample cavity (t_{cavity}) = 5.5 nm, the length of source (L_s) and drain (L_s) regions =45nm and gate metal work function is 4.1eV.

The dimensions for the charge plasma dopingless TFET biosensor are same as that of the conventional DG-TFET biosensor except for the fact that the P$^+$ source and N$^+$ drain regions are formed by depositing the metals with appropriate

978-1-5386-1357-3/17 $31.00 © 2017 IEEE

workfunctions. The primary requirements for charge plasma the concept are as follows: a) In order to form the P$^+$ source and N$^+$ drain regions the workfunctions of source and drain electrodes should be different than the workfunction of silicon in accordance to $\varphi_D < \chi_{Si} + (E_G/2)$ for drain contact and $\varphi_S > \chi_{Si} + (E_G/2)$ for source contact, where E_G is the bandgap of silicon, χ_{Si} (= 4.17eV) is the electron affinity of the bulk silicon. b) The thickness of the bulk silicon should be less than the Debye's length given by, $L_D = [(\varepsilon_{Si}.V_t)/(qN)]^{1/2}$ where ε_{Si} is the dielectric constant of the bulk silicon, V_t is the thermal voltage and N is the intrinsic carrier concentration in the bulk silicon [1]. In agreement with the charge plasma concept, 'n' type drain is formed by hafnium (φ_g = 3.9eV) electrode. Likewise to create the P$^+$ source region by induction of holes, platinum (φ_g = 5.93eV) metal has been considered at source side. The thickness of the bulk silicon is chosen to be 10 nm which is less than the approximated Debye's length to fulfill the second criteria of the charge plasma concept. To aid the process of tunneling silicon-germanium (SiGe) source have been used in both the devices with silicon to germanium composition ratio of 0.5 which improves the tunneling current. This is termed as energy band gap modulation of source region. A SiO$_2$ layer of thickness 1 nm has been inserted beneath source, drain contacts and also under the biosample nanocavity for two reasons. It prevents the formation of silicides under source and drain electrodes.

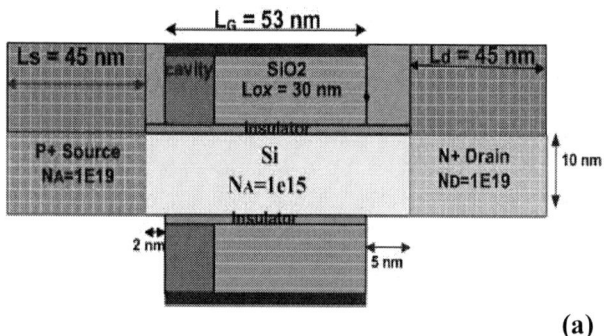

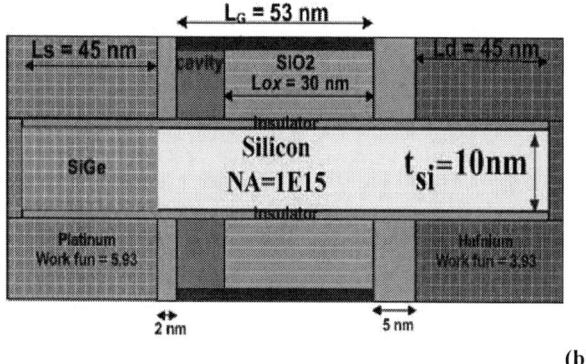

Fig. 1. Structures of (a) Conventional DG-TFET with SiGe as source material (b) Charge plasma based dopingless TFET with SiGe as source material. Note: Figures are not to the exact scale.

The double gated feature of both the devices enhances the control on the channel by enhancing the impact of dielectric constant or charge density of the biosample. The two gates operate simultaneously and thus a higher sensitivity can be achieved in the biosensors. A spacing of 2 nm and 5 nm is provided between the source electrode and nanogap cavity, and drain electrode and gate electrode, respectively. The spacing is of the order of depletion region width at source and drain junctions in the conventional DG-TFET. The gate control on the tunneling junction is dependent on the dielectric constant and charge density of the biosample in the nanogap cavity. The cavity for the biosample is considered towards the source junction as it has remarkable impact on the electrical characteristics (such as drain current) of the devices [13-15]. The n channel DG-TFET and CP-TFET bio-sensing devices are simulated in 2-D Atlas device simulator by Silvaco V5.19.20 [16]. The type of tunneling model chosen in the simulations determines the type and the mechanism of the band to band tunneling that will occur in the device. The non-local tunneling model is used in the simulations to calculate the tunneling rate and account the energy band profile throughout the devices. The local tunneling model calculates the tunneling rate at every point of the electric field where as the non-local model accesses the tunneling rate based on the spatial profile of energy band in the devices. The region (source channel interface) and the direction of tunneling have to be specified in a separate mesh when non local tunneling model is to be used. Alongside with the non-local BTBT model, charge concentration dependent SRH for recombination, Auger for minority carriers at higher current densities due to impact ionization, Fermi-Dirac statistics and field dependent mobility models have been included so that the devices can be accurately modeled to obtain near to exact results. The band-gap narrowing model has not been included in the simulations of CP-TFET biosensor as Atlas does not have a model that accounts for the band gap narrowing due to induced charge carriers [5].

III. RESULTS AND DISCUSSION

An extensive study on the effect of variations in charge density and dielectric constant of the biosample in the cavity on energy band profile, surface potential, tunneling rates and device characteristics has been carried out. For biosensing applications, it is important to analyze the sensitivity of the two devices which has been carried out in the latter sub-sections of this paper.

A. Energy Band Diagram:

To aid the mechanism of BTBT, a lower energy band gap source material has been used in both the devices which increases the tunneling rate and thereby increases the ON current of the device. By using SiGe as a source material, the valence band in source and conduction band get aligned in OFF condition thus channel has a considerable resistance between source and drain resulting in a very low OFF current as compared to thermionic devices.

Fig. 2 (a) and (b) shows the energy band profile of CP-TFET and DG-TFET biosensors in OFF and ON condition along the horizontal distance, respectively. In OFF state (V_{GS} = 0V, V_{DS} = 0.5V), the tunneling barrier width is large thus virtually no current flows through the junction. The non-local BTBT model accurately predicts the tunneling rate in ON and OFF conditions based on tunneling rate and energy band profile, hence, once gate voltage reaches to 1.5 V, band bending occurs that reduces the tunneling width as the conduction band of channel region moves below the valence band of drain region. We observed that the tunneling barrier width reduces more in CP-TFET than its counterpart device. Also vertical shift in energy bands can be observed at source side because deposition of platinum metal as a source electrode for P+ region formation in CP-TFET.

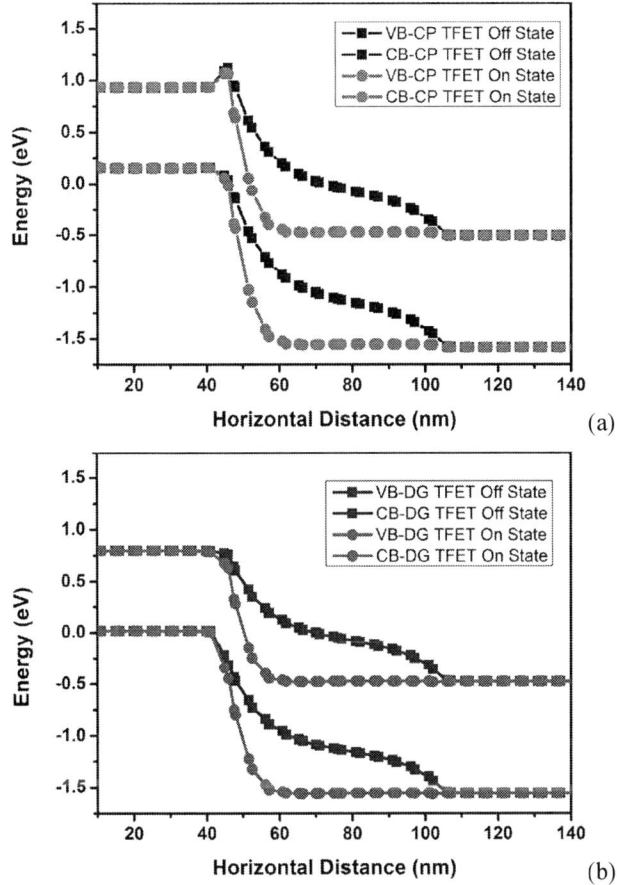

Fig. 2 Energy band profile along horizontal distance of CP-TFET and DG-TFET biosensor devices in ON-state ($V_{GS} = 1.5$ V, $V_{DS} = 0$V) and OFF-state ($V_{GS} = 1.5$ V, $V_{DS} = 0$ V) for K = 5.

Fig. 3 (a-b) show the band bending due to variation in dielectric constant (K) of biosample in cavity for DG-TFET and CP-TFET based biosensors in ON ($V_{GS}=1.5$V, $V_{DS}=0.5$V) condition. The tunneling barrier width decreases with an increase in dielectric constant and effect of dielectric constant is more prominent in charge plasma device than the conventional device. The higher dielectric constant bio-sample leads to improved coupling of gate electrode and source-channel junction thus the control over band bending and eventually the ON current is governed by the intrinsic properties of the biosample. The absence of biosample is simulated by K=1 in the cavity region.

The surface potential is an important parameter to analyze the effect of dielectric constant and charge density of biosample on band bending. The negative charge density of biomolecules influences flat-band band voltage and thus the tunneling rate of source-channel interface. Hence, it is important to analyze the surface potential below the nano-gap cavity to understand the effect of biomolecules on the electrostatic action of gate on the depletion region at source channel interface. One can clearly see most of the bending action takes place in the deletion region. Fig. 4 shows the variation of surface potential along the horizontal distance of the devices for different values of dielectric constant of biosample. In the absence of biosample (K=1), surface potential is lower due to increase in tunneling width at the source channel junction below the nanogap cavity. With the occupancy of cavity by the biomolecules with higher dielectric constant the tunneling barrier reduces, that leads to higher rate of the tunneling, hence, enhancing the electrostatic control of gate over the tunneling junction. The charge plasma TFET biosensor shows more variation in surface potential than in DG-TFET for different dielectric constants biosamples varying from 1 to 5 indicating a better control of gate over the channel

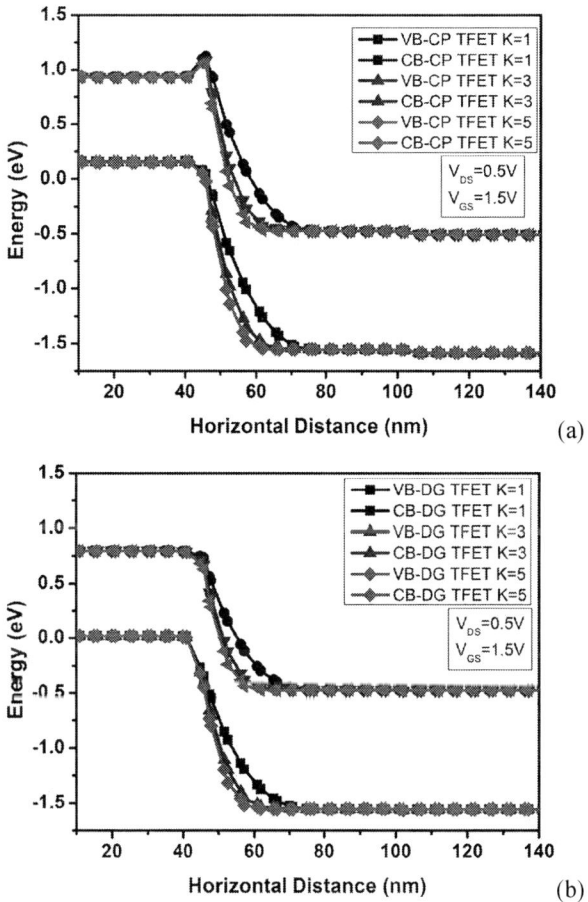

Fig. 3. Energy band profile along the horizontal distance of the device for different K in ON-state ($V_{GS} = 1.5$ V, $V_{DS} = 0.5$ V), (a) CP-TFET Biosensor (b) DG-TFET Biosensor.

B. Tunneling rate and ON current:

The fact that the tunneling length for CP-TFET and DG-TFET biosensor device architectures decreases with increasing dielectric constant of the biosample in the nanogap cavity has been already explained above.

978-1-5386-1357-3/17 $31.00 © 2017 IEEE

Fig. 5 (a-b) show the variation of e- tunneling rate at different values of charge density and dielectric constant of the biomolecules, respectively. As there is a gradual increment in the dielectric constant, the e- tunneling rate follows the same for both the devices. However the tunneling rate is more in DG-TFET biosensor than in CP-TFET suggesting that for a biosample of particular dielectric constant value the ON current is more in DG-TFET than CP-TFET biosensor.

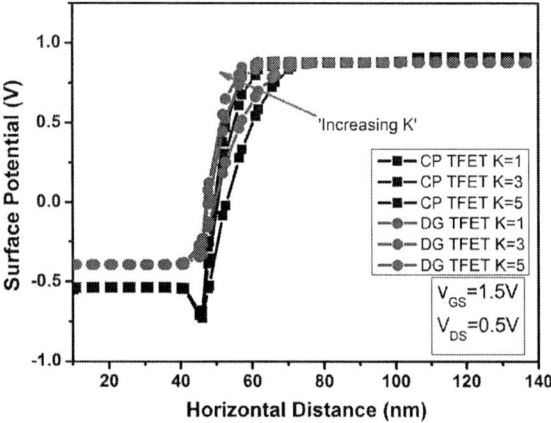

Fig. 4. Variation of surface potential horizontally along the device for different K in ON-state (V_{GS} = 1.5 V, V_{DS} = 0.5 V)

Since, in this work non-local BTBT model predicts the values for tunneling rates, the results are somewhat accurate as the tunneling rates are calculated for the energy band profile of the device. In earlier works [16] a local BTBT models were used which predicts the tunneling rates for various points according to the electric field, thus at V_{GS}=0V, it predicts a non-zero value of tunneling rate (drain current) leading to wrong estimation of the tunneling rate.

The tunneling rate variation has also been studied for the charge density contained by the biomolecules. The negative charge density modulates the flatband band voltage accordingly given by the classical equation of gate voltage required to invert the channel as [7]:

$$Vgs' = Vgs - (Vfb - \left(\frac{q\rho}{c}\right)) \qquad (1)$$

Where ρ represents the charge density of the biomolecules and C represents the total capacitance offered from the cavity. The tunneling rate change has been estimated for a decade change in charge density -1×10^{11}cm^{-3} to -1×10^{12}cm^{-3} as shown in Fig. 5 (a). The tunneling rate shows a monotonous detrimental change with an increase in negative charge density for both the devices. The rate predicted for a particular value of dielectric constant of biosample is more for DG-TFET than CP-TFET biosensor. The important thing to notice in this figure is that this decreasing rate is more prominent in CP-TFET than in DG-TFET indicating the sensitivities of the devices.

Fig. 6 (a-b) show the I_{on}/I_{off} for CP-TFET and DG-TFET biosensor devices for different values of charge density of the biomolecules in the cavity and dielectric constant of biosample is varied from 1 to 10 respectively. It can be visualized from the plot that the I_{on}/I_{off} is greater for CP-TFET than DG-TFET and the difference between I_{on}/I_{off} of the two devices keeps on

escalating as dielectric constant of biomolecules increases. This provides us a vital conclusion that CP-TFET has a better sensing capability than the conventional DG-TFET.

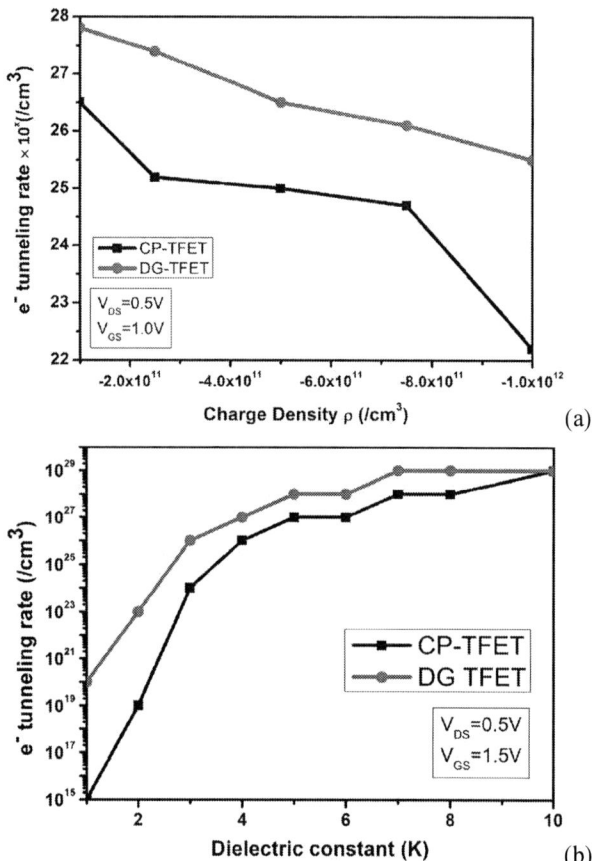

Fig. 5. e- tunneling rate for CP-TFET and DG-TFET biosensor devices in ON-state (V_{GS} = 1.5 V, V_{DS} = 0.5 V), (a) for different values of charge density of biomolecules (b) for various dielectric constants of the biosample in the nanogap cavity.

The I_{on}/I_{off} decreases with increase in negative charge densities and it follows the same trend as that decrease in CP-TFET similar to DG-TFET. The TFETs have limited ON current which causes a setback in its application as it results in smaller switching speeds. However, I_{on} can be increased by energy band gap modulation of source. The ON state current heavily depends on the gap between the source electrode and the gate electrode. The source-gate spacer thickness determines the closeness of the gate field to the tunneling path on the source side. The source spacer thickness is to be kept as minimum possible to achieve the maximum ON current, hence source to gate spacer is chosen as 2 nm in these structures.

C. Transfer and Output Characteristics:

The transfer characteristics of DG-TFET and CP-TFET biosensors for two different biosamples (simulated by dielectric constants of K=1, 5) are shown in Fig. 7. The plots follow the expected trend as that of any conventional TFET. The point worth noting is that CP-TFET has higher drain current than its counterpart DG-TFET and also OFF current for CP-TFET is lower than that of DG-TFET. The ON current for K = 5 is of the same order (1×10^{-7} A/μm) for both the devices

978-1-5386-1357-3/17 $31.00 © 2017 IEEE

for V_{DS} = 0.5 V and K = 5. The steepness of the transfer characteristics reveals good subthreshold swing of CP-TFET than DG-TFET.

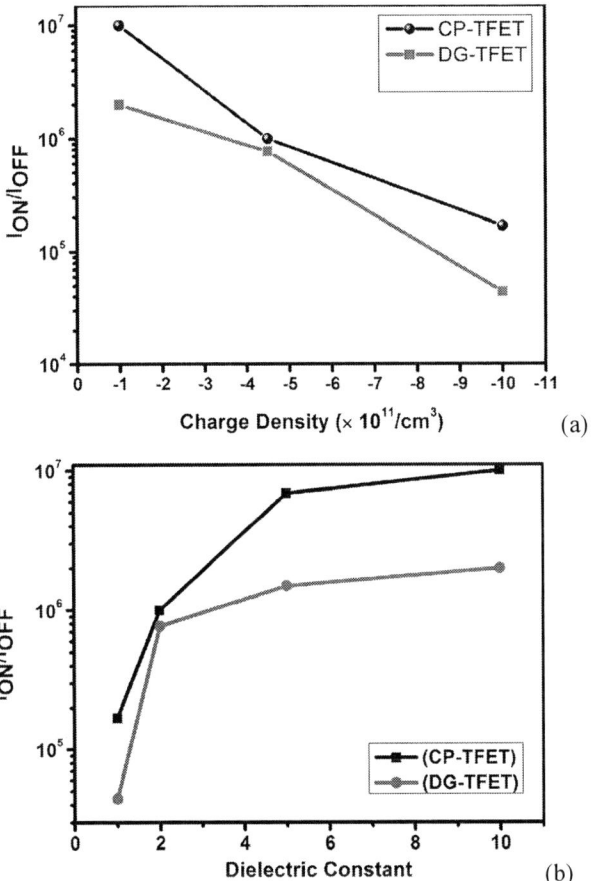

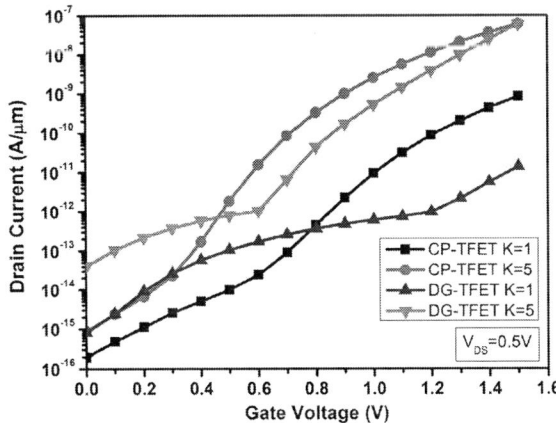

Fig. 6. Variation of I_{ON}/I_{OFF} ratio for CP-TFET and DG-TFET biosensor for (a) different values of charge density of biomolecules (b) different values of dielectric constants of the biosample in the nanogap cavity.

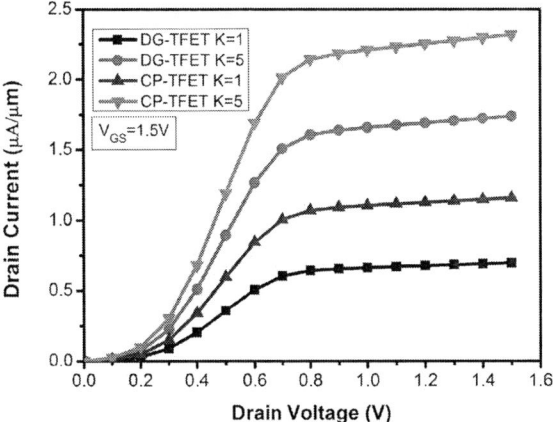

Fig. 8. Output characteristics of CP-TFET and DG-TFET biosensor devices for different dielectric constants at V_{GS} = 1.5 V.

Fig. 8 shows the output characteristics for CP-TFET and DG-TFET biosensors with different biosample dielectric constants. The output characteristics are exponential for low drain voltages and constant as drain voltages increase. The device physics for such plot can be explained as when V_{DS} is low accumulation mode of operation is dominant in the device thus increasing V_{DS} affects the horizontal electric field at the source channel interface increasing the tunneling probability. After a certain value of V_{DS}, the channel is fully depleted, and I_{DS} ceases to increase any further and attains a constant value thus saturating the drain current with respect to V_{DS}. The tunneling process becomes independent of drain to source voltage in the saturation region. It is evident from output characteristics with different dielectric constant that the saturation current is more for CP-TFET device than DG-TFET for any of the biosample.

D. Sensitivity Analysis:

In order to compare the performance of the two biosensor devices presented in this paper, it becomes important to analyze their relative behavior and simultaneously the relative sensing capability. For a biosensor to be efficient, the sensitivity towards a particular biosample should be high. The sensitivity in this work is the function of drain currents for various dielectric constants of the biomolecules and is given by:

$$S = (I_{bio} - I_{air})/I_{air} \qquad (2)$$

Where I_{bio} and I_{air} represent the drain currents when the cavity is filled with biosample and no biosample is present in the cavity.

Fig. 9 shows sensitivity against gate voltage for both the devices. The sensitivity plots are obtained for a dielectric constant (K = 5) for the biomolecules present in the cavity. It can be seen that the sensitivity has low values for smaller gate voltages and as gate voltage is increased sensitivity attains higher order of magnitude. It is worth noting that the sensitivity of CP-TFET biosensor is higher than DG-TFET biosensing device, it is mainly due to higher ON current in ON state when cavity is completely filled with biomolecules.

Fig. 7. Transfer characteristics for CP-TFET and DG-TFET biosensor for two different biosamples when V_{DS} = 0.5V

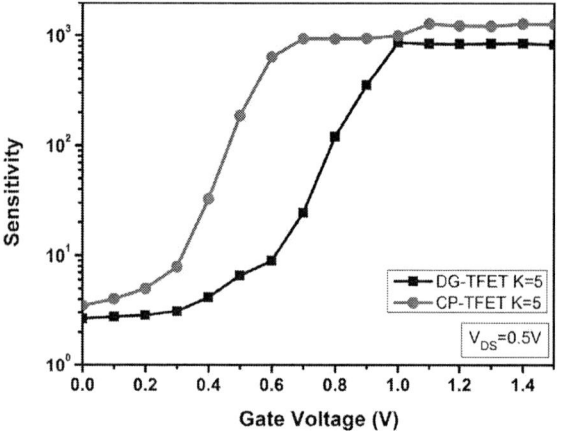

Fig. 9. Plots of sensitivity versus gate voltage (V_{GS}) at K=5 for CP-TFET and DG-TFET at V_{DS}=0.5V.

IV. CONCLUSION

The performance of SiGe source charge plasma based biosensors is evaluated in terms of ON current, I_{ON}/I_{OFF} ratio, and sensitivity. The incorporation of lower bandgap material such as SiGe in source region of charge plasma TFET achieves maximum tunneling probability at low gate voltages and shows reduced ambipolar effects under negative gate bias. The proposed device based on charge plasma simplifies the TFET fabrication procedure and lowers the thermal budget. The simulation results also favors the idea of development ultrasensitive biosensing device based on charge plasma. The variations in dielectric constants show excellent variation in various parameters such as ON current sensitivity of the device.

ACKNOWLEDGEMENT

The authors acknowledge the Department of Science and Technology (DST) and science and engineering research board (SERB) Government of India for financial support under early career research award (ECRA) project no. ECR/2017/000216. The authors would also like to acknowledge the assistance from VLSI LAB under special man power development programme (SMDP) MNIT Jaipur.

REFERENCES

[1] R. J. E. Hueting, B. Rajasekharan, C. Salm, and J. Schmitz, "Charge plasma p-n diode," IEEE Electron Device Lett., vol. 29, no. 12, pp. 1367–1368, Dec. 2008.J. Clerk Maxwell, A Treatise on Electricity and Magnetism, 3rd ed., vol. 2. Oxford: Clarendon, 1892, pp.68-73.

[2] K. Boucart and A. M. Ionescu, "Double gate tunnel FET with high-k gate dielectric," IEEE Trans. Electron Devices, vol. 54, no. 7, pp. 1725–1733, Jul.

[3] D. Singh, S. Pandey, K. Nigam, D. Sharma, D. S. Yadav and P. Kondekar, "A Charge-Plasma-Based Dielectric-Modulated Junctionless TFET for Biosensor Label-Free Detection," in *IEEE Transactions on Electron Devices*, vol. 64, no. 1, pp. 271-278, Jan. 2017.

[4] R. Narang, K. V. S. Reddy, M. Saxena, R. S. Gupta, and M. Gupta, "A dielectric-modulated tunnel-FET-based biosensor for label-free detection: Analytical modeling study and sensitivity analysis," IEEE Trans. Electron Devices, vol. 59, no. 10, pp. 2809–2817, Oct. 2012.

[5] . M. J. Kumar and S. Janardhanan, "Doping-less tunnel field effect transistor: Design and investigation," IEEE Trans. Electron Devices, vol. 60, no. 10, pp. 3285–3290, Oct. 2013.

[6] S. Kanungo, S. Chattopadhyay, P. S. Gupta, and H. Rahaman, "Comparative performance analysis of the dielectrically modulated full-gate and short-gate tunnel FET-based biosensors," IEEE Trans. Electron Devices, vol. 62, no. 3, pp. 994–1001, Mar. 2015.

[7] R. Narang, M. Saxena, R. S. Gupta, and M. Gupta, "Dielectric modulated tunnel field-effect transistor—A biomolecule sensor," IEEE Electron Device Lett., vol. 33, no. 2, pp. 266–268, Feb. 2012.

[8] C. Hu, "Green transistor as a solution to the IC power crisis," in Proc. 9th Int. Conf. Solid-State Integr.-Circuit Technol., Oct. 2008, pp. 16–20.

[9] A. Lahgere, C. Sahu, and J. Singh, "PVT-aware design of dopingless dynamically configurable tunnel FET," IEEE Trans. Electron Devices, vol. 62, no. 8, pp. 2404–2409, Aug. 2015.

[10] C. Sahu and J. Singh, "Charge-plasma based process variation immune junctionless transistor," IEEE Electron Device Lett., vol. 35, no. 3, pp. 411–413, Mar. 2014.

[11] C. Sahu and J. Singh,"Potential benefits and sensitivity analysis of dopingless transistor for low power applications," IEEE Trans. Electron Devices, vol. 62(3), 729-735, 3/2015.

[12] K. Cecil, J. Singh, Influence of germanium source on dopingless tunnel-FET for improved analog/RF performance Superlattices Microstruct., 101 (2017), pp. 244-2

[13] A. S. Verhulst, W. G. Vandenberghe, K. Maex, S. De Gendt, M. M. Heyns, and G. Groeseneken, "Complementary silicon-based hetero-structure tunnel-FETs with high tunnel rates," IEEE Electron Device Lett., vol. 29, no. 12, pp. 1398–1401, Dec. 2008.

[14] B. Jang and A. Hassibi, "Biosensor systems in standard CMOS processes: Fact or fiction?" IEEE Trans. Ind. Electron., vol. 56, no. 4, pp. 979–985, Apr. 2009.

[15] H. Im, X.-J. Huang, B. Gu, and Y.-K. Choi, "A dielectric-modulated field-effect transistor for biosensing," Nature Nanotechnol., vol. 2, pp. 430–434, Jul. 2007.

[16] K. Boucart and A. M. Ionescu, "Double gate tunnel FET with ultrathin silicon body and high-k gate dielectric," in Proc. ESSDERC, 2006, pp. 383–386

[17] ATLAS Device Simulation Software, Silvaco Int., Santa Clara, CA, USA, 2016.

Neutralization of the Effect of Hardware Trojan in SCADA system using Selectively placed TMR

Nagendra Babu Gunti and Karthikeyan Lingasubramanian
Department of Electrical and Computer Engineering
University of Alabama at Birmingham
Birmingham, Alabama, 35294
Email: nagsphd@uab.edu and klinga@uab.edu

Abstract—Supervisory Control And Data Acquisition (SCADA) systems are the primary technology used to automate our critical infrastructure and major industries in order to improve their efficiency. Their dependability is challenged by probable vulnerabilities in the core computing system. These vulnerabilities can be appear on both front (software) and back (hardware) ends of the computing system. While the software vulnerabilities are well researched and documented, the hardware threats are normally overlooked. However, with hardware-inclusive technological evolutions like Cyber-Physical Systems and Internet-of-Things, hardware vulnerabilities should be addressed appropriately. Hardware Trojans are malicious alterations in the systems that leak confidential information or disable the entire system. In order to achieve a countermeasure, we propose to neutralize the effect of Hardware Trojans through a Triple Modular Redundancy (TMR) based methodology. In order to address the inevitable overhead on area, TMR will be implemented only on select paths of the system i.e., equally probable output paths which are found vulnerable to Trojan placement. In this work, we propose to neutralize the effect of hardware Trojan based vulnerabilities in SCADA systems, with the use of selectively placed TMR methodology.

Keywords—Hardware Trojans, Neutralization, TMR, SCADA system

I. INTRODUCTION

Every major infrastructure and industry in modern world is operated by SCADA systems. Such autonomous functionality has enabled efficient operation and increased yield. However, the same autonomy also poses possible threats through vulnerabilities in the computing systems that form the heart of SCADA systems. One of the largest blackouts in American history, that happened in August 2003, might have been caused by a cyber-attack. The emergence of the usage of globalized business model for production of electronic devices has resulted in hardware security and trust issues. Without trusted foundries, the systems they support cannot necessarily be expected to perform as specified and may even be susceptible to attack by a malicious adversary. A Hardware Trojan is a covert malicious, modification of an electronic circuit or design, which results in undesired behavior of an electronic device [1]. These alterations can provide a back door entry to a SoC or an Embedded system. An adversary can also utilize Trojans to leak sensitive information from a system or can even deny providing the service during the execution of critical applications.

The primary channel for attacks in SCADA systems is through the computing system, both at the front-end (software) and back-end (hardware). While software based vulnerabilities are well documented, there has been very little emphasis on hardware. The computing hardware in SCADA systems are built using electronic components based on very large scale Integrated Circuits (ICs) technology. The incredible demand of ICs has globalized its industry, where the manufacturing foundries are situated all around the world. This provides a wider possibility for adversaries to infiltrate these foundries and get access to the hardware manufacturing process. Under such circumstances, these adversaries can add malicious circuitry, called hardware Trojans, to the existing design and make them stealthy. These hardware Trojans can provide back door entry for potential attacks, thus making the computing system vulnerable. In the advent of technological evolutions in the form of Cyber-Physical Systems and Internet-of-Things, such hardware vulnerabilities cannot be overlooked. Since SCADA systems are going to be the stepping- stone towards these advance technologies, it is important to understand hardware based vulnerabilities, like hardware Trojans, in them. However, such examination will be expensive if it relies on performing tests on actual SCADA systems. To handle such security vulnerability, neutralization of the threats without depending on its detection will be necessary.

In order to achieve such a countermeasure, we propose to use redundancy based strategy employing Triple Modular Redundancy (TMR) which can neutralize the effect of Trojans without the need of detection. Though, this methodology has significant area and power overhead, it will be a useful method for critical applications like Industrial Control Systems which are not constrained by them. Also, such application specific strategies for security are essential since it is implausible to devise a single countermeasure for every application [2], [3]. We have also diagnosed the outputs of the digital circuit to find the ones which are vulnerable to Trojan placement and Hence TMR can be implemented on such paths to reduce the area and power overheads. We achieve this by using a probabilistic model to realize security aware TMR scheme to narrow the options for Trojan insertion and to aid the detection of Trojans at the post-manufacturing phase.

In digital system, TMR is a fault tolerant redundancy scheme which has traditionally been used to mask the unpredictable malfunction of a system due to aspects like process variations. To utilize TMR, digital systems have three copies of same subsystems which perform identical functions. The outputs of these copies are fed to a majority voter which transmits the final output by neutralizing the effect of Trojan as

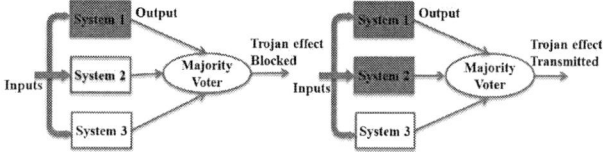

Fig. 1. (a) TMR with Trojan effect blocked (b) TMR with Trojan effect transmitted

TABLE I.   SIMULATION RESULTS OF C432

Output Gates	Golden Circuit	Circuit Under Test	TDS
G223	92.4915	91.6417	0.850
G329	75.9875	75.4677	0.520
G370	63.6604	63.3872	0.273
G421	85.3448	84.6379	0.707
G430	52.1914	52.1476	0.044
G431	49.0048	49.0247	0.020
G432	48.1379	48.1752	0.037

shown in Fig. 1(a). In a TMR induced system, the Trojan has to be placed in at least two copies in order to be effective as shown in Fig. 1(b). Such placement makes the Trojan size relatively big, therefore making it relatively easy to detect through side channel based techniques. Therefore, this technique forces the adversary to place the Trojan on non-TMR paths. Trojans placed in these non-TMR paths should be easier to detect through traditional logic testing methods. The objective of this work is to understand the behavior of the primary outputs in reference to Trojan detection sensitivity. Using the proposed probabilistic model it is shown that the Trojans placed in the path leading to predictable outputs have better detection sensitivity and therefore, in our security scheme TMR is performed on the paths which lead to random unpredictable outputs [10], [11].

Though we know that the presence of TMR is helpful in neutralizing the effect of Trojans, it is not beneficial to implement TMR on the whole system. Hence, in order to decide the placement of TMR, probability distribution of the primary outputs of the circuit is analyzed to decide on the outputs which are vulnerable to Trojans and transient errors. All the simulation results obtained with probabilistic modeling are validated using a functional simulator. In order to analyze the behavior of outputs in the presence of Trojan, the C432 is simulated with probabilistic model whose results are tabulated in Table I. It is observed that the predictable outputs such as G223, G421, G329 and G370 are having higher TDS than unpredictable outputs G430, G431 and G432. We can infer that TDS of unpredictable outputs is ten times lower than predictable outputs making them vulnerable to Trojan insertion. The unpredictability of equally probable outputs makes it difficult to detect transient error on these paths, so placing TMR on these paths will neutralize the effect of Trojan and transient errors. Every major infrastructure and industry in modern world is operated by SCADA systems. Such autonomous functionality has enabled efficient operation and increased yield. However, the same autonomy also poses possible threats through vulnerabilities in the computing systems that form the heart of SCADA systems. One of the largest blackouts in American history, that happened in August 2003, might have been caused by a cyber-attack. The

primary channel for attacks in SCADA systems is through the computing system, both at the front-end (software) and back-end (hardware). While software based vulnerabilities are well documented, there has been very little emphasis on hardware. The computing hardware in SCADA systems are built using electronic components based on very large scale Integrated Circuits (ICs) technology. Under globalization of manufacturing circumstances, an adversary can add malicious circuitry, called hardware Trojans, to the existing design and make them stealthy. These hardware Trojans can provide back door entry for potential attacks, thus making the computing system vulnerable. In the advent of technological evolutions in the form of Cyber-Physical Systems and Internet-of-Things, such hardware vulnerabilities cannot be overlooked. In this work, we propose to neutralize the effect of hardware Trojan based vulnerabilities in SCADA systems using the proposed TMR methodology.

The specific contributions of this work are as follows.

- We present a completely virtual platform for studying hardware Trojan based vulnerabilities in SCADA systems.

- We present 3 probable threat models based on the ability of a hardware Trojan to disable the controllability of the computing system.

- We have shown that even a seemingly harmless threat posed by a hardware Trojan, can completely destabilize the SCADA system.

- We were successful in neutralizing the effect of Hardware Trojans in SCADA systems thereby securing the system from threat models.

II. BACKGROUND

A Hardware Trojan is a covert, malicious modification of an electronic circuit or design, which results in undesired behavior of an electronic device [6]. These alterations can provide a back door entry to a computing system. An adversary can also utilize Trojans to leak sensitive information from a system or can even deny providing the service during the execution of critical applications. This can potentially have serious consequences in critical applications spanning the domains of communications, space, military and nuclear facilities. The research and investigation of hardware Trojans are primarily conducted by IC engineers in a distributed fashion, focusing mostly on their detection techniques at post manufacturing stages. They are classified into two categories: 1) logic testing based and 2) side-channel based. The logic testing based detection depends on rare conditions to activate Trojans occurring at internal nodes of the circuit under test [5], [6]. They have very large Hardware Trojan design space and an extremely large number of input- output combinations which are required for testing. This makes test generation computationally infeasible due to time constraints. Side-channel based techniques involve observing the effect of Hardware Trojan on one or more physical parameter(s) such as transient current, leakage current or delay [8], [9], [12], [13]. And, side channel based approaches are affected by large process-induced parameter variations [14]. Such inadequacies in detection techniques has made it important to understand the

978-1-5386-1357-3/17 $31.00 © 2017 IEEE

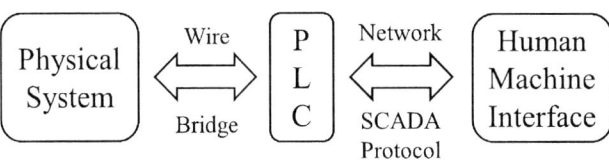

Fig. 2. SCADA System

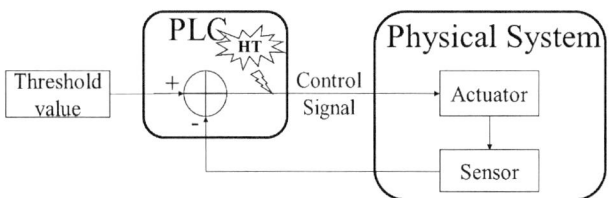

Fig. 3. Closed-loop control formed between PLC and Physical system; PLC hardware is vulnerable to presence of Hardware Trojan (HT) that can affect the control signal output.

presence on hardware Trojans, in SCADA systems, at run time in order to device countermeasures. Such studies obviously need the entire system to be considered instead of just the computing hardware in order to deal with hardware Trojans. In [10], it is discussed that TMR can be used to mask the effect of Trojans and that forces the adversary to place the Trojans in multiple copies which aids detection through side-channel analyses. In order to optimally use TMR it was suggested to be implemented only on equiprobable outputs since they are unpredictable and therefore vulnerable. However, this work did not discuss about the interplay between faults due to transient errors and Trojans. In this work, we propose to neutralize the effect of Trojan through a fault sensitive scheme using multi-level TMR to compensate the limitation of single level TMR on equiprobable paths.

III. SCADA SYSTEM

A supervisory control and data acquisition (SCADA) system is used to control industrial systems. Major components of a SCADA system include a human machine interface (HMI), which allows a user to monitor and manage the system; a programmable logic controller (PLC) which serve to control certain aspects of the physical infrastructure, such as sensors and actuators; a physical system which is being controlled. The communication between these components is achieved through a wire bridge which conveys the sensed data from the physical system to PLC; a Network SCADA protocol like MODBUS/TCP to transfer data between PLC and HMI. The PLC will be the computing system that forms the central core and plays a vital part in stabilizing the SCADA system as sown in Fig. 2.

IV. CASE STUDY: GAS PIPELINE SYSTEM

Gas pipelines are important infrastructures, not only for the power generation sector, but also for other sectors such as petrochemicals, chemicals, transportation, manufacturing, and district heating. The United States Patriot Act defines critical infrastructure as ?systems and assets, whether physical or virtual, so vital to the United States that the incapacity or destruction of such systems and assets would have a debilitating impact on security, national economic security, national public health or safety, or any combination of these matters.? [11]. By this definition, gas pipelines can be considered critical infrastructures, and therefore deserve special attention in terms of cyber-security. The Gas Pipeline Testbed is a small closed pipeline that tries to mimic the behavior of a real gas pipeline. An electrical pump increases the pressure inside the pipeline by pumping air into it. There are also pressure sensors constantly measuring the pressure inside the pipeline. A solenoid valve alleviates the internal pressure when opened. The pump, the solenoid valve and the digital pressure sensor are connected

to a Programmable Logic Controller (PLC) that controls the system. The role for the PLC is to maintain pressure between a high and a low setpoint by turning on and off the pump. There is also a manual operation mode that enables the user to manually control the pump and the valve from the HMI.

A. Hardware Trojan Implementation

The functional relation between the PLC and Physical system will form a closed-loop as shown in Fig. 3 . The PLC will control the physical system based on its dynamic changes, which are relayed through the sensor. The control mechanism is established by comparing this dynamic response from the physical system with a threshold value. This relational comparison will trigger the PLC to provide a control signal to the actuator of the physical system thus maintaining safe functionality. The relational comparison followed by subsequent arithmetic operation and production of control signal will be performed in the PLC and so it will be built as a digital computing hardware. As discussed before such hardware can be vulnerable to inclusion of hardware Trojans. As shown in Fig. 3, such Trojans can be designed to affect the control signal and damage the capability of PLC to control the physical system. This will destabilize the entire SCADA system leading to potential hazardous situations. In this work, we have tested the effect of such a vulnerability by designing the PLC as a digital logic circuit and implementing a hardware Trojan that will alter the digital control signal. In order to study the proposed threat models, we made the Trojan circuitry accessible and controllable.

B. Controller Circuit Design

The controller that controls the gas pipeline system is implemented as a digital circuit such that the study of Hardware Trojans is possible. This controller is designed using a greater than comparator, less than comparator and an OR gate as shown in Fig. 4 . The inputs to this controller are minimum pressure point, maximum pressure and the pressure in the pipeline sensed by the sensor whereas the output is the pump control. The output of greater than comparator goes ?High? when the minimum pressure point is greater than or equal to the pressure in the pipeline thereby turning ON the pump control, which increases the pressure in the system. The output of the less than comparator goes ?High? when the pressure in the pipeline is lesser than or equal to the maximum pressure point thereby turning ON the pump control, to increase the pressure in the system. Thus, the pressure in the pipeline system will be maintained between the minimum and maximum pressure points as inputted by the user.

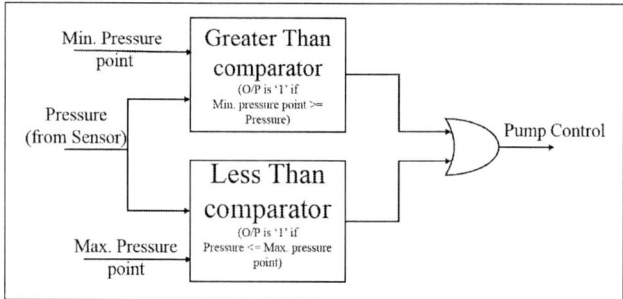

Fig. 4. Design of Controller to control pump in the pipeline

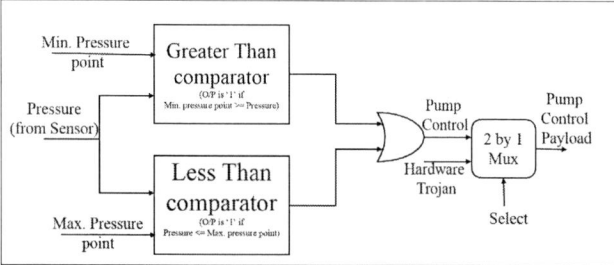

Fig. 5. Hardware Trojan in the Controller using multiplexer

C. Hardware Trojan Design

Now, we design a Hardware Trojan that can cause the threats to the system as discussed in the above sections. Here, we used a 2 by 1 multiplexer which outputs the Pump Control when Select is '0' and outputs Hardware Trojan when it is '1' as shown in Fig. 5. The output of 2 by 1 multiplexer, Pump Control Payload is the payload (signal that gets affected with Trojan) of the Hardware Trojan.

D. Hardware Trojan Taxonomy

This work focuses on Hardware threats which are inserted at design phase or fabrication phase, at gate level that are internally triggered which changes the functionality of the circuit as highlighted in Fig.6.

E. Threat Models

- Threat 1: Turn off PLC permanently. In this threat model, the controller is completely taken over by Hardware Trojan and the system will be destabilized.

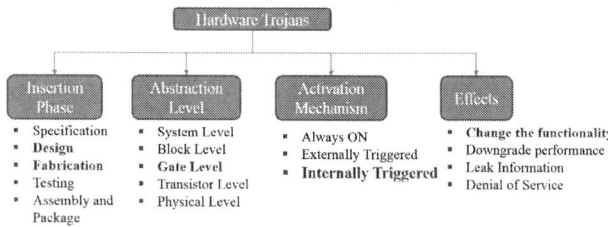

Fig. 6. Hardware Trojan Taxonomy

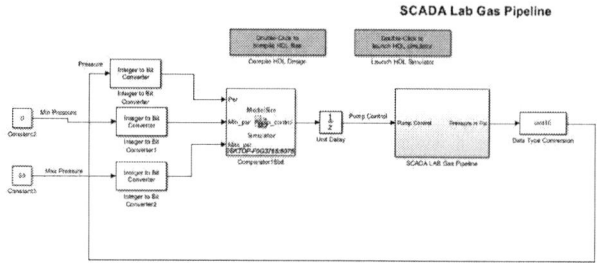

Fig. 7. Experimental setup to study the effect of Hardware Trojan on SCADA gas pipeline system

- Threat 2: Turn off PLC for long period. In this threat model, the Trojan gets activated after some period and remains active, which will cut off the controller.

- Threat 3: Turn off PLC for short period. In this threat model, Hardware Trojan is active just for a short period and hence system should get back to its normal state after the deactivation of Trojan This threat model will help us to study the time the system will need to regain its stability.

V. EXPERIMENTAL RESULTS

The experimental setup consists of a SCADA gas pipeline model created in a Simulink platform and the controller, which controls the pressure in the pipeline, is co-simulated using Modelsim in Simulink.

A. Experimental Setup and Procedure

The experimental setup consists of a SCADA gas pipeline model created in a Simulink platform and the controller, which controls the pressure in the pipeline, is co-simulated using Modelsim in Simulink. The controller is designed as digital circuit in VHDL language, which gives the flexibility to customize and study the effect of Hardware Trojans on the system. This controller circuit is simulated using Modelsim and its output is connected as input to the Simulink model of SCADA gas pipeline system as shown in the Fig. 7.

We have simulated the above setup to maintain the pressure in system between 0 and 50. The output of the system is shown in Fig. 8 in which the Trojan is inactive. It is observed that the pump control turns OFF when the pressure in the pipeline goes above 50 and turns ON when it goes below 50, thus maintaining the pressure within the minimum and maximum limits in the pipeline. In this case, the Trojan is deactivated and hence the effect of Trojan is not observed in spite of its presence which makes Hardware Trojans stealthy and difficult to detect.

B. Threat Model 1

In this threat model, we have studied the effect of Hardware Trojan on the system when it is active. In this case, the Trojan is active after 10 seconds and remains active for the rest of the time. We can observe from the simulation results shown in Fig. 10 that the Trojan makes the pump control to remain ON beyond the maximum limit (50) and thus the pressure in the

978-1-5386-1357-3/17 $31.00 © 2017 IEEE

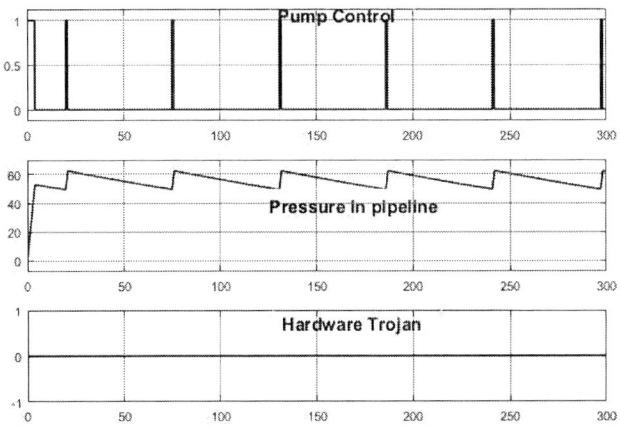

Fig. 8.   Pressure in the SCADA gas pipeline when Trojan is inactive

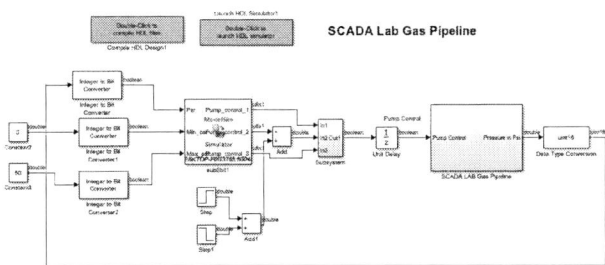

Fig. 9.   TMR implemented SCADA gas pipeline system

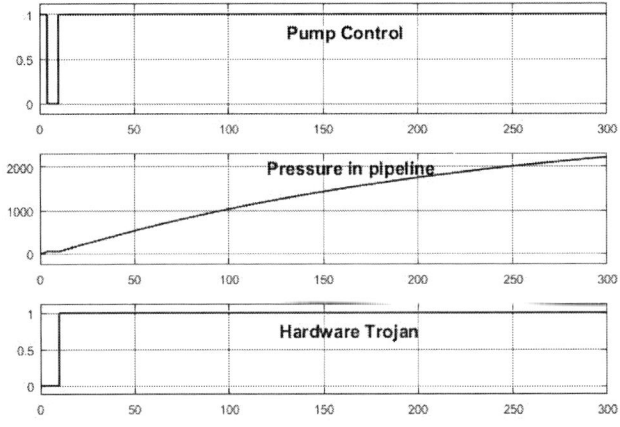

Fig. 10.   Pressure in the SCADA gas pipeline when Trojan is always active

pipeline keeps increasing rapidly. We have implemented TMR on the digital controller of the SCADA system as shown in Fig. 9. We have implemented TMR on SCADA system and neutralized the effect of Hardware Trojan and thus securing the system from Threat model 1 as shown in Fig.11.

C. Threat Model 2

In this threat model, the Trojan is activated after 175 seconds and remains active. The simulation result is shown in the Fig. 12 , it is inferred that the pressure is maintained

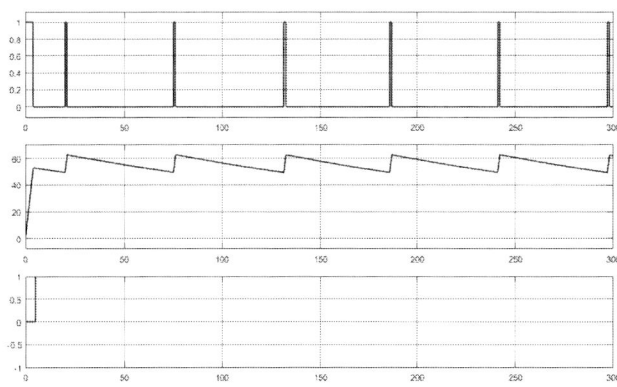

Fig. 11.   Neutralizing the effect of Hardware Trojan (Threat Model 1)

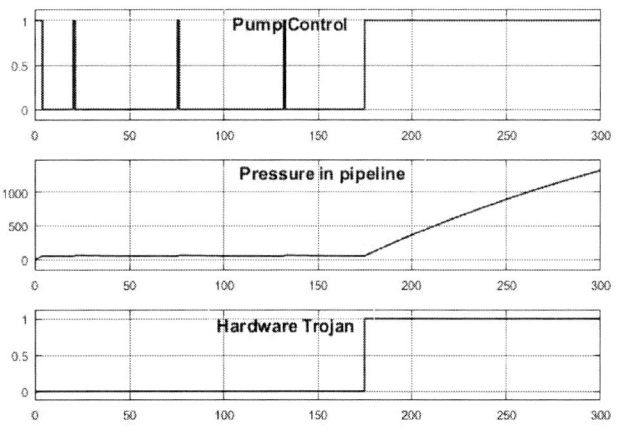

Fig. 12.   Pressure in the SCADA gas pipeline when Trojan is activated after 175 seconds

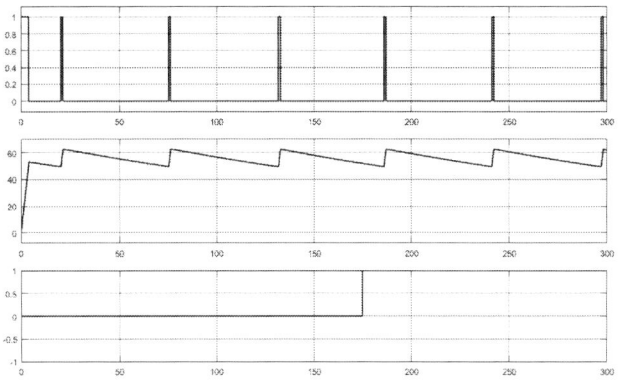

Fig. 13.   Neutralizing the effect of Hardware Trojan (Threat Model 2)

within the limits until the Trojan is active however once the Trojan is active, the system gets destabilized and the pressure keeps mounting on. We have implemented TMR on SCADA system and neutralized the effect of Hardware Trojan and thus securing the system from Threat model 2 as shown in Fig.13.

978-1-5386-1357-3/17 $31.00 © 2017 IEEE

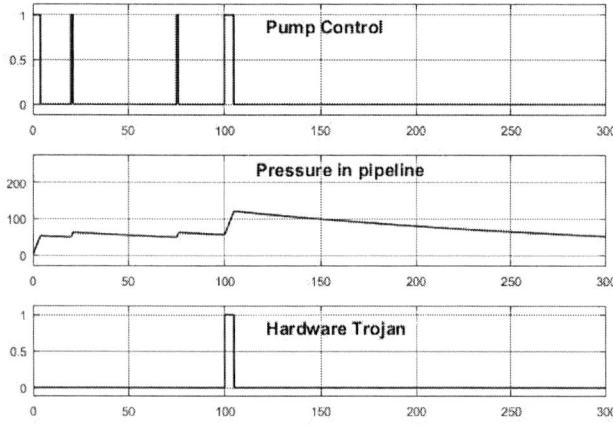

Fig. 14. Pressure in the SCADA gas pipeline when Trojan is active for a short period (5 seconds)

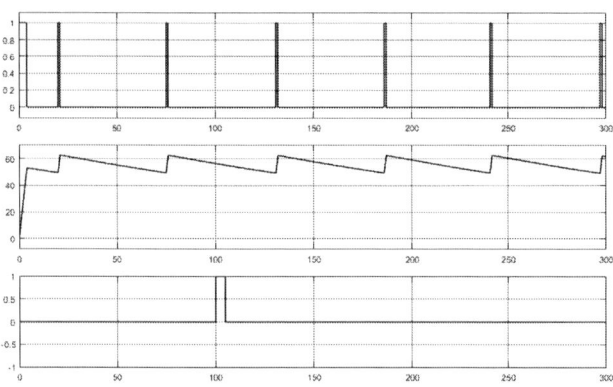

Fig. 15. Neutralizing the effect of Hardware Trojan (Threat Model 3)

D. Threat Model 3

In this threat model, the Trojan is active for a small period of time and we will study how the system stabilizes once the Trojan is deactivated. The Trojan is active from 100 to 105 seconds and during this period the system is unstable as the pressure keeps increasing beyond maximum limit. However, once the Trojan is inactive the system regains its stability as shown in Fig. 14 . This threat model clearly shows the typical stealthy and rare function of the Trojan which can create a hazardous situation. We have implemented TMR on SCADA system and neutralized the effect of Hardware Trojan and thus securing the system from Threat model 3 as shown in Fig.15.

VI. CONCLUSION

The inevitable usage of third party foundries by the globalized industry has resulted in hardware security and trust issues in ICs used in critical infrastructure. In this work, we have devised a security scheme which can neutralize the effect of hardware Trojans. We have shown that the control in design is needed for the circuit paths that lead to unpredictable outputs with random probability distribution. Using this synopsis we proposed a TMR based scheme where redundancy is selectively introduced on the unpredictable paths thereby enabling

neutralization of hardware Trojans. Also, such a design will limit the options for the adversary to insert Trojans. We presented 3 threat models based on this capability of the Trojan and implementing the proposed TMR methodology can neutralize their effects.

REFERENCES

[1] R. Chakraborty, S. Narasimhan, and S. Bhunia, "Hardware Trojan: Threats and emerging solutions," in *High Level Design Validation and Test Workshop, 2009. HLDVT 2009. IEEE International*, Nov 2009, pp. 166–171.

[2] M. Tehranipoor and F. Koushanfar, "A survey of Hardware Trojan taxonomy and detection," *Design Test of Computers, IEEE*, vol. 27, no. 1, pp. 10–25, Jan 2010.

[3] N. Jacob, D. Merli, J. Heyszl, and G. Sigl, "Hardware Trojans: current challenges and approaches," *Computers Digital Techniques, IET*, vol. 8, no. 6, pp. 264–273, 2014.

[4] N. B. Gunti, A. Khatri, and K. Lingasubramanian, "Realizing a security aware triple modular redundancy scheme for robust integrated circuits," in *Very Large Scale Integration (VLSI-SoC), 2014 22nd International Conference on*, Oct 2014, pp. 1–6.

[5] N. B. Gunti and K. Lingasubramanian, "Effective usage of redundancy to aid neutralization of hardware trojans in integrated circuits," *Integration, the VLSI Journal*, vol. 59, no. Supplement C, pp. 233 – 242, 2017. [Online]. Available: http://www.sciencedirect.com/science/article/pii/S0167926017300421

[6] R. Chakraborty, F. Wolff, S. Paul, C. Papachristou, and S. Bhunia, "MERO: A Statistical approach for Hardware Trojan Detection," in *Cryptographic Hardware and Embedded Systems - CHES 2009*, ser. Lecture Notes in Computer Science, C. Clavier and K. Gaj, Eds. Springer Berlin Heidelberg, 2009, vol. 5747, pp. 396–410.

[7] F. Wolff, C. Papachristou, S. Bhunia, and R. Chakraborty, "Towards Trojan-free trusted ICs: Problem analysis and detection scheme," in *Design, Automation and Test in Europe, 2008. DATE '08*, March 2008, pp. 1362–1365.

[8] D. Agrawal, S. Baktir, D. Karakoyunlu, P. Rohatgi, and B. Sunar, "Trojan detection using IC fingerprinting," in *Security and Privacy, 2007. SP '07. IEEE Symposium on*, May 2007, pp. 296–310.

[9] J. Aarestad, D. Acharyya, R. Rad, and J. Plusquellic, "Detecting trojans through leakage current analysis using multiple supply pads," *Information Forensics and Security, IEEE Transactions on*, vol. 5, no. 4, pp. 893–904, Dec 2010.

[10] B. Cha and S. K. Gupta, "Trojan detection via delay measurements: A new approach to select paths and vectors to maximize effectiveness and minimize cost," in *Design, Automation Test in Europe Conference Exhibition (DATE), 2013*, March 2013, pp. 1265–1270.

[11] J. Li and J. Lach, "At-speed delay characterization for IC authentication and Trojan Horse detection," in *Hardware-Oriented Security and Trust, 2008. HOST 2008. IEEE International Workshop on*, June 2008, pp. 8–14.

[12] S. Borkar, T. Karnik, S. Narendra, J. Tschanz, A. Keshavarzi, and V. De, "Parameter variations and impact on circuits and microarchitecture," in *Design Automation Conference, 2003. Proceedings*, June 2003, pp. 338–342.

2017 IEEE International Symposium on Nanoelectronic and Information Systems

Fault Sensitive Neutralization of Hardware Trojans using Multi-level Triple Modular Redundancy Scheme

Nagendra Babu Gunti and Karthikeyan Lingasubramanian
Department of Electrical and Computer Engineering
University of Alabama at Birmingham
Birmingham, Alabama, 35294
Email: nagsphd@uab.edu and klinga@uab.edu

Abstract—Hardware Trojans are malicious alterations in Integrated Circuits (ICs) that leaks confidential information or disables the entire IC. The detection of these Trojans is performed through logic or side channel based testing. Under sub-nm technologies this kind of detection will face more problems due to process variations. Therefore there is a need to address these vulnerabilities without relying on their detection. In this work we propose to neutralize the effect of Trojans by leveraging fault tolerant design methodologies like Triple Modular Redundancy (TMR) for security. Under this context a Fault can also be caused by a man made Trojan as compared to natural sources. To comprehensively address all these faults single TMR implementation is not helpful. In this work, a more efficient TMR approach to neutralize Trojans in the presence of natural faults is accomplished through multiple levels of TMR implemented after logic partitioning. We have diagnosed this approach for ISCAS'85 benchmark circuits of nodes within 900 and observed an improved Trojan Neutralization Rate (TNR). Our experiments show that the interplay of Trojan and natural faults actually reduces TNR. We have also shown that the Implementation of 3 levels of TMR considerably improves TNR.

Index Terms—Hardware Trojans, Neutralization, TMR, Transient errors

I. INTRODUCTION

The daily life is aided by the automation, monitoring or computational power provided by Electronic systems. These systems which are based on digital circuits are expected to be highly dependable and trustworthy. The extensive usage of them in almost all critical sectors including finance, military, and industry has only increased the expectations of their security. To be a part of a versatile infrastructure these systems involve integration of multiple components as System On Chip (SoC) or an Embedded system. Such applications contain several hardware elements which are manufactured in global foundries. The emergence of the usage of globalized business model for production of electronic devices has resulted in hardware security and trust issues. Without trusted foundries, the systems they support cannot necessarily be expected to perform as specified and may even be susceptible to attack by a malicious adversary. A Hardware Trojan is a covert malicious, modification of an electronic circuit or design, which results in undesired behavior of an electronic device [1]. These alterations can provide a back door entry to a

SoC or an Embedded system. An adversary can also utilize Trojans to leak sensitive information from a system or can even deny providing the service during the execution of critical applications.

A hardware Trojan can be classified into three main categories according to their physical, activation and action characteristics [2]–[4]. The physical characteristics category describes the various hardware manifestations of Trojans according to their shape and size; the activation characteristics describe the conditions which activate the Trojans, and action characteristics refer to the behavior of the Trojans. There are several techniques to detect Trojans but it is difficult to device a single Trojan detection technique that is applicable to all the varieties of Hardware Trojans. The detection of Trojans at the post-manufacturing test and validation phase, of the supply chain, is based on the physical and activation characteristics. Trojan detection techniques can be classified in two categories: 1) logic testing which focuses on activation characteristics and 2) side-channel which focuses on both physical and activation characteristics. The logic testing based detection depends on rare conditions to activate Trojans occurring at internal nodes of the circuit under test [5], [6]. Whereas, side-channel based techniques involve observing the effect of Hardware Trojan on one or more physical parameter(s) such as transient current, leakage current or delay [7]–[9]. Typically, the adversary would design a Trojan to evade detection by ensuring that 1) the rare activation of Trojan goes undetected by logic testing and 2) the physical characteristics, like size, are small enough to evade side channel based testing. Moreover, the inherent process variations due to device scaling will also make the detection process unsuccessful. To handle such a security vulnerability neutralization of the threats without depending on its detection will be necessary.

To achieve such a countermeasure we propose to use Triple Modular Redundancy (TMR) for security. Since TMR can provide fault-tolerant computation the same strategy can be used to mask the faults created by Trojans. However, the presence of transient errors due to process variations can prove to be a bottle neck in this scheme as shown in Fig. 2. In order to address this issue we propose a multi-level TMR scheme that can efficiently improve the neutralization of Hardware Trojans. The number of levels of TMR will depend

978-1-5386-1357-3/17 $31.00 © 2017 IEEE   105

on the complexity of the circuit and the application in which the circuit is used. Our experimental results on ISCAS'85 benchmark circuits have shown better neutralization with 3 levels of TMR for 8-bit ALU (C880) as compared to Interrupt Controller (C432) which has more interdependent nodes.

The rest of the paper is organized in the following manner. Section II gives the related work which will discuss different methods used for Trojan detection. In Section III, we present the limitation of single TMR and the proposed model of multi-TMR through logic partitioning. In Section IV, we discuss the results of experiments and Section V draws some important conclusions.

II. BACKGROUND

Trojan detection techniques can be classified into two categories: 1) logic testing based and 2) side-channel based. The logic testing based detection depends on rare conditions to activate Trojans occurring at internal nodes of the circuit under test [5], [6]. In this approach, the outputs of circuit under test are compared with the outputs of golden circuit. In this method random set of input patterns are applied to the circuit under test and the corresponding probability of logic 1 of primary output is compared with the design of the circuit. Side-channel based techniques involve observing the effect of Hardware Trojan on one or more physical parameter(s) such as transient current, leakage current or delay [7]–[11]. These parameter(s) from circuit under test are compared with the pre-characterized value(s) of the parameter(s) obtained from golden circuit. Both the methods of detecting Hardware Trojans have positive and negative aspects. The logic testing based approach has very large Hardware Trojan design space and an extremely large number of input-output combinations which are required for testing. This makes test generation computationally infeasible due to time constraints. And, side channel based approaches are affected by large process-induced parameter variations [12]. In [14], it is discussed that TMR can be used to mask the effect of Trojans and that forces the adversary to place the Trojans in multiple copies which aids detection through side-channel analyses. In order to optimally use TMR it was suggested to be implemented only on equiprobable outputs since they are unpredictable and therefore vulnerable. However, [14] this work did not discuss about the interplay between faults due to transient errors and Trojans. In this work, we propose to neutralize the effect of Trojan through a fault sensitive scheme using multi-level TMR to compensate the limitation of single level TMR on equiprobable paths.

III. MULTI-TMR FOR TROJAN NEUTRALIZATION IN THE PRESENCE OF SEU

A. Use of single TMR for neutralization of Trojan

In digital systems, TMR, a form of multiple modular redundancies, is used to improve the reliability. In TMR, three systems (S1, S2 and S3) perform a process and the result is compared by a majority voting system to produce a single correct output. In this work, we are leveraging TMR to mask the error induced due to Trojans. This TMR scheme can be used to mask the effect of Trojan, if it is placed in just one of

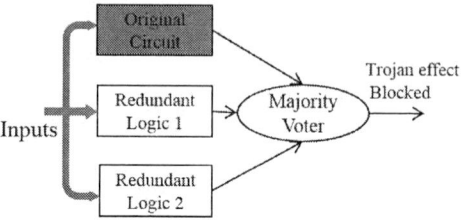

Fig. 1. (a) TMR with Trojan effect blocked (b) TMR with Trojan effect transmitted

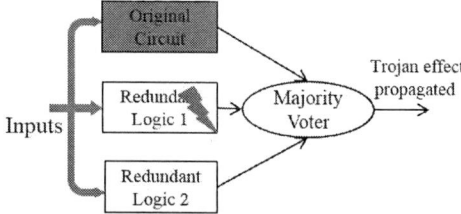

Fig. 2. Trojan effect propagated even in the presence of TMR

the systems as in Fig. 1. So in order to be effective, the Trojans have to be placed in at least two systems. The use of three redundant logics along with majority voter eliminates the error induced by Trojans as single point of failure and neutralizes their effect. In [15], it is presented that the optimized way of placing the TMR is on unpredictable output paths as they suffer with low Trojan Detection Sensitivity. So, TMR will be more useful on paths leading to equally probable outputs than on paths leading to biased outputs. Eventually, this forces the placement of Trojans on the paths leading to biased outputs. To neutralize the Trojan induced error and SEU, it is good to incorporate TMR in paths leading to such unpredictable equally probable outputs thereby improving both security and reliability.

B. Limitation of single TMR to mask Trojan in the presence of error

The majority voter performs an important task in the TMR approach. When a Trojan is inserted in TMR designs, the majority voter could neutralize the effect from the final output. However, there are natural faults occurring known as SEU [16], [17] in processor execution due to electrical noise or external radiation rather than design or manufacturing defects. So, when there are both Trojan and SEU appearing simultaneously in two distinct redundant logic blocks, such as original circuit and redundant logic 1 in Fig. 2, the majority voter may not vote the correct output. In this kind of scenario, when SEU and Trojan causing faults in distinct redundant blocks, single TMR is not helpful in mitigating either the error or the effect of Trojan. There could be a case where SEU can occur in one of the redundant logics (Redundant logic 1) whose effect is propagated and Trojan is inserted in another redundant block (Redundant logic 2) which makes the majority voter to fail in propagating the correct output as in Fig. 4(a). Hence, we have to reduce the amount of error with the help of more number of majority voters so that the effect of Trojan is neutralized.

978-1-5386-1357-3/17 $31.00 © 2017 IEEE

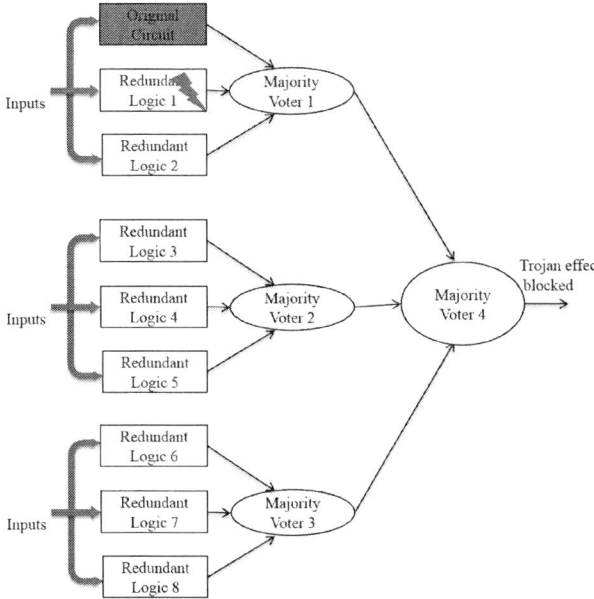

Fig. 3. Basic Multi TMR method to neutralize Trojan in the presence of error

C. Efficient use of TMR to neutralize Trojan even in the presence of error

1) Basic Multi-TMR Method: One traditional way to enhance TMR to handle multiple SEUs is multi-TMR in which single TMR is copied 3 times generating 3 outputs which are fed to a majority voter to produce the final output. The implementation of basic Multi-TMR is shown in Fig. 3. In this method, the Trojan can be blocked even in the presence of SEU in a distinct block. It is shown that there is Trojan in original circuit and SEU in Redundant Logic 1 causing an error at the output of majority voter 1. However the outputs produced by the other copies are non-erroneous which helps in generating non-erroneous outputs by majority voter 2 and majority voter 3. These two non-erroneous inputs will make the majority voter 4 to vote the correct output thus blocking the Trojan. This multi-TMR can help in blocking the Trojan in the presence of SEU, however, this way of implementation requires large area and power overhead which is a major limitation in the present day applications.

2) Multi-TMR Method through Logic Partitioning: A more efficient TMR approach to neutralize Trojan in the presence of SEU is the logic partitioning TMR, as shown in Fig. 4(b), where the TMR design is partitioned into 3 logic blocks and each of them has redundant blocks performing the same task in tandem whose outputs are compared and the majority is voted as the final output by majority voter. When there is a Trojan along with two SEUs occurring simultaneously in this TMR design, there can be three different situations according to their positions, i.e., SEU 'a', 'b' as presented in Fig. 4(b). Apparently, both upset 'a' and 'b' will not provoke any faults in the final partition, because of the presence of majority voters 1 and 2 respectively. So the effect of Trojan on the final partition is neutralized by the majority voter 3. This way, the

effect of the Trojan can be neutralized even in the presence of multiple SEUs.

In order to reduce the probability that SEU affect two distinct redundant logic blocks driving the same voter, we can partition the redundant logic parts into smaller logic blocks with majority voters. Then two simultaneous SEU from distinct redundant logic blocks can be voted by different majority voters, just like SEU 'a' and 'b' in Fig. 4(b).

IV. RESULTS

For analysis purposes, the simulations were performed on gate level circuits of ISCAS85 benchmarks. In [14], it is shown that TMR has to be implemented on equally probable paths to ensure the security of ICs. So, the proposed methodology is validated by implementing TMR on equally probable output paths by replicating the entire path from primary input to primary output 3 times, which are fed to a perfect majority voter to generate the correct output. Hence, in these experiments we have chosen equiprobable outputs of C17 (G16, G17), C432 (G430, G431, G432) and C880 (G767, G768) circuits and implemented TMR on those output paths. In all these experiments a 2-bit comparator which is shown in Fig. 5(b) is used as Trojan whose activation is controlled. This Trojan introduced in the given circuits is controlled with two newly introduced inputs to ensure their rare activation conditions. The probability of the inputs to Trojan is chosen in such a way that the activity of Trojan is 0.01, which translates to 10,000 times in a million runs. From here-on, in the results shown the Trojan free circuit will be referred to as 'Golden circuit' and the Trojan prone circuit will be referred to as 'Circuit Under Test'. This TMR implemented circuits are simulated in which 2-bit comparator based Trojan is inserted. For analysis purpose, the Trojan is inserted close to the primary output so that its effect is not masked by the error. Once, the probability distribution of primary outputs of golden circuit is obtained, 2-bit comparator is introduced as Trojan whose activation is 0.01 in one of the paths leading to primary output. Now, the probability distribution of that primary output of the circuit under test is determined and the difference of probability of golden circuit and circuit under test produces the resultant Trojan Neutralization Rate (TNR). All the results represented are probabilities of '1' expressed in terms of percentages.

A. Case A: Blocking of Trojan in the absence of error

In this case we present how TMR can neutralize the effect of Trojan in the absence of error. In Table I, it is shown that a single TMR is able to block the effect of Trojan on the outputs G16, G17 of the benchmark circuit C17. It is observed that the probability of '1' in both Golden Circuit and Circuit Under test (with Trojan) is same even if the Trojan is active and hence TNR of 100.00% which represents neutralizing the effect of Trojan completely. Also, we present the results of bigger benchmark circuits C432 and C880 in Table II and III respectively. It is observed that even in bigger circuits single TMR is able to completely block the effect of Trojan in the absence of error.

978-1-5386-1357-3/17 $31.00 © 2017 IEEE

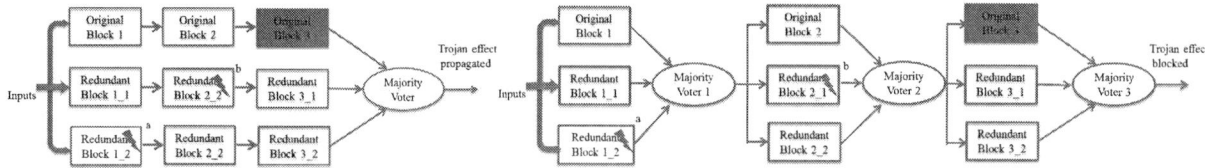

Fig. 4. (a) TMR with Trojan effect transmitted (b) TMR with Trojan effect blocked

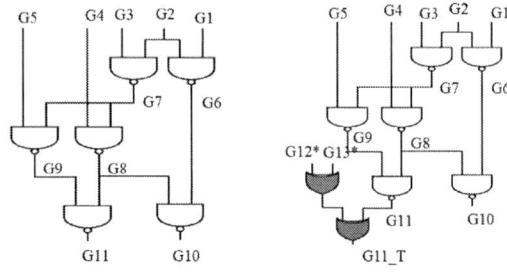

Fig. 5. (a) Gate level structure of C17 (b) C17 with trojan inserted

TABLE I
TNR OF C17 IN THE ABSENCE OF ERROR

Output	Golden Circuit	Circuit Under Test	TNR (%)
G16	56.2318	56.2318	100.00
G17	56.2256	56.2256	100.00

TABLE II
TNR OF C432 IN THE ABSENCE OF ERROR

Output	Golden Circuit	Circuit Under Test	TNR (%)
G430	52.1546	52.1546	100.00
G431	49.0505	49.0505	100.00
G432	48.1187	48.1187	100.00

TABLE III
TNR OF C880 IN THE ABSENCE OF ERROR

Output	Golden Circuit	Circuit Under Test	TNR (%)
G767	49.9888	49.9888	100.00
G768	49.9942	49.9942	100.00

TABLE IV
TNR OF C17 WITH ERROR IN THE SAME COPY

Output	Golden Circuit	Circuit Under Test	TNR (%)
G16	56.2318	56.2318	100.00
G17	56.2256	56.2256	100.00

B. Case B: Both Trojan and error in the same copy

Here, we present a case where SEU occurs in the same copy in which the Trojan is inserted. This case is interesting as the error caused by the SEU will be flipped by the Trojan or vice-versa. So, the error caused by the Trojan is masked by SEU and irrespective of this, the majority voters votes the correct output as they are majority. The simulation results of benchmark circuits C17, C432, C880 presented in Table IV V and VI respectively. It is observed that when there is SEU occurring in the same copy in which Trojan is inserted, its effect is completely blocked by the majority voter.

TABLE V
TNR OF C432 WITH ERROR IN THE SAME COPY

Output	Golden Circuit	Circuit Under Test	TNR (%)
G430	52.1546	52.1546	100.00
G431	49.0505	49.0505	100.00
G432	48.1187	48.1187	100.00

TABLE VI
TNR OF C880 WITH ERROR IN THE SAME COPY

Output	Golden Circuit	Circuit Under Test	TNR (%)
G767	49.9888	49.9888	100.00
G768	49.9942	49.9942	100.00

TABLE VII
TNR OF C17 WITH ERROR IN A DISTINCT COPY

Output	Golden Circuit	Circuit Under Test	TNR (%)
G16	56.2318	55.9202	69.00
G17	56.2256	56.033	81.00

TABLE VIII
TNR OF C432 WITH ERROR IN A DISTINCT COPY

Output	Golden Circuit	Circuit Under Test	TNR (%)
G430	52.1546	52.6568	50.00
G431	49.0505	49.5891	46.00
G432	48.1187	48.6595	46.00

TABLE IX
TNR OF C880 WITH ERROR IN A DISTINCT COPY

Output	Golden Circuit	Circuit Under Test	TNR (%)
G767	49.9888	50.4827	51.00
G768	49.9942	50.4894	50.00

C. Case C: Trojan and error present in distinct copies

In this section, we present a case which explains about the limitation of single TMR and the requirement of multi-TMR to mask the effect of Trojan in the presence of error. There is a probability that an SEU can occur in a copy other than in which the Trojan, in this situation the majority voter votes the error as the final output as the majority of its inputs are erroneous (one due to SEU and the other due to Trojan). Table VII shows the result when the SEU occurs in a distinct copy than the one with Trojan, it is observed that the effect of Trojan is transmitted on G16 and G17 with a sensitivity of 69% and 81% respectively. Also, the simulation results of C432 and C880 in Table VIII and Table IX show a lower TNR than the absence of TMR. The occurrence of SEU helps in transmitting the effect of Trojan. This is the limitation of single TMR, where the occurrence of SEU worsens the masking of Trojan. The occurrence of SEU is a major issue with single TMR, hence the mitigation of SEUs should be considered so that the occurrence such cases can be minimized.

TABLE X
BLOCKING OF TROJAN IN THE PRESENCE OF MULTIPLE ERRORS USING
BASIC MULTI-TMR METHOD IN C17

Output	Golden Circuit	Circuit Under Test	TNR (%)
G16	56.1941	56.1941	100.00
G17	56.244	56.244	100.00

TABLE XI
BLOCKING OF TROJAN IN THE PRESENCE OF MULTIPLE ERRORS USING
BASIC MULTI-TMR METHOD IN C432

Output	Golden Circuit	Circuit Under Test	TNR (%)
G430	52.2353	52.2353	100.00
G431	49.1463	49.1463	100.00
G432	48.1701	48.1701	100.00

TABLE XII
BLOCKING OF TROJAN IN THE PRESENCE OF MULTIPLE ERRORS USING
BASIC MULTI-TMR METHOD IN C880

Output	Golden Circuit	Circuit Under Test	TNR (%)
G767	50.0269	50.0269	100.00
G768	49.9707	49.9707	100.00

TABLE XIII
BLOCKING OF TROJAN IN THE PRESENCE OF MULTIPLE ERRORS USING
LOGIC PARTITIONING IN C17

Output	Golden Circuit	Circuit Under Test	TNR (%)
G16	56.3095	56.3095	100.00
G17	56.3025	56.3025	100.00

TABLE XIV
BLOCKING OF TROJAN IN THE PRESENCE OF MULTIPLE ERRORS USING
LOGIC PARTITIONING IN C432

Output	Golden Circuit	Circuit Under Test	TNR (%)
G430	52.1009	52.1009	100.00
G431	49.0184	49.0184	100.00
G432	48.1363	48.1363	100.00

TABLE XV
BLOCKING OF TROJAN IN THE PRESENCE OF MULTIPLE ERRORS USING
LOGIC PARTITIONING IN C880

Output	Golden Circuit	Circuit Under Test	TNR (%)
G767	50.0556	50.0556	100.00
G768	50.0329	50.0329	100.00

D. Solution A: Basic Multi-TMR method

Here we use the basic multi-TMR method which is traditionally used to handle multiple SEUs to deal with the neutralization of Trojan in the presence of SEU in a distinct copy. This method is tolerant to higher error rate and is reliable. The simulation results of C17 circuit with basic multi-TMR implemented are shown in Table X. In this case, Trojan is inserted in the 1st copy and SEU in the 2nd copy. It is observed that the TNR of G16 and G17 is 100.00% implying that the Trojan is masked completely even in the presence of SEU in a distinct copy. This is also true if there is Trojan and SEU in distinct copies driving different majority voters. However the major limitation of this method is that it requires 9 times the area and power dissipation to implement, which are a concern in the modern day applications.

E. Solution B: Multi-TMR using Logic Partitioning

As discussed, this method is based on partitioning the circuit into blocks and implementing a TMR on each and every partition using a majority voter. This way of implementation with little overhead ensures reliable and secure integrated circuits. Here, we have SEU occurring in one of the blocks and Trojan in other block which are driving distinct majority voters. The simulation results of C17 with logic partitioning implemented on it are shown in Table XIII. It is observed that just with 2-partitions we are able to neutralize the effect of Trojan in the presence of SEU. Also, Logic partitioning is implemented on complex bigger benchmark circuits such as C432 and C880 whose results are shown in Tables XIV and XV respectively. All these results prove that the short coming of single TMR can be compensated using multiple levels of TMR through logic partitioning.

The requirement of the number of levels of TMR depends on the complexity of the circuit. We have analyzed the behavior of C880 and C432 under the influence natural errors and Trojan after implementing different levels of TMR. The results of equiprobable outputs G767 and G768 of C880 circuit are

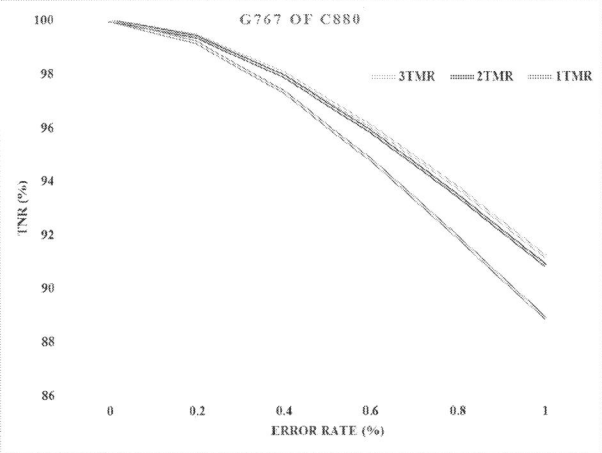

Fig. 6. G767 of C880

shown in Fig. 6 and Fig. 7 respectively. It is observed that 3 levels of TMR produces better TNR compared with 1 TMR implying better neutralization of the effect of Trojan even in the presence of higher amounts of errors. The results of equiprobable outputs G430 and G431 of C432 circuit are shown in Fig. 8 and Fig. 9 respectively. It is observed that 3 level TMR has not produced significant improvement in TNR because of the complexity of the circuit. Hence, such circuits will need more number of levels of TMR since higher dependency of the logic gates on their previous levels.

V. CONCLUSION

Our experiments show implementation results of using TMR to neutralize Hardware Trojans thereby eliminating the need for their detection. We have also shown the possible issue of interference between faults created by transient errors and Trojans and addressed it through multi-level TMR scheme which improves TNR. As a discussion point we have also shown that such a countermeasure will depend on circuit

978-1-5386-1357-3/17 $31.00 © 2017 IEEE

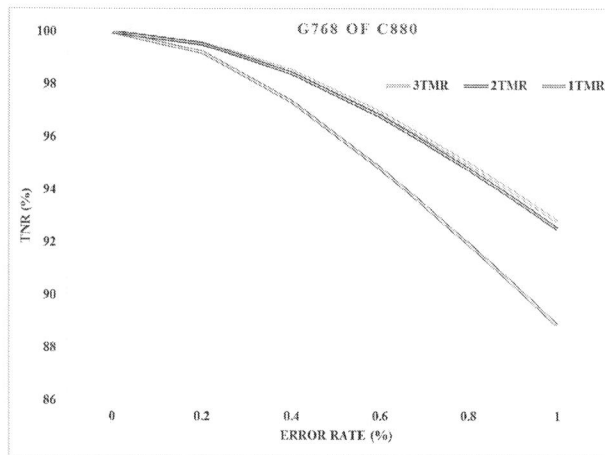

Fig. 7. G768 of C880

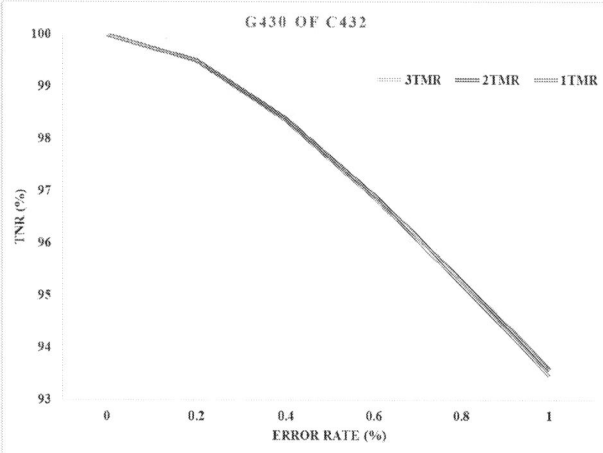

Fig. 8. G430 of C432

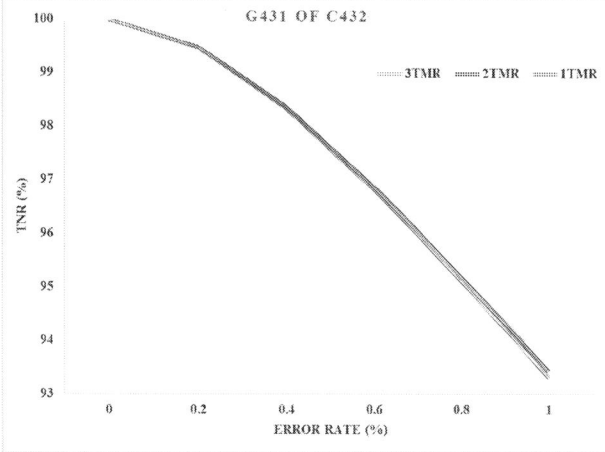

Fig. 9. G431 of C432

complexity. While we have shown that 3 levels of TMR is able to produce small improvement of TNR in a complex circuit

like Interrupt controller (C432) the future work will focus on identifying the optimal number of levels needed for effective neutralization of Hardware Trojans in such circuits.

REFERENCES

[1] R. Chakraborty, S. Narasimhan, and S. Bhunia, "Hardware Trojan: Threats and emerging solutions," in *High Level Design Validation and Test Workshop, 2009. HLDVT 2009. IEEE International*, Nov 2009, pp. 166–171.

[2] M. Tehranipoor and F. Koushanfar, "A survey of Hardware Trojan taxonomy and detection," *Design Test of Computers, IEEE*, vol. 27, no. 1, pp. 10–25, Jan 2010.

[3] N. Jacob, D. Merli, J. Heyszl, and G. Sigl, "Hardware Trojans: current challenges and approaches," *Computers Digital Techniques, IET*, vol. 8, no. 6, pp. 264–273, 2014.

[4] R. Karri, J. Rajendran, K. Rosenfeld, and M. Tehranipoor, "Trustworthy Hardware: Identifying and Classifying Hardware Trojans," *Computer*, vol. 43, no. 10, pp. 39–46, Oct 2010.

[5] R. Chakraborty, F. Wolff, S. Paul, C. Papachristou, and S. Bhunia, "MERO: A Statistical approach for Hardware Trojan Detection," in *Cryptographic Hardware and Embedded Systems - CHES 2009*, ser. Lecture Notes in Computer Science, C. Clavier and K. Gaj, Eds. Springer Berlin Heidelberg, 2009, vol. 5747, pp. 396–410.

[6] F. Wolff, C. Papachristou, S. Bhunia, and R. Chakraborty, "Towards Trojan-free trusted ICs: Problem analysis and detection scheme," in *Design, Automation and Test in Europe, 2008. DATE '08*, March 2008, pp. 1362–1365.

[7] S. Narasimhan, D. Du, R. Chakraborty, S. Paul, F. Wolff, C. Papachristou, K. Roy, and S. Bhunia, "Hardware Trojan Detection by multiple-parameter side-channel analysis," *Computers, IEEE Transactions on*, vol. 62, no. 11, pp. 2183–2195, Nov 2013.

[8] D. Agrawal, S. Baktir, D. Karakoyunlu, P. Rohatgi, and B. Sunar, "Trojan detection using IC fingerprinting," in *Security and Privacy, 2007. SP '07. IEEE Symposium on*, May 2007, pp. 296–310.

[9] J. Aarestad, D. Acharyya, R. Rad, and J. Plusquellic, "Detecting trojans through leakage current analysis using multiple supply pads," *Information Forensics and Security, IEEE Transactions on*, vol. 5, no. 4, pp. 893–904, Dec 2010.

[10] B. Cha and S. K. Gupta, "Trojan detection via delay measurements: A new approach to select paths and vectors to maximize effectiveness and minimize cost," in *Design, Automation Test in Europe Conference Exhibition (DATE), 2013*, March 2013, pp. 1265–1270.

[11] J. Li and J. Lach, "At-speed delay characterization for IC authentication and Trojan Horse detection," in *Hardware-Oriented Security and Trust, 2008. HOST 2008. IEEE International Workshop on*, June 2008, pp. 8–14.

[12] S. Borkar, T. Karnik, S. Narendra, J. Tschanz, A. Keshavarzi, and V. De, "Parameter variations and impact on circuits and microarchitecture," in *Design Automation Conference, 2003. Proceedings*, June 2003, pp. 338–342.

[13] N. B. Gunti, A. Khatri, and K. Lingasubramanian, "Realizing a security aware triple modular redundancy scheme for robust integrated circuits," in *Very Large Scale Integration (VLSI-SoC), 2014 22nd International Conference on*, Oct 2014, pp. 1–6.

[14] N. B. Gunti and K. Lingasubramanian, "Effective usage of redundancy to aid neutralization of hardware trojans in integrated circuits," *Integration, the VLSI Journal*, vol. 59, no. Supplement C, pp. 233 – 242, 2017. [Online]. Available: http://www.sciencedirect.com/science/article/pii/S0167926017300421

[15] K. Lingasubramanian and S. Bhanja, "Probabilistic maximum error modeling for unreliable logic circuits," in *Proceedings of the 17th ACM Great Lakes Symposium on VLSI 2007, Stresa, Lago Maggiore, Italy, March 11-13, 2007*, 2007, pp. 223–226. [Online]. Available: http://doi.acm.org/10.1145/1228784.1228842

[16] K. Lingasubramanian, S. M. Alam, and S. Bhanja, "Maximum error modeling for fault-tolerant computation using maximum a posteriori (MAP) hypothesis," *Microelectronics Reliability*, vol. 51, no. 2, pp. 485–501, 2011. [Online]. Available: http://dx.doi.org/10.1016/j.microrel.2010.07.156

2017 IEEE International Symposium on Nanoelectronic and Information Systems

A Novel Intrusion Detection Algorithm: An AODV Routing Protocol Case Study

Gurveen Vaseer[1], Garima Ghai[2] and Pushpinder Singh Patheja[3].
Department of Computer Science and Engineering, Oriental University, Indore.[1,2]
Department of Computer Science and Engineering, VIT University, Bhopal.[3]
Email-ID: gurveenv@orientaluniversity.in[1], garimaghai@orientaluniversity.in[2]
and pspatheja@gmail.com[3].

Abstract— **Mobile ad-hoc network (MANET) is a collection of movable nodes capable of self-routing, constraining energy and decentralized handling of nodes. It faces many challenges due to uncertainty of network topology i.e. security and congestion. In this paper we propose a novel algorithm for intrusion detection against attacks such as probing, Denial-of-service (DoS), vampire and User-To-Root (U2R) in a MANET environment. The attack detection has been carried out using a profile (behavior) analysis and a confusion matrix (True positives, True negatives, False positives, False negatives). The performance of a standard Ad hoc On-Demand Distance Vector (AODV) routing protocol has been reported for all 4 types of attack in a network simulator-2 (ns-2) environment. To the best of authors' knowledge, this is the first paper reporting a novel intrusion detection algorithm using behavior analysis for an AODV protocol in a MANET environment.**

I. INTRODUCTION AND CONTRIBUTIONS

Mobile ad-hoc network (MANET) is the most promising and rapidly growing technology that is primarily based on a self-organized and speedily deployed network [1]. As a result of its features, MANET is more suitable for real world applications in which the network topology changes quickly. Nodes in MANETs join and leave the network dynamically exhibiting their independent and self-deployable behavior. No mounted set of infrastructure and centralized administration is required in this kind of a network. Nodes are interconnected through wireless interfaces. The dynamic nature of such networks makes it extremely vulnerable to varied link attacks. The basic requirements for secured networking are protocols that guarantee confidentiality, handiness, legitimacy and integrity of network [2].

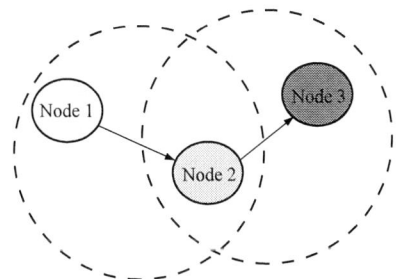

Fig. 1. Representation of a mobile ad hoc network (MANET)

Figure 1 shows the composition of MANET where 3 different, independent nodes communicate with each other in different environments through another independent node. Several existing security solutions for wired networks are ineffective and inefficient for MANET environment because the transmission takes place in an open space making the MANETs more susceptible to security attacks. There are 4 major kinds of security attacks viz. Denial of Service(DoS), probing, vampire and User-to-Root (U2R) [3]. In DoS attack, the system is flooded with unwanted traffic so that legitimate users cannot it hence crashing the server [4]. In probing, a device or an activity can be introduced in the system to gain access of the system hence damaging it. In U2R, the attacker gains local access to the victim machine and tries to gain super user privileges [5]. Vampire attack is a type of DoS attack, but it consumes the node's energy and decreases the network efficiency and reliability. In the presence of a security protocol, effectiveness of various attacks will be reduced. The mobile hosts dynamically establish paths among one another in order to communicate. Therefore, the success of MANET communication depends a lot on the collaboration of the concerned mobile nodes [1]. Ad hoc On-demand Distance Vector (AODV) routing protocol is an on-demand protocol i.e. it discovers routes on and as needed basis using route discovery process [6].

The AODV protocol is initiated when a node wants to communicate with other nodes outside its range. In this, first unicast routes are determined to the destination. Then, messages are sent to nodes with the following message types: RREQ (route request), RREP (route reply) and RERR (route error). RREQ is generated when a node needs to find a route to the destination and has the sequence number of the destination. RREP is generated if the destination is reached or a path has been traced by the node. RERR is generated for broken links in the path. Each node maintains a route table that contains information about reaching destination nodes. Destination information is assured by comparing the sequence number of the incoming AODV message with the sequence number for that destination. The process continues till the route is complete and source nodes and destination nodes are well routed.

The *novel contributions* of this paper are as follows:

978-1-5386-1357-3/17 $31.00 © 2017 IEEE   111

1) A novel intrusion detection algorithm for a MANET environment is proposed.
2) The algorithm is designed to detect probing, Denial-of-service (DoS), vampire and User-To-Root (U2R) attacks.
3) The proposed algorithm uses profile (behavior) analysis and confusion matrix for detection.
4) The impact of the attacks is studied using AODV protocol.

The remainder of the paper is organized as follows: Section II presents the related research. Section III presents the proposed research and discusses the proposed algorithm. Section IV discusses the output responses of the network under normal (no attack) and abnormal (under attack) conditions. This is followed by conclusions and future research in section V.

II. RELATED RESEARCH

In this section we discuss the existing research in the area of MANET security against probing, vampire, DoS and U2R attacks. Authors in [7] present a real-time intrusion detection system (IDS) using the Multi-agent System (MAS-IDS) to reduce the time of process traffic information network. Furthermore, the analysis of enormous amounts of information in the system within the shortest possible time has been achieved. Swati Paliwal et. al.[8] propose a methodology that supports Genetic formula for detection of inquisitor, Denial of Service(DoS) and Remote to User (R2L) attacks. The planned approach aims at gaining detection of the inquisitor, R2L and DoS attacks with minimum false positive rate. Out of the total intrusions in testing dataset, detection of more than 97% of the intrusions is anticipated by this approach. Distributed intrusion detection architecture proposed by Jaydip Sen [9] supports autonomous and cooperating agents with no centralized analysis parts. The agents collaborate by using a gradable communication of interests and information, and therefore the analysis of intrusion information is formed by the agents at the bottom level of the hierarchy. S. T. Sheu et.al. [10] present a secured routing methodology for police work that prevents network attacks like false reports and Gray-hole attacks in wireless device networks. Authors in [11] discuss the energy efficient protocols that divide the network to efficiently maintain the energy consumption of sensing element nodes. D.R. Raymond et al. [12] discuss the denial-of-sleep attacks at the MAC layer. Authors in [13] propose a methodology that aims to extend the lifetime of power-constrained networks by using less energy to transmit and receive packets. S. S. George et. al. [14] present a routing protocol to bind the harm caused by vampires within the forwarding phase. Table I summarizes the comparison of the proposed research in this paper with existing literature.

III. PROPOSED WORK

The proposed work detects the four types of attacks i.e. probing, DoS, vampire and U2R in MANETs. All the above listed attacks have been detected through specific behavior analysis based methodology. The detection engine creates two tables, one containing the normal profile such as TCP, UDP,

AODV related formats and the other is abnormal table containing the behavior of abnormality such as probing, vampire, DoS and U2R. If the abnormality matches with a particular attack, the attacks are classified in that particular class which helps in detection of each attack through this analysis. In our simulation engine, the packet format for TCP, UDP are fixed i.e. in the header part TCP/UDP will be specified. If the header part is missing or does not match with the standard TCP/UDP/AODV formats implies that it is a hampered packet. For simulation purposes, separate environments have been created for the 4 types of attacks. For a multiple attack scenario, we may utilize the unique categorization techniques of each attack for detection [15]. The complete algorithm is discussed in subsection III-A.

A. Proposed Algorithm

In this subsection we describe the proposed detection algorithm which detects all four types of attacks (vampire, probing, DoS, U2R). The algorithm is divided into three sections viz. input, procedure and output shown in algorithm 1.

In the proposed algorithm 1, data is tested using 50 nodes in a range of 800×800 m. Here, I is a collection of addresses of intermediate nodes between sender and receiver. It is compared with R (receiver node's address) until there is a match. Once there is a match, rpkt, which is a routing packet, is received by I and forwarded until it reaches the destination node. Because network related communication is possible with both Transport, Communication and Application layers, we have grouped TCP/UDP and AODV. Data is classified as normal and abnormal depending on the packet header that the network generates; those with standard TCP/UDP/AODV headers will be classified as normal and the rest can be presumed as abnormal. Attack detection is done through behavioral analysis; when data is passed through the simulation engine if data $== b_h(n)$ then no traces of abnormal data are found hence no attack has occurred. If data $== b_h(ab)$ then an attack has occurred which can be of four types:

1) If data is being captured then it is a probing attack.
2) If junk messages are getting transmitted and they do not match TCP,UDP standards then attack type is DoS.
3) If there is abnormal energy consumption and the path is disabled, it is vampire attack.
4) If IP is being modified then it is a U2R attack.

Figure 2 shows the flowchart for the proposed algorithm. Once the attacks are detected, we can start counterattack measures. Counter attack measures can be designed by analysis of respective behavior of the 4 attacks. This can be done by abolishing suspicious nodes and hence ensuring secure communication between sender and receiver nodes [15].

Table II shows the behavior table, where S: send, R: Receives, F: Forward, D: Drop, N: Normal, H: High. When there is absence of attacks, all the protocol values (TCP,UDP,AODV) are satisfied hence generate a positive acknowledgement to the receiver and the queue utilization is normal. Whereas in any of the attack scenarios based on behavioral conditions, there is drop in packets, no acknowledgement is sent and queue

978-1-5386-1357-3/17 $31.00 © 2017 IEEE

TABLE I

COMPARISON WITH RELATED WORKS

Reference	Work done	This work
Yaseen [7]	MAS-IDS to reduce time complexity of network	Detects 4 major attacks
Paliwal [8]	Attack free environment using genetic formula	Attack free environment using behavior analysis
Sen [9]	Distribution intrusion detection architecture minimizes intrusions	Remove intrusions
Sheu [10]	Prevention of grey hole attacks	Detection of 4 attacks
Doshi [13]	Optimization of energy using power constrained networks	Removal of attacks
George [14]	Prevent vampire attacks through routing protocol	Detection through behavior analysis

TABLE II

BEHAVIOR TABLE

Parameters	Normal	Abnormal			
		Vampire	U2R	Probe	DoS
Packet Type	TCP, UDP, AODV	Energy = 0	IP modification	Loop	Message
Event Type	S, R, F	D	R, D	D	F
Acknowledgement	Yes	No	No	No	No
Queue Utilization	N	H	H	H	H

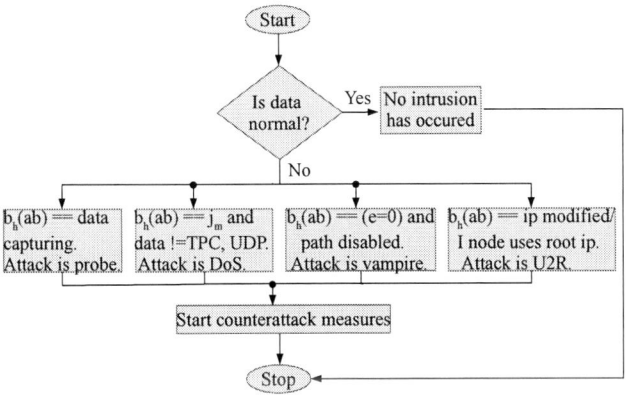

Fig. 2. Flowchart for proposed algorithm

TABLE III

NETWORK SIMULATION PARAMETERS

Parameter	Value
Number of nodes	50
Simulation area	800 × 800 m
Routing Protocol	AODV
Attack types	Probing, DoS, vampire, U2R
Simulation time	100 seconds
Transport layer	TCP, UDP
Traffic type	CBR, FTP
Packet size	512 bytes
Mac Standard	802.11
Antenna type	Omni-Antenna
Number of Traffic connections	10
Node speed	Random

utilization is high which proves there are more malicious nodes in the queue.

The parameters considered for network modeling are number of nodes, simulation area, standard routing protocol type, transport protocol, MAC standards, application protocols and antenna type as shown in Table III. All these parameters are useful for network architecture design and analyzing the behavior of the MANET.

IV. NETWORK OUTPUT RESPONSES

This section describes the response of the simulated network under normal conditions (no attack) as well as abnormal conditions (under attack). The responses reported are Throughput, Normal Routing Load (NRL), End-to-End Delay, Accuracy, Confusion Matrix i.e. True positives (t_p), True negatives (t_n), False positives (f_p), False negatives (f_n). For each output response, we compare the behaviour of AODV routing under no attack with AODV routing under DoS attack, probing attack, U2R attack and vampire attack. So, for each output response, there are 5 different scenarios.

A. Normal Routing Load

Normal routing load (NRL) is the ratio between number of routing packets (N_{rpkt}) to the total data received by the receivers (Data$_{received}$), as shown in equation 1.

$$NRL = \frac{N_{rpkt}}{Data_{received}} \times 100. \tag{1}$$

When the ratio is minimum it means network overhead is low. We report that AODV routing under normal conditions requires low overhead and U2R attack requires 8% more overhead because U2R gives wrong route frequently in order to lower the requirement of routing packets in this scenario. The other three attacks spread junk packets in the network so the network is heavily congested and thus increase the

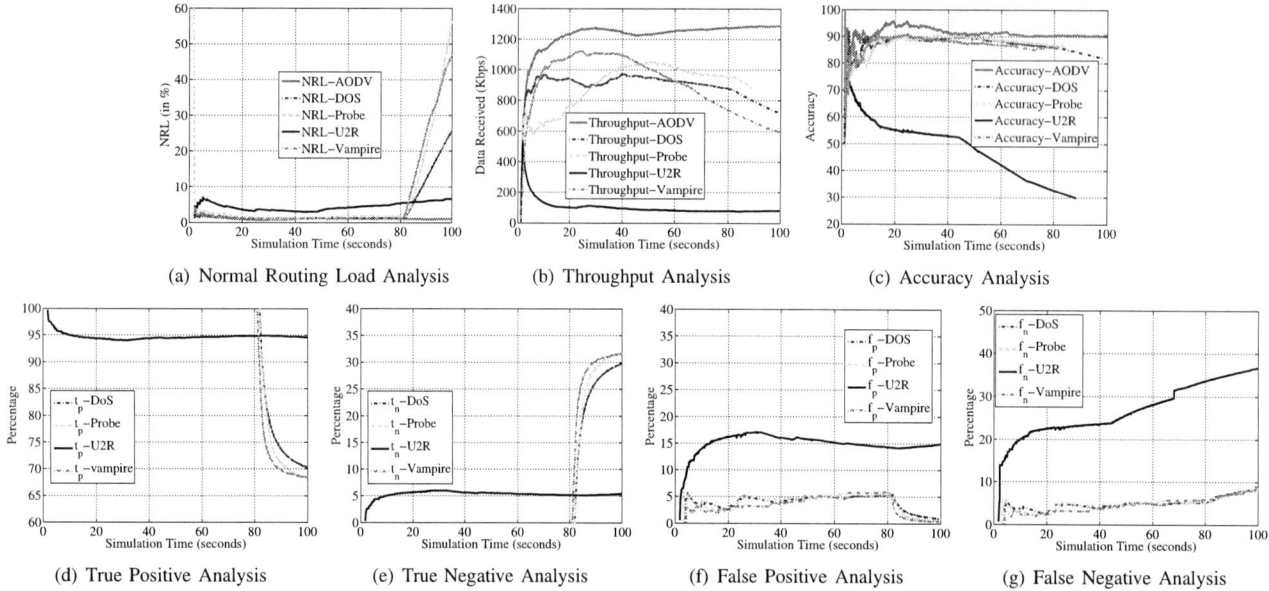

(a) Normal Routing Load Analysis   (b) Throughput Analysis   (c) Accuracy Analysis

(d) True Positive Analysis   (e) True Negative Analysis   (f) False Positive Analysis   (g) False Negative Analysis

Fig. 3. Network Output Responses.

overhead of the network. Figure 3(a) shows the NRL in the 5 different scenarios where x-axis shows the simulation time in seconds and y-axis shows the NRL (in %).

B. Throughput Analysis

Figure 3(b) shows the throughput for 5 different cases where x-axis shows the simulation time in seconds and y-axis shows the data received per second (Kbps). Throughput is defined as the ratio between data received by the receivers ($Data_{received}$) and Time (in seconds) as shown in equation 2.

$$Throughput = \frac{Data_{received}}{Time(sec.)}.\qquad(2)$$

AODV distance vector routing in no attack environment provides higher throughput, whereas the 4 scenarios under attack environment capture the genuine data through attacker nodes and decrease the throughput of the network. For the probing attack scenario, the throughput is 0 from 90th second onwards, meaning no data is received by the genuine receiver, all the data is received by the attacker resulting in damage to the whole network.

C. Accuracy Analysis

Accuracy in terms of network communication is the percentage of accurate data received by the receivers. When the attacks are not present in the network its performance is better. Figure 3(c) shows that all 4 types of attacks degrade the network performance, because each attacker module modifies the data packets and corrupt it. This corrupt data is treated as inadequate data. In the case of U2R attack, the accuracy degrades from 70% to 30% implying that U2R attack is more harmful than other attacks. AODV routing under no attack shows above 90% accuracy in receiving data.

D. End-to-End Delay Analysis

End to end delay depends on various factors of the network i.e. congestion status, queue, utilization, number of available paths and channel capacity. From Table IV, we observe that AODV under normal conditions generates minimum delay for data transmission and in other 4 cases, when the network is under attack, the end-to-end delay increases because unwanted junk messages are transmitted by the attacker node that increase the network delay from the unnecessary queue and bandwidth utilization.

TABLE IV
AVERAGE END-TO-END DELAY

Network	End-to-End Delay (ms)
Normal	0.33
DoS	0.52
Vampire	0.45
Probe	0.52
U2R	0.65

E. True Positive analysis

True positive is the total set of normal data (TCP, UDP) which are detected by the detection algorithm. When the transmitted data is passed through the detection algorithm, the data is compared with the respective format of particular data and if it is 100% accurate, it means it is true positive data. In Figure 3(d), DoS attack, probing, U2R and vampire attack are considered for the analysis and it is observed that for U2R the true positive is nearly 95%. However, for the other 3 cases true positive is 100% till 80 seconds and after that result degrades because the nodes move out of the radio zone.

978-1-5386-1357-3/17 $31.00 © 2017 IEEE   114

Algorithm 1 Intrusion Detection Algorithm

1: **Input Factors:**
2: M: Mobile node, S: Source node, R: Receiver node, Ψ : Radio range = 550m, b_h(n,ab): behavior table containing normal and abnormal behaviors, I: Set of intermediate nodes, A: Attack types (probing, vampire, DoS, U2R), j_m: Junk message, e: Energy of nodes, R_p: AODV routing protocol, rpkt: Routing Packet
3: **Output Responses:** Throughput, Normal Routing Load (NRL), End-to-End Delay, Accuracy, Confusion Matrix i.e. True positives (t_p), True negatives (t_n), False positives (f_p), False negatives (f_n).
4: **Procedure:**
5: S ← broadcast (AODV, S, R)
6: **if** I $\neq$ R and I in Ψ **then**
7:      I ← receive rpkt
8:      I ← forward rpkt to next hop
9: **else if** I == R **then**
10:      R ← receives rpkt
11:      Select shortest path
12:      R ← create reverse route for ack
13:      S ← receives ack
14:      Send data (S, R, data)
15: **else**
16:      Node out of range
17:      Node unreachable
18: **end if**
19: **Attack detection module:**
20: Data passes into detection engine, Compare data and b_h(n,ab).
21: **if** data == b_h(n) **then**
22:      data is normal TCP, UDP or AODV
23: **else if** data ==b_h(ab) **then**
24:      Data shows abnormal activity: A
25:      **if** b_h(ab) == data capturing **then**
26:          A ← probe
27:      **else if** b_h(ab) == j_m and data $\neq$ TCP, UDP, AODV **then**
28:          A ← DoS
29:      **else if** b_h(ab) == (c=0) and path disabled **then**
30:          A ← vampire
31:      **else if** b_h(ab) == ip modified or I node uses root ip **then**
32:          A ← U2R
33:      **end if**
34: **end if**

F. True Negative analysis

True negative is the total set of abnormal data which is detected by detection algorithm. If the data detected does not belong to the actual data group it means the data is abnormal and based on abnormality it can be classified as a particular attack. When the attacker node retrieves credentials of normal user and gains access to the root node, it signifies that it is a U2R attack. Similarly, if a node consumes network energy and degrades the network actual performance, then it is a vampire attack. In Figure 3(e), the true negative is shown for 4 attack scenarios.

G. False Positive Analysis

False positive also known as false alarm, is the total set of normal data which are detected but should actually be abnormal data. If the value is low or zero, it signifies that the proposed detection algorithm is accurate in measuring the abnormality. In the data transmission some data properties are depicted as normal data but they are actually unusual data which are not detected under true negative. All this data belongs to the category of false positive that creates confusion for detection algorithm. In Figure 3(f), all 4 types of attack are analyzed under MANET environment. The percentage is not greater than 20% meaning that a maximum 20% confusion for attack is detected from detection algorithm.

H. False Negative Analysis

False negative is the total number of abnormal instances detected which should be normal data. That abnormality in network is due to some reason i.e. either some packet has been dropped by the MAC, collision, route or queue based drop. However, the system detects the dropped data as attack symptoms and the data is treated as unusual by the detection algorithm. In Figure 3(g), all four type of attacks are measured.

The overall performance summary is presented in Table V.

TABLE V

PERFORMANCE SUMMARY

Parameter (units)	DoS	Vampire	Probe	U2R	Normal
SEND (no. of packets)	5142	3901	4756	1586	8238
RECV (no. of packets)	4196	3203	3901	423	7458
PDF (% data received)	81.6	82.11	82.02	26.67	90.53
NRL	25.91	46.99	56.42	6.82	1.01
End-to-End Delay (ms)	0.52	0.45	0.52	0.65	0.33
Dropped packets (no. of packets)	946	698	855	1163	780

V. CONCLUSIONS AND FUTURE RESEARCH

In this paper we have presented an intrusion detection algorithm to detect the probing, vampire, DoS and U2R attacks. We have shown that all the four type of attacks degrade the network performance with respect to throughput, accuracy, NRL and End-to-end delay. We have shown the efficiency and accuracy of our proposed algorithm using confusion matrix. True positive gives the minimum output nearly 70% implying that 70% is truly detected normal data by detection algorithm. Other analysis (True negatives, False positives, False negatives) also show promising results. For future research, this work will be extended for prevention of detected attacks through lightweight techniques. Also, the algorithm may be revised to handle multiple types of attacks occurring simultaneously in the network.

REFERENCES

[1] S. E. Khediri, N. Nasri, A. Benfradj, A. Kachouri, and A. Wei, "Routing protocols in MANET: Performance comparison of AODV, DSR and DSDV protocols using NS2," in *Proceedings of the IEEE International Symposium on Networks, Computers and Communications*, 2014, pp. 1–4.

[2] M. L. Rajaram, E. Kougianos, S. P. Mohanty, and U. Choppali, "Wireless Sensor Network Simulation Frameworks: A Tutorial Review," *IEEE Consumer Electronics Magazine (CEM)*, vol. 6, no. 2, pp. 63–69, April 2016.

[3] I. Butun, S. D. Morgera, and R. Sankar, "A survey of intrusion detection systems in wireless sensor networks," *IEEE communications surveys & tutorials*, vol. 16, no. 1, pp. 266–282, 2014.

[4] D. R. Raymond and S. F. Midkiff, "Denial-of-service in wireless sensor networks: Attacks and defenses," *IEEE Pervasive Computing*, vol. 7, no. 1, 2008.

[5] S. Ganapathy, P. Yogesh, and A. Kannan, "An intelligent intrusion detection system for mobile ad-hoc networks using classification techniques," *Communications in Computer and Information Science*, vol. 148, pp. 117–122, 2011.

[6] S. R. Das, E. M. Belding-Royer, and C. E. Perkins, "Ad hoc on-demand distance vector (aodv) routing," 2003.

[7] Al-Yaseen, W. Laftah, Z. A. Othman, and M. Z. A. Nazri, "Real-time intrusion detection system using multi-agent system," *IAENG International Journal of Computer Science*, vol. 43, no. 1, pp. 80–90, 2016.

[8] S. Paliwal and R. Gupta, "Denial-of-service, probing and remote to user (r2l) attack detection using genetic algorithm," *International Journal of Computer Applications*, vol. 60, no. 19, pp. 57–62, 2012.

[9] J. Sen, "A distributed intrusion detection system using cooperating agents," *arXiv preprint arXiv:1111.0382*, 2011.

[10] S. T. Sheu, M. Kao, Y. Hsu, and Y. en Cheng, "Secure routing protocol for detecting grayhole attack and false report along with elliptic curve cryptography in wireless sensor network," in *Proceedings of the IEEE Students Conference on Electrical, Electronics and Computer Science*, 2014.

[11] A. Vincy and V. U. Devi, "Maximizing lifetime of nodes in wireless ad hoc sensor network by preventing vampire attack," in *Proceedings of the IEEE International Conference on Innovations in Engineering and Technology*, 2014.

[12] D. R. Raymond, R. C. Marchany, M. I. Brownfield, and S. F. Midkiff, "Effects of denial-of-sleep attacks on wireless sensor network mac protocols," *IEEE transactions on vehicular technology*, vol. 58, no. 1, pp. 367–380, 2009.

[13] S. Doshi, S. Bhandare, and T. X. Brown, "An on-demand minimum energy routing protocol for a wireless ad hoc network," *ACM SIGMO-BILE Mobile Computing and Communications Review*, vol. 6, no. 3, pp. 50–66, 2002.

[14] S. S. George and R. Suma, "Attack-resistant routing for wireless ad hoc networks," *International Journal of CS & IT*, vol. 5, no. 3, 2014.

[15] S. Bose and A. Kannan, "Detecting denial of service attacks using cross layer based intrusion detection system in wireless ad hoc networks," in *Proceedings of the IEEE International Conference on Signal Processing, Communications and Networking*, 2008, pp. 182–188.

978-1-5386-1357-3/17 $31.00 © 2017 IEEE

Security Evaluation of MTJ/CMOS Circuits Against Power Analysis Attacks

S. Dinesh Kumar and Himanshu Thapliyal
Department of Electrical and Computer Engineering
University of Kentucky, Lexington, KY, USA
Email: hthapliyal@uky.edu

Abstract—Research explorations in new devices, new architectures and algorithms are being performed to reduce leakage power dissipation. As a solution to reduce the leakage power in CMOS based designs, Magnetic Tunnel Junction (MTJ) devices are being investigated to design MTJ/CMOS Logic-In-Memory (LIM) circuits. The MTJ/CMOS circuits have advantages such as near-zero leakage power and non-volatility which make them useful to design sudden power-outage resilient non-volatile processors. However, the security of the existing MTJ/CMOS circuits against power analysis based side-channel attacks need to be evaluated before deploying these circuits in real world applications. Therefore, in this paper, we are performing the security evaluation of the existing MTJ/CMOS circuits against power analysis attacks for the first time in the literature. From the simulations, it is shown that the existing MTJ/CMOS circuits consume high current during the switching of MTJs thereby leaking the information and becoming vulnerable to power analysis based attacks. Further, to thwart power analysis attacks in MTJ/CMOS circuits, we propose a novel secure MTJ/CMOS logic (SMCL) which consumes uniform current irrespective of switching of MTJs. Simulations are performed using 45nm CMOS technology with perpendicular anisotropy CoFeB/MgO MTJ model using Cadence Spectre simulator. Calculated values of Normalized Energy Deviation (NED) and Normalized Standard Deviation (NSD) show that the proposed SMCLL gates consume uniform energy for every cycle of operation irrespective of their input transition. The uniform energy rate and low power operation shows that the proposed SMCL gates are energy-efficient in nature and resistant to power analysis attacks.

I. INTRODUCTION

CMOS based computing is reaching its limits and high leakage power has become one of the major concerns in CMOS logic designs. One approach for designing low-power circuits is to employ non-silicon emerging devices in the design. Among the various emerging devices, Spin Transfer Torque Magnetic Tunnel Junction (STT-MTJ) is considered as a promising candidate for designing low-power circuits. STT-MTJ has the advantages such as near-zero leakage power, non-volatility, high integration density and easy compatibility with CMOS devices [1], [2], [3]. Hybrid MTJ/CMOS based Logic-In-Memory (LIM) architecture based circuits show high potential in designing low power circuits [4] and sudden power-outage resilient non-volatile processors [5]. However, the security of the MTJ/CMOS circuit against side-channel attacks must be thoroughly verified before implementing in commercial devices.

A Side-channel attack uses the unintentional information leaked by a cryptographic device to retrieve the secret key.

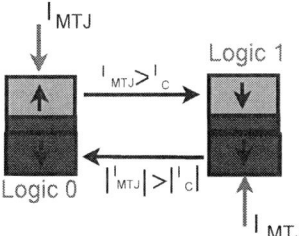

Fig. 1. Vertical Magnetic Tunnel Junction (MTJ) nanopillar structure with Spin Transfer Torque (STT) switching mechanism

Power analysis attack is one type of side-channel attack where the attacker monitors the power consumption of the cryptographic device without making any physical changes to the device [6], [7]. In recent years, researchers have focused on the security evaluation of Spin Transfer Torque Magnetic Random Access Memory (STT-MRAM) against power analysis attacks [8], [9], [10]. However, there is no security evaluation of hybrid MTJ/CMOS circuits against power analysis attacks. To the best of our knowledge, this is the first work to evaluate the security of the MTJ/CMOS circuits against power analysis attacks.

A. Motivation

MTJ/CMOS based LIM circuits have nearly zero leakage power dissipation and they are very appropriate to design low-power hardware and sudden power-outage resilient non-volatile processors [5]. However, the security of the MTJ/CMOS circuit against side-channel attacks must be thoroughly verified before implementing in commercial devices. Power analysis attack is one type of side-channel attack where the attacker monitors the power consumption of a cryptographic device without making any physical changes to the device. In this paper, we are evaluating the security of the existing MTJ/CMOS based LIM circuits against power analysis attacks. Further, we are also proposing a novel MTJ/CMOS circuit which consumes 50% less energy than the existing MTJ/CMOS based LIM circuits and has lower Normalized Energy Deviation (NED) and Normalized Standard Deviation (NSD) values. In this paper, we have made the existing LIM based MTJ/CMOS circuits secure by masking the MTJ's during the writing of data in the MTJ.

978-1-5386-1357-3/17 $31.00 © 2017 IEEE

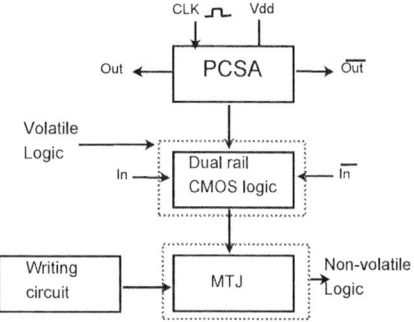

Fig. 2. Structure of existing LIM based MTJ/CMOS circuits

B. Contribution of the paper

In this paper, we are evaluating the existing Logic-In-Memory (LIM) based MTJ/CMOS circuits against power analysis attacks. From our simulations, we have shown that the existing MTJ/CMOS circuits are prone to power analysis attacks. In this paper, we have also proposed a novel Secure MTJ/CMOS Logic (SMCL) circuits which have uniform energy consumption for various input transitions. Further, the proposed SMCL circuits have 50% of energy savings as compared to the existing MTJ/CMOS circuits. In this paper, we have also proposed a novel SMCL based full adder circuit. The security evaluation of the proposed SMCL circuit is evaluated by calculating the NED and NSD values of the proposed SMCL circuit.

C. Organization of the paper

Section II discusses the background of the MTJ device, MTJ/CMOS based LIM architecture and Differential Power Analysis (DPA) attack. Section III analyzes the information leakage in the existing MTJ/CMOS LIM circuits. Section IV presents the circuit design and operation of the proposed low-power and secure MTJ/CMOS circuit. Section IV also presents the theoretical analysis of the energy consumption of the proposed MTJ/CMOS circuit as compared to the existing MTJ/CMOS based LIM circuits. Section V presents the simulation results of the proposed MTJ/CMOS circuit along with the existing PCSA based MTJ/CMOS circuit.Section VI concludes the paper.

II. BACKGROUND

A. Magnetic Tunnel Junction (MTJ)

Magnetic Tunnel Junction (MTJ) is a vertical nanopillar strcuture which consists of two ferromagnetic (FM) layers and an oxide barrier [11]. In the standard application of MTJ devices, the magnetization of one of the FM layers is fixed, while the other FM layer is free to take one of two orientations (parallel and antiparallel) as shown in Fig. 1 [12]. Based on the orientation of the FM layers, parallel (P) or antiparallel (AP), the MTJ device shows either a low resistance (RP) or high resistance (RAP) characteristic [13]. The resistance difference between the two configurations of MTJ device is given by the tunnel magnetoresistance ratio $TMR = (R_{AP} - R_P)/R_P$.

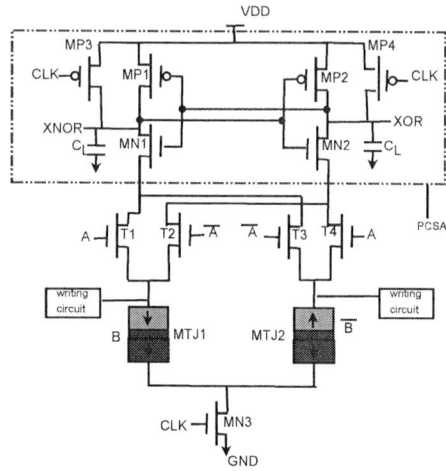

Fig. 3. Existing PCSA based MTJ/CMOS XOR gate [14] [1]

B. MTJ/CMOS based Logic-In-Memory (LIM) circuits

Fig. 2 shows the structure of the existing Logic-In-Memory (LIM) based MTJ/CMOS circuits. The LIM architecture consists of a Pre-Charged Sense Amplifier (PCSA) circuit which is used for sensing the outputs. The dual rail CMOS logic tree is used to evaluate the inputs and the MTJs are used to store the non-volatile data.

C. Differential Power Analysis attack

Differential Power Analysis (DPA) attack is one of the most widely used hardware attacks to reveal the secret key stored in a cryptographic device. DPA attack is used to reveal the secret key stored in a cryptographic device by correlating the instantaneous power consumed by the device with the input data. To guess the key, DPA uses statistical methods to evaluate the power traces with uniform plain texts.

III. INFORMATION LEAKAGE IN PRE-CHARGE SENSE AMPLIFIER BASED MTJ/CMOS CIRCUITS

This section explains the information leakage in the Pre-Charge Sense Amplifier (PCSA) based MTJ/CMOS circuit. As an example, the information leakage in the PCSA based MTJ/CMOS circuit is illustrated by PCSA based MTJ/CMOS XOR gate.

The operation of PCSA is explained through the existing PCSA based MTJ/CMOS XOR gate (Fig. 3) [1] [14]. The PCSA works in two phases depending on CLK: (i) When CLK is set to "0", the outputs (XOR, XNOR) are precharged to "1" (ii) when CLK is set to "1", the output voltages start discharging to ground. However, due to the difference in resistances of the different configuration of the MTJ (parallel and anti-parallel), the discharge speed will be different for each branch. For example, if MTJ1 is configured in anti-parallel configuration and MTJ2 is configured in parallel configuration, then $R_{MTJ1} > R_{MTJ2}$. Due to the difference in resistances between R_{MTJ1} and R_{MTJ2}, the discharge current through MTJ2 will be greater than MTJ1. When XNOR becomes less

978-1-5386-1357-3/17 $31.00 © 2017 IEEE   118

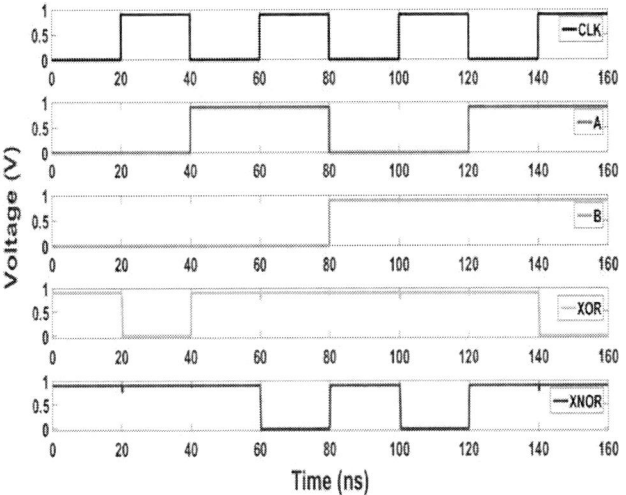

Fig. 4. Transient waveform of existing PCSA based XOR gate

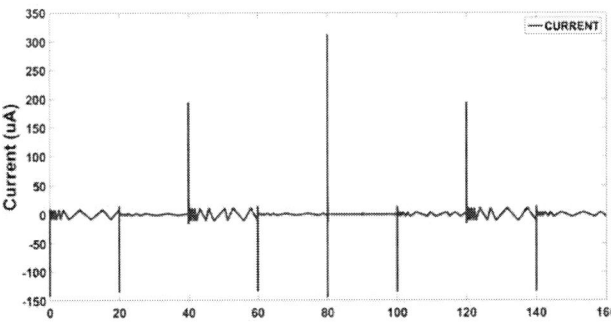

Fig. 5. Current consumption of PCSA based MTJ/CMOS XOR gate for input in Fig. 4

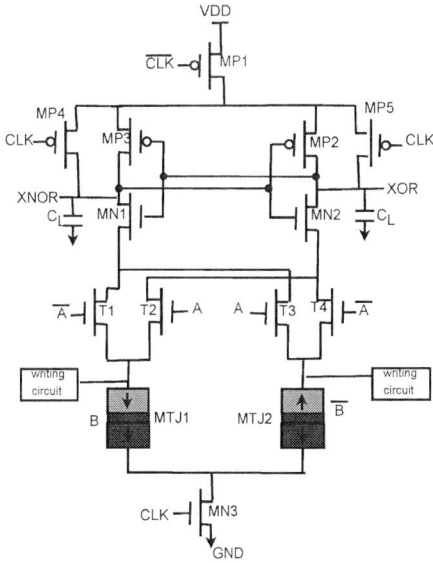

Fig. 6. Schematic of the proposed SMCL based XOR gate

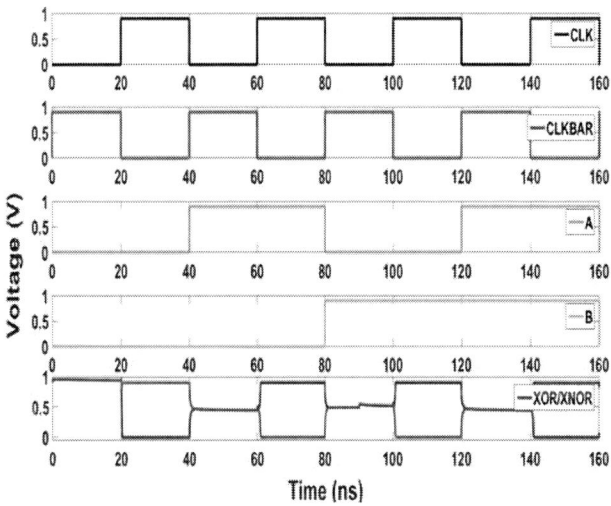

Fig. 7. Transient analysis of the proposed SMCL based XOR gate

than the threshold switching voltage of the inverter composed by MP2 and MN2, XOR will be charged to "1" and XNOR will be discharged to "0".

As we can see in Fig. 5, during the input change in MTJ, PCSA based MTJ/CMOS circuit consumes huge current. The current consumption during the writing of data to the MTJ can reveal the data stored in the MTJ which can be vulnerable to DPA attack.

IV. PROPOSED SECURE MTJ/CMOS LOGIC (SMCL) CIRCUITS

This section explains the operation of the proposed Secure MTJ/CMOS Logic (SMCL) circuits. In the proposed SMCL circuit, the MTJ's are masked from the power supply during the data are written to the MTJ.

A. Operation of the proposed SMCL circuit

This section explains the operation of the proposed SMCL circuit. The circuit operation of the proposed SMCL circuit is explained by the operation of an XOR gate. Fig. 6 shows the schematic diagram of the proposed SMCL based XOR gate.

Transistor MP1 is used to disconnect the MTJ from V_{dd} when the data is written in it. Transistors MP2, MP3, MN1 and MN2 are used to stabilize the outputs. Transistors MP4 and MP5 are used for charge sharing the outputs.

Let us understand the operation of the proposed SMCL based XOR gate through each phase of the clock.

Charge-sharing phase: During the charge-sharing phase, CLK=0, $\overline{CLK}$=1. When CLK=0, transistors MP4 and MP5 will be turned ON. Since the proposed SMCL XOR gate is dual rail in nature, the outputs will be shared between the output nodes during the charge-sharing phase. During the charge-sharing phase, MP1 is turned OFF to mask the MTJ while writing the data in the MTJ's. Moreover, in the proposed SMCL circuits, the outputs are precharged to $V_{dd}/2$ unlike the

978-1-5386-1357-3/17 $31.00 © 2017 IEEE

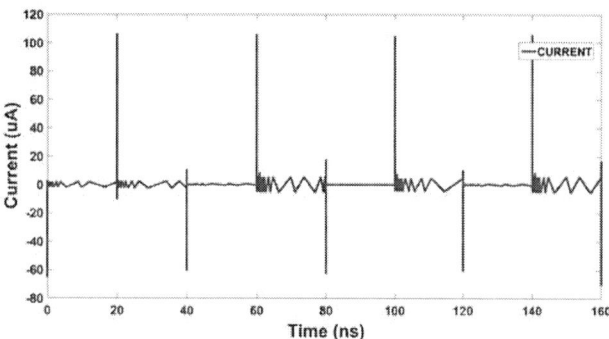

Fig. 8. Current consumption of the proposed SMCL based XOR gate

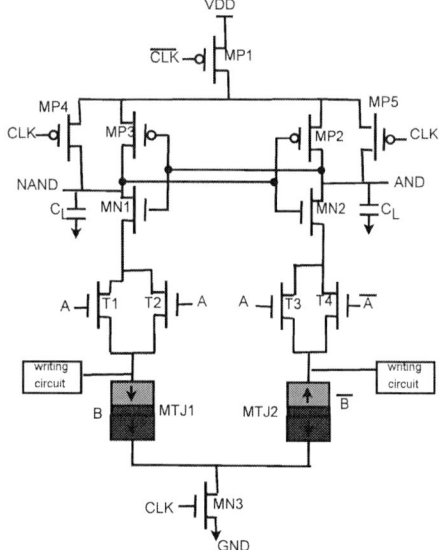

Fig. 9. Schematic of the proposed SMCL AND gate

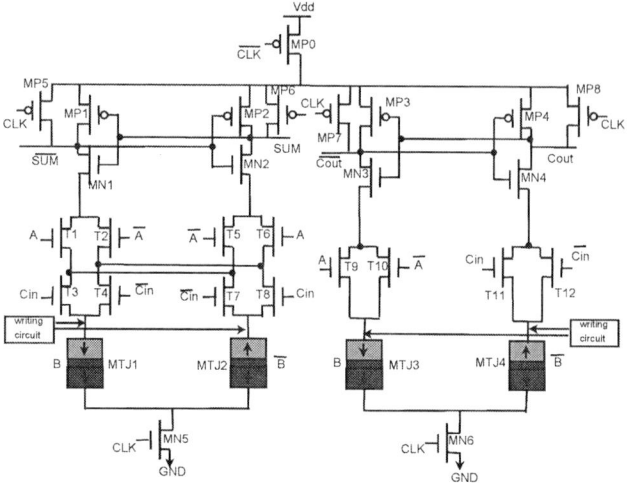

Fig. 10. Schematic of the proposed SMCL based full adder circuit

B. Proposed MTJ/CMOS full adder circuit

Fig. 10 shows the schematic of the proposed SMCL based Magnetic Full Adder (MFA). The inputs for the full adder are "A", "B" and "Cin" and the outputs are SUM and Cout. In the proposed magnetic full adder circuit, MP1 to MP8 and MN1 to MN4 are used as the sense amplifier. MP0 is used to mask the MTJ during the write operation. Rest of the transistors are used for evaluating SUM and Cout outputs. The MOS tree structure of the SMCL magnetic full adder is based on the following equations(4-5):

$$Sum = A.B.Cin + \overline{B}.A.\overline{Cin} + \overline{A}.B.\overline{Cin} + \overline{A}.Cin.\overline{B} \quad (1)$$

$$Cout = A.B + A.Cin + B.Cin \quad (2)$$

Let us assume that B="1" and $\overline{B}$="0". Let us assume that A=1 and Cin=1. During the charge-sharing phase of the CLK, all the inputs are passed to the circuit and the non-volatile data is stored in the MTJs. uring the charge-sharing phase of the CLK, all the inputs are passed to the circuit and the non-volatile data is stored in the MTJs. At this phase, all the ouput nodes are charge shared and will be at $V_{dd}/2$. During the evaluate phase of the clock, inputs are evaluated. In this illustrative example for A=1 and Cin=1, $\overline{Sum}$ will be discharged faster through the transistors T1 and T3. and SUM output will be charged to Vdd. Similarly, $\overline{Cout}$ will be discharged to ground through T9 transistor and Cout will be charged to Vdd.

C. Theoretical analysis of energy consumption in the proposed SMCL circuit

This section theoretically analyzes the energy consumption in the proposed SMCL circuit with the existing PCSA based MTJ/CMOS circuit. For analysis let us consider the PCSA based MTJ/CMOS XOR gate (Fig. 3) and the proposed SMCL based XOR gate (Fig. 6).

conventional PCSA MTJ/CMOS circuit where the outputs are precharged to V_{dd}. Since the outputs are precharged to $V_{dd}/2$, the proposed SMCL circuits consume low power as compared to the existing PCSA based MTJ/CMOS circuits.

Evaluate phase: During the evaluate phase, CLK=1, $\overline{CLK}$=0. In this phase, transistors MP1 and MN3 will be turned ON and MP4 and MP5 will be turned OFF. For analysis, let us assume that the input A=0, B=1. When A=0, transistors T2 and T3 will be turned OFF while T1 and T4 will be turned ON. The resistance of MTJ1 will be less as compared to resistance of MTJ2. When CLK=1, the charge stored in XNOR output will be discharged to ground through T1 and MTJ2 which makes transistor MN2 to turn OFF. Since, the transistor MN2 is turned OFF, the XOR output will be charged to V_{dd}. The transient waveforms of the proposed SMCL XOR gate is shown in Fig. 7. Fig. 8 shows the uniform current consumption of the proposed SMCL XOR gate. Fig. 9 shows the schematic of the proposed SMCL AND gate.

978-1-5386-1357-3/17 $31.00 © 2017 IEEE

1) Energy consumption in PCSA based MTJ/CMOS XOR gate: The energy dissipated to charge a capacitor is given by,

$$E_{diss} = \frac{1}{2}CV_{dd}^2 \qquad (3)$$

where, C is the capacitance value and V_{dd} is the voltage swing.

During the pre-charge phase of the CLK, both XOR and XNOR outputs are charged to V_{dd}. Assuming that one of the outputs will be charged to V_{dd} in the previous cycle, the total energy dissipated to charge the load capacitors is given by,

$$E_{diss,pre-charge} = \frac{1}{2}CV_{dd}^2 \qquad (4)$$

Similarly during the evaluate phase of the CLK, one of the two outputs (XOR or XNOR) will be discharged to ground while the other output will be at V_{dd}. So, the energy dissipated during the evaluate phase of the CLK is given by,

$$E_{diss,eval} = \frac{1}{2}CV_{dd}^2 \qquad (5)$$

So, the total energy dissipated in one clock cycle of the PCSA based MTJ/CMOS XOR gate is given by,

$$E_{diss,PCSA} = E_{diss,pre-charge} + E_{diss,eval} \qquad (6)$$

$$E_{diss,PCSA} = CV_{dd}^2 \qquad (7)$$

The total energy dissipated in one clock cycle of the PCSA based MTJ/CMOS XOR gate is given as CV_{dd}^2.

2) Energy consumption in proposed SMCL XOR gate: Let us assume that one of the outputs is already charged to V_{dd} during the previous cycle. During the charge sharing phase of the proposed SMCL XOR gate (Fig. 6), output voltages will be pre-charge to $V_{dd}/2$. During this phase, the charge and voltage are divided equally between the two load capacitors (XOR and XNOR). So, the total energy dissipated in the charge sharing phase of the proposed SMCL XOR gate is given by,

$$E_{diss,charge-sharing} = \frac{1}{4}CV_{dd}^2 \qquad (8)$$

Similarly during the evaluate phase of the CLK, one of the two outputs (XOR or XNOR) will be discharged to ground while the other output will be charged to V_{dd}. So, the energy dissipated during the evaluate phase of the CLK is given by,

$$E_{diss,eval} = \frac{1}{2}C(V_{dd}/2)^2 + \frac{1}{2}C(V_{dd}/2)^2 \qquad (9)$$

$$E_{diss,eval} = \frac{1}{4}CV_{dd}^2 \qquad (10)$$

So, the total energy dissipated in one clock cycle of the proposed SMCL XOR gate is given by,

$$E_{diss,proposed} = E_{diss,charge-sharing} + E_{diss,eval} \qquad (11)$$

$$E_{diss,proposed} = \frac{1}{2}CV_{dd}^2 \qquad (12)$$

The total energy dissipated in one clock cycle of the proposed SMCL XOR gate is given as $0.5CV_{dd}^2$ which is 50% less than the existing PCSA based MTJ/CMOS XOR gate.

V. SIMULATION RESULTS

This section presents the simulation results of the proposed SMCL circuits. Simulations are performed using Cadence Spectre simulator with 45nm standard CMOS technology with perpendicular anisotropy CoFeB/MgO MTJ model. Table I shows the MTJ device parameters used in this work [15]. The simulations are performed at 12.5MHz with V_{dd}=0.9V and load capacitor is 1fF.

TABLE I
MTJ DEVICE PARAMETERS USED FOR SIMULATIONS [15]

Parameter	Description	Value
t_{sl}	Thickness of the free layer	1.3nm
a	Length of surface long axis	40nm
b	Width of surface short axis	40nm
t_{ox}	Thickness of the Oxide barrier	0.85nm
TMR	Tunnel Magneto Resistance ratio	150 %
RA	Resistance Area Product	5ohmμm^2
Area	MTJ layout surface	$40nm \times 40nm \times \pi/4$
I_{co}	Critical switching current	Min. 40 μ A

The size of all the transistors are W/L=120nm/45nm. Table II gives the comparison of the PCSA based XOR gate and the proposed SMCL XOR gate. From Table II, we can see that the proposed SMCL XOR gate has 50% of energy savings as compared to the existing PCSA based XOR gate which is same as the results obtained in theoretical analysis in section IV B.

Table III shows the comparison of the PCSA based AND gate and the proposed AND gate. From Table III, we can see that the proposed SMCL AND gate has 42.7% and 50% of power and energy savings as compared to the existing PCSA based AND gate. From Table IV, we can see that the proposed SMCL based FA has 47.27% less energy consumption as compared to the existing PCSA based FA.

TABLE II
PERFORMANCE COMPARISON OF PCSA BASED XOR GATE AND PROPOSED SMCL XOR GATE

	PCSA based XOR [1]	Proposed SMCL XOR gate	% impr.
Avg. energy (fJ)	3.604	1.871	50
Avg. power (nW)	40.34	23.1	42.7
Device count	11MOS +2MTJ	12MOS+2MTJ	-

TABLE III
PERFORMANCE COMPARISON OF PCSA BASED AND GATE AND PROPOSED SMCL AND GATE

	PCSA based AND [1]	Proposed SMCL AND gate	% impr.
Avg. energy (fJ)	3.414	1.768	50
Avg. power (nW)	38.24	21.37	44.11
Device count	10MOS +2MTJ	12MOS+2MTJ	-

978-1-5386-1357-3/17 $31.00 © 2017 IEEE

TABLE IV
PERFORMANCE COMPARISON OF PCSA BASED FULL ADDER AND
PROPOSED FULL ADDER CIRCUIT

	PCSA based full adder [1]	Proposed SMCL full adder	% impr.
Avg. energy (fJ)	115.4	60.84	47.27
Avg. power (nW)	360.5	208.9	42.05
Device count	25MOS +4MTJ	26MOS+2MTJ	-

A. Security metrics analysis of the MTJ/CMOS gates

This section discusses the security metric analysis of the MTJ/CMOS gates. The parameter Normalized Energy Deviation (NED), defined as $(E_{max} - E_{min})/E_{max}$, is used to indicate the percentage difference between minimum and maximum energy consumption for all possible input transitions. Normalized Standard Deviation (NSD) indicates the energy consumption variation based on the inputs and it is calculated as $\frac{\sigma_E}{\bar{E}}$. $\bar{E}$ denotes the average energy dissipation for various input transitions. In general, a 'n' input gate will have 2^{2n} possible input transitions. For example, a 2 input gate will have 16 input transitions. σ_E denotes the standard deviation of the energy consumed dissipated by the circuit and it is shown as $\sqrt{\frac{\sum_{i=1}^{n}(E_i - \bar{E})^2}{n}}$.

TABLE V
SIMULATED AND CALCULATED RESULTS FOR MTJ/CMOS XOR GATE

Logic family	PCSA based XOR gate	Proposed SMCL XOR gate
E_{min}(fJ)	2.9	1.431
E_{max}(fJ)	6.3	1.85
NED (%)	69.3	2.63
NSD(%)	61.27	1.23

TABLE VI
SIMULATED AND CALCULATED RESULTS FOR MTJ/CMOS AND GATE

Logic family	PCSA based AND gate	Proposed SMCL AND gate
E_{min}(fJ)	2.77	1.25
E_{max}(fJ)	6.56	2.9
NED (%)	75.2	5.4
NSD(%)	64.33	3.22

From Table V and VI, we can see that the NED and NSD values for the proposed SMCL circuit are very less than the existing PCSA based MTJ/CMOS circuit. The lower the values of NED and NSD, the higher the resilience of the circuit towards power analysis attack.

VI. CONCLUSION

In this paper, we have shown the susceptibility of MTJ/CMOS circuits to power analysis attacks where the attacker uses the power traces to reveal the secret key. In order to improve the security of the existing MTJ/CMOS circuits against power analysis attacks, we have proposed a novel Secure MTJ/CMOS Logic (SMCL) circuit which has

uniform power consumption. SMCL gates and one bit full adder consumes 50% and 47% less energy, respectively, than the existing PCSA based MTJ/CMOS circuits. The proposed SMCL consumes uniform power by masking the MTJ during the write operation from the power supply thereby thwarting the power analysis based side-channel attacks. The proposed SMCL will find applications in the design of sudden power-outage resilient non-volatile secure processors.

REFERENCES

[1] E. Deng, Y. Zhang, J.-O. Klein, D. Ravelsona, C. Chappert, and W. Zhao, "Low power magnetic full-adder based on spin transfer torque mram," *IEEE transactions on magnetics*, vol. 49, no. 9, pp. 4982–4987, 2013.

[2] W. Kang, W. Lv, Y. Zhang, and W. Zhao, "Low store power high-speed high-density nonvolatile sram design with spin hall effect-driven magnetic tunnel junctions," *IEEE Transactions on Nanotechnology*, vol. 16, no. 1, pp. 148–154, 2017.

[3] W. Kang, Y. Zhang, Z. Wang, J.-O. Klein, C. Chappert, D. Ravelosona, G. Wang, Y. Zhang, and W. Zhao, "Spintronics: Emerging ultra-low-power circuits and systems beyond mos technology," *ACM Journal on Emerging Technologies in Computing Systems (JETC)*, vol. 12, no. 2, p. 16, 2015.

[4] W. Zhao, M. Moreau, E. Deng, Y. Zhang, J.-M. Portal, J.-O. Klein, M. Bocquet, H. Aziza, D. Deleruyelle, C. Muller *et al.*, "Synchronous non-volatile logic gate design based on resistive switching memories," *IEEE Transactions on Circuits and Systems I: Regular Papers*, vol. 61, no. 2, pp. 443–454, 2014.

[5] N. Onizawa, A. Mochizuki, A. Tamakoshi, and T. Hanyu, "Sudden power-outage resilient in-processor checkpointing for energy-harvesting nonvolatile processors," *IEEE Transactions on Emerging Topics in Computing*, vol. 5, no. 2, pp. 151–163, 2017.

[6] P. Kocher, J. Jaffe, and B. Jun, "Differential power analysis," in *Advances in cryptologyCRYPTO99*. Springer, 1999, pp. 789–789.

[7] S. Mangard, E. Oswald, and T. Popp, *Power analysis attacks: Revealing the secrets of smart cards*. Springer Science & Business Media, 2008, vol. 31.

[8] A. Iyengar, S. Ghosh, N. Rathi, and H. Naeimi, "Side channel attacks on sttram and low-overhead countermeasures," in *Defect and Fault Tolerance in VLSI and Nanotechnology Systems (DFT), 2016 IEEE International Symposium on*. IEEE, 2016, pp. 141–146.

[9] T. Winograd, H. Salmani, H. Mahmoodi, K. Gaj, and H. Homayoun, "Hybrid stt-cmos designs for reverse-engineering prevention," in *Proceedings of the 53rd Annual Design Automation Conference*. ACM, 2016, p. 88.

[10] A. Chakraborty, A. Mondal, and A. Srivastava, "Correlation power analysis attack against stt-mram based cyptosystems." *IACR Cryptology ePrint Archive*, vol. 2017, p. 413, 2017.

[11] J. S. Moodera, L. R. Kinder, T. M. Wong, and R. Meservey, "Large magnetoresistance at room temperature in ferromagnetic thin film tunnel junctions," *Physical review letters*, vol. 74, no. 16, p. 3273, 1995.

[12] R. Zand, A. Roohi, S. Salehi, and R. F. DeMara, "Scalable adaptive spintronic reconfigurable logic using area-matched mtj design," *IEEE Transactions on Circuits and Systems II: Express Briefs*, vol. 63, no. 7, pp. 678–682, 2016.

[13] B. Behin-Aein, J.-P. Wang, and R. Wiesendanger, "Computing with spins and magnets," *MRS Bulletin*, vol. 39, no. 08, pp. 696–702, 2014.

[14] Y. Gang, W. Zhao, J.-O. Klein, C. Chappert, and P. Mazoyer, "A high-reliability, low-power magnetic full adder," *IEEE Transactions on Magnetics*, vol. 47, no. 11, pp. 4611–4616, 2011.

[15] Y. WANG, Y. ZHANG, J.-O. Klein, T. Devolder, D. Ravelosona, C. Chappert, and W. Zhao, "Compact model for perpendicular magnetic anisotropy magnetic tunnel junction," Aug 2017. [Online]. Available: https://nanohub.org/publications/56/2

Quantum Circuit Designs of Integer Division Optimizing T-count and T-depth

Himanshu Thapliyal[*], T. S. S. Varun[*], Edgard Muñoz-Coreas[*], Keith A. Britt[†] and Travis S. Humble[†]

[*]Department of Electrical and Computer Engineering
University of Kentucky, Lexington, KY
Email: hthapliyal@uky.edu
[†]Quantum Computing Institute
Oak Ridge National Laboratory, TN

Abstract—Quantum circuits for basic mathematical functions such as division are required to implement scientific computing algorithms on quantum computers. In this work, we propose two designs for quantum integer division. The designs are based on quantum Clifford+T gates and are optimized for T-count and T-depth. Quantum circuits that are based on Clifford+T gates can be made fault tolerant in nature but the T gate is very costly to implement. As a result, reducing T-count and T-depth have become important optimization goals. Existing quantum hardware is limited in terms of number of available qubits. Thus, ancillary qubits are a circuit overhead that needs to be kept to a minimum. We propose two quantum integer division circuits. The first quantum integer division circuit is based on the non-restoring division algorithm. The proposed non-restoring division circuit is optimized for total quantum hardware (T-count and T-depth) cost but requires $2 * n + 1$ ancillary qubits. We also propose a quantum integer division circuit based on the restoring division algorithm. The proposed restoring division circuit is optimized for total qubits. The design requires only n ancillary qubits but will need more quantum hardware than the non-restoring division circuit. Both proposed quantum circuits are based on (i) a new quantum conditional addition circuit, (ii) a new quantum adder-subtractor and (iii) a new quantum subtraction circuit. Further, both designs are compared and shown to be superior to existing work in terms of T-count and T-depth. The proposed quantum non-restoring integer division circuit has a 96% improvement in terms of T-count and a 93% improvement in terms of T-depth compared to existing work. The proposed quantum restoring integer division circuit has a 91% improvement in terms of T-count and a 86% improvement in terms of T-count compared to the existing work.

I. INTRODUCTION AND BACKGROUND

Quantum circuits of arithmetic operations are needed to design quantum hardware for implementing quantum algorithms such as Shor's factoring algorithm, the discrete log problem, class number algorithm and triangle finding algorithm [1] [2]. Dividers are one of the major computational units in quantum arithmetic and have applications in circuit designs of quantum algorithms [3] [1].

Quantum circuits that are based on Clifford+T gates can be made fault tolerant in nature permitting reliable and scalable quantum computation [4] [5]. The Clifford+T gate family is illustrated in [6]. The T gate is very costly to implement compared to the Clifford gates making reducing T-count and T-depth important optimization goals [5] [7]. Existing quantum

hardware is limited in terms of number of available qubits [8]. Thus, ancillary qubits are a circuit overhead that needs to be kept to a minimum.

In the existing literature, there are a handful of integer divider designs based on reversible gates targeting mostly reversible computing [9] [10] [11]. Among these designs we found only [12] to be suitable for quantum computing. The quantum integer division circuit in [12] implements the restoring division algorithm and uses the quantum Fourier transform to perform the division operation. However, the design in [12] is not optimized for T-depth and T-count. The quantum division circuit in [12] uses controlled phase shift gates. It is known that the controlled phase gates required by the design in [12] can only be approximated by Clifford+T gates [13]. The Clifford+T based approximations of the controlled phase gates have a high T gate cost [13]. Further, the T gate cost increases as the accuracy of the controlled phase gate approximation is improved [13]. Thus, implementing all the controlled phase gates required by the design in [12] with a high degree of accuracy will result in a design with high T-count and T-depth [13].

This paper presents two designs for quantum circuit integer division based on Clifford+T gates. The first quantum circuit is based on the non-restoring division algorithm and the second quantum circuit is based on the restoring division algorithm. Both proposed quantum integer division circuits are based on (i) a new quantum conditional ADD operation circuit, (ii) a new quantum adder-subtractor and (iii) a new quantum subtraction circuit. The proposed non-restoring division circuit is optimized for total quantum hardware (T-count and T-depth) cost. The trade off for reducing the quantum hardware of the design is the need to use more ancillary qubits. The non-restoring division circuit requires $2 * n + 1$ ancillary qubits. The proposed quantum restoring division is designed with the aim to minimize total qubits. We reduce the number of ancillary qubits to n but must use more quantum hardware than the proposed quantum non-restoring division circuit. Both the proposed restoring quantum integer division circuit and proposed non-restoring quantum integer division circuit are compared and shown to be superior to existing work in terms of T-depth and T-count.

978-1-5386-1357-3/17 $31.00 © 2017 IEEE

123

This paper is organized as follows. Section II presents the design of the (i) new quantum conditional addition circuit, (ii) a new quantum adder-subtractor and (iii) a new quantum subtraction circuit used in the proposed quantum division circuits. In section III the design of the proposed quantum non-restoring integer division circuit is discussed. The design of the proposed quantum restoring integer division circuit is presented in section IV.

II. Design of Quantum Circuits Used In Proposed Integer Division Circuits

The quantum circuits that are required for developing the proposed non-restoring and restoring integer division circuits are: (i) controlled adder-subtractor, (ii) quantum subtractor and (iii) conditional ADD operation circuit. The quantum circuit designs of the quantum adder-subtractor, quantum subtractor and the conditional ADD operation circuit are discussed in the following sections.

A. Design of Quantum Subtractor

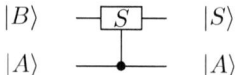

Fig. 1. Graphic symbol of quantum subtractor. S represents the quantum subtraction operation

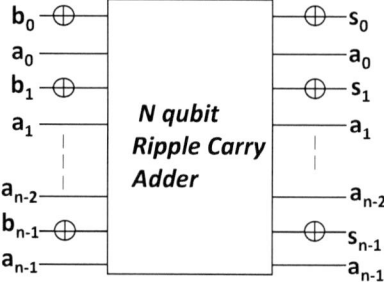

Fig. 2. Circuit design of N qubit quantum subtractor based on N qubit quantum ripple carry adder

Fig.1 shows the symbol of the quantum subtractor circuit. The subtractor circuit takes two n qubit inputs $|A\rangle$ and $|B\rangle$. The input a is regenerated at the output. The n-qubit output $|S\rangle$ has the result of the subtraction of b and a. Fig.2 shows the circuit design of N qubit subtractor based on N qubit quantum ripple carry adder. As shown in Fig.2, a quantum ripple carry adder is required to develop a quantum subtractor circuit. We use the quantum ripple carry adder proposed in [14] for developing the quantum subtractor circuit. To perform subtraction, we use the design approach presented in [15]. Thus, the input qubits $|B\rangle$ are complemented before being applied to the quantum ripple carry adder. Then, the ripple carry adder calculates $\bar{b} + a$. At the end of computation, the

input qubits $|B\rangle$ are complemented again. As a result, the quantum subtractor calculates $(\overline{\bar{b} + a})$ which is equivalent to $b - a$ [15].

B. Design of Quantum Adder-Subtractor

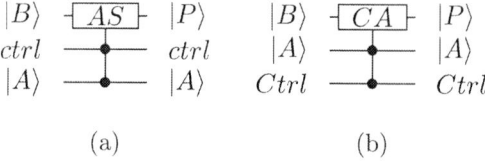

Fig. 3. Graphic symbols of (a) Adder-Subtractor (b) Conditional ADD operation circuit. AS represents add or subtract operation. CA represents conditional add operation

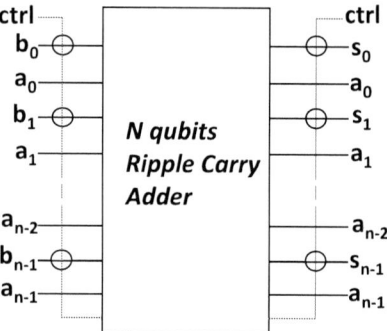

Fig. 4. Circuit design of N qubit quantum adder-subtractor based on N qubit quantum ripple carry adder

Fig. 3(a) shows the graphic symbol of the quantum controlled addition or subtraction circuit. The quantum adder-subtractor circuit operates as follows: (i) when the input labeled $ctrl$ is high (refer Fig. 3(a)), the circuit output is $|P\rangle = |B - A\rangle$, (ii) when the $ctrl$ input is low, the circuit output is $|P\rangle = |B + A\rangle$.

The complete working circuit of the quantum adder-subtractor circuit is shown in Fig. 4. The quantum adder-subtractor circuit is based on the design presented in [15] and uses the ripple carry adder in [14]. The quantum adder-subtractor calculates $(\overline{\bar{b} + a})$ when $ctrl$ is high. The expression $(\overline{\bar{b} + a})$ is equivalent to $b - a$.

C. Design of Quantum Conditional ADD Operation Circuit

Fig. 3(b) shows the graphic symbol of the quantum conditional ADD operation circuit. The quantum conditional ADD operation circuit operates as follows: (i) when the input labeled $ctrl$ is high (refer Fig. 3(b)), the circuit output is $|P\rangle = |B + A\rangle$, (ii) when the $ctrl$ input is low, the circuit output is $|P\rangle = |B\rangle$.

The complete working circuit of quantum conditional ADD operation circuit is shown in Fig.5 for 4 qubit operands. The quantum conditional ADD circuit uses a modified version of the ripple carry adder proposed in [14]. We were able

978-1-5386-1357-3/17 $31.00 © 2017 IEEE   124

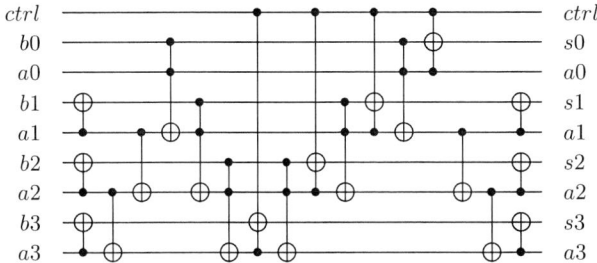

Fig. 5. Circuit design of quantum conditional ADD operation circuit

---

Algorithm 1: *Proposed quantum non-restoring division algorithm*

function $Non-Restore$ $(|Q_n\rangle, |R_n\rangle, |D_n\rangle)$
  for $i = 0$ *to* $n-1$ **do**
    /* **Start Core Engine Phase** */
    if($|R_{[0:n-1]}\rangle > 0$) **then**
      $(|Q_{[1:n-1]}\rangle, |R_{[0:n-1]}\rangle) = $ Leftshift $(|Q_{[0:n-1]}\rangle, |R_{[0:n-1]}\rangle)$;
      $|R_{[0:n-1]}\rangle = |\dot{R}_{[0:n-1]}\rangle + |D_{[0:n-1]}\rangle$;
    else
      $(|Q_{[1:n-1]}\rangle, |R_{[0:n-1]}\rangle) = $ Leftshift $(|Q_{[0:n-1]}\rangle, |R_{[0:n-1]}\rangle)$;
      $|R_{[0:n-1]}\rangle = |\dot{R}_{[0:n-1]}\rangle - |D_{[0:n-1]}\rangle$;
    end if;
    if($|R_{[0:n-1]}\rangle > 0$) **then**
      $|Q_{[0]}\rangle = 1$;
    else
      $|Q_{[0]}\rangle = 0$;
    end if;
    /* **End Core Engine Phase** */
  end for;
  //after n iterations//
    /* **Start Supplementary Restoring Phase** */
    if($|R_{[0:n-1]}\rangle > 0$) **then**
      $|R_{[0:n-1]}\rangle = |R_{[0:n-1]}\rangle$;
    else
      $|R_{[0:n-1]}\rangle = |R_{[0:n-1]}\rangle + |D_{[0:n-1]}\rangle$;
    end if;
    /* **End Supplementary Restoring Phase** */
  return R;
end function

TABLE I
PROPOSED QUANTUM NON-RESTORING DIVISION ALGORITHM

---

to remove the qubit that performs the carry out for the adder in [14] as we do not need the carry out qubit in the proposed integer dividers. The addition architecture in [14] uses Peres gates to perform the addition. The Peres gate can be decomposed into a Feynman and a Toffoli gate. By replacing the Feynman gate with a Toffoli gate, we can use the control line ($ctrl$) to determine whether the conditional ADD circuit will perform addition or no operation. Although, Fig.5 is just shown for 4 qubit operands, it can easily be extended to any operand size.

III. Design of Non-Restoring Quantum Integer Division Circuit

The quantum circuits that are required for developing the hardware implementation of the proposed non-restoring division algorithm are: (i) Leftshift operation circuit, (ii) controlled adder-subtractor, and (iii) conditional ADD operation circuit. We observed that we can eliminate the LeftShift operation circuit by combining $|R_{[0:n-2]}\rangle$ and $|Q_{[n-1]}\rangle$ to form an n qubit register there by saving the quantum resources.

The proposed non-restoring division algorithm for quantum circuits is shown in Table I. In Table I, the inputs to be given are: (a) $(|Q_{[0:n-1]}\rangle$, n qubit register in which the dividend is loaded; (b) $|D_{[0:n-1]}\rangle$, n qubit register in which the divisor is loaded; (c) $|R_{[0:n-1]}\rangle$, n qubit remainder register which is initiated to 0 at the start. At the end of computation, we get the quotient at $|Q_{[0:n-1]}\rangle$ and remainder at $|R_{[0:n-1]}\rangle$. The divisor is retained at the output. Also, $n+1$ garbage qubits are produced. The methodology to design our proposed quantum non-restoring integer division circuit is developed from the non-restoring division algorithm shown in Table I. The Steps of the methodology are presented below.

A. Design Methodology for Quantum Non-Restoring Integer Division Circuit

From Table I, we can see that the algorithm is divided into two phases. (i) Core Engine Phase and (ii) Supplementary Restoring Phase. The Core Engine Phase is iterated n times. Supplementary Restoring Phase takes place after the end of n iterations of the Core Engine Phase. The Supplementary Restoring Phase is repeated once. A quantum circuit is developed for each of these phases. The final circuit that performs the integer division using the non-restoring integer division

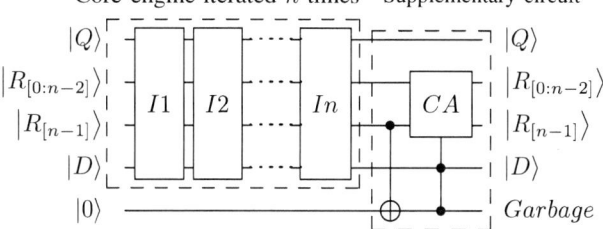

Fig. 6. Quantum non-restoring integer divider circuit design

algorithm is shown in Fig. 6. In Fig. 6, *I1* represents the first iteration of the Core Engine Phase, *I2* represents the second iteration and *In* represents the final iteration.

1) Core Engine Phase: Fig. 7 represents the quantum circuit that does the operations that are marked under the Core Engine Phase in the algorithm in Table IV. We now elaborate on how the information moves in Fig. 7 .

- Step 1. $|D_{[0:n-1]}\rangle$ holds the divisor, $|R_{[0:n-1]}\rangle$ is initialised to zero, and $|Q_{[0:n-1]}\rangle$ holds the dividend.
- Step 2. We consider, $|Q_{[n-1]}\rangle$ and $|R_{[0:n-2]}\rangle$, as one combined register.
- Step 3. The combined register of Step 2 and $|D_{[0:n-1]}\rangle$ are applied as two n qubits inputs to the quantum adder-subtractor circuit. In Fig. 7, AS represents the adder-subtractor circuit. At the end of computation, register $|D_{[0:n-1]}\rangle$ emerges unchanged and the combined register

978-1-5386-1357-3/17 $31.00 © 2017 IEEE   125

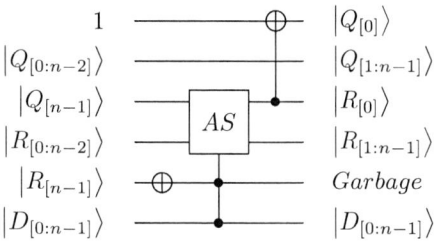

Fig. 7. Quantum non-restoring integer divider circuit design for first iteration(core engine)

now holds the sum or difference of the combined register and D.

- Step 4. Qubit $|R_{[n-1]}\rangle$ is complemented and applied as the $ctrl$ qubit to quantum adder-subtractor circuit.
- Step 5. The $ctrl$ qubit is left out as garbage.
- Step 6. An ancillary qubit set to 1 and qubit $|Q_{[n-1]}\rangle$ are applied to a CNOT gate. $|Q_{[n-1]}\rangle$ is the control qubit and 1 is the target qubit.

The Steps from 1 to 6 constitute the operations of the Core Engine Phase. From the algorithm in Table I, it can be seen that Steps 2 to 6 of the Core Engine Phase are iterated n times. So, the circuit in Fig. 7 that represents the Core Engine Phase is also iterated n times (see Fig. 6). The outputs of the first iteration as inputs to the second iteration and so on for all n iterations.

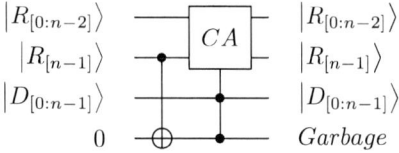

Fig. 8. Quantum circuit implementation of the Supplementary Restoring Phase(refer Table 1)

2) Supplementary Restoring Phase: After the end of n iterations of the Core Engine Phase, $|R_{[0:n-1]}\rangle$ might be negative at the end of n iterations. In that case, it has to be restored by adding the divisor. This restoration of the negative remainder is carried out by the Supplementary Restoring Phase quantum circuit shown in Fig. 8. The quantum circuit shown in Fig. 8 is the quantum implementation of the Supplementary Restoring Phase marked in the algorithm in Table I. We now elaborate on how the information moves in the supplementary circuit.

- Step 1. The qubit $|R_{[n-1]}\rangle$ and an ancillary qubit set to 0 are applied as inputs to a CNOT gate. $|R_{[n-1]}\rangle$ is the control qubit and the ancillary qubit is the target qubit. The target now holds the value of $|R_{[n-1]}\rangle$.
- Step 2. The ancillary qubit is used as $ctrl$ qubit to the conditional ADD operation quantum circuit.
- Step 3. Registers $|R_{[0:n-1]}\rangle$ and $|D_{[0:n-1]}\rangle$ are applied as inputs to conditional ADD operation quantum circuit.

In Fig. 8, CA represents the conditional ADD operation circuit. $|D_{[0:n-1]}\rangle$ emerges unchanged and $|R_{[0:n-1]}\rangle$ will contain either the sum or emerge unchanged.

- Step 4. The control qubit $|R_{[0:n-1]}\rangle$ is left out as garbage.
- Step 5. After Step 4, we have the Quotient in $|Q_{[0:n-1]}\rangle$, and the remainder in $|R_{[0:n-1]}\rangle$. The divisor $|D_{[0:n-1]}\rangle$ is unchanged.

B. Cost Comparison With Existing Work

TABLE II
RESOURCE COUNT OF PROPOSED NON-RESTORING ALGORITHM
DIVISION CIRCUIT

Designs	Adder-Subtractor	conditional ADD operation circuit	Non-Restoring Divider
T-count	$(14n-14)$	$(21n-14)$	$14n^2+21n-28$
T-depth	8	16	$8*n+7$
Ancilla qubits	0	0	$2*n+1$

TABLE III
COMPARISON OF RESOURCE COUNT BETWEEN PROPOSED AND EXISTING
WORK

	1	Proposed	% impr. w.r.t. 1
T-count	$\approx 400n^2$	$14n^2+21n-28$	$\approx 96\%$
T-depth	$130*n$	$8*n+7$	$\approx 93\%$
Ancilla qubits	$2n$	$2*n+1$	$\approx$ -

1 is the work in [12]

The resources used in the design of the proposed quantum non-restoring integer division circuit is presented in Table II. As shown in Table II, the proposed design will require $2*n+1$ ancillary qubits. n ancillary qubits are used during initialization of remainder register and the remaining $n+1$ are transformed to garbage output. The T-count required by the design is given by summing the cost of adder-subtractor and conditional ADD operation quantum circuit at each stage. T-count of the proposed quantum non-restoring integer division circuit is $14n^2+21n-28$. The T-depth required by the design is given as $8*n+7$.

Comparison of resource costs between the proposed quantum non-restoring integer division circuit and the existing work is shown in Table III. To calculate the T-count and T-depth for [12] we use T-count and T-depth values from approximate phase gate implementations reported in [13]. The implementations with the poorest accuracy are used. This is because the T gate cost increases significantly as a function of accuracy. Table III shows that the proposed quantum circuit of integer division has an improvement ratio of 93% in terms of T-depth, and 96% in terms of T-count.

IV. DESIGN OF RESTORING QUANTUM INTEGER DIVISION CIRCUIT

The quantum circuits that are required for developing the hardware implementation of the proposed restoring division

978-1-5386-1357-3/17 $31.00 © 2017 IEEE

algorithm are (i) Leftshift operation circuit, (ii) n qubit quantum subtractor and (iii) Conditional ADD operation circuit. We observed that we can eliminate the LeftShift operation circuit by combining $|R_{[0:n-2]}\rangle$ and $(|Q_{[n-1]}\rangle$ to form an n qubit register which is actually equal to performing an left shift operation. By combining the qubits in this way, we do not have to use a separate left shift operation circuit.

The proposed restoring division algorithm is shown in Table IV. In Table IV, the inputs to be given are: (a) $(|Q_{[0:n-1]}\rangle$, n qubit register in which the dividend is loaded ; (b) $|D_{[0:n-1]}\rangle$, n qubit register in which the divisor is loaded; (c) $|R_{[0:n-1]}\rangle$, n qubit remainder register which is initiated to 0 at the start. The algorithm repeats n times. At the end of n iterations, we get the quotient at $(|Q_{[0:n-1]}\rangle$ and the remainder at $|R_{[0:n-1]}\rangle$. The divisor is retained at the output. The methodology to design our proposed quantum restoring integer division circuit is developed from the restoring division algorithm shown in Table IV. The Steps of the methodology are presented below.

Algorithm 1 : *Proposed Restoring division algorithm*

---

function $Restore$ $(|Q_n\rangle, |R_n\rangle, |D_n\rangle)$
  for $i = 0$ to $n-1$ **do**
    $(|Q_{[1:n-1]}\rangle, |R_{[0:n-1]}\rangle) = $ Leftshift $(|Q_{[0:n-1]}\rangle, |R_{[0:n-1]}\rangle)$;
    $(|R-D_{[0:n-1]}\rangle = |R_{[0:n-1]}\rangle - |D_{[0:n-1]}\rangle)$;
    if$(|R_{[0:n-1]}\rangle > 0)$ **then**
      $|Q_{[0]}\rangle = 1$
      $|R_{[0:n-1]}\rangle = |R-D_{[0:n-1]}\rangle$;
    else
      $|Q_{[0]}\rangle = 0$;
      $|R_{[0:n-1]}\rangle = |R-D_{[0:n-1]}\rangle + |D_{[0:n-1]}\rangle$;
    end if;
  end for;
//repeat for n iterations//
return R;
end function

TABLE IV
PROPOSED RESTORING DIVISION ALGORITHM FOR QUANTUM CIRCUITS

A. Design Methodology for Quantum Restoring Integer Division Circuit

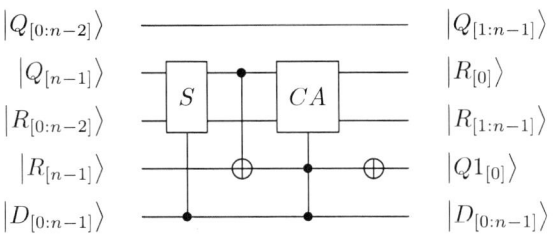

Fig. 9. Quantum restoring integer divider circuit design for a single iteration

Fig.9 shows the quantum circuit generated for the quantum restoring division circuit after 1 iteration of our design methodology. The Steps of the proposed methodology are repeated n times. Hence, the circuit in Fig. 9 is also iterated

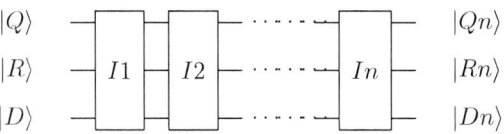

Fig. 10. Quantum restoring integer divider circuit design(for n iterations)

n times. This is done by using the outputs of the first iteration as inputs for the next iteration. Fig. 10 shows the complete quantum restoring division circuit where $I1$ represents the first iteration, $I2$ represents second iteration and In represents the final iteration. We now elaborate on how information moves through the circuit shown in Fig. 9.

- Step 1. The $|D_{[0:n-1]}\rangle$ holds the divisor, $|R_{[0:n-1]}\rangle$ is initialised to zero, and $|Q_{[0:n-1]}\rangle$ holds the dividend.
- Step 2. We consider, $|Q_{[n-1]}\rangle$ and $|R_{[0:n-2]}\rangle$, as one combined register.
- Step 3. The combined register mentioned above in Step 2, and $|D_{[0:n-1]}\rangle$ are given as inputs to the quantum subtractor circuit. Register $|D_{[0:n-1]}\rangle$ emerges unchanged. The combined register now holds the result of subtraction of R and D registers. Let us call this result as $|R-D_{[0:n-1]}\rangle$.
- Step 4. Qubits $|R-D_{[n-1]}\rangle$ and $|R_{[n-1]}\rangle$ are supplied to a CNOT gate. $|R-D_{[n-1]}\rangle$ is the control qubit and the $|R_{[n-1]}\rangle$ is the target qubit. The target now holds the value of $|R-D_{[n-1]}\rangle$ because $|R_{[n-1]}\rangle$ is always zero throughout the computation.
- Step 5. Qubit $|R_{[n-1]}\rangle$ is the control qubit to the conditional ADD operation circuit.
- Step 6. Registers $|R-D_{[0:n-1]}\rangle$ and $|D_{[0:n-1]}\rangle$ are the two n qubit inputs to the conditional ADD operation circuit. Register $|D_{[0:n-1]}\rangle$ emerges unchanged. The combined register will contain either the sum or emerge unchanged..
- Step 7. $|R_{[n-1]}\rangle$ is complemented.

Steps 2 through 7 are repeated n times. At the end of n iterations, the Quotient will be in $|Q_{[0:n-1]}\rangle$, the remainder in $|R_{[0:n-1]}\rangle$ and the divisor emerges unchanged.

B. Cost Comparison With Existing Work

TABLE V
RESOURCE COUNT OF PROPOSED RESTORING DIVISION CIRCUIT

	Subtractor	conditional ADD operation circuit	Restoring Divider
T-count	$(14n-14)$	$(21n-14)$	$35n^2 - 28n$
T-depth	8	16	$18*n$
Ancilla qubits	0	0	n

The resources used in the design of the proposed quantum restoring integer division circuit is presented in Table V. As shown in Table V, the proposed design will require n ancillary qubits during initialization of the remainder register. The T-count required by the design is given by summing the cost of

978-1-5386-1357-3/17 $31.00 © 2017 IEEE   127

TABLE VI
COMPARISON OF RESOURCE COUNT BETWEEN PROPOSED AND EXISTING WORK

	1	Proposed	% impr. w.r.t. 1
T-count	$\approx 400n^2$	$35n^2 - 28n$	$\approx 91\%$
T-depth	$130*n$	$18*n$	86.15%
Ancilla qubits	$2n$	n	50%

1 is the work in [12]

subtractor and conditional ADD operation quantum circuit at each stage. T-count of the proposed quantum restoring integer division circuit is $35n^2 - 28n$. The T-depth required by the design is given as $18*n$.

Comparison of resource estimation between proposed quantum circuit of integer division and the existing quantum circuit of integer division in [12] is shown in Table VI. To calculate the T-count and T-depth for [12] we use T-count and T-depth from approximate phase gate implementations reported in [13]. The implementations with the poorest accuracy were used. This is because the T-count increases significantly as a function of accuracy. Table VI showed that the proposed quantum circuit of integer division has an improvement ratio of 86.15% in terms of T-depth, and 91% in terms of T-count.

V. CONCLUSION

In this work, we have presented two designs for quantum circuit integer division based on Clifford+T gates. The first quantum circuit presented is based on the non-restoring division algorithm and the second quantum circuit presented is based on the restoring division algorithm. The design of subcomponents used in the proposed quantum integer division circuits such as the quantum conditional ADD operation circuit, quantum adder-subtractor and quantum subtraction circuit are also shown. The proposed quantum integer division circuits are shown to be superior to existing designs in terms of T-depth and T-count. We conclude that the proposed non-restoring division circuit can be integrated in a larger quantum data path system design where T-count and T-depth are of primary concern. We also conclude that the proposed restoring division circuit can be integrated in a larger quantum data path system design to implement quantum algorithms where qubits are limited and T-count and T-depth must be kept to a minimum.

Existing quantum circuit implementations do not include the additional qubit transformations that account for the available instruction set architecture, the hardware connectivity and layout constraints of a particular technology [16], [17]. For example, in trapped ion quantum computers (such as those presented in [18] and [19]) offer different methods to implement multi-qubit gates. These methods include piece-wise, nearest-neighbor interactions that address individual qubits as well as global interactions that apply coherent rotations uniformly to all available ions. The choice of which method to use depends on the layout of the device architecture and the relative complexity of the different instructions. Such

constraints will significantly impact how quantum circuits are implemented in practice. The proposed quantum integer division circuit designs do not take into account technology constraints. However, the T-count and T-depth cost savings of our quantum integer division circuits are unaffected by these hardware considerations. To efficiently implement quantum algorithms, new designs need to be investigated for integer division that minimize the overhead imposed by technology constraints.

REFERENCES

[1] P. Selinger et. al., *The Quipper System*, 2016, available at: http://www.mathstat.dal.ca/ selinger/quipper/doc/.

[2] S. Beauregard, "Circuit for Shor's algorithm using 2n+3 qubits," *Quantum Information & Computation*, vol. 3, no. 2, pp. 175–185, Mar 2003.

[3] M. A. Nielsen and I. Chuang, "Quantum computation and quantum information," 2002.

[4] A. Paler, I. Polian, K. Nemoto, and S. J. Devitt, "Fault-tolerant, high-level quantum circuits: form, compilation and description," *Quantum Science and Technology*, vol. 2, no. 2, p. 025003, 2017. [Online]. Available: http://stacks.iop.org/2058-9565/2/i=2/a=025003

[5] X. Zhou, D. W. Leung, and I. L. Chuang, "Methodology for quantum logic gate construction," *Phys. Rev. A*, vol. 62, p. 052316, Oct 2000. [Online]. Available: https://link.aps.org/doi/10.1103/PhysRevA.62.052316

[6] M. Amy, D. Maslov, M. Mosca, and M. Roetteler, "A meet-in-the-middle algorithm for fast synthesis of depth-optimal quantum circuits," *IEEE Transactions on Computer-Aided Design of Integrated Circuits and Systems*, vol. 32, no. 6, pp. 818–830, 2013.

[7] S. J. Devitt, A. M. Stephens, W. J. Munro, and K. Nemoto, "Requirements for fault-tolerant factoring on an atom-optics quantum computer," *Nature Communications*, vol. 4, p. 2524, Oct. 2013.

[8] IBM, *Quantum Computing - IBM Q*, 2017, available at: https://www.research.ibm.com/ibm-q/.

[9] N. M. Nayeem, A. Hossain, M. Haque, L. Jamal, and H. M. H. Babu, "Novel reversible division hardware," in *2009 52nd IEEE International Midwest Symposium on Circuits and Systems*, Aug 2009, pp. 1134–1138.

[10] S. V. Dibbo, H. M. H. Babu, and L. Jamal, "An efficient design technique of a quantum divider circuit," in *2016 IEEE International Symposium on Circuits and Systems (ISCAS)*, May 2016, pp. 2102–2105.

[11] F. Dastan and M. Haghparast, "A novel nanometric fault tolerant reversible divider," *International Journal of the Physical Sciences*, vol. 6, no. 24, pp. 5671–5681, October 2011.

[12] A. Khosropour, H. Aghababa, and B. Forouzandeh, "Quantum division circuit based on restoring division algorithm," in *Information Technology: New Generations (ITNG), 2011 Eighth International Conference on*. IEEE, 2011, pp. 1037–1040.

[13] V. Kliuchnikov, D. Maslov, and M. Mosca, "Fast and efficient exact synthesis of single-qubit unitaries generated by clifford and t gates," *Quantum Info. Comput.*, vol. 13, no. 7-8, pp. 607–630, Jul. 2013. [Online]. Available: http://dl.acm.org/citation.cfm?id=2535649.2535653

[14] H. Thapliyal and N. Ranganathan, "Design of efficient reversible logic-based binary and bcd adder circuits," *ACM Journal on Emerging Technologies in Computing Systems (JETC)*, vol. 9, no. 3, p. 17, 2013.

[15] H. Thapliyal, "Mapping of subtractor and adder-subtractor circuits on reversible quantum gates," in *Transactions on Computational Science XXVII*. Springer, 2016, pp. 10–34.

[16] K. A. Britt and T. S. Humble, "High-performance computing with quantum processing units," *ACM Journal on Emerging Technologies in Computing Systems (JETC)*, vol. 13, no. 3, p. 39, 2017.

[17] K. A. Britt and T. S. Humble, "Instruction set architectures for quantum processing units," *arXiv preprint arXiv:1707.06202*, 2017.

[18] N. M. Linke, D. Maslov, M. Roetteler, S. Debnath, C. Figgatt, K. A. Landsman, K. Wright, and C. Monroe, "Experimental comparison of two quantum computing architectures," *Proceedings of the National Academy of Sciences*, p. 201618020, 2017.

[19] E. A. Martinez, T. Monz, D. Nigg, P. Schindler, and R. Blatt, "Compiling quantum algorithms for architectures with multi-qubit gates," *New Journal of Physics*, vol. 18, no. 6, p. 063029, 2016.

2017 IEEE International Symposium on Nanoelectronic and Information Systems

High Performance Sense Amplifier based Flip Flop for driver applications

Anoop D
Dept. of Electronics and
Communication Engineering
National Institute of Technology Goa
Email: anoopdelamp@gmail.com

Dr. Nithin Kumar Y. B.
Dept. of Electronics and
Communication Engineering
National Institute of Technology Goa
Email: nithin.shastri@nitgoa.ac.in

Dr. Vasantha M. H.
Dept. of Electronics and
Communication Engineering
National Institute of Technology Goa
Email: vasanthmh@nitgoa.ac.in

Abstract—The presence of memory elements is in the rise in digital systems. The need for faster memory elements like latches and flip-flops are increasing day by day. A new sense amplifier based flip-flop architecture is presented in this paper. The proposed flip-flop provides ratioless design and faster rise operation. The simulation results, obtained for the $0.18\mu m$ technology, show improvements in the clock-to-output delay ($95\ ps$) and the power delay product of $39\ fJ$ when the outputs are loaded with $200fF$ load capacitance.

Keywords—Sense Amplifier, Flip Flop, Delay, Latches, SAFF

I. Introduction

The memory elements are more common than ever in the digital systems. The need for faster memory elements is the requirement of the day while consuming less power and occupying a small area. The clocked latches or flip-flops are mostly used in the peripheral circuits and I/Os. A significant portion of the system on the chip, digital integrated circuits are covered by memory elements like latches and flip-flops. So the need of faster, low power and small area memory elements are essential for the better performance of systems.

Timing elements of digital systems are the critical part for high-performance [1]. The faster flip-flops and latches are needed to improve the overall speed of systems. Commonly used architectures of latches and flip-flops in high-performance systems are compared and examined in the paper [2]. This paper analyzes the trade-off between speed and power.

A lot of structures are presented for high speed flip-flops and latches. Hybrid latch flip-flop (HLFF) structure is proposed in [3]. The paper [4] introduces a family of semi dynamical flip-flop (SDFF) and dynamic flip-flops which provide lower latency than the HLFF.

The sense amplifier based flip-flop (SAFF) proposed in [5] consists of a sense amplifier and a latch. It exhibits smaller setup time and hold time [8]. These features make the SAFF a good candidate for high-speed flip-flop design.

This paper is further organized as follows. The Section II reviews the existing architectures of SAFF. The Section III explains the Proposed latch architecture and its operation. Section IV discusses the results of the Proposed latch and its comparison with the existing SAFF. Finally, the paper is concluded in Section V.

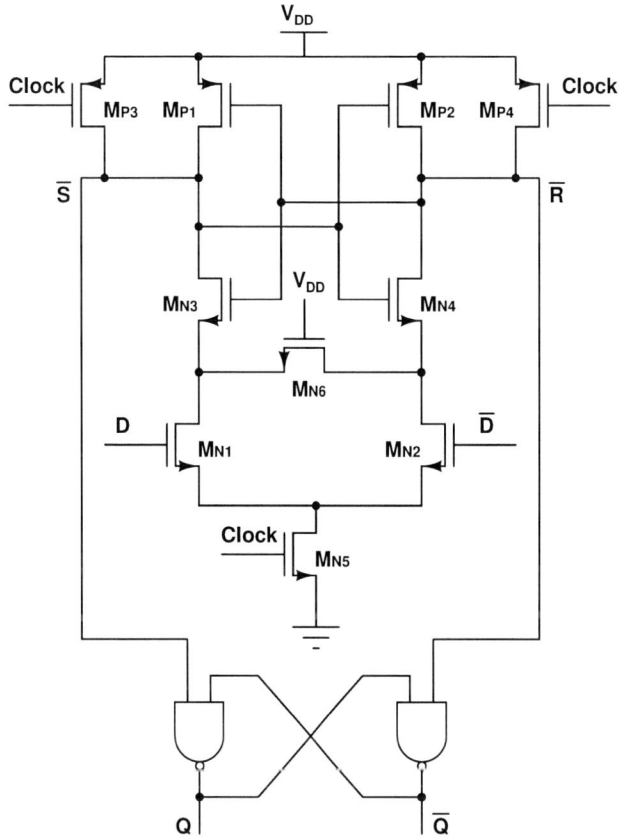

Fig. 1: Schematic of conventional SAFF with NAND latch [5]

II. Sense Amplifier based Flip-Flops

The SAFF is made up of two parts. One a sense amplifier and another a latch. The sense amplifier has two outputs $\overline{S}$ and $\overline{R}$ presented along with the NAND latch in Fig. 1. From these two outputs, a latch is to be created to construct the flip-flop.

Simplest solution for the latch is to make use of NAND latch. The forbidden state to it is when both of the inputs is logic zero. Since the signals from sense amplifier do not occur at logic zero simultaneously, the NAND latch provides required output.

978-1-5386-1357-3/17 $31.00 © 2017 IEEE   129

TABLE I: Truth Table of the Sense Amplifier based Flip-Flop

Clock	D	$\overline{S}$	$\overline{R}$	S	R	Q	$\overline{Q}$
0	X	1	1	0	0	Q	$\overline{Q}$
1	0	1	0	0	1	0	1
1	1	0	1	1	0	1	0

The output of NAND is equal to logic one when either of the inputs is zero and logic zero when both of the inputs is one. From the Fig. 1, it can be noted that the output Q depends on the two inputs $\overline{S}$ and $\overline{Q}$. Suppose initially $\overline{S}$ was logic zero and the output Q was logic one and then $\overline{S}$ changed to logic high and $\overline{R}$ to logic low, the output Q will change from logic one to logic zero after the output $\overline{Q}$ changes its value from logic low to logic high.

The output high to low transition takes one gate more delay than the output low to high transition. This introduces huge delay if both the outputs are driving large load. This dependency between the outputs is slowing down the flip-flop operation.

In order to counter this dependency between the outputs, a number of latch architectures are proposed in the literature [6], [7], [8]. In all these architectures, the sense amplifier remains the same.

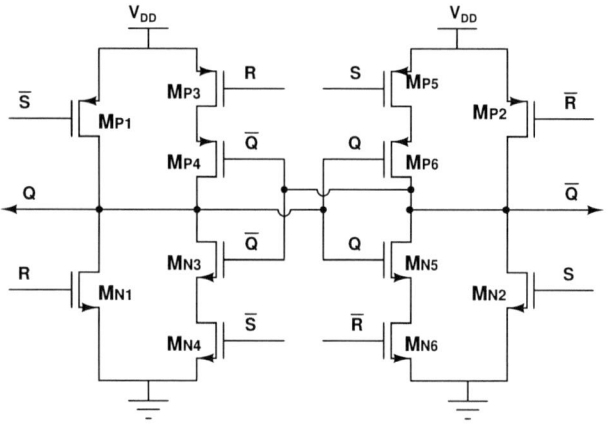

Fig. 2: Schematic of Nikolic's latch [6]

The Fig. 2 illustrates the circuit of Nikolic's latch [6]. The output Q in circuit is driven by the PMOS transistor M_{P1} and NMOS transistor M_{N1} when clock is logic high. The PMOS transistor is driven by $\overline{S}$ and NMOS transistor by R. Similarly for $\overline{Q}$, the PMOS transistor M_{P2} and NMOS transistor M_{N2} driven by $\overline{R}$ and S correspondingly. Inverters are required to produce S and R signals. They are the complement signals of $\overline{S}$ and $\overline{R}$ signals respectively.

The output low-to-high transition of Nikolic's latch has 2 gate delays and high-to-low transition takes 3 gate delays [8]. The architecture proposed in [7] and [8] reduces the high-to-low output delay.

In this architecture as well, the output low-to-high transi-

tion is driven by signal $\overline{S}$ when clock is logic high. But for high-to-low transition, a novel approach is introduced in [7]. From the Table I it can be observed that, the output Q is low when both clock as well as $\overline{S}$ equal to logic high. The low-to-high transition after the clock edge still takes 2 stage delay. One $Clock - \overline{S}$ another $\overline{S} - Q$. However high-to-low transition takes only 1 stage delay i.e., $Clock, \overline{S} - Q$. This is because when clock is logic low, $\overline{S}$ node is recharged to logic high. If input D is logic low, then there is no delay between $Clock - \overline{S}$ delay as $\overline{S}$ has already settled onto logic high value. So the only delay is in the latch stage i.e., the discharge path through the NMOS transistors M_{N1} and M_{N2}. This reduces the high-to-low transition of output significantly.

During refresh stage, the transistors M_{P1}, M_{N1} and M_{N2} are turned OFF. For the latching action back-to-back connected inverters are used. This inverter pair should be designed carefully as it should not pose problem during the active clock phase level. It should allow the changes in the Q node directed by the M_{P1}, M_{N1} and M_{N2} transistors. The advantage of this structure is it can be used as single-ended flip-flop i.e., there is no need to build latch for $\overline{Q}$ for Q to exist.

The Kim's architecture [7] suffers from glitches at the output. This glitch is present when the input data value is logic high and occurs at every positive edge of the clock. This will not occur at the transition of the output from high-to-low but rather when it is supposed to hold the output value logic value high. To understand this glitch, consider the situation when the output is already settled at logic high and the input D is logic high. When the clock is logic low, the the $\overline{S}$ is refreshed to logic one and in this half-period, the back-to-back inverters maintain the output. While the clock does the low-to-high transition, and D which is still at logic high, the $\overline{S}$ signal at the moment is still at logic high. It takes a certain amount of time, albeit small, to discharge to zero, is still enough to produce glitch at the output. It is because at the positive edge of clock, the transistors M_{N1} and M_{N2} are ON in the Fig. 3. The $\overline{S}$ signal changes to logic low after a certain time after the occurrence of clock edge. In this small window of time, the output discharges and later it starts to charge back to V_{DD}.

To overcome this glitches, one approach is to include an extra NMOS transistor in series with the existing two

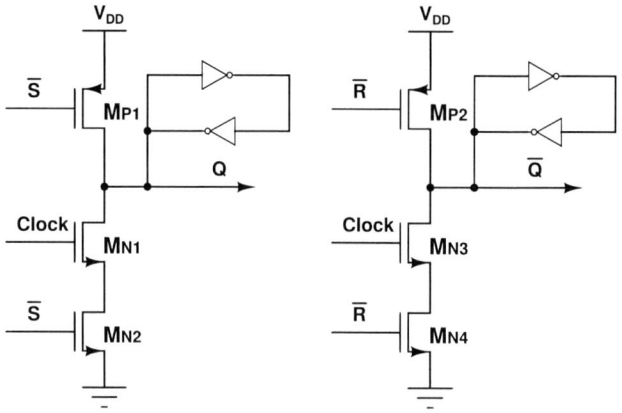

Fig. 3: Schematic of Kim's latch [7]

NMOSFETs as pictured in Fig. 4 presented in the paper [8]. The added transistor M_{N2} will be driven by input signal $\overline{D}$. This will guarantee that the discharging path will be OFF during clock low-to-high transition stage given D is logic high. During refresh stage, latching action is provided by the cross-coupling of the outputs. This architecture did not reduce the speed of operation compared to [7] and still was able to remove the glitch issue faced by it.

III. PROPOSED LATCH

The Proposed latch structure is shown in Fig. 5. From the Table I, it can be noted that the signals $\overline{S}$ and $\overline{R}$ are refreshed to supply voltage V_{DD} when clock is low. Thus the signals S and R are always logic low when clock is low.

When clock is logic high depending on the data input, one of the nodes $\overline{S}$ and $\overline{R}$ discharges to zero and the other node stays at logic high. By referring to the Table I, it can be noted that when the output Q discharges the signal $\overline{S}$ remains at logic high. So as mentioned in the paper [8], the transistors M_{N1}, M_{N2} and M_{N3} starts to discharge as soon as the clock transitions from low-to-high. Similarly, the $\overline{Q}$ discharges through the transistors M_{N4}, M_{N5} and M_{N6}. So for discharging takes only one gate delay.

From the Table I, it can be observed that when the output Q charges to logic high the signal $\overline{R}$ remains in logic high or the signal R remains in logic low. Although there is an inverter delay between the signals $\overline{R}$ and R, the transition occurs when the clock is low and is well settled before the occurrence of the rise-edge of the clock. The transistor M_{P3} controlled by $\overline{Clock}$ makes sure that the path exists only when the clock is logic high. The transistor M_{P2} is added in order to avoid the glitches at the output faced by the [7]. As soon as the rise-edge of the clock occurs, the output Q starts to charge to supply voltage V_{DD} through the transistors M_{P1}, M_{P2} and M_{P3}. Likewise, the charging path for the $\overline{Q}$ is controlled by the transistors M_{P4}, M_{P5} and M_{P6} controlled by signals S, D and $\overline{Clock}$ respectively.

When the clock is low, the PMOS transistors M_{P3}, M_{P6} and NMOS transistors M_{N1}, M_{N4} are OFF. The transistors M_{P1}, M_{P4} and M_{N3}, M_{N6} are turned ON. Based on the

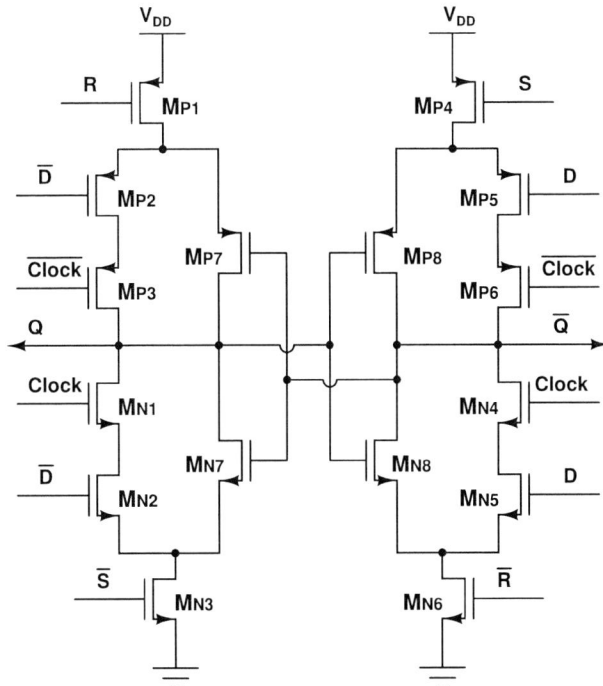

Fig. 5: Schematic of the Proposed Latch

existing value of the outputs, the transistors M_{P7}, M_{P8}, M_{N7} and M_{N8} turn ON and maintain the output logic value.

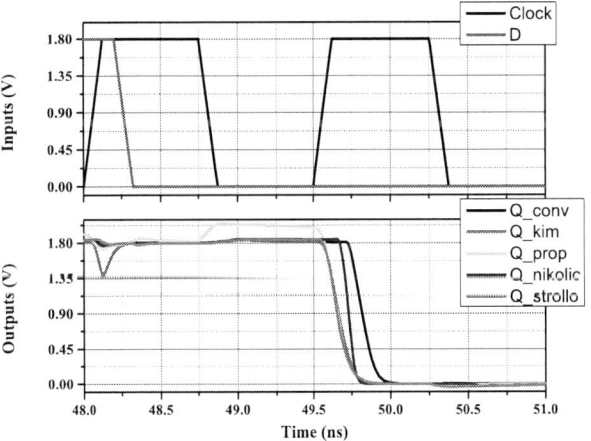

Fig. 6: Comparison of falling edge of Output waveform

IV. RESULTS

The circuits are simulated using $0.18\mu m$ technology node. All the Flip-Flops are clocked at period of $1.5\ ns$, the transition times are set at $125\ ps$. Both the outputs Q and $\overline{Q}$ are loaded with $200fF$ capacitor. The input data period is $32\ ns$ (16 times slower than the clock) and its transition time is also set at $125\ ps$.

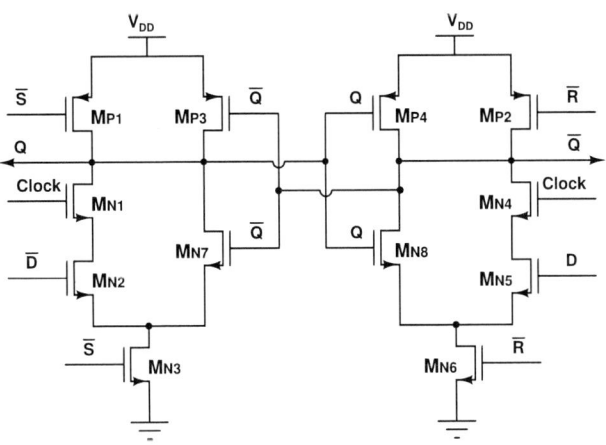

Fig. 4: Schematic of Strollo's latch [8]

978-1-5386-1357-3/17 $31.00 © 2017 IEEE   131

TABLE II: Performance measures of SAFFs

Parameter	Conventional [5]	Nikolic's [6]	Kim's [7]	Strollo's [8]	Proposed Structure
Rise delay (ps)	157.82	133.83	151.76	152.23	96.17
Fall delay (ps)	254.48	166.71	101.03	91.17	95.09
Power (μW)	236.9	265.1	378.2	246.6	408.9
Power Delay Product (fJ)	48.84	39.84	47.8	30.01	39.1

The Table II compares the results obtained from the flip-flops, the Proposed structure performs fairly well against the existing architectures. The Strollo's latch is slightly faster than the Proposed structure when the output transitions from high-to-low. The conventional nand-based latch performs poorly in this category. The comparison of different latches' output Q can be observed in Fig. 6. In the figure outputs of the Proposed latch (colour green), Kim's latch and Strollo's latch are very close to each other.

When it comes to the rise-delay, all the latches are close to each other, except the Proposed structure. It fairly leads in this category as it is evident in the Fig. 7 with the rise delay of 96.17 ps. The Nikolic's latch too performs well with the rise-delay of 133.83 ps. However, the power consumption of the Proposed latch is 408.9 μW, is the highest among all the latches compared.

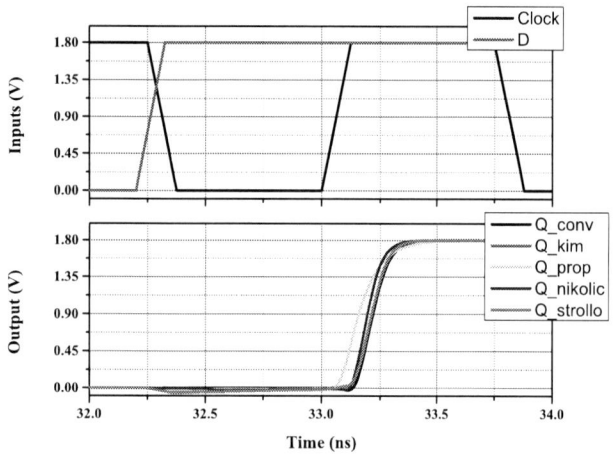

Fig. 7: Comparison of rising edge of output waveform

The Power Delay Product (PDP) is the figure of merit usually used for the digital circuits. The lower the value, higher the energy efficiency of the circuit. The conventional NAND based latch has the highest power delay product of 48.84 fJ while the Proposed latch's PDP is 39.1 fJ. The PDP of Proposed structure, Nikolic's latch are close to each other. The Strollo's latch has the lowest PDP of 30.01 fJ.

V. CONCLUSION

A new structure for sense amplifier based flip-flop is proposed. The rise-delay is the lowest among the existing latches and is 39 % faster than the conventional nand based flip-flop. The fall-delay is very close to the Strollo's latch and

Kim's latch and is 64 % faster than the conventional flip-flop. The power delay product of 39 fJ is comparable to the Nikolic's latch.

ACKNOWLEDGEMENT

This publication is an outcome of the Research work undertaken in the project under SMDP-C2SD, Department of Electronics and Information Technology, Ministry of Communication IT, Government of India.

REFERENCES

[1] G. Gerosa, S. Gary, C. Dietz, Dac Pham, K. Hoover, J. Alvarez, H. Sanchez, P. Ippolito, Tai Ngo, S. Litch, J. Eno, J. Golab, N. Vanderschaaf, J. Kahle, "A 2.2 W, 80 MHz superscalar RISC microprocessor", *IEEE Journal of Solid-State Circuits*, vol. 29, pp. 1440-1454, 1994.

[2] V. Stojanovic, V. G. Oklobdzija, "Comparative analysis of master-slave latches and flip-flops for high-performance and low-power systems", *IEEE Journal of Solid-State Circuits*, vol. 34, no. 4, pp. 536-548, April 1999.

[3] H. Partovi, R. Burd, U. Salim, F. Weber, L. D. Gregorio, D. Draper, "Flow-through latch and edge-triggered flip-flop hybrid elements", in *Proc. Int. Solid-State Circuits Conf.*, pp. 138-139, 1996.

[4] F. Klass, C. Amir, A. Das, K. Aingaran, C. Truong, R. Wang, A. Mehta, R. Heald, G. Yee, "A new family of semidynamic and dynamic flip-flops with embedded logic for high-performance processors", in *IEEE Journal of Solid-State Circuits*, vol. 34, no. 5, pp. 712-716, May 1999.

[5] M. Matsui, H. Hara, Y. Uetani, L. Kim, T. Nagamatsu, Y. Watanabe, A. Chiba, K. Matsuda, T. Sakurai, "A 200 MHz 13 mm ^{2} 2-D DCT macrocell using sense-amplifying pipeline flip-flop scheme", in *IEEE Journal of Solid-State Circuits*, vol. 29, no. 12, pp. 1482-1490, Dec. 1994.

[6] B. Nikolic, V.G. Oklobdzijia, V. Stajanovic, W. Jia, J. K. Chiu, M. M. Leung, "Improved sense-amplifier based flip-flop: Design and measurements", in *IEEE Journal of Solid-State Circuits*, vol. 35, no. 6, pp. 876-883, 2000.

[7] J. Kim, Y. Jang, H. Park, "CMOS sense amplifier-based flip-flop with two $N - C^2 MOS$ output latches", in *Electronics Letters*, vol.36, no.6, pag.498-500, Mar. 2000.

[8] A. G. M. Strollo, D. De Caro, E. Napoli, N. Petra, "A novel high-speed sense-amplifier-based flip-flop", in *IEEE Transactions on Very Large Scale Integration (VLSI) Systems*, vol. 13, no. 11, pp. 1266-1274, 2005.

978-1-5386-1357-3/17 $31.00 © 2017 IEEE

A novel low power high speed BEC for 2GHz sampling rate Flash ADC in 45nm technology

Sarfraz Hussain
Department of ECE
NERIST, Nirjuli
Arunachal Pradesh, India
Email: s.hussainec@gmail.com

Rajesh Kumar
Department of ECE
NERIST, Nirjuli
Arunachal Pradesh, India
Email: rk@nerist.ac.in

Gaurav Trivedi
Department of EEE
IIT Guwahati
Assam, India
Email: trivedi@iitg.ernet.in

Abstract—This paper depicts the idea of a novel bubble error corrector for removing the bubble error of order 1 and consuming less power. The earlier bubble error corrector (BEC) needed large number of transistors thus requiring more power. 3-input NAND gate with two inverted inputs is also used as a BEC but it requires more power than the proposed one as it requires more number of transistors. With a supply of 1V in 45nm technology, the BEC consumes 4.14 pico-Watt of dc power and 9.62 micro-Watt of average power. The maximum delay is calculated to be 20 pico-seconds. When used with Fat tree encoder it consumes 0.3 nano-Watt of dc power and 27.34 micro-Watt of average power and has a maximum delay of 74.76 pico-seconds.

Keywords-bubble error; encoder; Flash ADC; thermometer code; 45nm technology.

I. Introduction

In today's modern technology, ADC is a hot topic and has been an interesting topic of research way back to the invention of the Morse code. Flash ADC is the fastest among the other ADCs due to its parallel computing nature. Researchers are working on flash ADCs to make it work faster. But everything comes with a price and in this case it is mainly power dissipation. As the technology is evolving and transistor sizes are reducing many other factors are added to the design of a flash ADC like low power supply. Due to the reduction in transistor sizes, the flash ADC can now work faster but power dissipation is more and power supply is low. Therefore, low power flash ADCs are designed to overcome this problem. There are issues related to the design which includes bubble error correction and metastability.

A flash ADC comparator block generates a thermometer code. Thermometer-to-binary encoder converts the thermometer code to a binary code. Thermometer code is generated by a set of comparators used in Flash ADC. In this code, '0' or '0s' is followed by all '1s'. This resembles like a thermometer filled with mercury, thus the name thermometer code. If a '0' arises in between the '1s' it is termed as a bubble error. It is a look-alike of a bubble in a mercury filled thermometer. This bubble error can be of order one or higher. The bubble error is caused either due to fault in comparators or long delay in the comparator output. Comparator error

affects the conversion rate of an ADC therefore ADC has to be tolerant of such errors [3]. Generally to correct bubble errors two methods are used: gray encoding and thermometer bubble error suppression. Gray encoding is used so that only one bit error is introduced and approximate values can be achieved as described in [2]. Gray encoder do not correct the bubble error. There are several thermometer bubble error correction methods but they require more number of transistors [1]. One of the error reduction technique used in [4] uses a twin encoding scheme for higher efficiency but it adds delay and a large number of transistors.

Section II compares the four different encoders used in the recent decade for use in flash ADC design. Section III describes bubble error and its correction methods along with the proposed method. Simulation and results are discussed in section IV. Simulation is done in Cadence virtuoso using spectre simulator and 45nm technology.

II. Recent encoders used in Flash ADC

A flash ADC consists of two major component: comparator and encoder. For a n-bit flash ADC 2^n-1 number of comparators are required. The comparators converts the continuos time continus signal (analog domain)into continuous time discrete signal and the encoder converts the

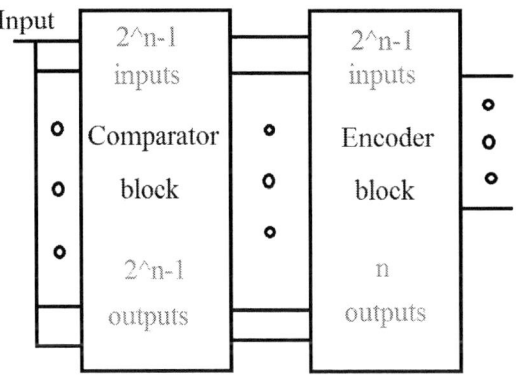

Figure 1.   Flash ADC block

continuous time discrete signal into discrete time discrete signal (digital domain). Four different types of encoders are discussed in this paper: Wallace tree, MUX based, Fat tree and pseudo NMOS. Block diagram of flash ADC is shown below in fig.1.

A. Wallace tree encoder

Wallace tree encoder is very efficient encoder but the problem lies with the speed [5], [7]. It is also called ONE's counter because it counts the number of '1s' and accordingly gives the output. For an efficient and low power ADC it can be used. But for high speed applications it is limited to MHz range sampling rate. It comprises of full adders and bubble error suppression is automatically done as the inputs are added. The main drawback of this encoder is speed. Given below in fig.2 is the block diagram of a Wallace tree encoder [11].

B. Fat Tree encoder

Fat tree is a high speed encoder and consumes less power [8]. The circuit signal delay is $O(\log_2 N)$ for fat tree whereas the signal delay is $O(\log_{1.5} N)$ for Wallace tree encoder . Fat tree is fast but it suffers from bubble error therefore modified Fat tree encoder is required which is explained in section III. Fig.3 shows a fat tree encoding structure [11]. Fat tree can be made to work faster by employing NAND and NOR gates instead of OR gates.

C. MUX based encoder

Another type of encoder is the MUX based encoder which consumes very less power [9], [10], [11]. The critical path is less than the others which make it to work in high speed

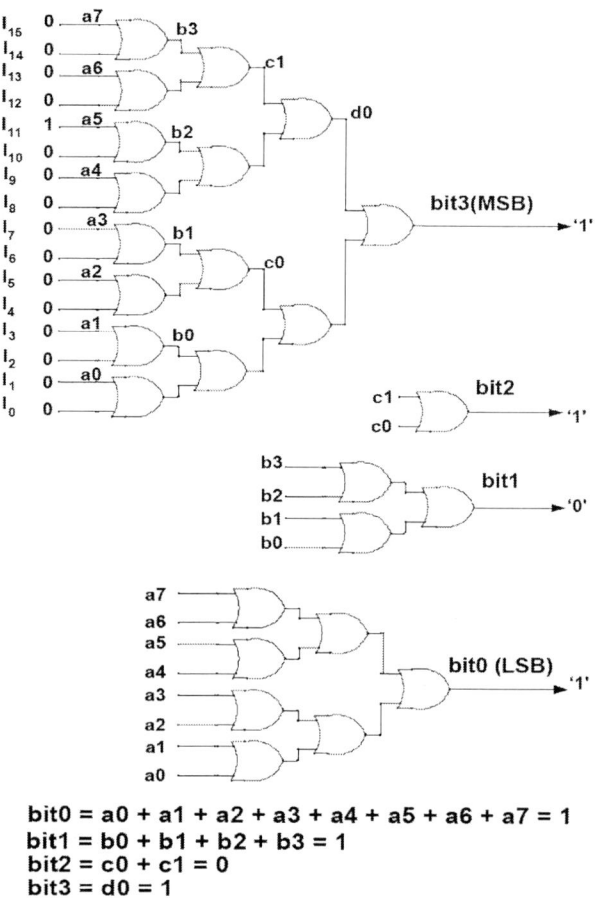

bit0 = a0 + a1 + a2 + a3 + a4 + a5 + a6 + a7 = 1
bit1 = b0 + b1 + b2 + b3 = 1
bit2 = c0 + c1 = 0
bit3 = d0 = 1

Figure 3.   4-bit Fat tree encoder   [11]

applications. It is fast but Fat tree is faster than the MUX based encoder. It is suitable for low power applications. Fig.4 shows MUX based encoder [11].

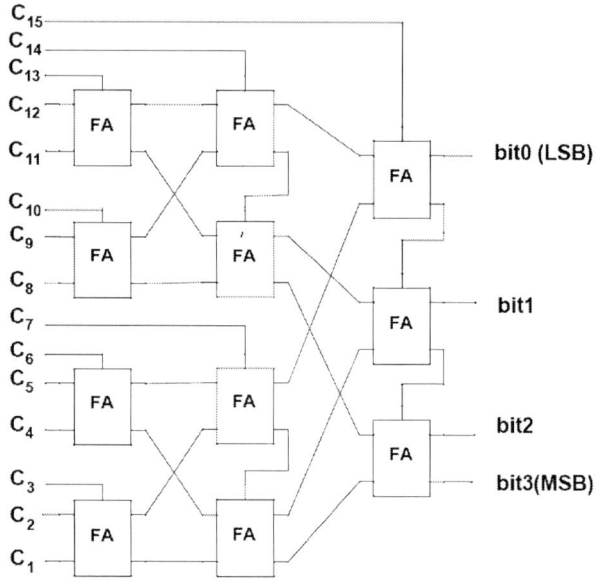

Figure 2.   4-bit Wallace tree encoder   [11]

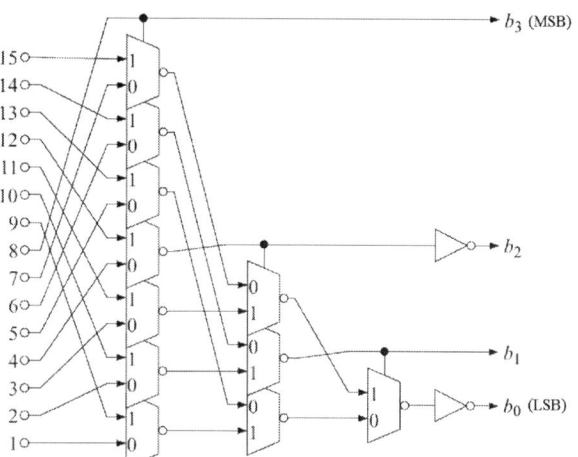

Figure 4.   4-bit MUX based encoder   [11]

978-1-5386-1357-3/17 $31.00 © 2017 IEEE   134

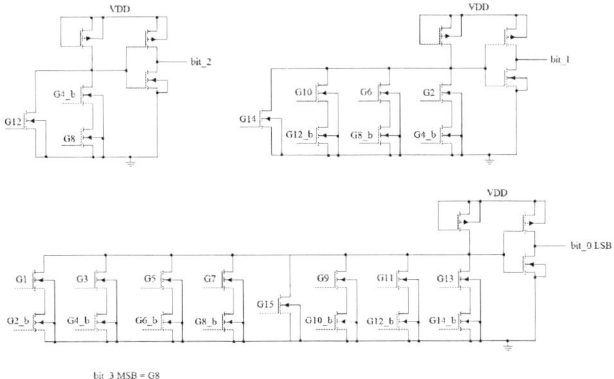

Figure 5. 4-bit pseudo NMOS based encoder [15]

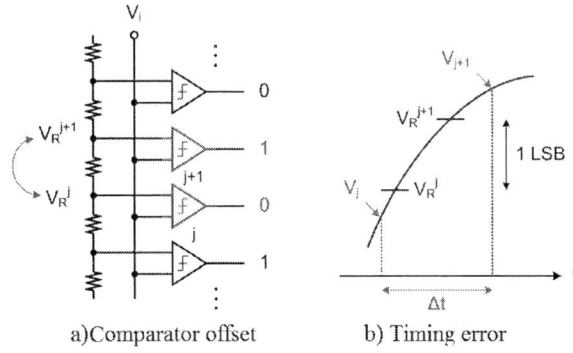

Figure 6. Causes of Bubble error [14]

Table I
COMPARISON OF ENCODERS IN 45NM TECHNOLOGY

Encoder type	DC Power (n Watt)	Average Power (μW)	Delay best case (pico-s)	Delay worst case (pico-s)	PDP (10⁻¹⁵)
Wallace tree	0.16	54.58	170.2	204.5	11
Fat tree	0.23	32.95	25.34	27.45	0.9
MUX based	0.05	20.11	14.18	66.42	1.3
Pseudo NMOS	0.10	59.47	14.18	32.14	1.9

D. Pseudo NMOS encoder

Pseudo NMOS encoder comprises of NMOS logic and a PMOS load [15]. It is faster as compared to all the other encoders but dc power consumption is more as the PMOS load is always in 'ON' state. Fig.5 shows pseudo nmos encoder [15]. The comparison between the encoders is shown in Table.I. The table shows the comparison in terms of dc power, average power, worst case delay, best case delay and power delay product (PDP). Fat tree encoder has a good PDP over the rest and has equal rise and fall time therefore it is preferred for use in low power and high speed encoder circuit.

III. BUBBLE ERROR AND ITS CORRECTION METHODS

Bubble error is caused due to large delay in comparator output or due to offset error in comparator. It can be of order 1 or 2 or of higher order. This paper corrects bubble error of order 1 and can be modified to correct higher order bubble. Fig.6 shows the cause of bubble error [14]. Bubble error correction methods include an additional circuitry along with the encoder. Some uses the concept of duplicating the thermometer code by using a set of OR gates [1] and some uses the concept of converting the thermometer code to one shot code by using NAND gate [13] and converting it back to binary. One of the method used earlier was to convert thermometer code to gray code and then converting it back to binary [6]. Efficiency was considered at first but now power dissipation and speed is an important factor in

high speed applications like in flash ADC. In Table.I, fat tree is preferred to be used in flash ADC due its low PDP therefore the proposed efficient and low power bubble error corrector is used with the fat tree encoder to provide better results. Table.II shows bubble errors of different types. The proposed bubble error corrector can correct one zero bubble effectively.

A. Bubble error corrector using OR Gates

Bubble error correction is done by majority voting in this case. The problem with this technique is that duplicate of the corrector circuit has to be added for every increase in the order of bubble error [1]. The nth output of the first encoder part is added with the $(n-1)^{th}$ output to give $(n-1)^{th}$ output of the second encoder part. The demerit of this encoder is speed and power dissipation. With the addition of more blocks the delay increases along with the power. Fig.7 shows the bubble error correction upto an order of 2.

B. Bubble error corrector using gray coding

Bubble error correction can be done by implementing gray coding to reduce the number of error bits. But converting a thermometer code to gray code and then again converting it to binary code is a gruesome process. In [6], [2], bubble

Table II
BUBBLE ERRORS OF DIFFERENT TYPES

	No bubble	1 bubble	2 bubble	1 Zero bubble	2 Zero bubble
	0	0	0	0	0
	0	0	0	0	0
Encoding	0	1	1	0	0
point	1	0	0	1	1
	1	1	0	1	1
	1	1	1	1	0
	1	1	1	0	0
	1	1	1	1	1
	1	1	1	1	1

error detection and correction circuit is used along with gray coding which reduces the speed and also increases the power consumption. Fig.8 shows the encoder block diagram [6].

C. Bubble error corrector using NAND Gates

A 3-input NAND gate with 2 inverted inputs can be used to correct bubble error of order 1 (1 zero bubble) and consume less power compared to the previous encoder [13]. For correcting higher order bubble error higher input NAND Gates can be used. Fig.9 shows NAND gate based bubble error correction. It converts the thermometer code into one shot code and then converts to binary code. The output is '0' only when the input is '001' else it is always '1'. Therefore single zero bubble can be corrected easily.

D. Proposed Bubble error corrector

The proposed bubble error corrector is a novel architecture which can be used in low power and high speed encoders. The circuit diagram is shown in fig.10. It can correct bubble

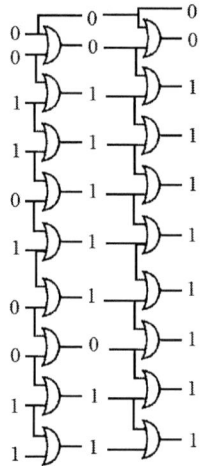

Figure 7.   OR gate based BEC

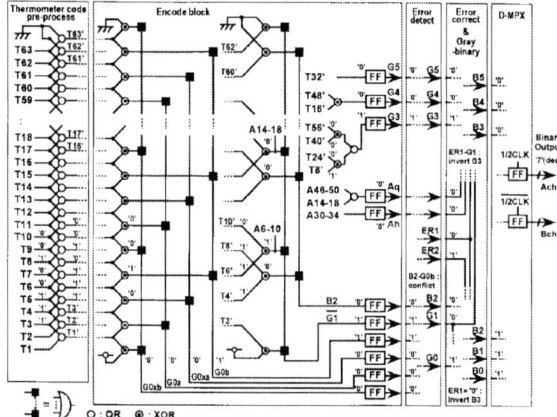

Figure 8.   Gray code based BEC [6]

Table III
NOVEL BEC TRUTH TABLE

A	B	C	OUT
0	0	0	0
0	0	1	1
0	1	0	0
0	1	1	0
1	0	0	0
1	0	1	0
1	1	0	0
1	1	1	0

error having 1 zero bubble efficiently and with a high speed. To increase the order of bubble error correction one more input can be added. This circuit uses less transistors as compared to the other encoders and dissipates less power. Truth Table is shown in Table. III.

IV. SIMULATION AND RESULTS

Simulations have been done in Cadence Virtuoso in 45nm technology using spectre simulator. The supply voltage is taken as 1 V. Input frequency is 2GHz. A rise time and fall time of 10ps is given in the input. All the transistors have the 45nm length and 120nm width. The circuit has been simulated in 27° C with nominal NMOS and PMOS. The results show that the novel bubble error corrector circuit consumes 4.14 pico-Watt of DC power, 9.62 μWatt of average power and has a delay of 20 pico-seconds. The transient analysis of the BEC is shown in fig.11.

A. Comparison between different BECs

The comparison tables. IV, V shows the comparison between different BEC with respect to 4 different types of bubble error. All the types are considered to have three inputs

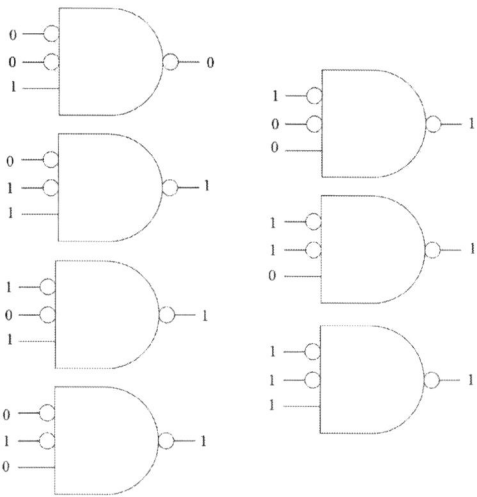

Figure 9.   NAND gate based BEC

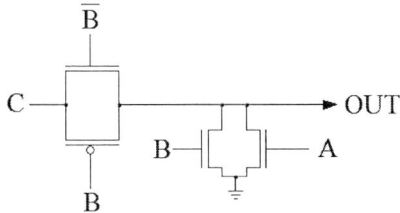

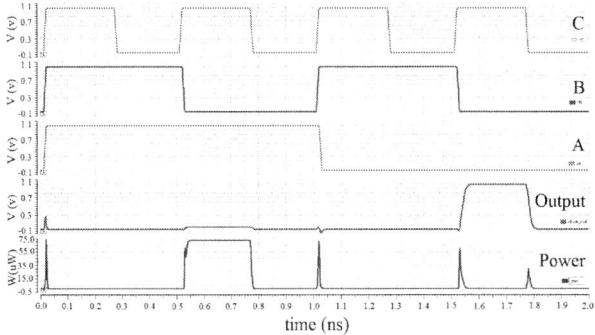

Figure 10. Proposed BEC circuit

Figure 11. Transient analysis of novel BEC circuit

and one output. The novel BEC is faster than the rest and consumes less power because of its circuitry. But the 3-input OR based BEC is more efficient.

Table IV
COMPARISON OF BEC FOR 1 BUBBLE ERROR

1 Bubble	OR Gate	NAND Gate	NOVEL BEC
0	0	1	0
0	0	1	0
1 encoding	1	0	1
0 point	1	1	0
1	1	1	0
1	1	1	0
1	1	1	0
1	1	1	0
1	1	1	0

Table V
COMPARISON OF BEC FOR 1 ZERO BUBBLE ERROR

1 Zero Bubble	OR Gate	NAND Gate	NOVEL BEC
0	0	1	0
0	0	1	0
0 encoding	0	1	0
1 point	1	0	1
1	1	1	0
0	1	1	0
0	1	1	0
1	1	1	0
1	1	1	0

Table VI
COMPARISON OF ENCODERS WITH AND WITHOUT BEC

Encoders	Average Power (µW)	Delay best case (pico-s)	Delay worst case (pico-s)	PDP (10^{-15})
Fat Tree Encoder				
Without BEC	32.95	25.34	27.45	0.9
With BEC	27.34	74.76	121	2.04
Pseudo NMOS based Encoder				
Without BEC	59.47	14.18	32.14	1.9
With BEC	72.31	17.1	73.66	5.32

Table VII
MONTE CARLO ANALYSIS OF ENCODERS

Encoders	Min	Max	Mean	Std. Dev.
Fat Tree Encoder				
Delay (in pico-s)	70.93	80.46	74.88	1.88
Avg Pow (in µW)	24.90	27.56	26.58	0.51
PDP (in femto)	1.82	2.21	1.99	0.061
Pseudo NMOS based Encoder				
Delay (in pico-s)	59.86	102.90	74.82	8.84
Avg Pow (in µW)	67.33	75.10	71.32	1.55
PDP (in femto)	4.14	7.55	5.34	0.67

Table VIII
PVT ANALYSIS OF PDP OF ENCODERS AT 27 °C

Encoders	ff	fs	nominal	sf	ss
Fat Tree Encoder					
With BEC (in femto)	1.76	2.21	2.04	–	2.59
Pseudo NMOS based Encoder					
With BEC (in femto)	4.62	3.96	5.32	–	7.08

Table IX
PVT ANALYSIS OF PDP OF ENCODERS AT 10 °C

Encoders	ff	fs	nominal	sf	ss
Fat Tree Encoder					
With BEC (in femto)	1.65	2.04	1.89	–	2.44
Pseudo NMOS based Encoder					
With BEC (in femto)	4.45	3.80	5.06	–	6.64

B. Encoders with and without BEC

From Table. I it is clear that PDP of Fat tree encoder is minimum. Therefore, it is preferred over the other types of encoder. Table. VI shows the comparison between fat tree and pseudo NMOS based encoder with and without bubble error corrector.

When bubble error circuit is connected, the overall power dissipation is reduced because the circuit elements are reduced but the delay is increased to 3 times. Worst case delay is 121 pico-seconds. Hence, this encoder can be used in efficient encoder designs with a sampling rate of 2 GHz and for low power applications.

C. Monte Carlo and PVT analysis of Encoders with BEC

Table. VII shows the Monte Carlo simulation of Fate tree and Pseudo NMOS based encoder with BEC. The overall result shows that the Fat tree encoder when used with BEC gives good results as compared to Pseudo NMOS based encoder with BEC. Table. VIII and Table. IX shows the PVT analysis of PDP of two encoders with BEC at 27 °C and 10 °C respectively. The variation over temperature is less in both the encoders. The overall performance of Fat tree encoder with BEC is good. In PVT analysis ff stands for fast NMOS fast PMOS, fs stands for fast NMOS slow PMOS, sf stands for slow NMOS fast PMOS, and ss stands for slow NMOS and slow PMOS.

V. CONCLUSION

A flash ADC requires a high speed comparator and encoder. Researchers are designing high speed ADCs but there is a catch. High speed costs power dissipation. As the technology is growing, power dissipation has become a major concern for low power supply. To balance both the perspectives we calculate the PDP. Fat tree encoder is chosen over other encoders because it has equal rise and fall times and its PDP is less than the rest. The results show that with the novel BEC a low power design is possible with the ability to correct 1 zero bubble error. The transient analysis in 45nm technology show that the proposed BEC consumes 9.62 μWatt of power with a delay of 20 pico-seconds. Fat tree encoder with novel BEC circuit has a worst case delay of 121 pico-seconds which makes it feasible to be used in 2 GHz sampling rate ADCs.

ACKNOWLEDGMENT

This work has been done from the Grant Received from Visvesvaraya PhD Scheme for Electronics and IT.

REFERENCES

[1] C. W. Mangelsdorf, "A 400-MHz input flash converter with error correction," *IEEE Journal of Solid-State Circuits*, vol. 25, pp. 184-191, Feb. 1990.

[2] A. Matsuzawa, Y. Kitagawa, I. Hidaka, S. Sawada, M. Kagawa,M. Kanoh, "An 8b 600 MHz flash A/D converter with multistage duplex Gray coding," *Symp. VLSI Circuits Dig. Tech. Papers*,Oiso, Japan. pp. 113-114, May. 1991.

[3] J. Van Valburg and R. J. Van de Plassche, "An 8-b 650-MS/s folding ADC," *IEEE Journal of Solid-State Circuits*, vol. 27, pp. 1662-1666, Dec. 1992.

[4] M. Ito, T. Miki, S. Hosotani, T. Kumamoto, Y. Yamashita, M. Kijima, T. Okada, "A 10 bit 20 MS/s 3 V supply CMOS A/D converter," *IEEE Journal of Solid-State Circuits*, vol. 29, pp. 1531-1536, Dec. 1994.

[5] Kaess, F., Kannan, R., Hochet, B. and Declercq, M., "New Encoding Scheme For High-Speed Flash ADC's," *30th IEEE International Symposium on Circuits and Systems (ISCAS'97)*, Hong Kong. pp.5-9, June. 1997.

[6] S. Tsukamoto, W. G. Schofield, T. Endo, "A CMOS 6-b, 400-MSample/s ADC with Error Correction," *IEEE Journal of Solid State Circuits*, Canada, USA. Vol. 33, No. 12, Dec. 1998.

[7] P. Pereira, J. R. Fernandes, and M. M. Silva, "Wallace tree encoding in folding and interpolation ADCs," *in proceedings of 2002 IEEE International Symposium on Circuits and Systems*, Arizona, USA. Vol. 1, pp. I-509-I-512, May. 2002.

[8] D.Lee, J.Yoo, K.Choi, J.Ghaznavi, "Fat Tree encoder design for Ultra high speed Flash A/D converters," *in Proceedings Of 2002 MWSCAS*, Oklahoma, USA. Vol. 2,PP. II-87-II-90, Aug. 2002.

[9] Erik Sall, M. Vesterbacka, "A Multiplexer based decoder for Flash Analog-to-Digital converter," *2004 IEEE Region 10 Conference TENCON*, Chiang Mai, Thailand. Vol. 4, pp. 250-253, Nov. 2004.

[10] E. Sall and M. Vesterbacka,"Comparison of two thermometer-to-binary decoders for high-performance flash ADCs," *2005 NORCHIP*, Oulu, Finland. pp.- 253-256, Feb. 2006.

[11] E. Sall and M. Vesterbacka,"Thermometer-to-binary decoders for flash analog-to-digital converters," *2007 18th European Conference on Circuit Theory and Design*, Seville, Spain. pp. 240-243, May. 2008.

[12] M. Rahman, K. l. Baishnab, and F. A. Talukdar, "A novel ROM architecture for reducing bubble and metastability errors in high speed flash ADCs," *20th International Conference on Electronics, Communications and Computer*, Cholula, Mexico. pp. 15-19, 2010.

[13] B. V. Hieu, S. Choi, J. Seon, Y. Oh, C. Park, J. Park, H. Kim, and T. Jeong, "A New Approach to Thermometer-to-Binary Encoder of Flash ADCs- Bubble Error Detection Circuit," *IEEE MWSCAS 2011*, Seoul, South Korea. pp. 1-4, Aug. 2011

[14] Prof. Y. Chiu, "*Lecture notes on Data converter - EECT 7327*," Semester Fall 2014.

[15] L. Nazir, B. Khurshid, R. N. Mir, "A 7GS/s, 1.2V Pseudo logic encoder based flash ADC using TIQ Technique," *2015 Annual IEEE India Conference (INDICON)*, New Delhi, India. pp. 1-6, Dec. 2015.

[16] G. T. Varghese, K. Mahapatra, "A Low Power Reconfigurable Encoder for Flash ADCs," *Global Colloquium in Recent Advancement and Effectual Researches in Engineering, Science and Technology (RAEREST 2016)*, Elsevier Procedia Technology, vol. 25, pp. 574-581, April 2016.

2017 IEEE International Symposium on Nanoelectronic and Information Systems

Comparison and design of dynamic comparator in 180nm SCL technology for low power and high speed Flash ADC

Sarfraz Hussain
Department of ECE
NERIST, Nirjuli
Arunachal Pradesh, India
Email: s.hussainec@gmail.com

Rajesh Kumar
Department of ECE
NERIST, Nirjuli
Arunachal Pradesh, India
Email: rk@nerist.ac.in

Gaurav Trivedi
Department of EEE
IIT Guwahati
Assam, India
Email: trivedi@iitg.ernet.in

Abstract—A modified dynamic comparator is proposed and compared in this paper. A dynamic comparator consists of a low gain amplifier connected to a latch circuit. The inputs are amplified during the evaluation period and the outputs are latched during the regeneration time. The proposed dynamic comparator is fast and consumes less power. At a clock frequency of 1.25GHz and 100mV $\triangle$Vin, the delay is 176.71ps and average power consumption is 119.81μW for a supply voltage of 1.8V. The calculated maximum PDP is 24.53 f. The proposed dynamic comparator is suitable for an efficient low power and high speed Flash ADC. The circuits are simulated in cadence virtuoso spectre with 180nm SCL technology.

Keywords-dynamic comparator; high speed; latch; low-power; flash ADC.

I. INTRODUCTION

ADC converts an analog signal to a digital domain. Flash ADC is the fastest of all the other ADCs. Comparator is one of the important part of a Flash ADC. 2^n-1 number of comparators are required for n-bit Flash ADC. For high speed aplications dynamic comparators are used instead of static comparators. Dynamic comparators are used for high speed applications because they provide positive feedback and therefore charging of output node is faster as compared to static comparator. Another advantage of dynamic comparators is low power consumption. Dynamic comparators are the comparators with low gain differential amplifier and a latch circuit. Dynamic comparator has two phases: evaluation phase and regeneration phase. In the evaluation phase, the input is compared with the reference voltage and the output node is set 'HIGH' or 'LOW' accordingly. All the transistors are kept in saturation region. When one of the differential pair nmos transistor is turned 'ON', the maximum current flows through that transistor and charges the capacitor. During the regeneration phase, the latch tries to maintain proper balance between the outputs.

Some comparators are current comparators [1], [11] and some are voltage comparators [2], [3], [5], [6], [7], [8], [9], [10]. Power consumption is however low in case of current comparators but implementation is cumbersome. Another current to voltage

converter is required as an additional circuit. Use of high resolution techniques mentioned in [2] and offset reduction techniques in [6], [7] gives rise to efficient designing techniques. The analysis of static and dynamic random offset in [4] shows the way of reducing offset and making the comparator robust. The proposed technique is not much affected by technological transistor sizing.

In this paper, five different comparators have been compared with the proposed comparator. In [5], there is static power dissipation which is a drawback of this comparator. Because of static power dissipation the overall power consumption of the circuit increases. In [7], [8], [9], [10] static power dissipation is negligible therefore saving energy and power. The proposed dynamic comparator is a modified form mentioned in [8] which has comparitively low power dissipation and has less delay. The output swing is more as compared to the rest. PDP of the proposed dynamic comparator is better than the others. Dynamic comparator circuit of [10] consumes the least power. The proposed comparator is the fastest and has a wide output swing than the rest. The proposed comparator also has a large output common mode voltage due to its dual rail property. Simulations show variations in average power and delay with respect to change in common mode voltage and input voltage.

Section II describes different types of dynamic comparators. It briefs the working of the comparators and compares with the rest. Proposed dynamic comparator is explained in the last part of the section. Simulation and results are shown in Section III. Simulations show that the proposed comparator can be used in an efficient low power flash ADCs as its PDP is less than the rest.

II. DYNAMIC COMPARATORS

A. Circuit Implementation and Comparison

For every output the comparator depends upon the clock input. This section explains the working of different dynamic comparator circuits. The total delay of the comparator depends upon the delay of the latch plus the delay of the differential pair amplifier. The comparator of [5] is shown

978-1-5386-1357-3/17 $31.00 © 2017 IEEE

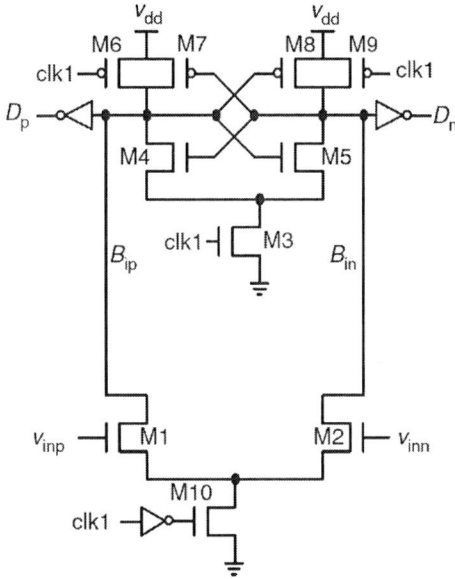

Figure 1. Comparator [5]

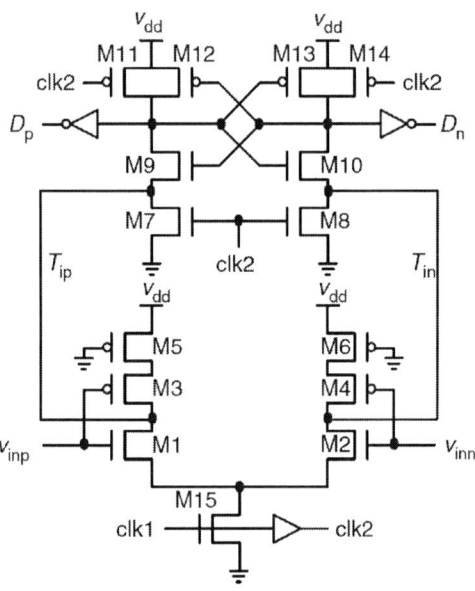

Figure 2. Comparator [7]

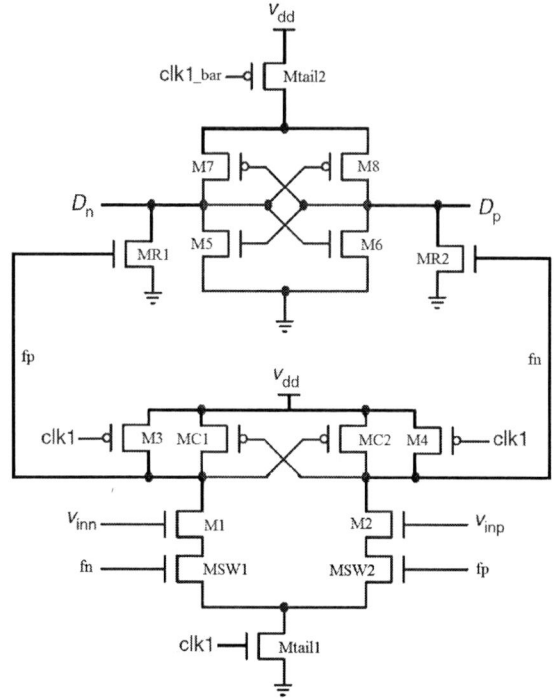

Figure 3. Comparator [8]

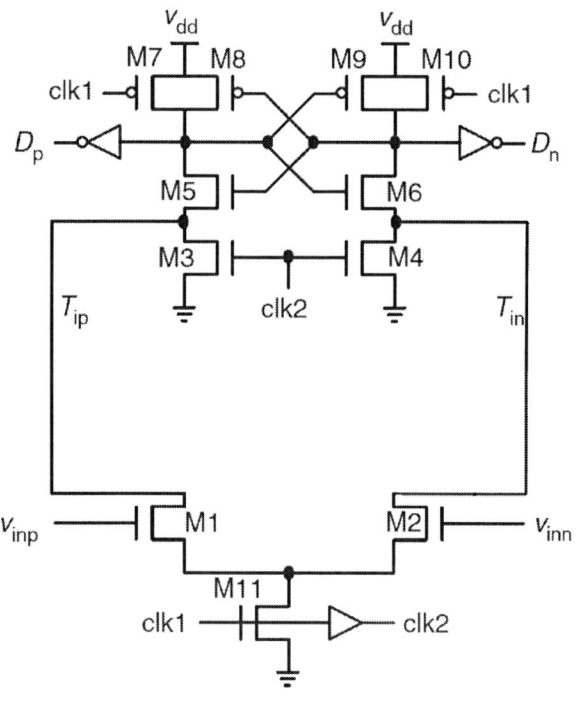

Figure 4. Comparator [9]

in fig.1. Static power consumption is a demerit of this comparator. When the clock clk1 is LOW, the transistors M6, M9 and M10 are turned ON and the transistor M3 is turned OFF. The comparator is in reset mode and the outputs are '0'. The transistors M6 and M9 pulls up the drain potential to VDD. If the input Vinp > Vinn, then the transistor M1 tries to discharge the potential to ground. But the outputs remain at their zero potential since the transistors M6 and M9 are ON.

After sometime when the clock clk1 is HIGH, the tran-

sistors M6, M9 and M10 are OFF and the transistor M3 is ON. This is the evaluation phase of the comparator. The drain potential of M2 is already at higher potential and M3 is

978-1-5386-1357-3/17 $31.00 © 2017 IEEE   140

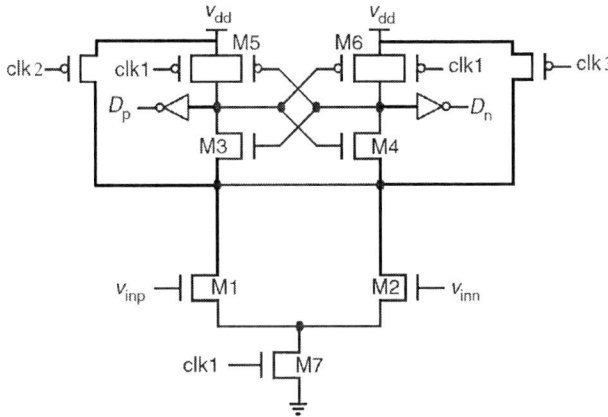

Figure 5.   Comparator [10]

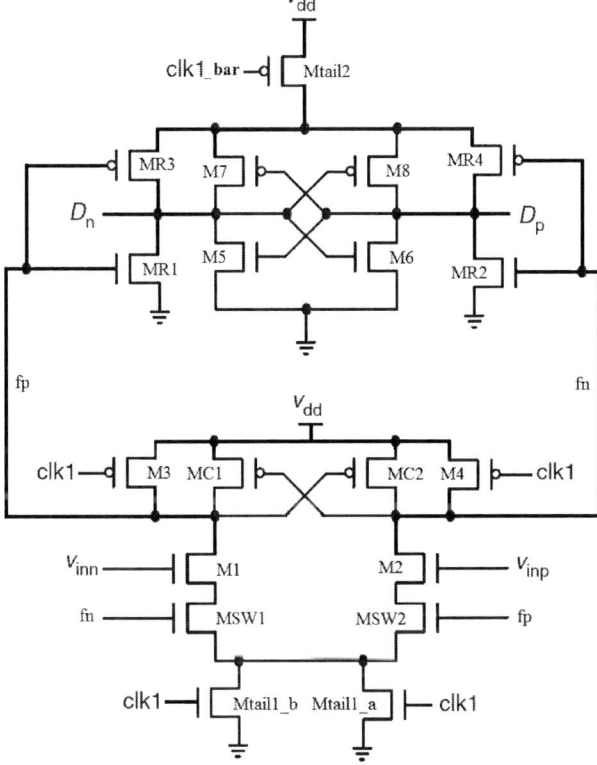

Figure 6.   Proposed Comparator

ON therefore M4 turns ON and output Dp becomes HIGH. The latch helps in holding the outputs in a particular level. In comparator circuit of [7] static power dissipation is very less and it is faster than [5]. Fig.2 shows the comparator of [7]. During the reset mode, clock clk2 is LOW and clk1 is also LOW. The clock pulse clk2 is the delayed pulse of clk1. PMOS and NMOS input drivers provide large swing in the output. A PMOS load is added to the differential pair to provide stability and higher gain. When the clock clk1 is

HIGH, there is a fast increase in current in the transistors. When the input Vinp is higher than Vinn, the current in the transistor M1 increases and pulls down the drain voltage to logic LOW and transistor M4 pulls up the drain voltage to VDD - Vtp, where Vtp is the threshold voltage of M6 transistor. Transistor M9 source potential becomes LOW and M10 source potential becomes HIGH. When clk2 goes HIGH, it pulls down the source potentials to LOW. Since the source potential of M9 is already LOW, the CMOS transistor logic formed by M9 and M12 initiates first and provides a stable output.

Comparator [8] is an efficient comparator having a good rise and fall time as well as large common mode voltage. Due to the dual rail configuration of this comparator the output swing is more than the other comparators. Single tail comparator is unable to draw much power required for a high speed comparator. Dual rail however provides sufficient power to the comparator circuit: one rail to latch and one rail to differential pair. Due to the high power consumption by the dual rail comparator a low power dual rail comparator was designed [8]. The new comparator consumes less power and provides high output swing. MC1 and MC2 provides pre-charges the nodes fn and fp. M3 and M4 act as PMOS loads. M1 and M2 provides low gain amplification. MSW1 and MSW2 acts as switches to lower the power dissipaion. M5, M6, M7 and M8 forms the latch. Two tail transistors are used Mtail1 and Mtail2. When fn is 'HIGH' MR2 is 'ON' and Dp becomes zero. Similarly, when fp is 'HIGH' MR1 is 'ON' and Dn becomes zero. When one node becomes zero the other node is charged to VDD. Fig.3 show compartor [8].

The comparator [9] is similar to [7] except its PMOS load transistors. The clock pulse given to transistors M7 and M10 is clk1 instead of clk2. The PDP of [9] is better than [7]. When clk1 is LOW, the output goes LOW. The clock pulse clk2 is the delayed pulse of clk1. When clk1 become HIGH and clk2 is still LOW for 20pico-s, the potential at the source of M5 pulled down to LOW when Vinp > Vinn. Now when clk2 goes HIGH, M3 and M4 tries to pull down the source potential to LOW. Since M5 source potential is already LOW, therefore, M5 acts before M6. The latching starts and reaches the HOLD state where outputs remain at their respective states for some time. Fig.4 shows the circuit diagram of [9]. The comparator [10] is a single rail dynamic comparator. A conventional dynamic comparator is not able to drive large transistors due to only one supply rail. The dynamic comparator consists of both differential pair and latch circuit which needs large amount of power in order to process quickly. Therefore it consumes less power but the delay is more. A low power design is mentioned in [10] for low power dissipation and offset calibration. Clk2 and CLK3 are delayed output of CLK1 and they are fed as an input to the latching circuit so that one node is charged first then the other. Such a latching technique provides better output. Fig.5 shows the comparator [10].

978-1-5386-1357-3/17 $31.00 © 2017 IEEE

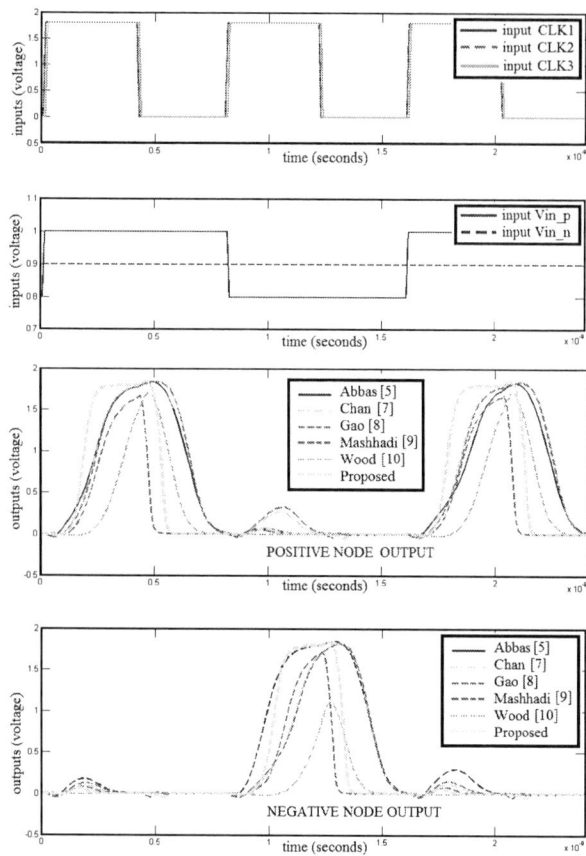

Figure 7.   Comparator outputs

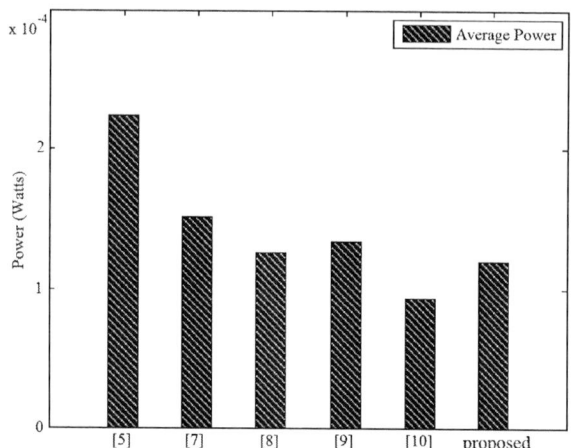

Figure 8.   Comparison of Average Powers

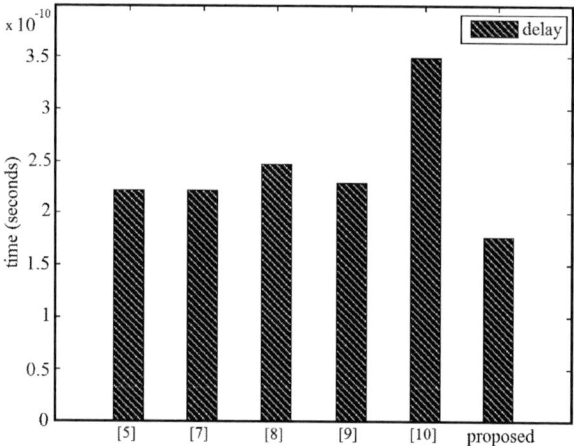

Figure 9.   Comparison of Delays

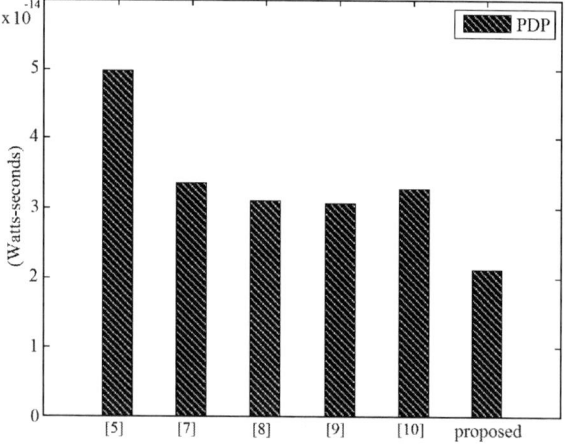

Figure 10.   Comparison of PDPs

B. Proposed Comparator

Proposed comparator is a modified form of [8]. The new comparator is not only fast but also consumes less power. The mean PDP is 22.94×10^{-15} which is less than the rest. The common mode voltage range is more as compared to all the other comparators. By adding a PMOS transistor to the overall circuit the output swing is increased. A PMOS and NMOS transistor at the output of the differential pair gives a better amplification and swing. When MR1 is turned 'OFF'

Table I
COMPARISON OF COMPARATORS IN 180NM TECHNOLOGY

Comparator	Vcm	DC Power	Average Power	Delay
	(V)	(nano-W)	(µW)	(pico-s)
Abbas [5]	0.9 to 1.4	88300	224.1	222.11
Chan [7]	0.1 to 0.9	0.066	151	222.17
Mashhadi [8]	0.7 to 1.2	0.012	125.5	247.13
Gao [9]	0.7 to 0.9	0.070	133.6	228.9
Wood [10]	0.9 to 1.5	0.060	93.5	349.23
Proposed	0.6 to 1.5	0.020	119.81	176.71

and MR2 is turned 'ON' i.e. it pulls down the node to zero potential then MR3 pulls the output node to VDD whereas

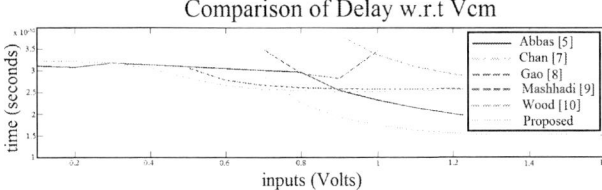

Comparison of Delay w.r.t Vcm

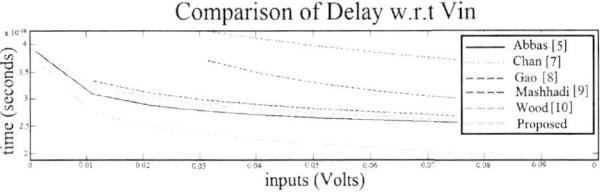

Comparison of Delay w.r.t Vin

Figure 11.   Delay with respect to Vcm and Vin

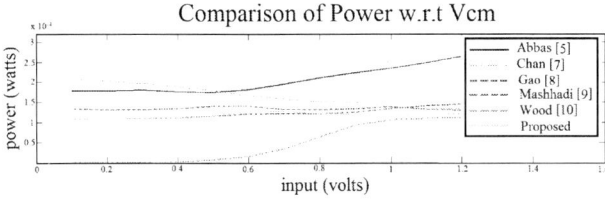

Comparison of Power w.r.t Vcm

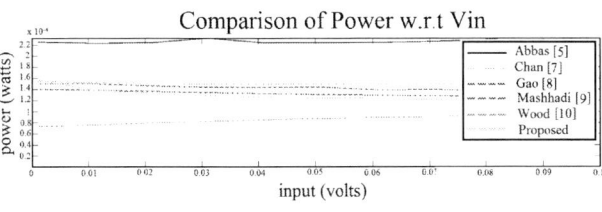

Comparison of Power w.r.t Vin

Figure 12.   Power with respect to Vcm and Vin

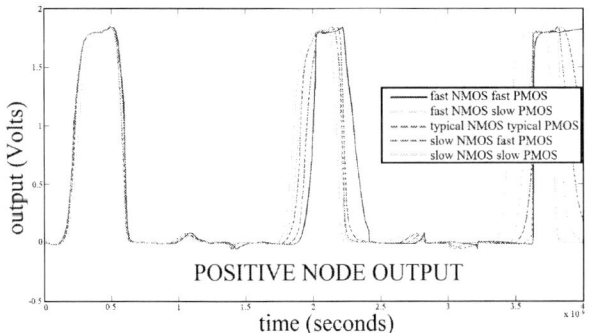

POSITIVE NODE OUTPUT

Figure 13.   Corner Analysis of Proposed Dynamic Comparator

MR4 is in 'OFF' state. Mtail1-a and Mtail1-b provides faster pull down which can also be achieved by just increasing the width of Mtail1 transistor. The width of Mtail1 is kept large and width of Mtail2 is kept small for faster operation. The proposed comparator is shown in Fig.6. At a clock frequency of 1.25GHz and 100mV ΔVin, a delay of 176.71 pico-seconds is achieved. DC power consumption is 0.02 nano-Watts and average power consumption is 119.81 μ-

Table II
MONTE CARLO ANALYSIS OF COMPARATORS

Comparators	Min	Max	Mean	Std. Dev.
Abbas [5]				
Delay (in pico-s)	193.7	352.3	237.8	33.23
Avg Pow (in μW)	210.6	227.7	219.5	3.894
PDP (in femto)	42.33	78.91	52.58	7.724
Chan [7]				
Delay (in pico-s)	227.6	330.4	253.3	19.37
Avg Pow (in μW)	137.1	157	148.6	3.773
PDP (in femto)	31.22	50.5	37.61	3.486
Mashhadi [8]				
Delay (in pico-s)	255.8	368.2	285.3	18.45
Avg Pow (in μW)	123.8	130	125.8	1.224
PDP (in femto)	31.93	47.85	35.9	2.65
Gao [9]				
Delay (in pico-s)	225.8	372.6	267.8	30.06
Avg Pow (in μW)	117.9	140.2	131.5	5.16
PDP (in femto)	26.62	50.35	35.31	4.967
Wood [10]				
Delay (in pico-s)	331.6	368.4	350.2	8.796
Avg Pow (in μW)	89.02	98.07	93.69	1.832
PDP (in femto)	29.83	35.07	32.81	1.051
Proposed				
Delay (in pico-s)	181.5	203.8	191.3	4.425
Avg Pow (in μW)	119.4	121.1	119.9	0.291
PDP (in femto)	21.71	24.53	22.94	0.548

Watts. The input common mode range is 0.6 to 1.5 V which is greater than the others and provides a better output swing.

The comparison between the different dynamic comparators and the proposed comparator is shown in Table.I. The common mode voltage range is the maximum in the proposed one. Comparator [8] consumes least dc power. Average power consumption is least in [10]. Proposed comparator is the fastest and has the least delay. Overall PDP is less which makes it suitable for applications in low power and high speed ADCs.

III.   SIMULATION AND RESULTS

The dynamic comparators are compared by using 180nm SCL technology. NMOS and PMOS lengths are taken as 180nm and widths are taken as 420nm. Clock freuncy is 1.25 GHz and input frequency is 625MHz. The voltage supply is 1.8 V with common mode input voltage of 0.9 V and 100mV ΔVin. The outputs are shown in fig.7. The comparison for average power is shown in fig.8 and for delay it is shown in fig.9. The power-delay product (PDP) is shown in fig.10. In addition to these simulations, the variation of power and delay with respect to common mode voltage Vcm and input

978-1-5386-1357-3/17 $31.00 © 2017 IEEE   143

Vinp is shown in fig.12 and fig.11 respectively. Finally, The corner analysis of the proposed comparator is done to check the reliability which is shown in fig.13. Corner analysis result is shown for the positive output node. In this analysis it is seen that when a slow NMOS and a slow PMOS is used, rise time of the output is less which shows that the latch is faster than the amplifier circuit. Overall analysis of the proposed comparator shows that it is reliable and efficient to use in low power and high speed applications. Table II shows the Monte Carlo analysis of different comparators.

IV. CONCLUSION

Dynamic comparators provide high speed comparison with the analog voltage. Single tail dynamic comparators requires large power supply to drive the circuit. One rail supply is therefore not sufficient. Although, single tail dynamic comparators consumes less power but cascoding reduces the output voltage swing. This gives rise to another type of comparator which is called dual tail comparator. Dual tail comparator provides large output swing and it provides sufficient power supply to the circuits. The proposed comparator a dual tail dynamic comparator and it is the fastest as compared to the others whereas comparator [10] consumes the least power among the rest. Low power design in [8] shows that the DC power consumption is less as compared to other recents designs. A PDP of 22.94f is achieved by designing the proposed dynamic comparator which is the least among others. This shows that it can be used in an efficient Flash ADC where low power and high speed is required. Encoder is said to be the bottleneck of a Flash ADC, so if encoder part is also fast then overall ADC performance will be high.

ACKNOWLEDGMENT

This work has been done from the Grant Received from Visvesvaraya PhD Scheme for Electronics and IT.

REFERENCES

[1] H. Traff, "Novel approach to high speed CMOS current comparators," in *Electronics Letters*, vol. 28, no. 3, pp. 310-312, Jan. 1992.

[2] B. Goll and H. Zimmermann,"A Low-Power 4GHz Comparator in 120nm CMOS Technology with a Technique to tune Resolution," *2006 Proceedings of the 32nd European Solid-State Circuits Conference*, Montreux, Switzerland. pp. 320-323. Feb. 2007.

[3] M. C. Huang and S. I. Liu,"A Fully Differential Comparator-Based Switched-Capacitor $\Delta\Sigma$ Modulator," *IEEE transactions on Circuits and Systems-II: Express Briefs*, Vol. 56, No.5. pp. 369-373. May 2009.

[4] J. He, S. Zhan, D. Chen, "Analyses of Static and Dynamic Random Offset Voltages in Dynamic Comparators," *IEEE Transactions on Circuits and Systems-I: Regular Papers*, Vol. 56, No.5. pp. 911-919. May 2009.

[5] M. Abbas, Y. Furukawa, S. Komatsu, J. Y. Takahiro and K. Asada, "Clocked comparator for high-speed applications in 65nm technology," *IEEE Asian Solid-State Circuits Conference 2010*, Beijing, China. pp. 1-4. Nov 2010.

[6] Y. Jung, S. Lee, J. Chae, G.C. Temes, "Low-power and low-offset comparator using latch load," *Electronics Letters*, Vol. 47 No. 3. Feb. 2011.

[7] C. H. Chan, Y. Zhu, U. F. Chio, S. W. Sin, U. Seng-Pan and R. P. Martins, "A reconfigurable low-noise dynamic comparator with offset calibration in 90nm CMOS," *IEEE Asian Solid-State Circuits Conference 2011*, Jeju, South Korea. pp. 233-236. Jan 2012.

[8] S. Babayan-Mashhadi and R. Lotfi, "Analysis and Design of a Low-Voltage Low-Power Double-Tail Comparator," in *IEEE Transactions on Very Large Scale Integration (VLSI) Systems*, vol. 22, no. 2, pp. 343-352, Feb. 2014.

[9] J. Gao, G. Li and Q. Li, "High-speed low-power common-mode insensitive dynamic comparator," in *Electronics Letters*, vol. 51, no. 2, pp. 134-136, Jan. 2015.

[10] S. H. Wood Chiang, "Comparator offset calibration using unbalanced clocks for high speed and high power efficiency," in *Electronics Letters*, vol. 52, no. 14, pp. 1206-1207. June 2016.

[11] W. I. I. Restu, B. W. M. Nasir, M. M. Bin Ibne Reaz, "Low power and high speed CMOS current comparators," in *2016 International Conference on Advances in Electrical, Electronic and Systems Engineering (ICAEES)*, Putrajaya, Malaysia. pp. 539-543. Nov. 2016.

[12] S. J. Kim, D. Kim, M. Seok, "Comparative study and optimization of synchronous and asynchronous comparators at near-threshold voltages," in *2017 IEEE/ACM International Symposium on Low Power Electronics and Design (ISLPED)*, Taipei, Taiwan. pp. 1-6. July. 2017.

An Automated Game Theoretic Approach for Co-operative Road Traffic Management in Disaster

Samya Muhuri
Computer Science & Technology
Indian Institute of Engineering
Science & Technology, Shibpur
Howrah, India
samyamuhuri.rs2015@cs.iiests.ac.in

Debasree Das
Computer Science & Technology
Indian Institute of Engineering
Science & Technology, Shibpur
Howrah, India
debasreedas1994@gmail.com

Susanta Chakraborty
Computer Science & Technology
Indian Institute of Engineering
Science & Technology, Shibpur
Howrah, India
sc@cs.iiests.ac.in/susanta.chak@gmail.com

Abstract—Transportation system gets paralyzed when a place affected by any catastrophic natural disaster. Some roads get blocked and vehicle density increases in all the other remaining routes. Each vehicle stuck at different position of the road network and tries to reach its destination in minimum time. A cooperative game theory based approach is proposed to regulate road traffic and minimize the waiting time of individual vehicles in any disaster situation. Each vehicle acts as a player and tries to increase its payoff value. Payoff value is based on the different parameters of the vehicle like its arrival time, velocity, priority and traffic density. Payoff value is adjusted according to the existing vehicle density in the road segment. Traffic density of each road is minimized simultaneously which increases the flow of the vehicles in the network cooperatively. Waiting time of a vehicle is directly proportional to the number of adjacent edges of the corresponding node and its density. The method also shows that any vehicle with higher priority will cross the road in minimum time. New alternative road creation process is also proposed based on graph theory for the worst situation. The alternative path will join the affected road with a minimum utilized node in the network to evacuate the affected traffic. Experimental results suggest that our proposed method achieve minimum average waiting time compare to the existing method. The shortest time travel path is found for each vehicle from its source to the desired destination in post disaster situation.

Keywords— road network; disaster management; game theory; clustering coefficient; traffic density.

I. INTRODUCTION

The vehicle population has been rapidly growing over the past several decades. The increasing number of vehicles on the roads frequently results in severe traffic congestion. Different natural and man-made disasters damage the existing road communication infrastructure partially or fully and make the traffic scenario even worst. Disaster Management can be categorized as the organization of the existing resources for dealing with all humanitarian aspects of emergencies to minimize the impact of calamities. Maintaining free-flows of the traffic in any disaster situation through traffic prediction and management is an impending challenge in the transportation system. The existing road network of any city can be viewed as a weighted network. Each edge represents

the road and vertex represents the road end points. Weight denotes the physical distance between two end points. The vehicles are moving with some velocity and priority. Each vehicle can only move through the open roads and try to reach its destination. In any catastrophe, some roads get blocked and accordingly the representing edges are removed from the existing network. As a result, the vehicle density in the neighbor roads increased and creates acute congestion. A novel disaster traffic management approach is designed in this paper for routing the vehicles through co-operative manner by increasing the traffic flow at each point. Alternative routes creation is proposed as a temporary solution to void the affected lanes if feasible in reality.

Routing multiple vehicles cooperatively by reducing the waiting time has been gaining increasing interest from researchers over the past several decades. The literature in this research domain can be roughly categorized into two main groups. It can be studied from the perspective of the whole road network [1-3] or the individual vehicle [4-5]. Kerner [6] shows that network breakdown minimization principle permits considerably greater network inflow rates at which no traffic breakdown occurs and frees flow of traffic remains in the whole network. Adaptive Kalman filtering based traffic prediction algorithm [7] estimates accurate travel time for any vehicle. A routing algorithm [8] minimizes the traffic jam occurrence through directing the paths of multiple vehicles cooperatively. The solution is presented for the conflict between all the crossing vehicles at road intersections using game theory in [9]. Crowdsourcing method is used in [10] to solve the problem of road congestion. Non-parametric approaches like machine learning [11], fuzzy logic [12], and neural network [13] is also used for easy flow of traffic. Some works have been reported on traffic management in disaster. In [14], authors describe some fundamental issues of the traffic management systems and the actual traffic conditions following the Hanshin-Awaji Earthquake. Kaviani et. al. [15] discusses management of road networks during bushfire and floods by Intelligent Disaster Decision Support System (IDDSS) method. Most of the existing disaster management methods ignore the car density in the neighbor roads of the affected area and do not suggest any method for alternative road creation. Our proposed method based on cooperative

978-1-5386-1357-3/17 $31.00 © 2017 IEEE

game theory can identify the best suitable shortest travel path for any individual vehicle from its source to the desired destination in any post disaster situation. The method efficiently reduces the average waiting time for all the individual vehicles. Traffic density minimized at each point in the road network which increases the flow of the vehicles cooperatively. Payoff value of individual vehicles is based on the different parameters like its arrival time, velocity, priority and traffic density. Payoff value is adjusted according to the existing traffic density of the corresponding road segment. The vehicle with a higher priority gets more payoff value than any random car and crosses the desired path in minimum time. The proposed method is effective in the real scenario as the reduction of waiting time in a queue for any individual vehicle can massively reduce the cost of road communication.

The rest of the paper is organized as follows: section II describes technological preliminaries, section III deals with proposed co-operative routing method in disaster with example. The experimental results are discussed in section IV and the conclusion is presented in the final section.

II. PRELIMINARIES

A. Road Network

The road network is denoted by an undirected graph $G(V,E,W,C)$, where V is the set of unordered vertices represents the road end points, E is the set of edges depicts the roads in between, the weight W is the set of length of each road and C is the set of maximum traffic capacity of the road segments. Traffic signals are present at each road crossing. A vehicle moving from one edge segment to other has to wait until it gets the green signal. It is assumed that the whole road network is enabled with internet of thing (IoT) infrastructure. Each vehicle moves in the road communicating with the centrally controlled IoT enabled device for collaborative traffic management. A vehicle is defined by $\alpha(T,V,P)$ where T denotes its arrival time in a road segment, V represents its velocity and P depicts its priority. Priority has a high impact on the travel time. An ambulance, vehicle of the police force or fire brigade etc. gets higher priority than any personal vehicle. Priority value always lies between 0 and 1 which denotes low priority and high priority respectively. A road network in normal condition is shown in Fig.1.a. It consists of 5 nodes and 10 edges. Let some roads get blocked (1-3,1-4,1-5) in disaster as shown in Fig.1.b. At this stage, each vehicle in the network has to reach its desired destination through the existing road infrastructure.

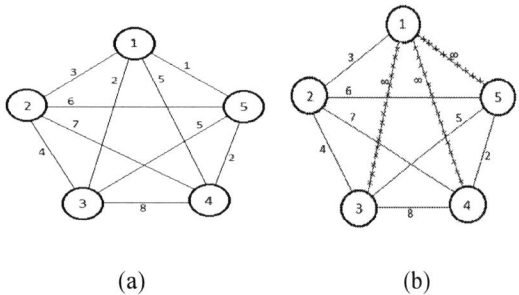

(a)   (b)

Fig. 1.   Road Network (a) before and (b) after disaster

B. Clustering Coefficient

Clustering co-efficient (CC) represents the association among the nodes [16]. In a small-world network, if the most number of nodes are neighbors of one another, then the nodes can be reached from every other by a small number of hops or steps. Clustering coefficient examines the neighborhood relationship according to their connectivity. Clustering Co-efficient can be measured by

$$CC(v_i) = \frac{no.of\ neighbor\ nodes\ of\ v_i\ connected\ by\ edges}{Max\ no.of\ edges\ may\ exist\ among\ neighbors} \ ...(i)$$

The maximum number of edges may exist among neighbors is $^nC_2$ where n is the number of neighbors. $CC(v_i)$ lies in between 0 and 1. It is equal to 1 if every neighbor connected to v_i is also connected to every other vertex within the neighborhood, and 0 if no neighbors are connected.

C. Game Theory

Game theory can be of two types which are cooperative and non-cooperative. The cooperative game theory calculates the gain of each player in cooperative manner, while non-cooperative game theory focuses on which moves players should rationally make to win individually [17]. The final result of a game is dependent on the strategies are chosen by all the players.

D. Traffic measuring parameters

i) Traffic Flow: [18]

Traffic flow describes the number of vehicles passing through a reference point in unit time. If average waiting time of all the vehicles decrease, then the flow of the traffic will increase. Traffic flow F of any road segment k can be denoted as

$$F_k = \frac{no.of\ vehicles\ passes}{waiting\ time} \ ...(ii)$$

In a disaster, we would like to reduce the waiting time of each vehicle to increase the traffic flow.

ii) Average Queue

Average queue measures the efficiency of the traffic management. It is the ratio between the duration of the congestion of an individual vehicle i, to the total waiting time for all the vehicles for any particular road segment. Average queue Q for a road segment k can be defined as

$$Q_k(i) = \frac{duration\ of\ congestion\ for\ i^{th}\ vehicle}{total\ wighting\ time\ for\ all\ vehicles} \ ...(iii)$$

Average queue value always remains between 0 and 1 representing low congestion and high congestion respectively.

III. PROPOSED CO-OPERATIVE TRAFFIC ROUTING METHOD IN POST DISASTER SITUATION

The urban road infrastructure can be represented as a connected graph. The vehicles are always looking for the shortest path from its source to the desired destination. The shortest geographical distance between any two points in a graph does not always represent the minimum travel time path. The vehicles are always looking for cross the less dense road to reach its destination in minimum time. Some roads get blocked and correspondingly detached from the original

978-1-5386-1357-3/17 $31.00 © 2017 IEEE   146

network in post disaster situation. Vehicles of the dense road should be cleared at first priority for smooth traffic movement. Otherwise, the road will be flooded and the vehicles have to wait for a long time. If the road reaches its maximum capacity, then the vehicles will go to the starvation state. In this section, we will describe our proposed cooperative game theoretic method to resolve the traffic congestion in disaster.

A. Routing Method based on game-theory

Any game consists of three basic elements which are players, player actions, and utilities or payoff function. We define these three elements and set up the game for a disaster affected road network. In this game, each vehicle is considered as a player. The vehicles are looking for the shortest travel time path to reach its destination and waiting for its turn to cross all the nodes ahead of it. Payoff value of each vehicle is calculated considering its arrival time, priority and velocity. Each player is tried to increase its payoff value which will reduce its waiting time. Payoff value of each vehicle represents its weightage in the road. It always lies between 0 and 1 which denotes low weightage and high weightage respectively. Each vehicle enters in the road segment with certain priority. The vehicle priority is adjusted with a weight factor which depends on the corresponding road situation. The weight depends on the arrival time of the vehicle and the traffic density. If vehicle density is high in a road segment and another vehicle arrives late with high priority, its payoff value will be reduced. If the difference in the arrival time between two consecutive vehicles is high, in the meantime the traffic density will reduce. The priority of the later vehicle will be increased to adjust the traffic flow. Algorithm 1 represents the method to adjust the payoff value of each vehicle.

Algorithm 1 Payoff value adjustment of a vehicle

---

1. **payoff** (*arrival_time* [ ], *vehicle_no* [ ], *priority* [ ])
2. **Begin**
3.   **for** each vertex $k \in G.V$
4.   **for** each vehicle α
5.   **if** (*vehicle_no* [α] == *first_vehicle*)
6.   *temp* ← *arrival_time* [α]
7.   **end If**
8.   **end for**
9.   **for** each vehicle $\alpha \notin$ *first_vehicle*
10.   *delay*[α] ← abs (*temp* - *arrival_time* [α])
11.   **for** i =1 to α
12.   *sum_delay*[α] ← $\sum$ *delay*[i]
13.   **end for**
14.   *weight* [α] ← *delay*[α] / *sum_delay*[α]
15.   *payoff* [α] ← *priority*[α] * *weight* [α]
16.   **return** *payoff* [α]
17.   **end for**
18.   **end for**
19. **End**

In the following example, we briefly describe the proposed method for adjustment of the payoff value of any vehicle. Let

nine vehicles came in the same road segment with some priority value at different time instant as shown in Table 1. Generally, for two consecutive vehicles, weight of the later one will be less as it increases the road density. Vehicle number 4 and 5 arrive at the same time. Priority of vehicle number 4 is greater than vehicle number 5 and correspondingly gets high payoff value. Arrival time delay between vehicle number 8 and 9 is high so the previous vehicles can cross the road within that specified time. The density of the road will decrease within that time and vehicle number 9 get high payoff value although the priority of both the vehicles is same.

TABLE I.   PAYOFF VALUE ADJUSTMENT

Vehicle no.	Priority	Arrival Time Stamp	Delay from the first arrived vehicle	weight $= \dfrac{arrival\ delay}{\sum arrival\ delay}$	Payoff (priority X weight)
1	0.5	0.10	0	1	0.5
2	0.2	0.12	0.02	1	0.2
3	0.6	0.15	0.05	0.71	0.43
4	0.8	0.16	0.06	0.46	0.37
5	0.5	0.16	0.06	0.32	0.16
6	0.7	0.18	0.08	0.30	0.21
7	0.4	0.19	0.09	0.25	0.10
8	0.6	0.20	0.10	0.22	0.13
9	0.6	0.38	0.28	0.38	0.23

Each vehicle tries to figure out its next hop station and seeks to increase its payoff. Maximum dense roads from the neighbors are to be found. This most dense road gets green signal alternatively until its density becomes less than its neighbor roads. All the vehicles will get it turn to cross the signal and density of the roads will be optimized. If the corresponding road is highly dense than others, then it will get green signal first and its neighbors have to wait. If density is equal for more than two roads then to break the conflict, the road with high summative payoff value will get the advantage. If there is no traffic rule violation, more than one road gets green signal in parallel. Actual vehicle speed v is determined by

$$v = v_f(1 - \frac{\rho}{\rho_{max}})\ldots(iv)$$

where v_f is the speed with which the vehicle is travelling, ρ is current road density whereas ρ_{max} is the maximum road capacity. If traffic density reaches its maximum capacity, the velocity of the individual car become zero and finds starvation. In post disaster situation, we try to reduce the value of ρ to maximize the actual vehicle speed.

If the density of the neighbors is high than the corresponding road k, then traffic flow time of the neighbours is treated as waiting time for each vehicle resides at k. Else, vehicles at k get its turn to move. The weighting time w_t for each vehicle of the road segment k is calculated as

$$W_t(k) = \begin{cases} t_{flow}(neighbour_k) \ if \ \rho_{neighbour_k} > \rho_k \\ 0 \qquad\qquad\qquad otherwise \end{cases} \ ...(v)$$

The actual travel time for a vehicle to reach its destination is the summation of the travel time through the shortest path and the waiting time in each signal point. Algorithm 2 describes the method to calculate the actual travel time of a vehicle in any disaster affected network.

Algorithm 2 Travel time of a vehicle

---

1. **travel_time** (*network G, vehicle_no, vel, arrival_time, priority, dest*)
2. **Begin**
3.   **for** each edge $e \in G.E$
4.   **if** *block* [e] = = true //*locate disaster*
5.   *weight* [e] = infinity
6.   **end for**
7.   **for** each vehicle α
8.   *source* $\leftarrow$ *G. V* [α]
9.   *dist* $\leftarrow$ *min_distance* [*source, dest*]
10.   *actual_vel* $\leftarrow vel\,[1 - \frac{no\_of\_vehicle[source]}{c[source]}]$
11.   *temp* $\leftarrow payoff$ (*arrival_time, vehicle_no, priority*)
12.   *travel_time* [α] $\leftarrow \frac{dist*temp}{actual\_vel}$
13.   **end for**
14.   **for** each vehicle α
15.   *hop_station* [ ] $\leftarrow$ *intermediate_nodes* [*source,dest*]
16.   **for** each *station* $\in hop\_station$ [ ]
17.   *adj* [*station*] $\leftarrow$ *station* $\cup$ *neighbour*[*station*]
18.   *hop_time* $\leftarrow$ 0
19.   **for** each *station* in *adj* [*station*]
20.   choose the max dense station
21.   **if** *max_dense_station* = = *station*
22.   **break**
23.   **else**
24.   *temp_dist* $\leftarrow$ min (*weight*[$e \leftarrow station \in G.V$])
25.   *clearing_time* $\leftarrow \frac{temp\_dist}{avg\_vel[\alpha \in station]}$
26.   *hop_time* $\leftarrow$ *hop_time* + *clearing_time*
27.   end for
28.   *waiting_time* [α] $\leftarrow$ *waiting_time* [α] + *hoptime*
29.   **end for**
30.   *actual_travel_time* $\leftarrow$ *travel_time* [α] +   *waiting_time* [α]
31.   **end for**
32. **End**

Lemma: Waiting time of a vehicle is directly proportional to the number of adjacent edges of the corresponding node and its vehicle density.

Proof: Case 1: Let the traffic density of each road is constant. If number of adjacent edges increases, then the vehicles have to wait for its turn for large number of cases and waiting time will increase. If there are no adjacent edges, then the corresponding vehicles can move without any resistance.

Case 2: Let the number of adjacent edges is constant. If the density at any road segment reaches its maximum point i.e. $\rho \cong \rho_{max}$ then from equation (iv) we can say that $v = 0$. The vehicle will go to the starvation state and waiting time increases. If density comes to minimum point i.e. $\rho \cong 0$ then $v = v_f$ and vehicles need not to stop. It will travel with its arrival speed.

B. Generation of alternative path

Sometimes the existing road connection is not sufficient for efficient traffic flow after any natural disaster. Some places may get fully disconnected and accordingly represent as an isolated vertex. The cars which are stuck in between the blocked path have to be transmit through an alternative way. Alternative path creation is dependent on the structure of the network, present traffic situation and feasibility. A novel alternative path creation method is proposed here for systematic road traffic management. If any one of the corresponding two crossings of a road segment has a clustering coefficient CC (v_i) large than a threshold, then no alternative path is necessary to clear the affected traffic. The traffic can find its way through the adjacent nodes as neighbors are strongly connected between themselves. Otherwise, an alternative path is generated between the blocked node and the minimum summative payoff node if it is feasible in real life. The node having minimum summative payoff value can bear the additional pressure of overhead traffic. A normal road network is shown with 5 nodes and 6 edges in the following example as shown in Fig. 2. a. We consider 0.3 as our clustering coefficient threshold value for alternative path creation. Let there are blocks in the edges (3,4) and (5,4). Node 4 is getting isolated and needs to be connected to the main network for smooth vehicle transmission.

TABLE II.   CC VALUE OF EACH NODE FOR THE GRAPH IN FIG.2.a.

Node number	1	2	3	4	5
CC	0.33	0	1	0	1

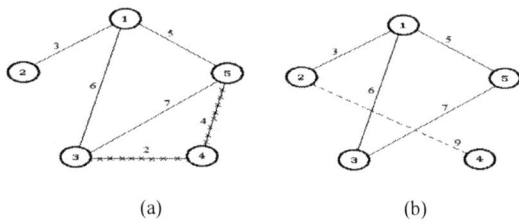

(a)   (b)

Fig. 2. Road Network (a) before disaster and (b) alternative path

The clustering coefficient of the node 4 is zero as shown in Table II. An alternative path likes to be created from node 4 to overcome the disaster. Let Node 2 has minimum payoff in the whole network and get connected with node 4 as depicted in Fig. 2.b. Node 3 and 5 are also directly affected by the disaster. Their high clustering coefficient value will manage the existing traffic load and will not be considered for creation of the alternative path. An alternative path is not the substitute of the previous path. It will just improve the situation and vacate the disaster affected traffic. In Fig.2.a. there is a direct path from node 3 to node 4. After the alternative path creation, vehicles from node 3 have to cross node 1 and 2 to reach its destination node 4.

Algorithm 3 Generation of alternative path

---

1. generate_new_edge(*Network G, block_start, block_end*)
2. Begin
3.   **if** (*weight*[*block_start, block_end*]= = ∞) //disaster
4.   **if** ((*CC (block_start)*≥ *T*) | | (*CC (block_end)* ≥ *T*))
5.   **do** nothing
6.   **else**
7.   find a node *v* ∈ *G.V* with min payoff
8.   **end if**
9.   *temp←v*
10.   **if** ((*weight*(*temp, block_start*)= =0) || (*weight*[*temp, block_start*]= =∞))
11.   create an edge *e₁*, *e₁*=(*temp, block_start*), *G.E* ∪ {*e₁*}
12.   **end if**
13.   **if** ((*weight*(*temp, block_end*)= =0) || (*weight*[*temp, block_end*]= =∞))
14.   create an edge *e₂*, *e₂*=(*temp, block_end*), *G.E* ∪ {*e₂*}
15.   **end if**
16.   **end if**
17. End

IV. EXPERIMENTAL RESULTS

In this section, we create a road network with 20 nodes and 23 edges to evaluate the performance of our proposed method. The experimental setup includes low-end machine with Intel(R) Core(TM) i5- 7600 processor, 4GB RAM and 500 GB hard disk. The experiment is simulated using Java platform. Two hundred random vehicles are generated in different nodes of the network at different time instance with different velocity and priority. Each vehicle chooses its destination and reaches there in a specified time. Now a disaster environment is created manually where six roads are blocked. The vehicles ask for the shortest travel path from the same source to destination. Proposed method calculate the traffic density of each road. According to the individual payoff value, each vehicle recognized its suitable shortest time travel path. Fig.3. depicts the difference between the travel time of each vehicle before and after the disaster situation. The method reduces the waiting time for individual vehicle as compared to the exiting method [19] as shown in Fig. 4. Average waiting time is 3.38 and 3.20 for exiting method [19] and proposed method respectively. Reduced waiting time decreases average queue value of vehicles as shown in Fig.5. All the results show that our proposed method based on cooperative game theory can identify the best suitable shortest travel path for any individual vehicle in any disaster situation. Traffic density gets minimized at each point in the network which increases the flow of the vehicles cooperatively. The

vehicle with higher priority gets more payoff value than any random vehicle and cross the desired path in minimum time.

CONCLUSION

This paper proposes a novel algorithm for urban traffic management for any post disaster situation. The proposed method is based on cooperative game theory. The algorithm finds the shortest time travel path for each vehicle from its source to the desired destination. The method works on co-operative manner and reduces vehicle density in each road segment. The results of the simulated experiments show that the proposed algorithm achieves minimum average waiting time and finds the minimum time travel path for each vehicle from its source to desired destination after disaster. In future, we aim to resolve the conflict of crossing vehicles at different road intersections like Diamond interchange, Trumpet interchange, Cloverleaf interchange, etc. Another future extension of the work is aiming to take the road advantage in front of some important places like hospitals, schools, rescue centre etc. and use separately in disaster.

REFERENCES

[1] Z. Cao, H. Guo, J. Zhang, F. Oliehoek, and U. Fastenrath, "Maximizing the probability of arriving on time: A practical q-learning method," in Proc. 31th AAAI Conf. Artif. Intell., 2017, pp.4481-4487.

[2] H. Rakha and A. Tawfik, "Traffic networks: Dynamic traffic routing, assignment, and assessment," in Encyclopedia of Complexity and Systems Science, R. A. Meyers, Ed. New York, NY, USA: Springer, 2009, pp. 9429–9470.

[3] N. Gartner and C. Stamatiadis, "Traffic networks, optimization and control of urban", in Encyclopedia of Complexity and Systems Science, R. A. Meyers, Ed. New York, NY, USA: Springer, 2009, pp. 9470–9500.

[4] L. Padgham and M. Winikoff, "Developing Intelligent Agent Systems: A Practical Guide", vol. 13, New York, NY, USA: Wiley, 2005.

[5] J. Auld and A. Mohammadian, "Activity planning processes in the agentbased dynamic activity planning and travel scheduling (ADAPTS) model", Transp. Res. Part A Policy Practice, vol. 46, no. 8, pp. 1386–1403, 2012.

[6] B. S. Kerner, "Optimum principle for a vehicular traffic network: Minimum probability of congestion," J. Phys. A Math. Theoretical, vol. 44, no. 9, pp. 41–66, 2011.

[7] L. Chu, S. Oh, and W. Recker, "Adaptive Kalman filter based freeway travel time estimation", in Proc. 84th Transportation Research Board (TRB) Annual Meeting, 2005, pp.1-21.

[8] H. Guo, Z. Cao, M. Seshadri, J. Zhang, D. Niyato, and U. Fastenrath, "Routing Multiple Vehicles Cooperatively: Minimizing Road Network Breakdown Probability", IEEE Transactions on Emerging Topics in Computational Intelligence, vol. 1, no. 2, 2017, pp. 112-124.

[9] M. Elhenawy, A. A. Elbery, A. A. Hassan and H. A. Rakha, "An Intersection Game-Theory-Based Traffic Control Algorithm in a Connected Vehicle Environment", International Conference on Intelligent Transportation Systems, 2015, pp. 343 – 347.

[10] T. T. Alam, A. N. Chowdhury and M. Z. Rahman, "An Intelligent Road Traffic Management System Using Nvidia GPU", 19th International Conference on Computer and Information Technology, December 18-20, 2016, North South University, Dhaka, Bangladesh, pp. 419-424.

[11] J. Rzeszotko and S. H. Nguyen, "Machine Learning for Traffic Prediction", Fundamenta Informaticae, vol. 119, no. 3-4, 2012, pp. 407-420.

[12] V. P. Vijayan, and B. Paul,"Multi Objective Traffic Prediction Using Type-2 Fuzzy Logic and Ambient Intelligence," in Proc. 2010 International Conference on Advances in Computer Engineering (ACE), pp.309-311, 2010.

978-1-5386-1357-3/17 $31.00 © 2017 IEEE   149

[13] D. Chen, L. Lü, M-S. Shang, Y. C. Zhang, T. C. van Hinsbergen, A. Hegyi, J. van Lint, and H. van, Zuylen, "Bayesian neural networks for the prediction of stochastic travel times in urban networks," IET Intelligent Transport Systems, vol. 5, no. 4, pp. 259-265, 2011.

[14] Y. Iida, F. Kurauchi and H. Shimada, "Traffic Management System Against Major Earthquakes", IATSS Research, Volume 24, Issue 2, 2000, pp. 6-17.

[15] A. Kaviani, R. G. Thompson, A. Rajabifard, G. Griffin and Y. Chen, "A decision support system for improving the management of traffic networks during disasters", in Proc. Australasian Transport Research Forum 2015, 2015, Sydney, Australia, pp.1-14.

[16] D. J. Watts, and S. H. Strogatz, "Collective dynamics of small-world networks", Nature, vol. 393, 1998, pp. 440-442.

[17] M. U. B. Niazi, A. B. Özgüler and A Yildiz, "Consensus as a Nash Equilibrium of a Dynamic Game", in Proc. International Conference on Signal-Image Technology & Internet-Based Systems (SITIS), 2016, pp. 365 – 372.

[18] H. Lieu, "Traffic-Flow Theory", Public Roads. US Dept of Transportation, vol. 62, No. 4, 1999.

[19] Ciyun Lin and Bowen Gong, "Transit-Based Emergency Evacuation with Transit Signal Priority in Sudden-Onset Disaster", Discrete Dynamics in Nature and Society, vol.2016, pp. 1-13, 2016.

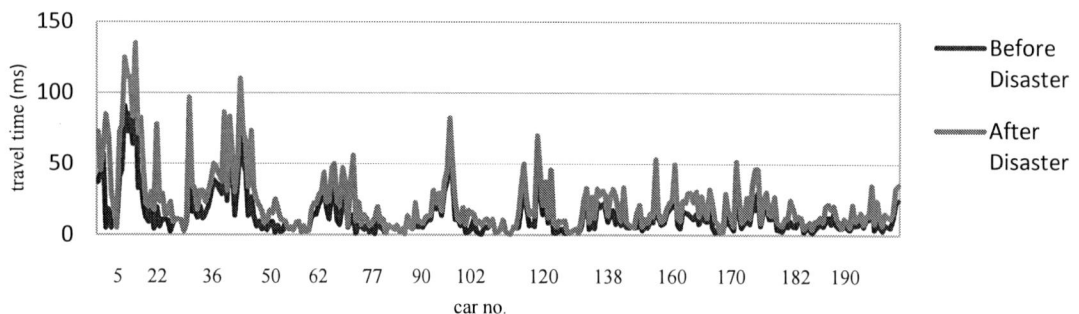

Fig. 3. Comparison of travel time before and after disaster

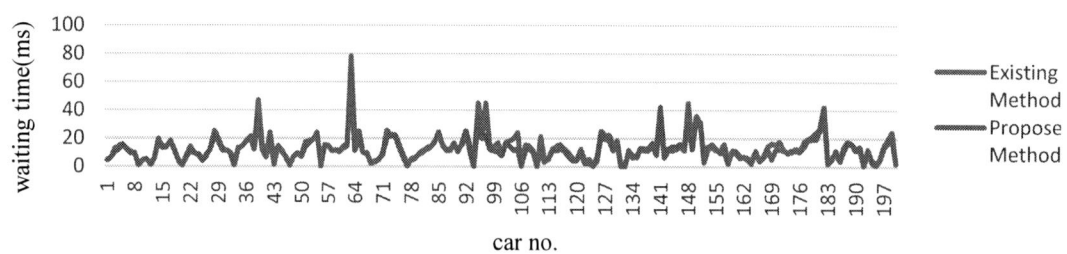

Fig. 4. Comparison of waiting time for proposed and existing method

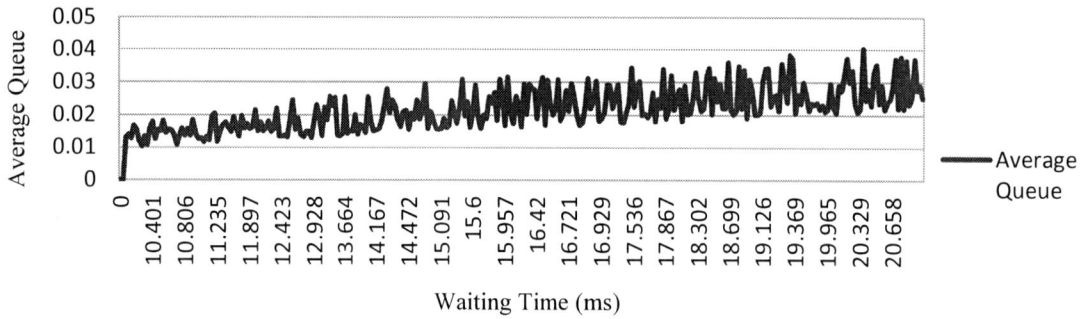

Fig. 5. Average queue of the vehicles

Security Enhancements to System on Chip Devices for IoT Perception Layer

Sudeendra kumar K, Sauvagya Sahoo, Abhishek Mahapatra, Ayas Kanta Swain, K.K.Mahapatra
kumar.sudeendra@gmail.com, sauvagya.nitrkl@gmail.com, kmaha2@gmail.com
National Institute of Technology, Rourkela

Abstract- Internet of Things (IoT) will drive the growth for semiconductor industry in next decade. In the era of IoT, millions of smart computing things are connected to solve customized applications. The tenets of IoT design are agility, scalability and security. Security is one of the important tenets for the success of IoT. In this paper, we discuss the challenges and possible solutions for IoT security that needs to be addressed at IoT perception layer/Edge node. The inevitable component of the IoT edge node is microcontroller/System on Chip (SoC). The microcontroller/SoC used in sensitive applications consists of Trusted Execution Environment (TEE), a hardware support for security. TEE's are not sufficient to address all the security issues in IoT systems. Hardware security issues like hardware Trojans, counterfeiting and debug security are tightly interlinked with the IoT perception layer security. There can be common solution to the hardware security issues and IoT perception layer security. In this paper, we briefly discuss the challenges in IoT design, IoT security, vulnerabilities of edge device, existing solutions and need for new security architecture for IoT edge nodes. And finally we present what security features, the next generation SoC/microcontrollers should incorporate to solve both hardware intrinsic security and IoT perception layer security more holistically.

Keywords: Internet of Things, Microcontroller/System-on-Chip Security, Hardware Security, Security Infrastructure IP

I. INTRODUCTION

The idea of Internet of Things (IoT) is to connect the digital and physical world seamlessly to create a network of objects which communicate with each other. There can be millions of objects in the network with a capability to take intelligent decisions. The success of IoT can raise the quality of human life to a new level. The complete IoT ecosystem consists of sensor and actuators, microcontrollers with modest processing and connectivity capabilities, network gateways and cloud computing. Generally, consumer of data is also a producer of data in IoT ecosystem. Smart devices enabled with microcontrollers suitable for IoT applications will interface with sensor, actuators and networks. The communication networks generally used are Wi-Fi, GPRS, 3G, Zigbee, Bluetooth, etc and finally cloud infrastructure to support large scale processing and data storage. The known challenges in an IoT ecosystem are: - identification or authentication for addressing an IoT node, choosing a right connectivity technique, maintaining data compliance across network and security. The permeation of IoT's will bring drastic change to every known sector including agriculture, manufacturing and services. The core tenets to follow in the IoT design are: - agility, scalability, cost and security. Security is of prime importance because it is a part of challenge and also it is one of the core tenets in the IoT design [1].

The simple architecture of IoT ecosystem contains edge nodes, gateways, network and cloud [2]. The edge nodes (devices) are found in industries, consumer electronics, and home appliances etc, which collect data from sensors and interact with physical objects in the ecosystem. The edge nodes are also called as perception layer devices in IoT ecosystem [2]. The edge nodes are generally powered by microprocessors/microcontrollers or System on Chip (SoC) chips. Edge nodes are connected to internet through gateways. Gateways are generally standalone devices, which support wireless and wireline communication protocols. In recent times, gateways are getting embedded with edge nodes [3]. Gateways are generally designed using microcontroller/SoC devices. The universal gateways can be designed with the right mix of hardware and software. IoT device connected to cloud through internet is a globally interconnected network used to share data, store data and used to extract the meaningful information out of data through large scale data crunching. The IoT value chain includes variety of solutions: - hardware (processors, chipsets, SoC's, gateways etc), custom designed software solutions, network services and cloud services [1].

There are many research problems associated with IoT. The well-known research issues in IoT are discussed briefly: -

Data interoperability: - Billions of devices generate disparate data with their own formats. Data transmitted from processor to gateways and to cloud servers should use data format, which is understood across the different IoT layers. The standardization of data and representation languages, with a scope for constant upgradation is a good research problem [2].

Low power device (edge node) support: - Most of the IoT edge devices are low power and battery operated. Choosing components to design an edge node is very important. The microcontroller/SoC used in edge device must support internet connectivity. Internet connection is not optimized for low power consumption [19]. Power management in IoT edge devices is crucial, as it should be powered on and connected to network continuously. Designing an edge device with low power components without affecting the performance is a challenge for designers [2].

Security and Privacy: - Success of IoT based product or solution highly depend upon its strong security features. Security of IoT system is complex and challenging because it is a network of many connected devices communicating with each other all around the globe. In this complex ecosystem, at any stage there can be malicious hardware or software which will compromise security. The security issues are serious,

978-1-5386-1357-3/17 $31.00 © 2017 IEEE   151

because data communicated in the IoT networks are sensitive ranging from personal health information, trade secrets of a commercial firm and sensitive government information. It is very critical to ensure data security and system security, which is a multidimensional research problem [2].

Product development and manufacturing: - Product development and manufacturing are highly diversified in the IoT ecosystem, due to large number of applications coming under IoT umbrella and high number of IoT hardware (mainly edge devices and gateways) designers and manufacturers. Lack of standards in edge device and gateway design will create compatibility issues across IoT edge nodes in future. There is a need for regulations and standards to follow in designing edge and gateway nodes. Developing standards and regulatory mechanisms needs serious research.

Analytics: - Data mining and crunching is a crucial component in IoT ecosystem, to extract meaningful information from raw data generated from different sensors and data collected from various sources including social media. Handling different patterns of data is a unique challenge in IoT. Each edge node will have some amount of processing capability, which is helpful in converting a data into standard pattern and small amount of processing. Based on this concept, fog computing technique is developed and discussed widely in recent IoT literature. The traditional data mining algorithms suitable for centralized computing infrastructure does not fit into IoT. There is a need to develop novel algorithms for data analytics which support fog computing and modern IoT requirements. Further, critical challenges are: - to decide on how much data to collect, aggregation of data through different edge nodes, routers, gateways and finally into cloud [4].

In paper [7], authors list the vulnerabilities and threats to the IoT edge node and paper [2] describes the categories of security issues ranging from hardware exploitation to power management in the perception layer of IoT ecosystem. SoC/Microcontroller is an inevitable part of perception layer and solutions to the vulnerabilities at IoT edge nodes should be addressed at SoC level. The SoC manufacturers are also facing hardware security issues like counterfeiting and debug security. This paper discusses solutions available to address security issues in current day microcontroller/SoC and necessary features in SoC to address the security challenges in IoT and hardware security.

In this paper, we mainly focus on security problems and solutions connected with IoT edge device (perception layer). The section II discusses the IoT security comprehensively and compares IoT with embedded systems and cyber physical systems security. Section III discusses the IoT edge device vulnerabilities and section IV discusses the existing solutions found in literature for IoT edge device security. Finally, in section V we present the possible solutions the SoC/microcontroller ODM's (Original Design Manufacturers) can incorporate in their products targeted to design IoT edge devices, which are useful in securing the IoT edge devices more holistically. Section VI concludes the paper.

II. IoT SECURITY

In the development of embedded systems used in sensitive applications, security was considered as important part of product development lifecycle. IoT is unique in several aspects in comparison with traditional embedded systems and cyber physical systems.

IoT and Traditional Embedded Systems: Embedded systems are designed to perform specific applications with the right mix of hardware and software. In embedded systems, depending upon the functionality connection to internet is an optional one. Most of the embedded systems are isolated devices without connectivity to internet. In the case of IoT, connection to internet is mandatory. Embedded systems will communicate with few devices nearby and in the case of IoT, billions of heterogeneous devices talk to each other in the ubiquitous network. Another difference between IoT and conventional embedded systems is the amount of data getting processed. IoT ecosystem process large scale data compared to embedded system. Tamper resistance and suitable encryption schemes to protect sensitive data are generally found security measures in embedded systems. In IoT, we have to deal with range of security issues from perception layer edge devices to cloud data centres, which will be discussed in this section [5].

IoT and Cyber Physical Systems (CPS): - CPS is a system which comprises of both cyber and physical components. Both cyber and physical components are bind together using computers and communication systems makes the CPS [6]. CPS is an integration of sensors, actuators, communication networks and control systems, in which status of physical system is monitored with the help of sensors and suitable control action is taken. Data transactions will occur through communication channels. The concept of IoT is very similar to CPS. CPS is well established domain which has got numerous application areas similar to IoT like smart grid etc. IoT is very different from CPS in terms of number of systems connected to communication link (internet), the amount of data generated and scope for scalability. CPS is a system infrastructure and IoT is networking infrastructure connected to large number of devices compared to CPS. For CPS, real-time control is a primary goal, while in IoT, resource sharing, data sharing and effective data crunching with minimum latency are important. The security aspect in CPS is much simpler challenge than security of IoT. Applying conventional cryptographic techniques, the data security is achieved in CPS and techniques used in embedded systems security are equally applicable in CPS.

The factors responsible for emergence of IoT are: Availability of high performance processors at low prices, improvements in data storage techniques (cloud) and efficient wireless communication systems (4G) to connect multiple layers in the system hierarchy at affordable prices. All these factors made a concept of IoT into reality, which is redefining the products and services in most of the application verticals of semiconductors.

978-1-5386-1357-3/17 $31.00 © 2017 IEEE

Factors affecting security of IoT:-

- Scalability: -The major challenge for security is scale, diversity and customization. The IoT system will have billions of devices connected in the network. Designing holistic data and system security using generic cryptographic techniques for a large system is difficult. Designing scalable security architecture in the presence of customization is a critical challenge [7].
- Diversity: - Diversity of IoT edge device platforms and gateways make generic security design challenging. Customized security architecture is deployed by edge device manufacturers (platform developers) which makes the designing security for whole IoT system more complex and challenging.
- Device life: - In few IoT applications, device life is very short and in few applications like automotive device life is very high. Security techniques needs to be designed based on application and device shelf life. For the devices with long life, there must be option to update security features through software/firmware updates to defend against new threats and attacks.
- Open-source systems: - Open source software is widely used in IoT systems. The usage of open source programs is advantageous for quick development of IoT devices. Developers should carefully choose the software and must have a framework to validate the open source software, that it does not contain any malicious functionality. At present, most of the IoT edge device manufacturers use open source software and most of the IoT edge devices use same open source software may have same type of vulnerabilities, which is an advantage, such that, same solution fits wide range of applications. At the same time, it is equally dangerous; the whole system will collapse if an adversary use the same vulnerability in all edge nodes and attack all devices in the network [2].
- Firmware/Software Update or Upgradation: - IoT systems can be upgraded remotely through the network. Most of the manufacturers send updates over the air (OTA). Supported by cloud, update distribution can be done using centralized infrastructure without much user involvement. In this case, security challenge is of two types: - proper authentication and identification of the edge device in the network of billions and to make sure no malicious drivers or software is not installed on edge devices in the name of updates and upgrades [2].
- Cloud Data centre: -All commercial activity in IoT is supported with cloud based data storage infrastructure. Cloud is an integral part of IoT ecosystem and enables most of the services. Cloud security and IoT edge device security are mutually dependent factors. In designing security, we have to look into complete ecosystem more holistically across different hierarchies from cloud to final edge devices to ensure there is no compromise in security at any layer.
- Authentication and pairing: - Large number of devices in the IoT network complicates the pairing of devices. Pairing is required to aggregate and segregate the data generated from edge nodes. It is important make pairing process secure. The adversary can tamper with data and node identification credentials. Compromise in the security at one node may open up communication channel among device to device and device to routers. And power consumption and security are closely interconnected in IoT [7]. Security architects must consider power consumption patterns and available power supply capacity at edge node in designing protection to edge nodes.

III. VULNERABILITIES OF IOT EDGE DEVICES (ATTACKS ON PERCEPTION LAYER)

The IoT edge devices collect data from sensors and control the physical components of the system. The possible attacks on IoT edge devices are discussed below [7] [2]: -

- Node Capture Attacks: - In this type of attack, an adversary will capture the node or replace the entire node, or tamper the hardware device. The confidential information like cryptographic keys, access keys and other assets related to digital rights management are exposed. The fake node can act like a malicious node in the network, which further compromise the security of entire IoT network.
- Malicious Code Injection Attack: - The adversary will inject the malicious code into the memory of the node, using debug modules of the node. Malicious code will not only perform unintended functions, but can also give the adversary access to complete IoT network. This happens mainly during firmware/software upgradation through OTA.
- False Data Injection Attack: - An adversary can pump erroneous data through a tampered node, which lead to malicious events in delivery of services to end-user/customer. It may also cause denial of service (DoS) attack.
- Side-channel attacks (SCA): - Different types of SCA based on power consumption, timing, test infrastructures, fault attacks, electromagnetic and laser based attacks leak the secret keys used in encryption of sensitive data. Different types of countermeasures against various SCA are implemented with cryptographic modules in modern chips. Edge device should have defense against all types of possible SCA.
- Eavesdropping and Interference: - Adversary can eavesdrop at any point in the communication channel both wireline and wireless to leak the data. Light weight encryption algorithms must be used in IoT edge device to keep the data protected from eavesdropping. Adversary can also interfere and

create denial of service attack by pumping noise and distort the data during transmission.

- Sleep Deprivation Attacks: - This attack is similar to Denial of Service (DoS) attack, by draining out the battery connected to edge device. Generally, edge devices are designed for low power consumption. Either by tampering hardware or pumping malicious into memory (code executes in infinite loop) can increase the power consumption of the edge device which will lead to DoS type of attack by draining out the battery.
- Booting vulnerabilities: - During boot process, most of the protection mechanisms are not enabled and an adversary will try to access the sensitive sections of both hardware and software to get secret keys used for encryption, digital rights management etc. Securing the boot process is crucial in IoT edge devices.
- Hardware Exploitation: - Adversary can access the debug ports, JTAG, on-chip instruments used for debug and diagnosis and get an access to confidential assets of the edge device, which are commercial sensitive. This attack is performed by dumping malicious firmware and accessing the right ports and analysing the data using sophisticated test equipments. IoT designer should choose the microcontroller/SoC with suitable debug security features or designer has to create protection by developing security firmware.
- Software exploitation: - Software vulnerabilities in IoT are very similar to conventional general computing systems and traditional embedded systems. Software architecture of IoT is an extension to the framework followed in embedded systems and all types of vulnerabilities applicable to embedded software are equally applicable to IoT also.

We discussed the possible attacks on IoT edge devices in this section which is a part of the large IoT ecosystem. Most of the vulnerabilities discussed here are applicable to gateways also. Multi-level attacks using the vulnerabilities in edge devices, gateways, routers and data centres are possible. The holistic security architecture from edge device to cloud based data centre working with tight cooperation can mitigate the attacks on IoT ecosystem.

IV. EXISTING SOLUTIONS

Most of the IoT literature treats the IoT from network perspective and propose specific security solution for a given application or for a given device, which is not a generic solution and incompatible in many aspects when it comes to practical implementation. In this paper, our focus is mainly on IoT edge device. The microcontroller or SoC is a heart of IoT edge device. The root of trust coming from hardware is always better and reliable than the root of trust coming from software implementation. The dedicated security module in the microcontroller/SoC used in the IoT edge device will help in designing a better protection mechanism. We primarily focus on the solutions the modern day SoC devices offer in designing the security of IoT edge nodes. The security

architectures defined for traditional embedded systems are currently used in IoT devices also. These architectures will solve few security issues. Further improvements to these architectures are required to address new variety of edge device vulnerabilities in IoT ecosystem. The literature towards standardizing the security architecture has led to developing a Trusted Execution Environment (TEE) [8]. TEE is a tamper resistant computing environment running a separation kernel. TEE guarantees the authenticity of program code, integrity of crucial assets of system (processor registers, secured memory) and confidentiality of code and data stored in persistent memory [8]. Apart from this, TEE is also useful in proving the authentication and identification of the system. The content of TEE (both code and data) should get securely updated time to time to resist all types of old and new attacks from adversary. For outside world, TEE is a module which guarantees the isolation between secure and non-secure environments for both code and data. Trusted program module (TPM) is a sub-set of TEE, which is a secure cryptographic processor designed to generate the cryptographic keys, authentication and remote attestation. Several TEE modules are designed in both industry and academic research. The TEE modules designed for microcontroller/SoC are: - ARM TrustZone [9], Intel Software Guard Extension (SGX) [10] and Samsung KNOX [11]. All these TEE modules are programmable. The security architecture can be reconfigured according to requirement of end-user.

ARM TrustZone is a virtualization scheme with hardware support for memory, I/O and interrupts. Due to virtualization, ARM core can provide two virtual cores; one for secure and another for non-secure operations [9]. Based on ARM TrustZone, companies define their own proprietary TEE schemes. The details of TEE of few companies are open to public, like Nokia (Microsoft) integrate TEE called ObC [12] in Lumia devices. Similarly, Samsung's TZ-RKP [13] TEE is deployed in its Galaxy series mobile phone. Documentation of TEE architectures of ObC and TZ-RKP is openly discussed. There are few other TEE techniques like SecuriTEE from Solacia [14], QSEE of Qualcomm for which documentation is not available for public. There are few more open source TEE's developed jointly by commercial organizations and open-source community like OP-TEE [15] from STMicroelectronics and TLK from NVidia.

The TEE techniques discussed above have more similarities than differences. All TEE's are software mechanisms supported by hardware (like key generation and operating modes etc). TEE's serve the security issues in traditional embedded systems, mobile phones and high-end electronic gazettes with a reasonable success. The capabilities of TEE are not enough to address all the IoT security related vulnerabilities discussed in the previous section. Significant improvements in TEE or security architectures are required to address the IoT security issues holistically. TEE does not address the modern hardware security issues like Hardware Trojans, using unreliable counterfeit parts and debug security. Current TEE provide protection to sensitive code, sensitive data and crucial assets like processor registers and software API's. Along with the security features TEE support, we need an enhanced version of TEE or Infrastructure IP (Intellectual

978-1-5386-1357-3/17 $31.00 © 2017 IEEE

Property) core for security which also address modern hardware security threats. Recent research literature in this direction can be found. Li et al in [16], propose a language and framework for implementing security policies in software. A. Basak et al in [17] propose a microcontroller based security infrastructure core for implementing security policies, which is known as E-IIPS (Extended Infrastructure IP for Security), which address modern security threats like hardware Trojans and counterfeiting. Further, there is a need to improve both TEE and IIPS to address all security issues. In a new programmable solution a firmware update should be enough to update the TEE or IIPS to protect the systems against new variety of threats and attacks. The E-IIPS or TEE does not support debug and on-chip instrumentation security, digital rights management and defense against all types of SCA.

V. FUTURE SOLUTIONS

The future microcontroller/SoC should support security features required in IoT domain and these features are also equally useful in other standalone embedded applications. Every semiconductor application vertical will change with the advent of IoT. It is always better, if root of trust comes from hardware. In the IoT era, what extra security features the microcontroller/SoC should incorporate than the current TEE offer to system developers. The Table I show the vulnerabilities of the IoT edge devices, existing security features in current day microcontrollers and possible security solutions future microcontroller/SoC offer.

TABLE I
VULNERABILITIES, EXISTING SOLUTIONS AND FUTURE SOLUTIONS FOR IOT PERCEPTION LAYER SECURITY

Name of the Vulnerability	Existing Solution in current microcontroller/SoC.	Possible Security Solutions microcontroller/SoC should support in IoT era
Node Capture Attacks: - Edge Node replaced with a fake node. [2] [7]	Cryptography based Attestation in TEE	Physical Unclonable Function (PUF) based authentication and identification scheme.
Malicious Code/Data Injection Attack: - Malicious code/data into the memory of the edge node. [2] [7]	Solution can be of two ways: - (1) Protection of porting malicious code/data into memory. (2) Defense against malicious activity of code through some security program. Existing TEE partially supports both protection and defense, which is not sufficient for IoT security. Updates are ported to IoT through OTA, so further enhancements to existing techniques are required. Defense against Hardware Trojan (HT) is not available in current schemes (TEE and other schemes).	Proper Identification/Authentication of edge nodes in the billions of nodes in the IoT network and defense against all types of malicious activity either from the firmware or inherent hardware Trojans in the microcontroller/SoC. Physical unclonable functions (PUF) or any other hardware security primitive may become mandatory in future microcontrollers to mitigate counterfeiting, identification of edge nodes in IoT, authenticating the device during updating the firmware in IoT edge node.
Side-channel attacks (SCA): - Leaking secret keys used in cryptography using different SCA based on power consumption, timing, test infrastructures, fault attacks, electromagnetic and laser based attacks. [2] [7]	Few microcontrollers used in sensitive applications having cryptographic modules consist of a countermeasure against power analysis SCA. Defense against other SCA schemes and for multi-level attacks is not available in current microcontrollers/SoC.	Cryptographic modules in microcontroller/SoC play a crucial role in IoT security. Defence schemes to safeguard the crypto modules against all kinds of SCA and multilevel attacks is mandatory in future microcontroller/SoC.
Eavesdropping and Interference: - Leaking secret data and denial of service type attack. [2] [7]	TEE support data encryption which prevents eavesdropping.	Along with TEE features, microcontrollers should have mechanism to detect DoS type attacks and security program should take suitable remedial measure.
Sleep Deprivation Attacks: - This attack is similar to Denial of Service (DoS) attack, by draining out the battery connected to edge device. [2] [7]	Current microcontrollers/SoC does not have any mechanism against this type of attack.	Same as above.
Booting vulnerabilities: - During boot process, most of the protection mechanisms are not enabled. [2] [7]	TEE or few microcontrollers support secure boot mechanisms. Access to confidential and sensitive assets is restricted during boot process.	In the current TEE, boot is secured through password and cryptography. The hardware based boot security can incorporated in future microcontrollers.
Hardware Exploitation: - Adversary can access the debug ports, JTAG, on-chip instruments get an access to confidential assets of the edge device [2] [7]	Currently, TEE or any concrete mechanism towards securing the debug instruments is not found in microcontroller/SoC devices.	Debug security will be most important feature in future microcontrollers targeted to use in IoT. A holistic Debug security will be mandatory feature in future microcontrollers.
Software exploitation: - Similar to virus, software Trojans etc. [2] [7]	Light weight anti-virus and other software defense mechanisms for embedded systems are already present.	Role of microcontroller at application level will be very less. Operating system and software level anti-virus suitable to IoT may solve the problem. Still current TEE and in IIPS protects crucial assets, even the threat comes from software applications.

Based on the discussion in Table I, we can list the few features, microcontrollers should incorporate for better security: -

- Physical Unclonable Functions (PUF): - PUF is a hardware security primitive, which is useful for microcontroller manufacturer, system developer and end user.

- Microcontroller manufacturers can mitigate IC counterfeiting and implement digital rights management (DRM) using PUF. System developers can use a PUF for accurate identification of edge nodes in IoT and authenticating the edge node, when edge node requests for upgradation of firmware or a new feature. End user can make use of PUF for cryptographic key generation [20].

- PUF circuits generate the unique set of large challenge response pairs (CRP), which are helpful in addressing the security issues. PUF derive their uniqueness in their CRP's from the process variation occurring during the fabrication of IC's. The details on PUF can be found in [18]. So future microcontrollers should have PUF circuits which are highly useful.

- Countermeasures against Side-Channel Attacks (SCA): - Defense against variety of SCA is a well-established research and microcontroller/SoC should include best possible countermeasures against all kinds of SCA, without affecting power consumption and performance of the microcontroller.

- Debug Security: - Large number of test structures and on-chip instruments will go into the chip, which are required at different times of microcontroller life-cycle, starting from production tests to in-field maintenance. Access to these instruments has become streamlined and easy after adopting the IEEE P 1687 on-chip instrument access standards [21]. So the microcontroller designers should include well defined, easy and secure access mechanism to on-chip instruments to safeguard the critical assets of system developer and end-user.

- Defense against Hardware/firmware Trojans: - A Security Controller to protect the critical assets of system against hardware Trojans and firmware Trojans.

- Finally, microcontroller should have light weight, efficient cryptographic module to encrypt the sensitive data.

The scalable infrastructure IP for security which addresses all the components discussed above will serve as centralized resource, which can provide more holistic security.

VI. CONCLUSION

In this paper, we have discussed the security challenges of overall IoT systems; with main focus on vulnerabilities of IoT edge devices/node (also known as perception layer). The solutions to the security problems in perception layer lies in designing efficient security architecture in microcontroller/SoC, which is inevitable component of the IoT edge node. We discuss the existing Trusted Execution Environment (TEE) in current day microcontrollers and their capabilities to solve security issues. And based on the review of vulnerabilities and existing solutions, we present the list of security features the next generation microcontroller/SoC should have to address the IoT era security issues in a more holistic way. The comprehensive solution may be an infrastructure IP core for security which implements all the features listed for future microcontrollers.

REFERENCES

[1] D.Evans, "The Internet of Things- How the next evolution of the internet is changing everything" Cisco Internet Business Solutions Group-white paper, 2011.

[2] Sandip Ray, et al "The Changing Computing Paradigm With Internet of Things: A Tutorial Introduction". IEEE Design & Test 33(2): 76-96, 2016.

[3] J.Folkens "Building Gateways for internet of things" Texas Instruments White Paper, www.ti.com/lit/wp/spmy013/spmy013.pdf.

[4] Z. Chen, G. Xu, V. Mahalingam, L. Ge, J. Nguyen, W. Yu, and C. Lu., "A cloud computing based network monitoring and threat detection system for critical infrastructures", *Big Data Research*, 3:10–23, 2016.

[5] J. Wu and W. Zhao. Design and realization of winternet: From net of things to internet of things. *ACM Transactions on Cyber-Physical Systems*, 1(1), November 2016.

[6] S. H. Ahmed, G. Kim, and D. Kim. Cyber physical system: Architecture, applications and research challenges. In *Proc. of 2013 IFIP Wireless Days (WD)*, November 2013.

[7] J. Lin, W. Yu, N. Zhang, X. Yang, H. Zhang, W. Zhao, "A survey on internet of things: Architecture enabling technologies security and privacy and applications", *IEEE Internet-of-Things Journal 2017*.

[8] Mohamed Sabt, et al," Trusted Execution Environment: What It is, and What It is Not", **IEEE** Trustcom/BigDataSE/ISPA -2015.

[9] ARM Limited, "Building a secure system using Trustzone technology," 2009.

[10] F. McKeen et al, "Innovative instructions and software model for isolated execution". In Hardware and Architectural Support for Security and Privacy (HASP). ACM,-2013.

[11] Samsung, "Samsung KNOX," www.samsungknox.com.

[12] K. Kostiainen, et al, "On-board credentials with open provisioning," in Proceedings of the 4th International Symposium on Information, Computer, and Communications Security, ser. ASIACCS '09. New York, NY, USA: ACM, 2009, pp. 104–115.

[13] A. M. Azab, et al, "Hypervision across worlds: real-time kernel protection from the arm trustzone secure world," in Proceedings of the 2014 ACM SIGSAC Conference on Computer and Communications Security, ser. CCS '14. New York, NY, USA: ACM, 2014, pp. 90–102.

[14] Solacia, "SecuriTEE." [Online]. Available: http://www.sola-cia.com/en/securiTee/product.asp.

[15] P. Brand, "Op-tee." [Online]. Available: https://github.com/OP-TEE.

[16] X. Li, et al, "Sapper: A Language for Hardware-Level Security Policy Enforcement," in International Conference on Architectural Support for Programming Languages and Operating Systems, 2014.

[17] Abhishek Basak, et al, "Security Assurance for System-on-Chip Designs With Untrusted IPs", IEEE Trans. Information Forensics and Security, 1515-1528, June-2017.

[18] C.Herder et al, "Physical Unclonable Functions and Applications: A tutorial", Proceedings of the IEEE, Volume: 102, Issue: 8, Aug. 2014.

[19] http://www.ieee802.org/15/pub/TG4.html.

[20] H. Kang et al, Cryptographic key generation from PUF data using efficient fuzzy extractors, 16th International Conference on Advanced Communication Technology (ICACT), 2014

[21] https://standards.ieee.org/findstds/standard/1687-2014.html.

An Efficient MapReduce-based Adaptive K-Means Clustering for Large Dataset

Tapan Chowdhury
Dept. of CSE
Techno India, Salt Lake
Kolkata, India
tapan2005cse@gmail.com

Arijit Mukherjee
Dept. of CSE
IIIT, Hyderabad
Hyderabad, India
mkarijit@gmail.com

Susanta Chakraborty
Dept. of CST
IIEST, Shibpur
Howrah, India
susanta.chak@gmail.com

Abstract— **MapReduce-based clustering of a large dataset is gaining rapid importance in the fields of data science. Its main task is to make groups of data from the large dataset such that all data points categorized into a single group are similar to one another. K-means clustering method is one of the most extensively used clustering methods but it minimizes clustering criteria by iteratively relocating data points between clusters until a locally optimal partition is attained, so convergence is local and globally optimal solutions cannot be guaranteed in case of a large dataset. Due to the random selection of K-initial seeds, it decreases the quality of clusters. This paper proposes a MapReduce-based adaptive K-means clustering approach. The adaptive nature of our approach, improves the efficiency based on two key concepts. First, it selects the initial seeds that are spread throughout the large dataset via statistical testing. Second, it reduces the impact of the outlier on a large dataset and clustering large dataset using MapReduce framework. Our approach is adaptive for large dataset by adapting the size and nature of the dataset to make the process more robust and efficient. The experimental result shows that our approach improves the performance of clustering compared to earlier works.**

Keywords—Clustering; K-means; MapReduce.

I. INTRODUCTION

A large amount of data is being produced everyday globally. This will expand exponentially over the next few years which require existing analytical methods suitable to handle such huge volume of data. And moreover, the volume of the data is not the only concern; variety is also a big concern. The data being generated is from different sources which are mostly unregulated. The challenge is that extract knowledge from a large amount of data using data mining techniques. In this regard, clustering plays an important role to categorize data into a group of similar data.

K-means [1] is one of the popular methods in data mining because of its simplicity. This method randomly chooses initial centers and repeatedly assigns each input points to its nearest cluster centers by calculating the Euclidian distance between two and then recalculates cluster centers. But in case of a large volume of data K-means is not a suitable clustering method in terms of initial seed selection. In K-means global optimal solutions cannot be achieved due to random initialization.

There are various clustering methods built on K-means. K-means++ [2], which improves the initialization to reach the optimal solutions but it is inherently sequential nature. K-means++ requires K-passes over the data to select the initial set of centers which is not good for clustering large datasets. A parallel version of K-means++ is Scalable K-Means++ [3], which reduce the number of passes for better initialization.

Randomly selecting 'K' initial seeds is not a good approach as it may compromise with the accuracy. And in cases where the dataset is very large and the number of Clusters needs is few, then selecting the 'K' initial centroid in random manner would not be a wise choice.

In this paper, we propose an efficient adaptive method to improve the methodology for choosing 'K' initial centroids. We select a sample size large enough to give an appropriate idea of the entire dataset. Then we select the 'K' best fit points as the initial seeds. The objective is that the initial seeds would be well spread throughout the dataset and it reduces the number of iterations to select the initial seeds.

Secondly, we propose an approach to reduce the impact of outliers or noisy data points on the efficiency of the method. Centroids are always very sensitive to outliers. The presence of such points in the dataset causes the centroid to shift a great extent towards the outliers in the iterative method. In this paper, certain outlier data points are ignored while updating the centroid. All the computations are done in parallel using MapReduce framework to improve the efficiency of clustering. It demonstrates the effectiveness of parallel computations on large dataset.

The rest of the paper is organized as follows: Section II describes the related works. Section III presents some preliminaries. The proposed method is explained in Section IV. Section V shows experimental results and analysis with three datasets which illustrate the effectiveness of our proposed approach. Finally, the conclusion is drawn in section VI.

II. RELATED WORKS

Several works on clustering technique have been proposed in different kinds of literature. In different surveys several clustering methods have been described [4,5,6], however, we will enter into the deep insight of particular approaches that are

relevant to our work: K-means [1], K-means ++ [2], Parallel K-means [7].

K-means [1] is one of the fundamental methods in data mining and machine learning. In practical it is most popular approach because of its simplicity and effectiveness in different applications. K-means approach belongs to the class of combinatorial optimization also known as iterative relocation technique. K-means approach falls under the class of NP-Hard problem even for $K=2$ [8]. This method minimizes a given clustering criterion by iteratively relocating data points between clusters until a locally optimal partition is attained. In this basic iterative approach, convergence is local and the globally optimal solution cannot be guaranteed. In case of large datasets convergence criterion takes more time for clustering and depending on dataset cluster quality can be low. In K-means++ [2] authors introduced different concept of initial seed selection. Their variation proposes to choose the first data point at random and then the following points are chosen with the weighted probability proportional to distance square from the closest point already chosen as a centroid. The problem with K-means++ is that for 'K' centroids there will be K-initial iterations over the entire dataset which is time consuming. Also, this approach is not suitable for parallel processing of large dataset.

Parallel K-means [7] is the method for the process of a large-scale dataset in parallel. This approach satisfies major three criteria of speedup, scale up and size up. But since it was basically a MapReduce-based version of the K-means method, it didn't deal with the problems exist for large-scale datasets, such as the presence of noisy data points or outliers in the large dataset, which would affect the quality of clustering. Also, there were no mechanisms to ensure selection of good initial centroids.

In our proposed solution we proposed an approach to overcome the problem of selecting suitable initial centroids by statistical testing and dealing with outliers by Chauvenet's criterion [9], in addition, to make the method suitable for Hadoop framework [10,11].

III. PRELIMINARIES

In this section, we introduce the elementary background of our proposed work: sample size selection, Chauvenet's criterion, and MapReduce framework.

A. Sample size selection

Statistically, selection of sample size depends on the level of precision, confidence level and degree of variability.

Definition 1. The level of precision e is the range in which the true value of the sample size is estimated. The range is expressed in terms of percentage of points, ±5% [16].

Definition 2. Confidence level is the percentage of all possible samples that can be expected to contain the real dataset parameter [16].

Definition 3. Degree of variability refers to the distribution of attribute value in the dataset. In case of more heterogeneous dataset, to obtain a given level of precision a larger sample size is required [16].

Definition 4. Z-score is the number of standard deviations from the mean of data point. It can be defined as $z=(x-\mu)/\sigma$, where z is the z-score, x is the value of point, μ is the mean and σ is the standard deviation [16].

A sample size n_0 can be defined as [16]

$$n_0 = \frac{z^2 \, P(1-P)}{e^2} \qquad (1)$$

where, z is the z-score, derived from the confidence level. P is the variability and e is the level of precision.

TABLE I.   Z-SCORE

Confidence level	z-score
90%	1.645
95%	1.96
98%	2.326
99%	2.576

z-score values for corresponding confidence level are shown in Table I.

A given sample size provides proportionately more information for a small dataset than for a large dataset. So the new sample size n' can be defined by Cochran's correction formula [16] as

$$n' = \frac{n_0}{1 + \frac{(n_0-1)}{N}} \qquad (2)$$

where, N is the size of the dataset.

Definition 5. Compactness is the average distance between every pair of data points in a cluster [5]. Compactness CP can be defined as

$$\overline{CP}_k = \frac{1}{|n_k|} \sum_{x_i \in n_k} |x_i - W_k| \qquad (3)$$

where, n_k refers to the points present in the kth cluster, W_k is the kth centroid and x_i is the point belongs to n_k.

Compactness in global measures that is the average of all clusters can be defined as

$$\overline{CP} = \frac{1}{K} \sum_{i=1}^{K} \overline{CP}_k \qquad (3a)$$

where, $\overline{CP}_k$ is the compactness of cluster k [5].

Definition 6. Fitness is equivalent to the sum of the distance of all points from their corresponding cluster centroid and then divided by the total number of points exist in the dataset. It can be defined as

$$\overline{FT} = \frac{1}{|N|} \sum_{i=1}^{|N|} |x_i - w_{ik}| \qquad (4)$$

where, N is the number of points in the dataset, x_i is the ith point in the dataset and w_{ik} is the centroid for the cluster which contains the point x_i.

B. Chauvenet's criterion

Chauvenet's criterion [9] is used to detect the outlier points in a dataset. Outlier points are detected for n number of observations. In this method mean μ and standard deviation σ are calculated. If $n*erfc (| point\ x_i - \mu | / \sigma) < \frac{1}{2}$ then it rejects the point x_i. A data point within a stipulated cluster would be identified as an outlier based on its Euclidean distance from the centroid.

C. MapReduce framework

To show the effectiveness of parallelism, MapReduce [12, 13, 14] paradigm is used in our proposed method. Google proposed MapReduce framework. It is a parallel programming model to process the large volume of structured, semi-structured and unstructured data. This programming model has three functions: mapper, combiner and reducer. Mapper function generates the intermediary *<key, value>* pair by dividing the dataset into a different number of blocks. Hadoop Distributed file system is used to partition the large dataset into different blocks. All the intermediary *<key, value>* pairs are combined and sorted by Combiner function according to the key. Reduction of the intermediary *<key, value>* pair is done by Reducer function. This function generates unique <key, value> pair as an output.

IV. MAPREDUCE-BASED ADAPTIVE K-MEANS CLUSTERING

In this paper, we propose MapReduce-based adaptive K-means clustering approach for clustering of a large dataset. Iteration 'K' times over a huge dataset of might fetch accurate result but will be very time-consuming.

Our solution aims to be more efficient and adaptive while working with large volume of data. Working with a scenario where we need a fixed number of clusters 'K' and where $K<N$, N is the total number of points in the dataset. Selection of 'K' initial seeds at random leaves a high probability that the initial centroids are not well spread throughout the dataset. This takes a higher number of iterations to reach the desired result and sometimes highly impacts the accuracy of cluster formation. Therefore, we take a larger number of points, say $K*M$ which would improve the probability of points being better spread throughout the dataset. Then we would select the best fit points as our initial centroids. Criteria for best fit points would be that the centroids should not lie much closer to one another. Therefore, we would start merging close points until we are left with 'K' seed points.

Another modification we propose is to make the procedure more adaptive to deal with outliers or noisy data points. The presence of outlier or noisy data points within a cluster can cause the centroid to shift towards the outlier. Our aim is to reduce the impact of outliers on the movement of cluster centers. This would be done by detecting outlier points, present within individual clusters and ignoring them while finding the next centroid of that cluster.

All the computations are done in MapReduce framework for the tasks and data parallelism of our proposed work. Our proposed method is divided into different tasks: M-factor initialization, selection of K best-fit points from $K*M$ points,

clustering with respect to K best-fit points and reducing the impact of outliers.

A. M-factor initialization

In our proposed approach instead of selecting K initial points, we select $K*M$ initial points uniformly at random. Selection of a large number of points will improve the probability of having well spread out the distribution. This will give a better precision of most suitable points for initial seeds. We select $K*M$ points from the entire dataset as our sample. With the increase of sample size, accurate decisions are taken from the sample.

Since the nature of the dataset is unknown, in our proposed method we consider variability P value is 0.5, which is the maximum value of proportion. Equation (1) can be redefined as

$$n_0 = \frac{z^2 (0.5)^2}{e^2} = \frac{z^2 (0.25)}{e^2} = \frac{1}{4}\left(\frac{z^2}{e^2}\right) \qquad (5)$$

By the equation (2) and (3), we can define n' as

$$n' = \frac{z^2 N}{z^2 + 4e^2 (N-1)} \qquad (6)$$

We consider n' to be a good sample size for the selection of initial seeds. In our proposed approach we increase 'K' number of initial centroids by 'M' times. Therefore, we select '$K*M$' points which are equal to n', and equation (6) can be redefined as

$$K*M = \frac{z^2 N}{z^2 + 4e^2 (N-1)} \qquad (7)$$

$$M = \frac{1}{K}\left(\frac{z^2 N}{z^2 + 4e^2 (N-1)}\right) \qquad (8)$$

*Lemma: Inference obtained from the large number of sample points 'K*M' becomes close to the expected result 'K' and try to become closer as many as sample points 'M' are chosen.*

Proof: Let $P_1, P_2, P_3,...,P_n$ are randomly taken value from sample space, where 'n' is the number of sample points. Consider μ is the expected result. According to the *law of large numbers*, $\bar{P}_n = \frac{1}{n}(P_1 + P_2 + \cdots + P_n)$ where, $\bar{P}_n$ is the mean or sample average. $\bar{P}_n$ converges to the expected result that is $\bar{P}_n \to \mu$ for $n\to\infty$. Therefore, for 'K' no. of randomly selected centroids, selection of $K*M$ no. of points will give a well-spread distribution of points. For example consider a large dataset with the value of parameters shown in Table II. By the equation (8), we can compute M is 77 for the cluster number K is 5.

TABLE II.   DATASET PARAMETER

Parameter	value
Size	500000
Confidence level P	95%
z-score	1.96
level of precision e	0.05
Required number K	5

B. Selection of K best-fit points from K*M points

We select only K number points to find the K-best fit points from '$K*M$' point. Therefore, we will start with merging the points which are close to one another and assign weight to the merging point. We repeat merging process based on weight, with the closest points until K number points are left.

Consider the two points, $P_i(x_{i1}, x_{i2}, ..., x_{ik})$ and $P_i(x_{j1}, x_{j2}, .., x_{jk})$. Initial weights of these two points are m and n respectively. After merging these two points by considering their weight, the merging point is P_{ij}. Therefore,

$$P_{ij} = \frac{(m(x_{i1}) + n(x_{j1})}{m+n}, \frac{(m(x_{i2}) + n(x_{j2})}{m+n}, , \frac{(m(x_{ik}) + n(x_{jk})}{m+n}$$

(9)

C. Reducing the impact of outlier

The presence of outlier or noisy data points within a cluster can cause the centroid to shift towards the outlier. Our solution would be done by ignoring outlier points present within individual clusters while finding the next centroid of that cluster. In our proposed approach, first we assign each point to its closest cluster center and then the outliers using Chauvenet's criterion. The points would be classified as an outlier based on its Euclidian distance from the present centroid. The points which are identified as an outlier within a cluster would be marked and the value of these points won't be considered while calculating the new cluster center. Therefore, the movement of the centroid would not be affected by points which are very far away from most other points.

D. Proposed Algorithm

We present a parallel implementation of our proposed clustering method using MapReduce framework.Algorithm 1 shows the initial seed selection algorithm. We would take a different approach for initialization. We would select $K*M$ initial seeds instead of selecting K initial seeds. Then we would keep merging the closer points till we reach K points. The intuition behind the algorithm is simply that a higher number of points gives us a better probability of well-dispersed initialization.

Algorithm 1 *K-initial seed selection*

Input: k, m /* k is the number of cluster and m is the multiplication factor */

1: Number of Initial Seeds ← k
2: Multiplication Factor ← m
3: Select k*m points from the Dataset uniformly at random
4: len←k*m
5: d←dimensions
6: store[len][d] /*Store all the selected points
7: update[len] /*Initialize to Zero*/
8: Count ←len
9: while count>k
a: Find the closest Points. Store[x] and store[y]are the closest points.
b: for k← 0 to d
store[x][k]← (store[x][k]*update[x])

+(store[y][k]*update[y])/(update[x]+update[y])
c: end for
d: update[x]++
e: remove store[y]
f: count--
10: end while
11: Remaining K points are the final initialized points.

Algorithm for *Mapper* function is shown in Algorithm 2. The main objective of this algorithm is to compute the distance of data points from the current centroids in a parallel manner. The entire dataset is stored in HDFS in *<key, value>* pair format. Every *Map* task create an array comprising of the presently selected centroids that are considered as global variables. Assign each data points to its closest centroid by calculating the distance between the present centroids and each data points. Output *<key', value'>* pair is the index of the assigned centroid (*key'*) and the data point (*value'*).

Algorithm 2 *Mapper Function*

Input: Centroids[k] /* An array consisting of 'k', selecting centroids from the datasets. <key, value>represents records in a dataset, where key is the offset of the record ID and value is the string comprising of all the attributes of the records.*/

1: Min_Dist. ← ∞
2: index ← -1
3: For i← 0 to n-1
4: Do for j← 0 to k
5:   Dis ←Euclid_Dist.[Record[i], Centroid[j]]
6: If dis<Min_Dist.{Min_Dist. = dis; index=j}
7:   End
8: Output <key', value'>pair.

Algorithm 3 describes the algorithm for *Combiner* function. The *Combiner* function is being used to work with intermediate data from each *Map* Task. It will give a partial sum of all the points which have been assigned to the same cluster and the total number of points allotted to every cluster center by the same *Map* task. Also, we create an array which would store the distance of every point from its assigned centroid.

Algorithm 3 *Combiner Function*

Input: <Key k, Value v> and Centroids []/* Value v stores all attributes of a point assigned to a particular Key k*/

1: array_points[n]; /*n number of attributes of the records*/
2: count← 0
3: while v.next() <> Null
4: do
5: fori← 0 to x-1
6: Add each attribute of v.next() value to array_points[i]
7: end for
8: count++
9: end while
10: Point_dist← Euclid(array_points,Centroids[key])
11: key'← key
12: value'← string of array_points andarray_dist
13: output <key', value'>

978-1-5386-1357-3/17 $31.00 © 2017 IEEE

Algorithm for *Reducer* function is shown in Algorithm 4. The *reducer* function accepts the list of data points and their distance from the centroid. The Chauvenet's function is called by the reducer task. Points which fail to meet the criteria will be considered as outliers. These points will not be considered at the time of new centroid calculation. Once the new centroids have been updated, they will be sent to the all the *Mapper* functions.

Algorithm 4 *Reducer Function*

Input: *<key, List(W)>* /*key* is the Index of the cluster, *List(W)* is the list of nodes and their distance from centroids. */

1: Count← 0, j← 0 /* n is the number of dimensions and j is the number of points satisfy Chauvenet's criterion */
2: while list.next() <> Null
3: do
4: len ← w.length() /* number of points assigned to that centroid */
5: for i← 0 to n-1
6: Add value of attributes of w.next() to array[Count][i]
7: dist[i] ←w.next()   /* distance from centroid */
8: end while
9: K[]←chauvnets(dist[],len)
10: if k[i] <> 'large value'
11: array_sum[i] += array[i]
12: j++
13: enf if
14: Divide array_sum[1…..n] by j which will give coordinated from new centroid
15: key'← key
16: value'← string of new centroids' co-ordinates
17: Output <key', value'>.

Algorithm 5 shows the algorithm for Chauvenet's criterion function. This function is invoked by the *Reducer* function. It receives the list of data points and their distance from the current centroid. Outliers would be determined based on the distance from the existing centroid. Only the points whose distances are greater than average distance would be checked because our aim is to eliminate the points which are very far from the centroid, not the ones extremely close. The points which are tested positive for outliers would be marked.

Algorithm 5 *Chauvenet's criterion Function*

Input:array_dist [], count

1: flag ← 0
2: do
3: m ← mean(array_dist[],count)
4: sd← stddev(array-dist[], count, m)
5: for i← 0 to count-1
6:    Double p← array_dist[i]
7:    x←Math.abs((p-m)/sd)
8:    Double cal← n*(Erf.erfc(x))
9:    if n.x< 0.5 && p>m
10:     reject array_dist[i]
11:     Index ←i
12:     flag←1
13:   end if
14:   else

15:   Keep array_dist[i]
16: end for
17: remove rejected data and update count value
18: end do
19: while flag <> 0
20: Output <array_dist[], count >

V. EXPERIMENTAL RESULTS ANALYSIS

The performance of our proposed approach is evaluated with two machine learning datasets that are collected from UCI Machine Learning repository [15].

Experiments are done in Hadoop clusters using Java version 7 and with PCs of Intel(R) Core(TM) i5- 7600processor @4.10GHZ and 8GB RAM in Ubuntu 16.04.3 platform.

We mainly analyse the performance of the method based on two different criteria: Cluster quality and computation time. Earlier work parallel K-means [7] is scalable and showed a linear speed up. Therefore, in order to measure performance with respect to time, we consider the number of iterations taken to reach the converging condition.

Table III shows that for all the three datasets, our proposed method takes the lower number of iterations to reach the convergence condition than that of the earlier parallel K-means method. For the *Breast Cancer* dataset previous parallel method takes 10 iterations to reach convergence condition whereas our method reaches the convergence condition in 7 iterations. The Diabetes dataset and Heart Disease also follow the same trend, while the previous method takes 16 and 18 iterations respectively to converge, but the proposed method only takes 9 and 10 Iterations respectively to reach a convergence condition. Thus from this result, we can establish that under similar given conditions and same data, the proposed solution will reach convergence condition more quickly.

In order to evaluate the quality of clusters, two measures are taken, Compactness and Fitness. Compactness [5] in global measures is used for the validation of cluster. Clustering method is effective if it creates the clusters with instances that are nearby or analogous to one another. Fitness is equivalent to the sum of the distance of all points from their corresponding cluster centroid and then divided by the total number of points exist in the dataset.

A low compactness symbolizes that clusters are compact in shape, and points are close to the center. Fitness is a measure of the distance of the points from its cluster centroid. The presence of outliers will have a greater impact on compactness than fitness. Table III shows that the compactness and fitness for Diabetes dataset are 50.5 and 59.67 respectively using the earlier parallel K-means method but our proposed approach gives a compactness and fitness of 47.79 and 56.63 respectively. Here we observe that proposed solution gives more compact and well-formed clusters. The almost similar trend can be observed for the Heart disease and Breast cancer datasets. The quality of clustering is better in three cases. Accurate decisions can be taken from the clusters to diagnose particular disease. Therefore, our proposed approach will give better results.

VI. Conclusion

Over the years there have been significant researches in the field of data clustering. With the advance of World Wide Web, we are generating a large amount of data every day. Existing clustering methods face problems while working with such large-scale data. In this paper, we proposed the MapReduce-based adaptive parallel approach for clustering of the large dataset. The aim is to make the algorithm more adaptive to deal with two problems which might arise in clustering of large-scale data: initial seed selection and rejection of outliers. The proposed method is evaluated on several datasets and it shows our method performs better than earlier works.

References

[1] S. Lloyd, "Least squares quantization in PCM," *IEEE Trans. Inf. Theory*,vol. IF-28, no. 2, pp. 129–137, Mar. 1982.

[2] D. Arthur and S. Vassilvitskii, "k-means++: The advantages of careful seeding," in Proc. 18th Annu. ACM-SIAM Symp. Discrete Algorithms *Soc. Ind. Appl. Math.*, New Orleans, LA, USA, pp. 1027–1035, 2007.

[3] B. Bahmani, B. Moseley, A. Vattani, R. Kumar, and S. Vassilvitskii,"Scalable k-means++," PVLDB, 5(7):622–633, 2012.

[4] E. Chandra and V.P. Anuradha, "A survey on clustering algorithms for data in spatial database management systems," International Journal of Computer Applications, 24(9):19-26, 2011.

[5] Fahad A, Alshatri N, Tari Z, Alamri A, "A survey of clustering algorithms for Big Data: Taxonomy and empirical analysis," IEEE Transactions on Emerging Topics in Computing, 2(3):267–279, 2014.

[6] Xu R, Wunsch D," Survey of clustering algorithms," IEEE Transactions on Neural Networks. 2005 May; 16(3):645–78.

[7] W. Zhao, H. Ma, Q. He, "Parallel K-Means Clustering Based on MapReduce," Cloud Computing, Lecture Notes in Computer Science,Chapter IV, vol. 5931, pp. 674-679, 2009.

[8] M. Mahajan, P. Nimbhorkar, K. Varadarajan. "The Planar k-Means Problem is NP-Hard," Lecture Notes in Computer Science 5431: 274-285, 2009.

[9] Barnett, Vic and Lewis, Toby, Outliers in Statistical Data, 3rd edition. Chichester: J.Wiley and Sons, 1994.

[10] Alam A, Ahmed J, " Hadoop Architecture and Its Issues," Proceedings of International Conference on Computational Science and Computational Intelligence, pp. 288–291,2014.

[11] Dittrich J, Quiani-Ruiz J, "Efficient Big Data Processing in Hadoop MapReduce," Proceedings of the VLDB Endowment 5(12):2014–2015, 2012.

[12] Dean J, Ghemawat S, "MapReduce: simplified data processing on large clusters". Commun ACM 51(1): 107–113, 2008.

[13] Dawei Jiang,Beng Chin Ooi,Lei Shi,Sai Wu,"The Performance of MapReduce: An In-depth Study", Proceedings of the VLDB Endowment, Vol. 3, No. 1,2010.

[14] A. Srinivasan, T. A. Faruquie, and S. Joshi, "Data and task parallelism in ILP using MapReduce," Machine Learning, vol. 86, pp. 141-168, 2011.

[15] UCI Machine Learning repository: < http://archive.ics.uci.edu/ml/>

[16] G. D. Israel, "Determining sample size," Program Evaluation and Organizational Development, IFAS, University of Florida, 1992. [Online]. Available: http://edis.ifas.ufl.edu/ pd006.

TABLE III.   Experimental Results

Dataset	Parallel K-Means [7]			Proposed Method		
	Compactness	Fitness	Iterations	Compactness	Fitness	Iterations
Breast Cancer [15]	3.92	5.69	10	2.78	4.52	7
Diabetes [15]	50.5	59.67	16	47.79	56.63	9
Heart Disease [15]	33.31	33.50	18	31.24	31.43	10

978-1-5386-1357-3/17 $31.00 © 2017 IEEE

2017 IEEE International Symposium on Nanoelectronic and Information Systems

Enhanced Look-up Table Approach for Modeling of Floating Body SOI MOSFET

Sitansusekhar Roymohapatra*, Ganesh R Gore*, Akanksha Yadav*, Mahesh B. Patil*,
Krishnan S Rengrajan**, Maryam Shojaei Baghini*
* Department of Electrical Engineering, Indian Institute of Technology, Bombay, India, 400076
** Global Foundries (India), Bangalore, India, 560045

Abstract—**The Look-Up Table (LUT) approach is an effective tool for modeling semiconductor devices in the early stages of technology development for the purpose of circuit simulation. In this paper, it is shown that the conventional LUT approach leads to significant errors for SOI circuits. The cause for this discrepancy is explained. An improved LUT approach is proposed in which an auxiliary circuit is added to the conventional LUT model in order to accurately handle SOI devices. The new approach is validated for different circuits like CMOS inverter and SRAM cells. The proposed model shows substantial improvement in accuracy with 3% error as compared to 27% error in conventional LUT model.**

Keywords—*Look-up Table, Small-Signal Model, Large-Signal Model, Auxiliary Circuit, FDSOI, PDSOI, Enhanced LUT*

I. INTRODUCTION

In order to achieve low power and high performance, CMOS SOI has gained popularity as a technology option. Compact models such as BSIM are mostly used for CMOS circuit simulations. However, as the device length reduces, compact models have become increasingly complex, requiring a large number of parameters. It takes a significant amount of time to develop complex models and benchmark them for new technologies. The LUT approach [1-6] offers an attractive alternative in this context since it can be applied readily for circuit design or simulation with emerging technologies.

In the conventional LUT approach for MOS transistors [5] one terminal of the device is taken as a reference whereas I-V and Q-V data for other terminals are stored in tables to describe the device behavior. The I-V data is extracted under DCconditions whereas Q-V data is obtained from y-parameters extracted at different bias points using AC simulation or measurement. The look-up table approach including only I-V data was first applied to simulation of digital circuits in the timing simulator MOTIS, which was developed at Bell Laboratories in 1975 [1]. In 1982, Shima et al. reported a three-dimensional look-up table MOSFET model that was accurate enough for analog circuit simulations [2]. The effect of temperature was included in the LUT approach by Graham et al. in 1993 [3]. LUT approach for SOI devices was proposed by D. Nadezhin in 2003 [4] where the device body is considered as an independent terminal which effectively makes it a four terminal device. A look-up table approach for a three-terminal (called 3T-LUT in this paper) planar FINFET device was first proposed by Thakker in 2009 [5].

For simulating PDSOI devices with LUT approach, the body region was considered as a separate terminal in [4]. In this paper the 3T-LUT approach [5] has been extended to floating body SOI transistors without considering the body as a separate terminal. The proposed approach has an advantage of modeling devices with inaccessible node. We observed that, in order to accurately model the device behavior, it is crucial to account for the nonlinear frequency dependency of the small signal parameters in case of SOI devices, as described in section II. A look-up table approach for 28nm FDSOI with the condition of zero body bias has been reported [6]. The proposed enhanced LUT in this paper precisely models the behavior of both FDSOI and PDSOI devices.

This paper is organized as follows. In section II, the 3T-LUT model is reviewed, and the difficulty involved in extending it to SOI MOSFETs is pointed out. In section III a new LUT approach using an auxiliary circuit is proposed. In section IV, the proposed LUT approach is validated for 90nm PDSOI and 28nm FDSOI devices with simulation reults for a few representative circuits.

II. REVISITING THE 3T-LUT BASED SIMULATION FOR MODELING OF SOI MOSFETS

The 3T-LUT simulation set-up for generating I-V and Q-V data for a three-terminal device is shown in Fig. 1 [5]. I-V data is obtained by setting all AC signals to zero and applying DC bias at all terminals. DC simulated terminal currents are recorded at different bias points. In order to extract the Q-V data, y-parameters are obtained by AC simulation for various bias points, by keeping only one of the AC sources in Fig. 1 and recording all AC terminal currents. Terminal charges are then computed using the small signal y-parameters data. For example the charge corresponding to the gate terminal 2 at a given bias point is calculated as follows [5], with indices 1, 2, 3 denoting source, gate, and drain respectively.

$$Q_2(V_{21}^*, V_{31}^*) - Q_2(V_{21min}, V_{31min})$$

$$= \int_{V_{21min}}^{V_{21}^*} C_{22}(V_{21}^*, V_{31}^*) \partial V_{21} + \int_{V_{21min}}^{V_{21}^*} C_{23}(V_{21}^*, V_{31}^*) \partial V_{31} \quad (1)$$

where $C_{22} = \frac{Im\{y_{22}\}}{\omega}$ and $C_{23} = \frac{Im\{y_{23}\}}{\omega}$

In (1) V_{21} and V_{31} are gate and drain voltages with respect to source respectively, (V_{21}^*, V_{31}^*) is the bias point, and V_{21min}

978-1-5386-1357-3/17 $31.00 © 2017 IEEE

163

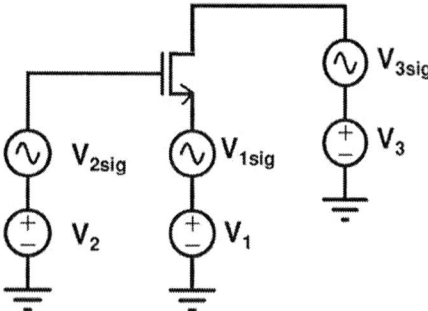

Fig. 1. Setup for extracting C-V data for standard 3T LUT

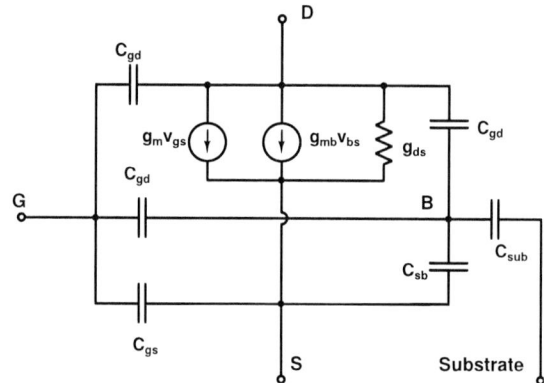

Fig. 3. Small-signal model of PDSOI MOSFET [7]

and V_{31min} are the starting bias voltages used in the LUT. In the 3T-LUT approach, the imaginary part of the small-signal y-parameter is a linear function of frequency for a FINFET device. However, in FBSOI devices, $Im\{y_{ij}\}$ is a nonlinear function of frequency for some of the small-signal parameters as shown in Fig. 2. This behavior can be related to the small-signal model of partially depleted SOI (PDSOI) shown in Fig. 3 [7] for which $Im\{y_{32}\}$ can be given by

$$Im\{y_{32}\} = \omega \frac{AC + \omega^2 BD}{C^2 + \omega^2 D^2} \qquad (2)$$

where A=$g_b c_{gd} + g_{mb} c_{gd}$, $B = c_{gb} c_{bb}$, $C = g_b$, $D = c_{bb}$

The nonlinear relation with frequency exists for y_{32}, y_{33}, y_{12} and y_{11} where 3, 2 and 1 denote the drain, gate and source terminal numbers, respectively. This effect cannot be captured by the 3T-LUT. To capture these effects we propose a new approach using another circuit (called auxiliary circuit in this paper) which is connected to the device presented using 3T-LUT as shown in Fig. 4. This auxiliary circuit will realize the frequency dependent part of y_{32} and y_{33} behavior. The auxiliary circuit is only used to match the difference between actual small-signal y-parameters of the device and y-parameters of the 3T-LUT.

The nonlinear relation of y-parameters with frequency can be eliminated in FDSOI devices with body bias. In this case, our enhanced LUT approach can also be used by simply disabling the auxiliary circuit.

III. DESIGN OF THE AUXLIARY CIRCUIT

As it was discussed in section II an auxiliary circuit with minimum possible number of components is included

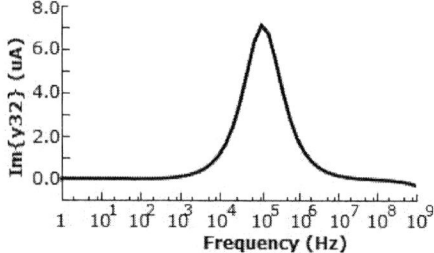

Fig. 2. $Im\{y_{32}\}$ of 90nm NMOS PDSOI for $V_{21} = V_{31} = 1V$

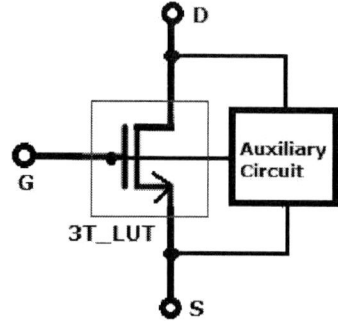

Fig. 4. Concept of the proposed model by combining the 3T-LUT and an auxiliary circuit

along with the 3T-LUT to model the nonlinear behavior of y-parameters with respect to the frequency. The auxiliary circuit is designed in such a way that it does not affect the already modeled DC behavior of the SOI device since there is no outwards DC current path from terminals of the auxiliary circuit. Therefore a novel approach is proposed in which a large-signal model is obtained by mapping its small-signal model to required y-parameters. Simulation data of 90nm floating body PDSOI NMOS device is used to discuss the procedure of designing the auxiliary circuit.

A. Large-Signal Model of the Auxiliary Circuit

First of all the required y-parameters of the auxiliary circuit are calculated from the difference between small-signal y-parameters of the FBSOI obtained from the compact model and the same parameters obtained from corresponding 3T-LUT model for the frequency up to 1GHz. Some properties of these y-parameters are listed as follows.

1) Source (1) terminal is taken as the reference and hence only y-parameters related to the drain (3) and gate (2) terminals need to be modeled by the help of the auxiliary circuit. It is also observed that y_{22} and y_{23} do not show nonlinear dependency with the frequency and hence they can be modeled by 3T-LUT. So, the auxiliary circuit is required to model y_{32} and y_{33} only.

2) The required small-signal y_{32} and y_{33} are nonlinear function of frequency as shown in Fig. 5 and 6

978-1-5386-1357-3/17 $31.00 © 2017 IEEE   164

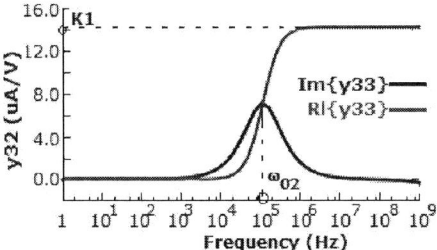

Fig. 5. Frequency characteristic of required y_{32} at $V_{21} = V_{31}$=1.0 V

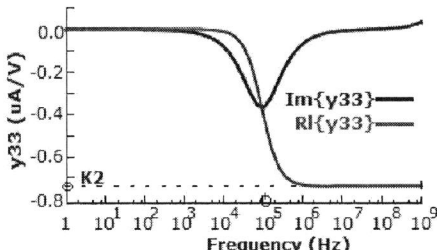

Fig. 6. Frequency characteristic of required y_{33} at $V_{21} = V_{31}$=1.0 V

respectively.

3) Small-signal y-parameters are function of bias voltages.

Based on the properties mentioned above, minimal large-signal model of the auxiliary circuit, the small-signal y-parameters of which can match above listed properties can be a simple nonlinear RC circuit as shown in Fig. 7a. Small-signal model shown in Fig. 7b is generated from large signal model shown in Fig. 7a and exhibits similar behavior to the impedance characteristics shown in Fig. 5 and Fig. 6. Large-signal current flowing through terminal T3 can be given by (3).

$$I_3 = \frac{\partial Q(V_{21}, V_{31}, V_C)}{\partial t} = p_2\frac{\partial V_{21}}{\partial t} + p_3\frac{\partial V_{31}}{\partial t} + p_c\frac{\partial V_C}{\partial t} \quad (3)$$

where $V_C = V_{31} - V_R$, $p_2 = \frac{\partial Q}{\partial V_{21}}$, $p_3 = \frac{\partial Q}{\partial V_{31}}$, $p_c = \frac{\partial Q}{\partial V_C}$. The nonlinear resistor of the large-signal minimal auxiliary circuit (Fig. 7a) is given by (4).

$$I_R = G(V_{21}, V_{31}) * V_R \quad (4)$$

B. Details of the Small-Signal Model of the Auxiliary Circuit

Small-signal model of the auxiliary circuit is shown in Fig. 7(b). Accordingly frequency domain small-signal representation of (3) at a given bias point (V_{21}^*, V_{31}^*) is given by (5).

$$i_3 = i_{a1} + i_{a2} + i_{a3}$$
$$= j\omega p_2(V_{21}^*, V_{31}^*)v_{21} + j\omega p_3(V_{21}^*, V_{31}^*)v_{31} + j\omega p_c(V_{21}^*, V_{31}^*)v_c \quad (5)$$

Similarly small-signal representation of (4) at the given bias point is given by (6).

$$i_r = \frac{\partial I_R}{\partial V_{21}}|_{(V_{21}^*, V_{31}^*)}v_{21} + \frac{\partial I_R}{\partial V_{31}}|_{(V_{21}^*, V_{31}^*)}v_{31} + \frac{\partial I_R}{\partial V_C}|_{(V_{21}^*, V_{31}^*)}v_c \quad (6)$$

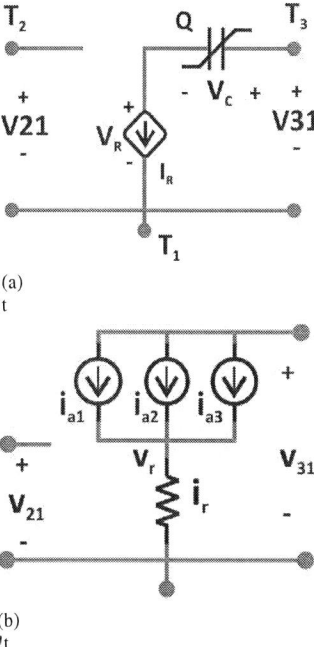

(a)
t

(b)
!t

Fig. 7. (a)Large signal model of the auxiliary circuit. (b)Small signal model of the auxiliary circuit

where $\frac{\partial I_R}{\partial V_{21}} = \frac{\partial G}{\partial V_{21}}V_R$ and $\frac{\partial I_R}{\partial V_{31}} = \frac{\partial G}{\partial V_{31}}V_R$ As no DC current flows through the non-linear resistor V_R=0 under DC condition and hence we get

$$\frac{\partial I_3}{\partial V_{21}} = 0 \;,\; \frac{\partial I_3}{\partial V_{31}} = 0 \quad (7)$$

Equation (6) can be rewritten as

$$i_3 = (\frac{\partial I_3}{\partial V_R}|_{(V_{21}^*, V_{31}^*)}v_r) = gv_r = \frac{v_{31} - v_c}{r} \quad (8)$$

Now from (5) and (8), and considering $i_r = i_3$ we get

$$i_3 = j\omega\frac{p_2v_2 + (p_3 + p_c)v_3}{1 + j\omega p_c r} \quad (9)$$

Values of p_2, $p_3 + p_c$ and $p_c r$ can be obtained from required y-parameters either analytically or by using transfer function estimation with one pole and one zero system. Analytically from (9) and with reference to Fig. 5 and 6 we get, $\omega_{02} = \frac{1}{rp_c}$, $p_2 = k_1\omega_{02}$ and $p_3 + p_c = k_2\omega_{02}$ where ω_{02} is the frequency correspond to local minima/maxima of $Im\{y_{32}\}$. k_1 and k_2 are constant asymptotic value of $Rl\{y_{32}\}$ and $Rl\{y_{33}\}$ at high frequencies respectively.

C. Large-Signal Model of the Auxiliary Circuit

After getting partial derivative values of Q and G, the next step is to get values of Q and G at different bias points and store them in the LUT. Integration method has been used in literature to get values of function from partial derivative. In our proposed model, Q and G are function of V_{21}, V_{31} and V_C. V_{21} and V_{31} are independent variable as they are applied bias voltages, but V_C is a dependent variable. In this case integration method cannot be used to get large-signal values of Q and G. Curve fitting method is another option to get approximate values of Q and G.

978-1-5386-1357-3/17 $31.00 © 2017 IEEE   165

Segment fitting approach is used to get values of Q. In this approach, range of gate and drain voltages are divided into smaller parts called segment as shown in Fig. 8. Each segment consists of some bias points. Generic third order polynomials as given by (10) and (11) are considered for Q and G respectively in each segment.

$$Q = b_{100}V_{31} + b_{010}V_{21} + b_{001}V_C + b_{200}V_{31}^2 + b_{110}V_{31}V_{21}$$

$$+ b_{101}V_{31}V_C + b_{020}V_{31}^2 + b_{011}V_{31}V_C + ... + b_{003}V_C^3 \quad (10)$$

$$G = a_{10}V_{31} + a_{01}V_{21} + a_{20}V_{31}^2 + a_{11}V_{31}V_{21} + ... + a_{03}V_{21}^3 \quad (11)$$

Expression of p_2, p_3 and p_C are derived by partial derivative of (10). Values of coefficients of polynomial in each segment are obtained by simultaneously mapping the polynomial expression of p_2 and $p_3 + p_C$ to the data obtained at various bias points of the corresponding segment as discussed in previous section. Numerical values of coefficients vary segment to segment. Coefficients of polynomial are then used in (10) to calculate values of Q at different bias point. Continuity of Q values across segments is also taken into account with boundary conditions during fitting. Lagrangian constrain optimization for least mean square error method is used for data fitting. Comparison of fitted derivative data with actual data for V_{31}=0.8 V and different values of V_{21} is shown in Fig. 9. Similar fitting were observed for other values of V_{31} with 6% maximum error and 2% root mean square error.

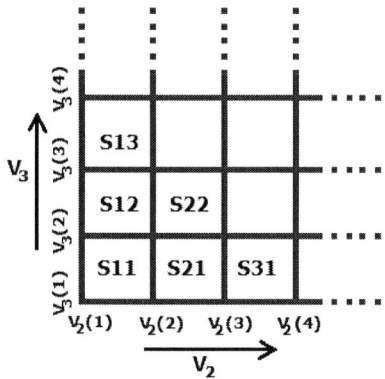

Fig. 8. Segmentation of supply range for data fitting.

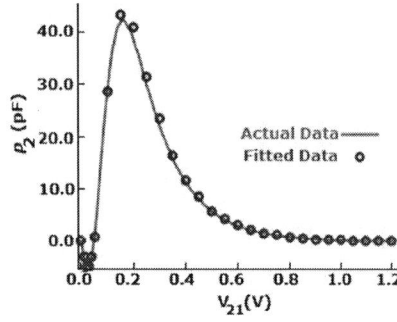

Fig. 9. Comparison of fitted derivative data with actual data for $V_{31} = 0.8V$ and different values of V_{21}.

IV. BENCHMARKING OF THE PROPOSED ENHANCED LUT MODEL

The proposed enhanced LUT (called E_LUT henceforth) approach is first benchmarked with 90nm PDSOI device. After getting large-signal components of the auxiliary circuit DC simulation, AC simulation and transient simulation results of 3T-LUT and the enhanced LUT model are compared with compact model (BSIM) based simulations for benchmarking in this section. BSIM model is used to benchmark the proposed model where as in practice experimental data or data extracted from TCAD simulation are used in absence of compact model. The proposed enhanced LUT approach is also evaluated with experimental data of 28nm FDSOI.

A. DC Simulation of PDSOI device

The DC simulation set-up and the I_D-V_{DS} characteristic of floating body PDSOI NMOS device for different values of V_{GS} is shown in Fig. 10. In section II and III we discussed that the 3T-LUT can match the DC behavior and we do not need any auxiliary circuit for modeling the DC behavior. Also in section III it is mentioned that the auxiliary circuit does not affect the DC behavior of the device and it can be verified from Fig. 10.

B. Small-signal Simulation of flaoting body PDSOI device

Biasing arrangement for AC simulation of NMOS device is shown in Fig. 10. As discussed in section III-A, y_{22} and y_{23} do not require an auxiliary circuit for modeling. Either y_{32} or y_{33} can be compared to benchmark the small-signal behavior. In

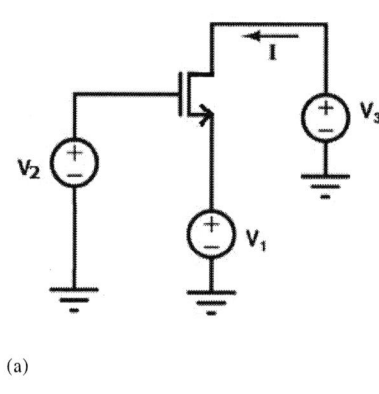

(a)

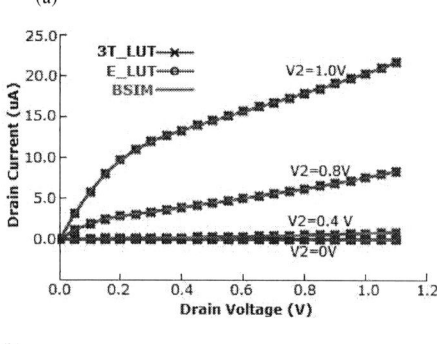

(b)

Fig. 10. Validatioin of DC characteristic of NMOS device, W = 200 nm, L=100 nm. (a)Circuit set up for simulation (b) $I_D - V_D$ characteristic of FBSOI device

this example the small signal y_{32} is extracted at $V_{31} = V_{21} = 1V$ for the comparison purpose. Comparison of the compact model simulation, 3T-LUT simulation and the enhanced LUT simulation results are shown in Fig. 11. As shown in Fig. 11, our novel modeling approach leads to a model which clearly takes care of non-linear variation of y-parameter (both real and imaginary parts) with frequency. In contrast 3T-LUT does not show the variation. Similar comparison was also observed for y_{32} and y_{33} for different bias points.

C. Transient Simulation of floating body PDSOI NMOS Device

Simulation setup of the NMOS device without any load is sown in Fig. 12. This scenario is selected in order to validate the modeling of the internal capacitances and resistances of the device. Drain terminal is biased at 1 V, source is connected to the ground and a square wave signal of 1 V peak to peak is applied at the gate terminal. The drain current is measured for the comparison purpose. As shown in Fig. 12 the enhanced LUT model gives around 2% error as compared to 27% in case of 3T-LUT when both are compared with BSIM results.

D. Simulation of CMOS Inverter

The output of a CMOS inverter for a square wave input is shown in Fig. 13. The improvement in simulation results due to the inclusion of the auxiliary circuit can be observed from Fig.13 and Table I. Maximum relative error of 3% in rise time is observed for enhanced LUT and in 3T-LUT the error is 26% when both are compared with BSIM results.

E. Simulation of SRAM Cell

The 6T SRAM cell simulated for reading data and comparing bit line voltage for 3T-LUT model and enhanced LUT

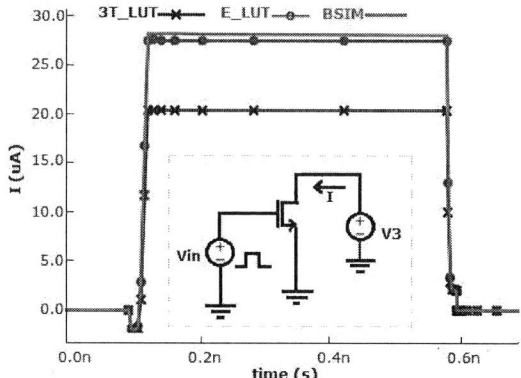

Fig. 12. Drain current comparison for square wave input with rise time = fall time = 20ps, pulse width = 450ps and period = 1ns. Maximum relative difference in the drain current observed in the enhanced LUT model is 2% and 3T_LUT is 27.4%, with BSIM based results for benchmarking.

model with reference to BSIM model is shown in Fig. 14(a). Ratio of device width of pull down transistor to that of access transistors is decided by comparing read noise margin and the voltage at node Q. This circuit is used only to validate the accuracy of the enhanced LUT model. Capacitors of value 10 fF are connected to bit line and bit line bar of the cell to represent the load due to all SRAM cells in a block. Before read operation, Q node is discharged to 0 V and QB, BL and BLB nodes are pre-charged to 1 V. Difference between bit line voltages, V_{BL}-V_{BLB}, is used for comparison as shown in Fig. 14(b).

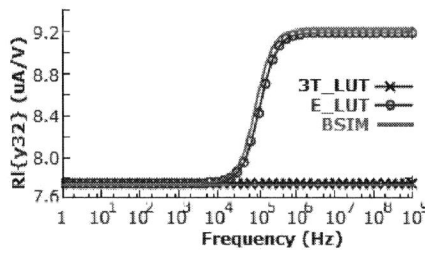

(a)

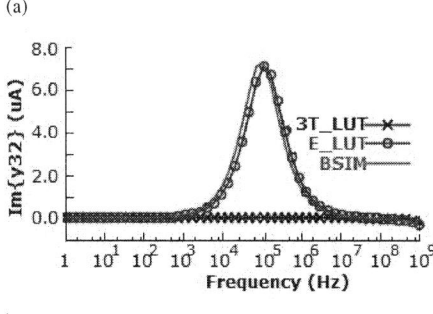

(b)

Fig. 11. AC simulation comparison(a) $Rl\{y_{33}\}$ (b) $Im\{y_{33}\}$. $V_{21} = V_{31} = 0V$. Small signal voltage is applied at the gate terminal and the current is measured at the drain.

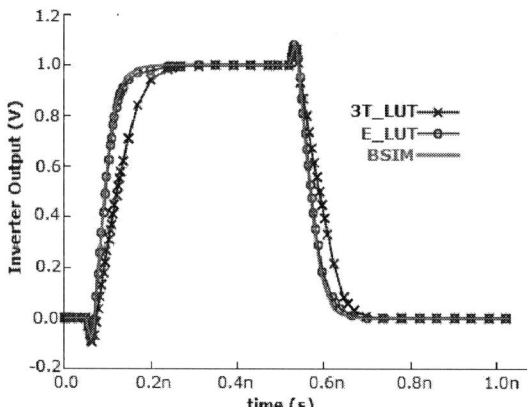

Fig. 13. CMOS inverter output voltage comparison for square wave input with rise time = fall time = 20ps, pulse width = 450ps and period = 1ns.

TABLE.I RISE AND FALL TIME OF INVERTER

	BSIM	E-LUT		3T-LUT	
		Values	% Difference	Values	% Difference
Rise time (ps)	53.8	55.4	3	67.9	26
Fall time (ps)	58.1	59.5	2.2	72.8	25

978-1-5386-1357-3/17 $31.00 © 2017 IEEE   167

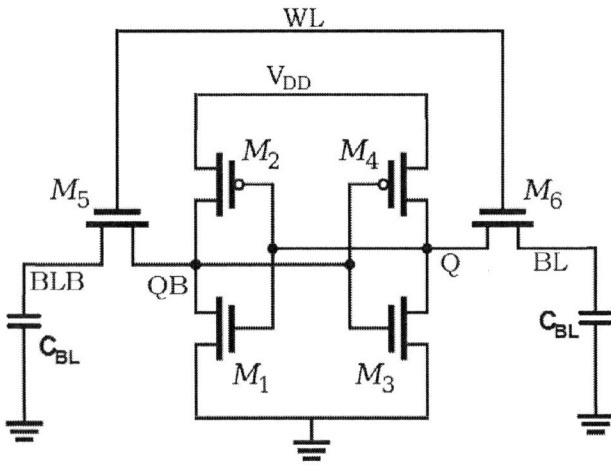

(a)

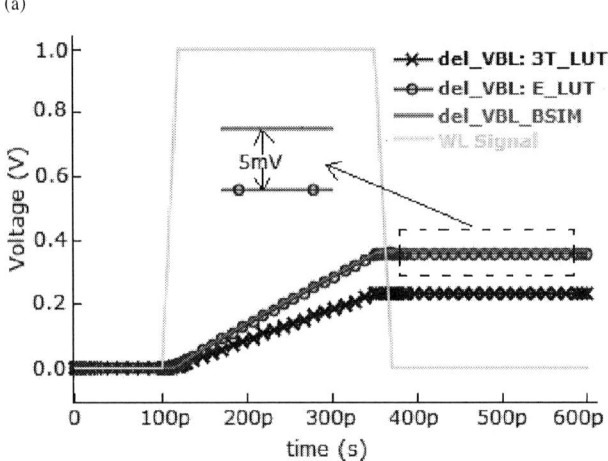

(b)

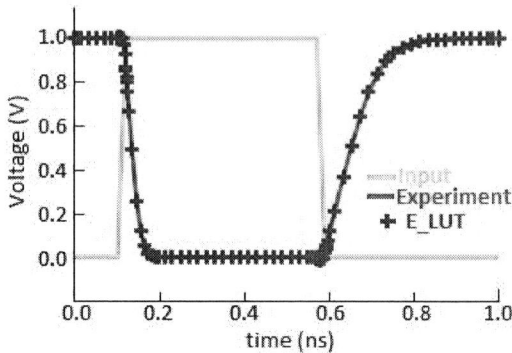

Fig. 15. Accuracy comparison of E-LUT for CMOS inverter with capacitive load of value 5 fF

Fig. 14. Read Mode simulation of 6T SRAM cell.(a) 6T SRAM cell schematic with W/L for M1 and M3 is 150n/100n. W/L for M2, M4, M5 and M6 is 200n/100n.(b)Variation of del_VBL during read operation with initial data Q=0 and QB=1. WL signal is a pulse input with rise time = fall time = 20ps, pulse width =250ps and period = 500ps

TABLE.II BITLINE DIFFERENCE AFTER READ

	BSIM	E-LUT		3T-LUT	
		Values	% Difference	Values	% Difference
V_{BL}-V_{BLB}	58.1	59.5	2.2	72.8	25

F. Simulation of 28nm FDSOI CMOS Inverter

The propsed E-LUT is used for simulation of 28nm FDSOI CMOS inverter. LUT is generated with body biased at VSS and VDD for NMOS and PMOS respectively.. The auxiliary circuit used in E-LUT is disabled during simulation of FDSOI device due to elimination of nonlinear dependency of frequency. Simulation result is compared with experimental data and is shown in Fig.15. The proposed E-LUT matches the results of FDSOI with a relative error of 0.6% in rise time.

V. CONCLUSION

This paper presents a novel look-up table approach for PDSOI and FDSOI devices. This requirement arises at the early stage of technology development where precise compact models are not available but designer would like to explore circuit level designs in the same technology. This paper then presents simulation results of different commnoly used circuits for logic and memory application in 90nm and 28nm SOI CMOS technologies using proposed Enhanced LUT. Benchmarking of the simulation results shows maximum relative error of 3%.

REFERENCES

[1] B. R. Chawla, H. K. Gummel, and P. Kozak, *MOTIS-An MOS timing simulator*, IEEE Trans. Circuits Syst., vol. CAS-22, no. 12, pp. 901910, Dec. 1975

[2] T. Shima, H. Tamada, R. Luong, and M. Dang, *Table look-up MOSFET modeling system using a 2-D device simulator and monotonic piecewise cubic interpolation*, IEEE Trans. Comput. Aided Design Integr. Circuits Syst., vol. CAD-2, no. 2, pp.

[3] M. G. Graham, J. J. Paulos, and D. W. Nychka, *Template-based MOSFET device model*, IEEE Trans. Comput.-Aided Design Integr.Circuits Syst, vol.14, no.8, pp.924-933, Aug.1995

[4] D. Nadezhin, S. Gavrilov, A. Glebov, Y. Egorov, V. Zolotov, D. Blaauw, R. Panda, M. Becer, A. Ardelea, and A Patel,*SOI transistor model for fast transient simulation*, International Conference on Computer Aided Design, pp.120-127, Nov. 2003

[5] R. A Thakker, C. Sathe, A. B. Sachid, M. S. Baghini, V. R. Rao, and M. B. Patil, *A new table based modelling of 28nm fully depleted silicon-on insulator (FDSOI)* IEEE Trans. Computer-Aided Design Integrated Circuits Syst., vol.28, no.7, pp.1061-1070, July 2009

[6] A. M.Abdalla, and J. Rodriguel,*A Novel Table-Based Approach for Design of FinFET Circuits*13th International Conference on Synthesis, Modeling, Analysis and Simulation Methods and Applications to Circuit Design (SMACD), Lisbon, 2016, pp. 1-4.

[7] V. Kilchytska, D. Levacq, D. Lederer, J.-P. Raskin, and D. Flandre, *Floating effective back-gate effect on the small-signal output conductance of SOI MOSFETs*, IEEE Electron Device Lett., vol. 26, no. 10, pp. 414416, Jun. 2003

Gap in pagination due to unavailable paper.

Pages 169-172

Basic CMOS Gate Design by Mixed-Mode Analysis of Step-Channel TMDG-MOSFET

Pankaj Kumar, Syed Samsuz Zaman, Manash Pratim Sarma
GUIST, Gauhati University, Guwahati-781014, India
Email: (cmnpps, mr.samsuzzaman,manashpelsc)@gmail.com

Ashok Ray, Dr. Gaurav Trivedi
EEE Dept., IIT Guwahati, Guwahati-781039, India
Email: (ashok.ray, trivedi)@iitg.ernet.in

Abstract—This paper introduces a novel MOSFET design having a Step-channel Triple Material with Double Gate (STMDG). Our work commences with the design of a STMDG MOSFET using Technology Computer Aided Design (TCAD) simulation tool Sentaurus. We have created two structures of STMDG-MOSFET. In first design (Type-I) the structure has a substrate whose thickness reduces from the source towards the drain. The second design (Type-II) has a tapering from both the source and drain ends wherein the central substrate region is the thinnest. The performance of these devices are contrasted against a typical Linear-channel Triple Material Double Gate MOSFET (LTMDG MOSFET). Both electrical and analog performances are used to analyze the effectiveness of the proposed structure. Using the IV characteristics, we obtain threshold voltage, on-off current ratio, subthreshold slope (SS) and drain induced barrier lowering (DIBL). Next we evaluate the electrostatic-potential and electric-field plots from electrical performance. Further the analog performance of the device is estimated from the transconductance, capacitance, gain and cut-off frequency versus gate voltage plots. Finally we have designed CMOS inverter, NAND and NOR gates using the STMDG Type-I and Type-II MOSFETs and optimized its voltage transfer characteristics (VTC) and switching characteristics to show the utility of our device.

Index Terms—CMOS; DIBL; LTMDG MOSFET; SS; STMDG MOSFET; TCAD; VTC.

I. INTRODUCTION

Since last five decades Moore's law has been enduring and MOSFET down-scaling has the main contribution in this regard. This integration results in the rapid growth of the performance of the device [1]. But now as the integration is going beyond 100 nm there is fair change of increase in the short channel effects (SCEs). In recent time researchers are giving attention to DG MOSFET device structures due to their natural property of suppressing the SCEs [3]. It shows high drive current, steep subthreshold slope(SS) and transconductance, due to the two-channel formation in symmetric DG-MOSFET. Though other devices like the tri-gate, quad-gate MOSFETs, FinFET, GAA [2] etc. are already being in utility, yet much sphere of double-gate MOSFET is unexplored.

Many authors came up with various innovative ideas to enhance the performance of the DG MOSFET. Some of them concluded that with increase in channel doping the threshold voltage (V_T) and I_{on}/I_{off} can be increased and DIBL can be decreased [4]. Further on implementing high-k dielectric as gate-oxide [5,6], increment in drain current (I_D), voltage gain and cut-off frequency and reduction in off-current (I_{off}),

DIBL and SS are observed. Strain can even be applied in the device by taking SiGe sandwiched between two thin layers of Si as substrate [7- 9]. A strained DG MOSFET shows lower DIBL and SS and higher on-current (I_{on}), capacitance and transconductance (g_m). Recently triple material DG MOSFET is the most favored DG MOSFET. The gate electrode of the TMDG MOSFET consists of three laterally contacting materials of degrading work-function from source to drain. This device demonstrates reduced SCEs and improved g_m and I_{off} [11]. They also show simultaneous transconductance enhancement and drain conductance reduction [10]. By innovating a DG MOSFET using the above techniques a strained linear-channel triple material double gate (LTMDG) MOSFET with high-k gate-oxide is obtained.

In [12], Jenal *et al* presents a new conical surrounding gate MOSFET with triple-material gate where the source diameter is greater than the drain diameter. It promises to exhibit higher I_D, minimum V_T, maximum I_{on}/I_{off} ratio and least SS for an optimized tapering ratio (source diameter/drain diameter) of 0.98. But it will be difficult to fabricate such a device due to its slanting dimension. We thus came up with the idea of decreasing the substrate thickness after every fixed length from source to drain in step format. By realizing this variation in the LTMDG MOSFET we obtained a new device and named it as step-channel triple material double gate (STMDG) MOSFET. Moreover we decided to make another device by tapering the LTMDG MOSFET at the central portion and keeping the source and drain at same thickness. This device shows further improvement in the performance. We titled this device as STMDG MOSFET Type-II and former device as STMDG MOSFET Type-I.

In this paper the drain-current characteristics, electrical performance and analog performance of the two new structures of STMDG-MOSFETs are compared with the conventional LTMDG-MOSFET. The innovative structures are found to show improved performance than the traditional LTMDG MOSFETs in all respect. They showed controlled SCEs, more I_{on}/I_{off} ratio and better electrical performance. The STMDG MOSFETs also have more gain and nearly equal maximum cut-off frequency with respect to the LTMDG MOSFET. With the three structures we have designed basic CMOS gates (NOT, NAND and NOR) to verify their functionality. The VTC and switching characteristics for the three gates have shown better results.

II. Performance Parameters

1) **Threshold Voltage(V_T):** [4] The gate voltage at which the device begins to conduct i.e. crosses a given arbitrary constant drain current, I_D is called threshold voltage. Mathematically,

$$V_T = \phi_M + 2\phi_F + (Q_D + Q_{SS})/C_{ox} + V_{in} \quad (1)$$

where, Q_{SS} implies the gate dielectric charge, Q_D implies the channel's depletion charge, C_{ox} refers to the gate capacitance, $\phi_M = \phi_1 + \phi_2 + \phi_3$ are work functions of the three gate materials, ϕ_F implies the fermi potential of the substrate, C_{ox} implies the capacitance of the gate-oxide and V_{in} implies the input voltage.

2) **Subthreshold Slope (SS):** [3] Subthreshold slope is the slope of $log(I_D)$ vs V_{GS} plot. It refers to the variation in gate voltage for per decade shift in drain current. Mathematically,

$$SS = [\delta log10(I_D)/\delta V_{GS}] \quad (2)$$

where, I_D implies drain current and V_{GS} implies gate-to-source voltage.

3) **Subthreshold Current (I_{off}):** [4] The undesirable leakage current (for n-type MOSFET) flowing between the drain and the source at $V_{GS} < V_{th}$ is called the subthreshold current or off-state current. It is the drain current measured at $V_{GS} = 0$ and $V_{DS} = V_{DD}$. It is desirable to keep I_{off} very small so as to minimize the static power and increase the circuit speed. Mathematically,

$$I_{off} = 100(W/L)10^{-(V_T/SS)} \quad (3)$$

where, W/L implies width to length ratio of the device.

4) **Drain Induced Barrier Lowering (DIBL):** [3] Drain Induced Barrier Lowering is a SCE which refers to the fluctuation of threshold voltage (V_T) by variation of drain-to-source voltage (V_{DS}). DIBL is enhanced at high drain voltages and shorter channel lengths. Mathematically,

$$DIBL = \Delta V_T/\Delta V_{DS} = (V_{T1} - V_{T2})/(V_{DS2} - V_{DS1}) \quad (4)$$

where, V_{T1} and V_{T2} are the threshold voltages for drain-to-source voltages V_{DS1} and V_{DS2}.

5) **Voltage Gain (A_v):** [3] Voltage gain is defined as the ratio of input transconductance to output transconductance. Mathematically,

$$A_v = V_{out}/V_{in} = g_m/g_d \quad (5)$$

where g_m implies input transconductance and g_d is the output transconductance. They are given by,

$$g_m = dI_D/dV_{GS} \quad (6)$$

$$g_d = dI_D/dV_{DS} \quad (7)$$

6) **Cut-off Frequency (f_c):** [3] Cut-off frequency is the threshold of frequency response beyond which the system's energy level begins to fall below 3 dB. Mathematically,

$$f_c = g_m/2\pi(C_{GS} + C_{GD}) \quad (8)$$

where, C_{GS} implies the gate-to-source capacitance, C_{GD} implies the gate-to-drain capacitance and C_G implies the gate capacitance.

7) **Noise Margin (NM):** [17] It is a parameter which determines the acceptable noise voltage at the gate so that the output is unaffected. Noise margin (or noise immunity) can be measured by two parameters- The LOW noise margin, NM_L, and the HIGH noised margin, NM_H. Mathematically,

$$NM_L = |V_{IL} - V_{OL}| \quad (9)$$

$$NM_H = |V_{OH} - V_{IH}| \quad (10)$$

where,V_{IH} implies the minimum input high voltage, V_{IL} implies the maximum input low voltage,V_{OH} implies the minimum output high voltage and V_{OL} implies the maximum output low voltage.

III. Device and Gate Structures

This paper presents three structures viz. LTMDG MOSFET, STMDG MOSFET Type-I and SMDG MOSFET Type-II as shown in Fig.1. These structures have channel length of 36 nm and its dimension ($W \times L \times t$) is $100nm \times 60nm \times 30nm$. The lengths of the three gates are taken to be $L_1, L_2, L_3 = 12nm$. The substrate is lightly doped with boron of doping concentration $1 \times 10^{16} cm^{-3}$ and the drain and source are doped with arsenic of doping concentration $1 \times 10^{20} cm^{-3}$ to a depth of 12 nm. Being a high-k material, HfO_2 is used as supporting gate-oxide over a thin layer (1 nm) of SiO_2 to reduce the tunneling effect. Strain is generated along the channel by epitaxial growth of silicon on either side of SiGe material where the mole fraction of Ge is fixed to 0.4 so as to increase the carrier mobility [7]. The device involves three gate materials i.e. Au(ϕ=5.1 eV), Cu(ϕ=4.65 eV) and Ag(ϕ=4.26 eV) of decreasing work function(ϕ) from source towards the drain. This leads to increased mobility in the channel and reduced electric field spike at the drain side thereby diminishing HCEs.

For the first structure (LTMDG MOSFET), the substrate is of constant depth (20 nm). In the second structure (STMDG MOSFET Type-I), the substrate depth is constantly diminished by 1 nm after every 12 nm substrate length. The substrate so obtained is 20 nm thick at the source end and 16 nm thick at the drain end. The third structure (STMDG MOSFET Type-II) on the other hand has thickness of 20 nm at both the source and drain ends. The thickness is diminished by 2 nm towards the middle from both ends after every 12 nm. Thus the middle portion of the third device is 16nm thick. All other parameters are similar for the three structures.

It has been estimated that on diminishing the substrate width the performance of the device can be enhanced. But if this width is scaled beyond a certain limit, HCEs and narrow width effects (NWEs) can come into play. Thus in order to enhance the carrier mobility and also to diminish HCEs and NWEs we diminished the substrate width in step form from the source to the drain. At the same time we kept on increasing the oxide thickness from source to drain and maintained a uniform device width. This makes C_{GS} more than C_{GD} in the saturation region which diminishes the leakage current and other SCEs.

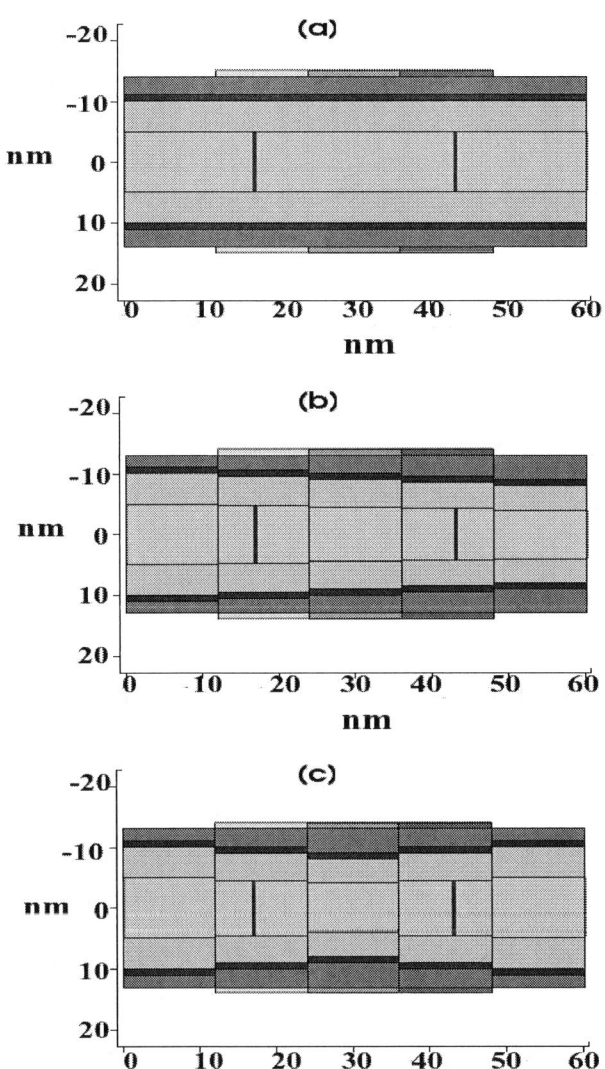

Fig. 1. Structure of (a)TMDG MOSFET (b)STMDG MOSFET Type-I (c)STMDG MOSFET Type-II

To verify the utility of our device we have implemented it in the design of a basic CMOS inverter and Universal gates (NAND and NOR). For designing these gates we have also created a p-type device. All the parameters involved in the design of the p-type device is same as that of the n-type device, but the n-type and p-type dopings are interchanged. Also the

W/L ratio of p-type device is doubled, remained same and quadrupled with respect to the n-type device to design CMOS Inverter, NAND and NOR gates respectively [13]. Fig. 2 shows the structures of CMOS Inverter, NAND and NOR gate using DG MOSFET.

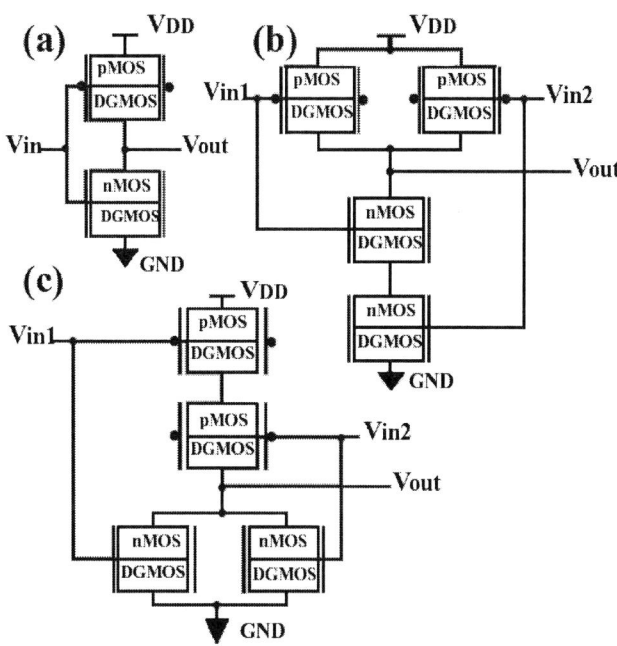

Fig. 2. CMOS (a)Inverter (b)NAND (c)NOR gates using DG MOSFET

IV. RESULTS AND DISCUSSION

This paper uses the TCAD device simulating tool Sentaurus for 2D simulation of our device [16]. The structures are designed, meshed and simulated using Sentaurus structure editor, Sentaurus mesh and Sentaurus device simulator respectively. The electron current continuity equation coupled to the 2D Poisson equation is solved fully using Drift-diffusion model. Mobility models used for this simulation includes doping dependence, high-field saturation and transverse field dependence. The silicon bandgap narrowing model 'Old slot boom' is used here to determine the effective intrinsic carrier concentration. Further recombination models Auger, avalanche and SRH are incorporated for device simulation.

A. Device Simulation

1) Drain Current Characteristics: The $log(I_D) - V_{GS}$ plot for $V_{DD} = 0.1, 1V$ and $V_{GS} = 1.5V$ and $log(I_D) - V_{DS}$ plot for $V_{DS} = 1V$ and $V_{GS} = 0.5, 1.5V$ for the three structures can be observed from Fig.3 and Fig.4 respectively. The result obtained from these plots is shown in the Table I. It can be observed from the table that the threshold voltage is approximately same for the three structures. The DIBL and SS for the STMDG MOSFETs are decreased, while the I_{on}/I_{off} ratio are increased by a factor of 72 and 144 for STMDG Type-I and Type-II MOSFET as compared to the LTMDG MOSFET.

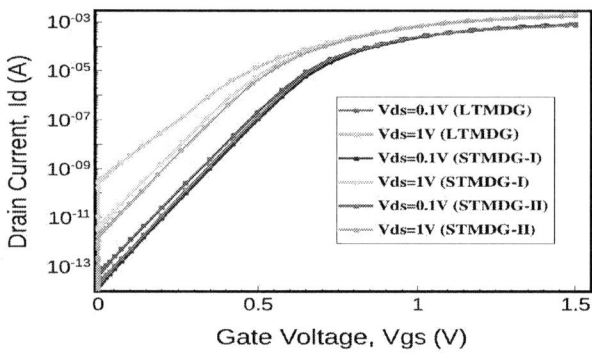

Fig. 3. $log(I_D) - V_{GS}$ Plot for $V_{DS} = 0.1, 1V and V_{GS} = 1.5V$

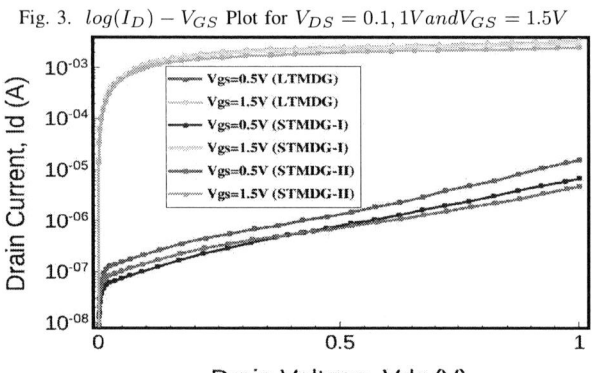

Fig. 4. $log(I_D) - V_{DS}$ Plot for $V_{DS} = 1V and V_{GS} = 0.5, 1.5V$

TABLE I
ELECTRICAL CHARACTERISTICS COMPARISON OF THE THREE
STRUCTURES

Parameters	LTMDG MOSFET	STMDG MOSFET Type-I	STMDG MOSFET Type-II
$V_T(V)$	0.5	0.55	0.5
$I_{on}(mA)$	2.2	2.1	1.8
$I_{off}(nA)$	0.3	4×10^{-3}	2×10^{-3}
I_{on}/I_{off}	7.33×10^6	5.25×10^8	1.05×10^9
$SS(mV/dec)$	75	72	70
$DIBL(mV/V)$	133	111	55

2) Electrical Performance: The electrical performance of
the devices is shown by the electrostatic potential and electric field plots along the channel length in Fig.5 and Fig.6
respectively. For obtaining all these plots the drain voltage
and gate voltage are fixed at 1 V and 1.5 V respectively. It
can be observed from the electrostatic potential plot that the
drain is at higher potential than that of the source and there
is a gradual decrement in potential from drain to source. This
allows a smooth flow of current from drain to source. The
magnitude of the electric field along the lateral position of
the devices is found to be minimum at the drain end which
confirms minimum HCEs and lesser DIBL. As the electric
field magnitude is minimum for STMDG MOSFET Type-II
structure, so it has a minimum DIBL and also least leakage
current.

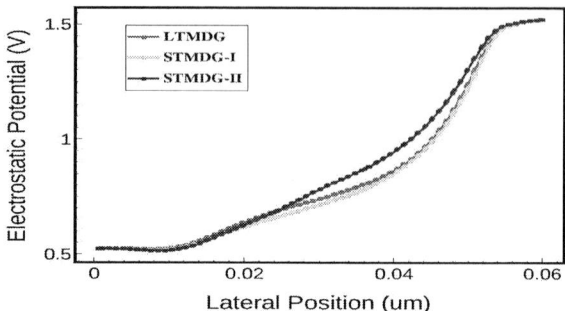

Fig. 5. Electrostatic Potential Plot for $V_{DD} = 1V and V_{GS} = 1.5V$

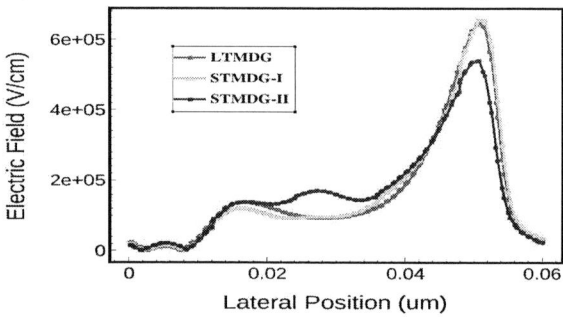

Fig. 6. Electric Field Plot for $V_{DD} = 1V and V_{GS} = 1.5V$

3) Analog Performance: To observe the analog performance of the devices the input/output transconductance, gate-
to-source/gate-to-drain capacitance and voltage gain/cut-off
frequency are plotted against the gate voltage as shown in
Fig.7, Fig.8 and Fig.9 respectively. The drain voltage is taken
to be 1 V and the gate voltage is varied from 0 to 1.5 V. The
frequency is fixed at 1 MHz for simulation.

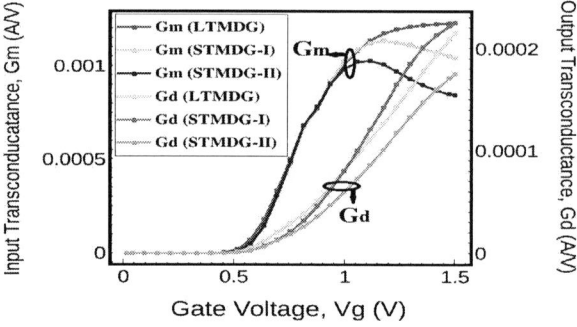

Fig. 7. G_m and G_d Vs V_{GS} Plot for $V_{DD} = 1V and V_{GS} = 1.5V$

The input transconductance increases exponentially till saturation and then began to decrease. On the contrary the output
transconductance keeps on increasing exponentially. G_m/G_d
ratio implies the voltage gain of a semiconductor device.
Thus it is desirable to obtain higher magnitude of G_m and
lower magnitude of G_d, thereby increasing the device's gain.
The voltage gain plot shows higher peaks at the linear and
saturation points. The STMDG MOSFETs show higher gain
than the LTMDG MOSFET. The gate-to-source capacitance

increases till saturation and thereafter becomes constant. On the contrary, the gate-to-drain capacitance keeps on increasing with varying slopes. The magnitudes of C_{GS} and C_{GD} are instrumental in obtaining the cut-off frequency plot. This plot shows a exponential increase in magnitude from few KHz to hundreds of GHz till it becomes constant. The STMDG MOSFETs show slightly lower cut-off frequency than the LTMDG MOSFET.

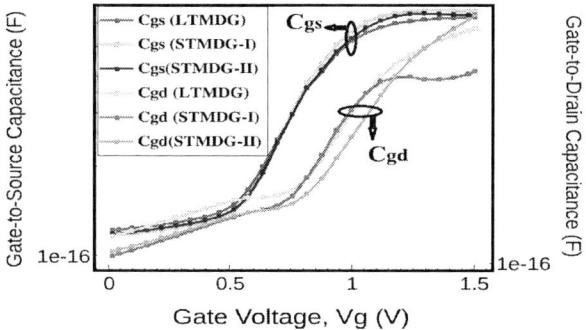

Fig. 8. C_{GS} and C_{DS} Vs V_{GS} Plot for $V_{DD} = 1V$ and $V_{GS} = 1.5V$

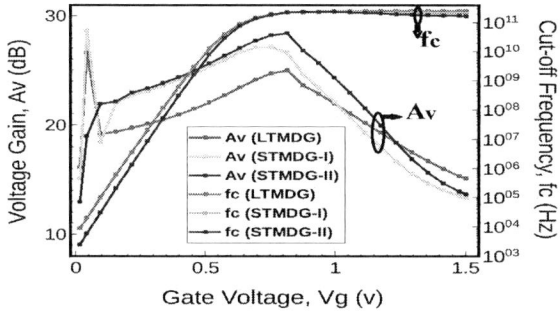

Fig. 9. Voltage gain and Cut-off frequency Vs V_{GS} Plot for $V_{DD} = 1V$ and $V_{GS} = 1.5V$

B. Gate Simulation

1) Voltage Transfer Characteristics: The voltage transfer characteristics (V_{out} vs V_{in} curve) of the three gates are plotted in Fig. 10. The values obtained from Fig.10(a) are shown in Table II. From the table we noted out that V_{OH} and V_{OL} are close to V_{DD} and 0 respectively and values of NM_L and NM_H are nearly equal. Also the threshold voltage of the inverter, (V_M) is very close to $V_{DD}/2$. This shows a good agreement with the symmetric CMOS inverter. The NAND gate shows a broader VTC (Fig.10(b)) and the NOR gate (Fig.10(c)) a narrower one as expected. To obtain all the VTC curves a capacitance of 10^{-17}F is connected at the load [14]. The VTCs of the STMDG MOSFETs show higher peaks at output high and lower peaks at output low as compared to the LTMDG MOSFET for the three gates. This verifies the superiority of the STMDG MOSFETs.

2) Switching Characteristics: To obtain the switching characteristics of the gates we have considered the following parameters [14]: Rise time (T_r)=10 psec, Fall time (T_f)=10 psec, On time (T_{on})=60 psec, Time period=140 psec, Load

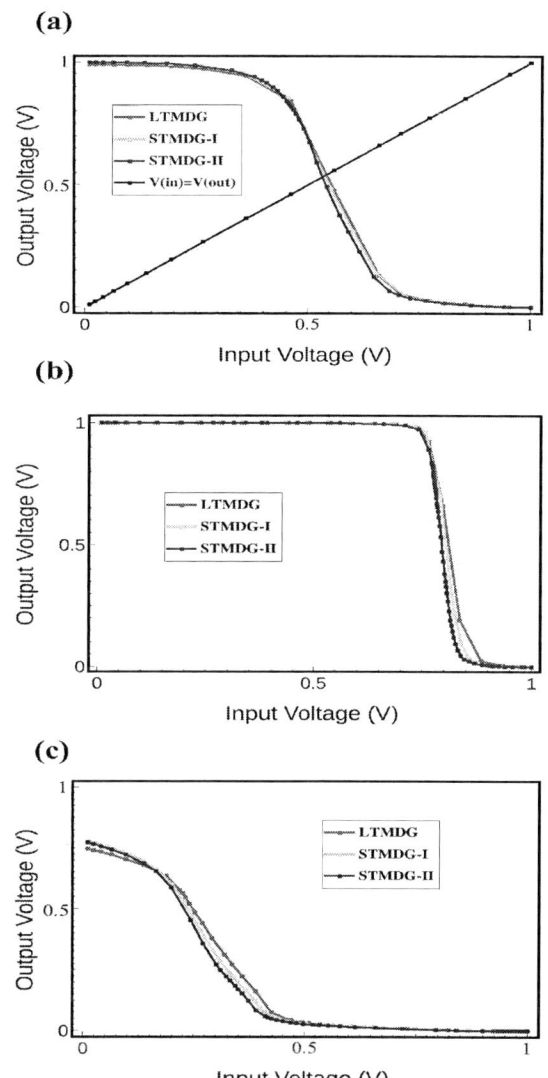

Fig. 10. VTC of (a)Inverter (b)NAND Gate (c) NOR Gate for $V_{DD} = 1V$

TABLE II
PERFORMANCE OF THE CMOS INVERTER

CMOS Inverter	LTMDG MOSFET	STMDG MOSFET Type-I	STMDG MOSFET Type-II
V_{DD}(V)	1	1	1
V_{OH}(V)	0.98	1	1
V_{OL}(V)	0.01	0.01	0.01
V_{IL}(V)	0.3	0.3	0.32
V_{IH}(V)	0.72	0.73	0.7
V_M(V)	0.54	0.535	0.53
NM_L(V)	0.29	0.29	0.31
NM_H(V)	0.26	0.27	0.3

capacitance (C_L)= 10^{-17}F for total time of 200 psec. Fig.11 shows the switching (timing) characteristics of the Inverter, NAND and NOR gates. V(in), V(in1) and V(in2) specifies the input voltages and V(out1), V(out2) and V(out3) specifies the output voltages of LTMDG, STMDG-I and STMDG-II MOSFETs. From Fig.11(a) we found that for input='0', output='1' and for input='1', output='0'. For NAND gate output='0', '1', '1', '1' are obtained for input='11', '00', '01', '10' as shown in Fig.11(b). Similarly for NOR gate output='0', '1', '0', '0' are obtained for input='11', '00', '01', '10' as shown in Fig.11(c). These outputs obtained for varying inputs verifies the truth table of the NOT, NAND and NOR gates. This shows that the gates are working in desirable way. Irrespective of this initially some spikes are obtained in the timing diagrams of STMDG MOSFETs because of the varying parasitic capacitance in step-channel caused by clock feedthrough phenomenon.

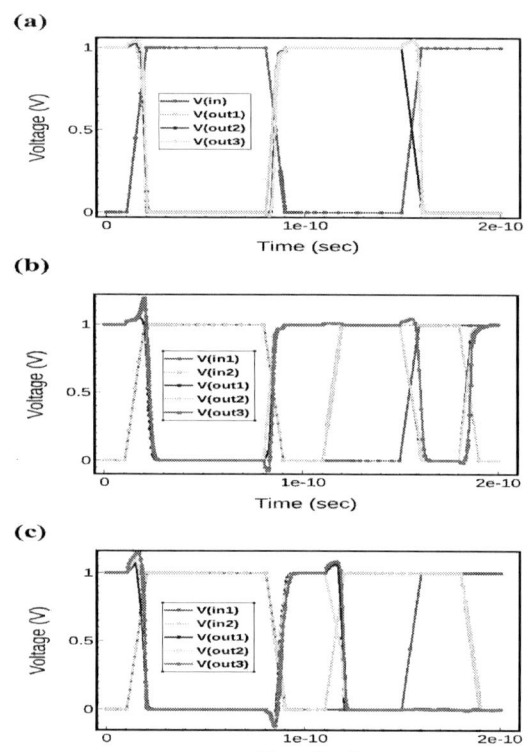

Fig. 11. Switching Characteristics of (a)Inverter (b)NAND (c) NOR Gates for $V_{DD} = 1V$

V. CONCLUSION

As estimated our device is capable of controlling SCEs and increasing the performance of double gate MOSFETs (DG-MOSFETs). The Step-channel TMDG MOSFET shows a decrease in SS, DIBL and more than hundred-fold increase in the on-off current ratio as compared to the Linear-channel TMDG MOSFET. The STMDG MOSFETs shows an elevation in electrical performance. They also have more gain and nearly equal maximum cut-off frequency when compared to the LTMDG MOSFET. The inverter shows good noise margin for the three MOSFETs. The NOT, NAND and NOR shows desired VTCs. Even the switching characteristics of the gates formed by these devices are found to follow there respective truth tables. These results estimates that the gates designed using our devices are working in its desired way. Hence it verified that the new device STMDG MOSFET is superior to TMDG MOSFET.

Further experiments can be done in future by varying the tapering ratio of the devices. Like the CMOS gates the devices can also be implemented to design other CMOS circuits and low power AC circuits like CMOS Amplifiers, Operational Transconductance Amplifiers, Schmitt Trigger, etc.

ACKNOWLEDGMENT

The authors like to thank CAC Lab, IIT Guwahati and all the Research Scholars for providing the resources and their support.

REFERENCES

[1] A. S. I. Association, *Itrs - international technology roadmap for semi-conductor*, 2012.
[2] B.Buvaneswari, *"A Survey of Multi Gate MOSFETs"*, International Journal of Innovative Research in Science, Engineering and Technology, Volume 3, Special Issue 3, March 2014.
[3] Ankita Wagadre, Shashank Mane, *"Design and Performance Analysis of DG-MOSFET for Reduction of Short Channel Effect over Bulk MOSFET at 20nm"*, Int. Journal of Engineering Research and Applications, ISSN : 2248-9622, Vol. 4, Issue 7(Version 1), July 2014, pp.30-34.
[4] Vinay Kumar Yadav, Ashwani K. Rana, *"Impact of Channel Doping on DG-MOSFET Parameters in Nano Regime-TCAD Simulation"*, International Journal of Computer Applications (0975 8887),Volume 37 No.11, January 2012.
[5] K.P.Pradhan, S.K.Mohapatra, P.K.Sahu, D.K.Behera, *"Impact of high-k gate dielectric on analog and RF performance of nano scale DG-MOSFET"*, Microelectronics Journa l45(2014)144151.
[6] Nour El Islam Boukortt, Baghdad Hadri, Salvatore Patan, *"Effects of High-k Dielectric Materials on Electrical Characteristics of DG n-FinFETs"*, International Journal of Computer Applications (0975 8887), Volume 139 No.10, April 2016.
[7] Diana Pradhan, Sanghamitra Das, Tara Prasanna Dash, *"Study of strained-Si p-channel MOSFETs with Hf_O2 gate dielectric"*, Superlattices and Microstructures 98 (2016) 203-207.
[8] Hossein Valinajad, Reza Hosseini and Mohhamad Esmael Akbari, *"Electrical Characteristics of Strained Double Gate MOSFET"*, IJRRAS Vol 13 Issue 2, November 2012.
[9] Arka Dutta, Kalyan Koley, Chandan K. Sarkar, *"Impact of underlap and mole-fraction on RF performance of strained-Si/$Si_{1-x}Ge_x$/strained-Si DG MOSFETs"*, Superlattices and Microstructures 75 (2014) 634646.
[10] Pedram Razavi and Ali A. Orouji, *"Nanoscale Triple Material Double Gate (TM-DG) MOSFET for Improving Short Channel Effects"*, International Conference on Advances in Electronics and Micro-electronics, 978-0-7695-3370-4/08 2008 IEEE DOI 10.1109/ENICS.2008.33.
[11] Santosh Kumar Gupta, Achinta Baidya and S. Baishya, *"Simulation and Analysis of Gate Engineered Triple Metal Double Gate (TM-DG) MOSFET for Diminished Short Channel Effects"*, International Journal of Advanced Science and Technology, Vol. 38, January, 2012.
[12] B Jenal, B S Ramkrishnal, S Dash and G P Mishra, *"Conical surrounding gate MOSFET: a possibility in gate-all-around family"*, Adv. Nat. Sci.: Nanosci. Nanotechnol. 7 (2016) 015009 (6pp).
[13] J. CharlesPravin, D.Nirmal, P.Prajoon, J.Ajayan, *"Implementation of nanoscale circuits using dual metal gate engineered nanowire MOSFET with high-k dielectrics for low power applications"*, Physica E 83(2016) 95100.
[14] Achinta Baidya, Trupti Ranjan Lenka, Srimanta Baishya, *"Mixed-mode simulation and analysis of 3D double gate junctionless nanowire transistor for CMOS circuit applications"*, Superlattices and Microstructures, 100 (2016) 14-23.
[15] Juncheng Wang, Gang Du, Kangliang Wei, Kai Zhao, Lang Zeng, Xing Zhang, and Xiaoyan Liu, *"Mixed-Mode Analysis of Different Mode Silicon Nanowire Transistors-Based Inverter"*, IEEE Transactions on Nanotechnology, Vol. 13, No. 2, March 2014.
[16] Synopsys, TCAD Sentaurus User Guide, Version Z-2013, 12 December 2016.
[17] Neil H.E. Weste, David Money Harris, "CMOS VLSI Design", 4th edition.

Modeling of Threshold Voltage and Subthreshold Current for P-Channel Symmetric Double-Gate MOSFET in Nanoscale Regime

Rekib Uddin Ahmed and Prabir Saha
Department of Electronics and Communication Engineering
National Institute of Technology Meghalaya
Shillong, Meghalaya, India
rekib@nitm.ac.in, sahaprabir1@gmail.com

Abstract—In this paper analytical threshold voltage and sub-threshold current model for lightly-doped p-channel symmetric Double-Gate (DG) MOSFET in nanoscale regime have been presented. Analytical equation of potential distribution in the channel has been derived by solving the Poisson's equation with the constraint of weak inversion region. Threshold voltage equation of the device has been derived from the inversion charge sheet density at maximum potential position. Subthreshold current model of the device has been derived by augmenting the core drain current model with one of the short-channel effect ie. threshold voltage roll-off. Physical effect like surface roughness scattering has been incorporated in the subthreshold current model. The obtained results from threshold voltage model as a function of channel length and silicon body thickness and subthreshold current model have been found in good agreement with simulation results obtained from device simulator Atlas.

Keywords—Threshold voltage, potential distribution, sub-threshold current, subthreshold slope, Atlas (Silvaco)

I. INTRODUCTION

In the prsesnt era, metal-oxide semiconductor field-effect transistor (MOSFET) sizes have entered in nanoscale regime. The purpose behind the downscaling of MOSFET is to increase the speed and to incorporate more transistors in the integrated circuits. The downscaling of complementary metal-oxide semiconductor (CMOS) devices is experiencing difficulties as the channel length is becoming smaller. The phenomena arising due to ultra-short channel length are called short-channel effect. The consequences of short-channel effects are threshold voltage roll-off, drain induced barrier lowering (DIBL), mobility degradation, velocity saturation and hot carrier effect [1]. Alternate devices have been proposed to extend the scalability of CMOS devices such as high-k gate dielectric, strained-silicon MOSFET, silicon on insulator (SOI) and multi gate MOSFET [2], [3].

Double-Gate (DG) MOSFET is one of the multi gate devices which has ability to show better immunity towards short-channel effects and it can be scaled to the shortest channel for a given oxide thickness [4]. DG MOSFET is a silicon on insulator device which is controlled by the two gates shown in Fig 1. The dual gate system and the undoped (or lightly doped) thin silicon body makes the device to show

better performance than conventional bulk MOSFET. The use of undoped (or lightly doped) silicon body eliminates the random dopant fluctuation effect and reduces the depletion charges which enhances the carrier mobility [4]. Two types of DG MOSFET have been found in literature [5], [6]: symmetric DG MOSFET in which gate work function of the two gates are same and asymmetric DG MOSFET is having the two gates are of different work functions.

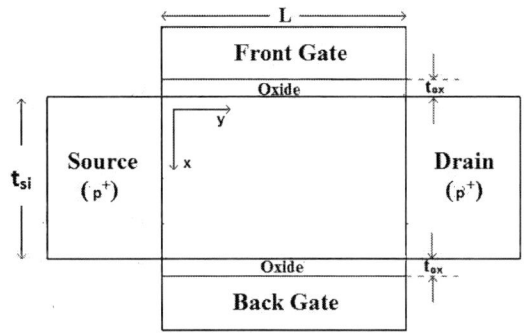

Fig. 1. Schematic cross section of p-channel symmetric DG MOSFET

To design any circuit and to perform simulation, at first model of the constituting devices are required. Goal of a device model is to accurately predict electrical characteristics of the device for a given applied bias. To perform circuit level simulation using DG MOSFET, drain current (I-V), charge (Q-V) and capacitance (C-V) model of the device are required. Several models for n-channel DG MOSFET are already been available in the literature [4]–[11]. The existing current, charge and capacitance model for n-channel DG MOSFET can be implemented using Verilog-A and can be used as a circuit simulating tool. But none of the above reported papers extended their work for p-channel DG MOSFET.

In order to simulate any circuits using DG-CMOS technology, modeling of p-channel DG MOSFET is important. Moers et al. [12] realized layout design of p-channel DG MOSFET with channel length of $50nm$ but the paper lacks analytical expression for drain current characteristics of the device. Song et al. [13] verified and calibrated the analytical potential model for symmetric DG MOSFET with experimental results of p-channel FinFET considering channel length upto $100nm$ which becomes inadequate for technology node below $45nm$. Kumari et al. [14] presented a charge based model for p-channel DG MOSFET considering Gaussian doped silicon body but the developed model does not satisfy for the uniform or lightly doped silicon body.

In this paper, analytical expression of threshold voltage for p-channel symmetric DG MOSFET is derived by adopting same approach used for modeling the n-channel device [8], [9]. Subsequently current equation for subthreshold region is modeled for p-channel considering device dimensions, channel length $L = 30nm$, silicon body thickness $t_{si} = 12nm$ and oxide thickness $t_{ox} = 1nm$. The proposed threshold voltage and subthreshold current model are verified by comparing with simulation results performed in Atlas (Silvaco).

II. PROPOSED THRESHOLD VOLTAGE MODEL

A schematic cross section of p-channel symmetric DG is shown in Fig 1 where x represents direction across the thickness of silicon body and y is direction along the channel. The source and drain regions are highly doped p-type and the body is lightly-doped n-type silicon material. One of the key equations governing the operation in MOS devices is Poisson's equation. In short p-channel devices, for operation in weak inversion region the 2D Poisson's equation is written considering the fixed charge concentration [8], [15]

$$\frac{\partial^2 \psi(x,y)}{\partial x^2} + \frac{\partial^2 \psi(x,y)}{\partial y^2} = -\frac{qN_{si}}{\epsilon_{si}} \quad (1)$$

where $\psi(x,y)$ is the potential distribution in the channel, q is the elementary charge, ϵ_{si} is the dielectric permittivity of silicon and N_{si} is the body doping concentration (lightly-doped). To develop a physical model for the DG MOSFET, potential distribution throughout the silicon body has to be determined under applied gate and drain bias. The Poisson's equation for weak inversion region can be solved to determine 2D potential distribution in the channel [8]. Potential distribution in the channel of p-channel symmetric DG MOSFET in weak inversion region is derived as

$$\psi(x,y) = V_{gs} - V_{fb_p} + \frac{1}{e^{\frac{2L}{\lambda_x}} - 1}[(V_{bi_p} + V_{ds} - V_{gs} + V_{fb_p})$$
$$\left(e^{\frac{L+y}{\lambda_x}} - e^{\frac{L-y}{\lambda_x}}\right) + \left(V_{bi_p} - V_{gs} + V_{fb_p}\right)\left(e^{\frac{2L-y}{\lambda_x}} - e^{\frac{y}{\lambda_x}}\right)] \quad (2)$$

where V_{gs} is the applied bias at front and back gate, V_{fb_p} is the flat band voltage which is approximately equal to work function difference between metal gate ϕ_m and lightly-doped silicon ϕ_s [4], n_i is the intrinsic charge concentration, $V_{bi_p} = -V_T ln\left(N_{si}N_{sd}/n_i^2\right)$ is the built-in voltage, V_T is the thermal voltage, N_{sd} is the source/drain doping concentration, V_{ds} is the applied bias across drain and source and ϵ_{ox} is the dielectric permittivity of the oxide. $\lambda_x = \sqrt{\frac{\epsilon_{si}t_{ox}t_{si}}{2\epsilon_{ox}}\left(1 + \frac{\epsilon_{ox}x}{\epsilon_{si}t_{ox}} - \frac{\epsilon_{ox}x^2}{\epsilon_{si}t_{ox}t_{si}}\right)}$ is the natural channel

length of the DG MOSFET. In short-channel DG MOSFETs, natural channel length is described as a function of channel depth x [8].

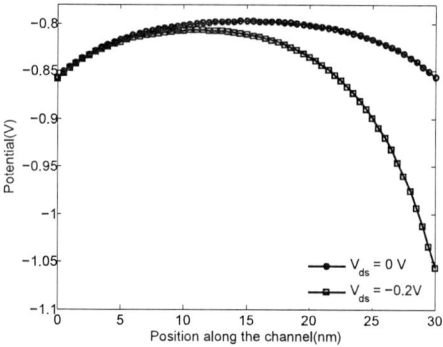

Fig. 2. Potential distribution along the effective conductive path with dimensions $L = 30nm$, $t_{ox} = 1nm$, $t_{si} = 12nm$ at bias condition $V_{gs} = -0.5V$ and different V_{ds}

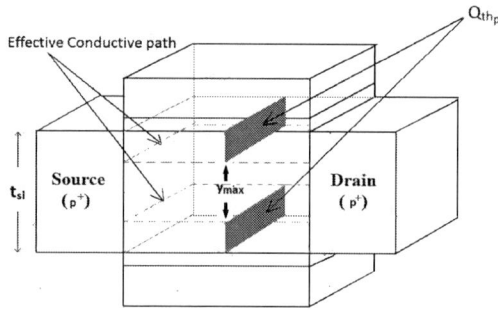

Fig. 3. Inversion charge sheet density at threshold condition $Q_{inv} = Q_{th_p}$

The effective conductive path of the device is located at a position $x = t_{si}/4$ from the surface [8]. Fig 2 shows the potential distribution along the effective conductive path for different applied V_{ds}. Maximum potential is observed at the center of the channel when $V_{ds} = 0V$. Maximum potential position shifts towards source when V_{ds} is applied. Maximum potential position in p-channel DG MOSFET implies less negative potential which means less hole density at that point. Maximum potential position along the effective conductive path in p-channel DG MOSFET is used to formulate the threshold voltage.

Threshold voltage for p-channel symmetric DG MOSFET is defined as the gate voltage at which the inversion charge (hole) sheet density Q_{inv} at the maximum potential position reaches a value Q_{th_p} which is sufficiently enough to turn on the device. Fig 3 shows the threshold condition to turn the device on. The existing threshold voltage model for n-channel DG MOSFET [9] is analyzed accordingly to derive the same for p-channel symmetric DG MOSFET. The analytical threshold voltage equation for lightly doped p-channel symmetric

DG MOSFET is derived as:

$$V_{th_p} = V_{fb_p} + A_p V_T \ln \left(\frac{Q_{th_p} N_{si}}{n_i^2 t_{si}} \right)$$

$$- B_p \left[V_{bi_p} - V_T \ln \left(\frac{Q_{th_p} N_{si}}{n_i^2 t_{si}} \right) \right]^{\frac{1}{2}} \quad (3)$$

$$\left[V_{bi_p} + V_{ds} - V_T \ln \left(\frac{Q_{th_p} N_{si}}{n_i^2 t_{si}} \right) \right]^{\frac{1}{2}}$$

$$- C_p (2V_{bi_p} + V_{ds})$$

where

$$A_p = -\frac{e^{\frac{4L}{\lambda}} - 2e^{\frac{2L}{\lambda}} + 1}{\left(e^{\frac{L}{\lambda}} - 1 \right)^4}$$

$$B_p = \frac{2e^{\frac{L}{\lambda}} \left(1 + e^{\frac{L}{\lambda}} \right)}{\left(e^{\frac{L}{\lambda}} - 1 \right)^2}$$

$$C_p = -\frac{2e^{\frac{3L}{\lambda}} - 4e^{\frac{2L}{\lambda}} + 2e^{\frac{L}{\lambda}}}{\left(e^{\frac{L}{\lambda}} - 1 \right)^4}$$

$$\lambda = \lambda_{x=t_{si}/4} = \sqrt{\frac{\epsilon_{si} t_{ox} t_{si}}{2\epsilon_{ox}} \left(1 + \frac{3\epsilon_{ox} t_{si}}{16\epsilon_{si} t_{ox}} \right)}$$

Inversion charge sheet density at threshold condition Q_{th_p} is expressed as a function of L and λ

$$Q_{th_p} = 2.064 \left(\frac{L}{\lambda} \right)^2 - 22.46 \left(\frac{L}{\lambda} \right) + 64.06$$

III. PROPOSED SUBTHRESHOLD CURRENT MODEL

For operation in near-threshold region, the potential along the channel is modulated by inversion charge carriers and eq (1) is no longer valid [8]. It is general practice in DG MOSFET that to model the drain current for short channel devices, the core model for long channel device is augmented with short-channel effects. The reported paper [10] extended the core model for long channel DG MOSFET [7] to present analytical drain current model for short n-channel DG MOSFET. For long p-channel DG MOSFET, the 1D Poisson's equation considering only inversion charge carriers (holes) can be written as [15]:

$$\frac{d^2 \psi(x)}{dx^2} = -\frac{q}{\epsilon_{si}} \frac{n_i^2}{N_{si}} e^{\frac{-(\psi(x) - V)}{V_T}} \quad (4)$$

where V is the quasi-fermi potential of the holes. In general, the concept of quasi-fermi potential comes into picture when drain to source voltage V_{ds} is applied. The value of V ranges from 0 at source and V_{ds} at drain end [10]. Eq (4) is integrated twice with respect to x and obtained the solution for electrostatic potential distribution [4].

$$\psi(x) = V + V_T \ln \left[\left(\frac{q n_i^2}{2\epsilon_{si} V_T N_{si}} \right) \left(\frac{t_{si}}{2\beta} \right)^2 \cos^2 \left(\frac{2\beta}{t_{si}} x \right) \right] \quad (5)$$

where

$$\beta = \frac{t_{si}}{2} \sqrt{\frac{q n_i^2}{2\epsilon_{si} V_T N_{si}}} e^{\frac{-(\psi_0 - V)}{V_T}} \quad (6)$$

$\psi_0 = \psi(x = 0)$ is electrostatic potential distribution at the center of p-channel DG MOSFET. Hole current density

considering the gradient in quasi-fermi potential can be written as [16]:

$$J_p(x, y) = -q\mu_p \frac{n_i^2}{N_{si}} e^{\frac{-(\psi(x) - V)}{V_T}} \frac{dV}{dy} \quad (7)$$

Integrating eq (7) with respect to x, the drain current expression is obtained.

$$I_{ds} = \mu_p \frac{2W}{L} \frac{4\epsilon_{si} V_T}{t_{si}} \int_0^{V_{ds}} \beta \tan \beta \, dV \quad (8)$$

where μ_p is hole mobility, W is the channel width. The boundary condition at the $Si - SiO2$ interface is:

$$\left. \frac{d\psi(x)}{dx} \right|_{x=t_{si}/2} = \frac{\epsilon_{ox}}{\epsilon_{si}} \frac{V_{gs} - \Delta V_{th} - V_{fb_p} - \psi(t_{si}/2)}{t_{ox}} \quad (9)$$

where $\Delta V_{th} = V_{th_p}(L = 1\mu m) - V_{th_p}(L = 30nm)$ is the threshold voltage roll-off. Threshold voltage roll-off is considered in the eq (9) to introduce the short-channel effect. Differentiating eq (5) with respect to x and substituting in eq (9) yields an implicit expression for β.

$$\ln(\beta) - \ln(\cos \beta) + 2r\beta \tan \beta = -\frac{V_{gs} - \Delta V_{th} - V_{fb_p} - V}{2V_T}$$

$$- \ln \left[\left(\frac{2}{t_{si}} \right) \sqrt{\frac{2\epsilon_{si} V_T N_{si}}{q n_i^2}} \right] \quad (10)$$

where $r = (\epsilon_{si} t_{ox})/(\epsilon_{ox} t_{si})$, dV is found by differentiating (10) with respect to β.

$$dV = 2V_T \left[\frac{1}{\beta} + (2r + 1) \tan \beta + 2r\beta \sec^2 \beta \right] d\beta \quad (11)$$

Substituting eq (11) in eq (8), drain current expression for p-channel symmetric DG MOSFET incorporated with threshold voltage roll-off effect is obtained.

$$I_{ds} = \mu_p \left(\frac{2W}{L} \right) \left(\frac{8\epsilon_{si}}{t_{si}} \right) V_T^2 \left[\beta \tan \beta - \frac{\beta^2}{2} + r\beta^2 \tan^2 \beta \right]_{\beta_s}^{\beta_d} \quad (12)$$

where β_s and β_d are the values of β at source and drain end respectively. β values are calculated from eq (10) using the computation method given in [17]. Inclusion of surface roughness scattering replaces the hole mobility μ_p with the function [18].

$$1/\mu = 1/\mu_{ac} + 1/\mu_{sr} + 1/\mu_b \quad (13)$$

where μ_{ac} is the mobility limited by acoustic phonons, μ_{sr} is the mobility limited by surface roughness scattering and μ_b is the mobility of holes in silicon body. In order to get best fit with the simulation data, subthreshold-factor parameter ss has been introduced which is obtained through Newton's Forward Interpolation method. The subthreshold current model for p-channel symmetric DG MOSFET considering threshold voltage roll-off and surface roughness scattering is written as:

$$I_{ds} = \mu \left(\frac{2W}{L} \right) \left(\frac{8\epsilon_{si}}{t_{si}} \right) V_T^2$$

$$\times \left[\beta \tan \beta - \frac{\beta^2}{2} + r\beta^2 \tan^2 \beta \right]_{\beta_s}^{\beta_d} ss \quad (14)$$

where
$$ss = (3.29V_{gs}^4 - 12.5V_{gs}^3 - 17.67V_{gs}^2 - 7.14V_{gs} - 0.989)V_{ds} + (31.31V_{gs}^4 + 21.01V_{gs}^3 + 7.232V_{gs}^2 + 1.834V_{gs} + 0.239)$$

IV. RESULTS AND DISCUSSIONS

A structure of p-channel symmetric DG MOSFET is designed in device simulator Atlas (Silvaco) with parameters: body doping concentration $N_{si} = 10^{15}cm^{-3}$, source/drain doping concentration $N_{sd} = 10^{20}cm^{-3}$, oxide thickness $t_{ox} = 1nm$ and metal gate of work function $q\phi_m = 4.71eV$. Threshold voltage is extracted from simulation for different values of channel length L and silicon body thickness t_{si} satisfying the relation $L/t_{si} \geq$ 2-3 [8]. The proposed threshold voltage model results are found by considering parameters: thermal voltage $V_T = 0.0259V$ at $300K$, electron affinity of silicon $q\chi = 4.17eV$, bandgap of silicon $E_g = 1.08eV$.

Table I shows the comparison of the proposed threshold voltage model with the results obtained from simulation. Model results are found close to the simulation results. Maximum error (1.78%) is observed in case of $L = 45nm$ and $t_{si} = 22nm$ and least error (0.09%) is found for $L = 30nm$, $t_{si} = 12nm$. The proposed threshold voltage model can be applied to sub-30-nm p-channel symmetric DG MOSFETs.

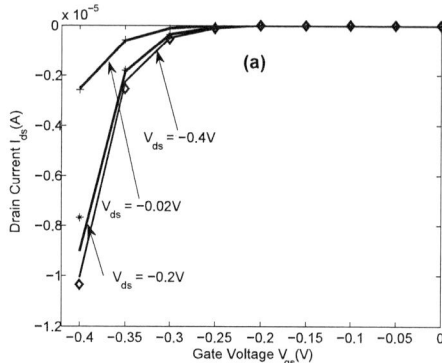

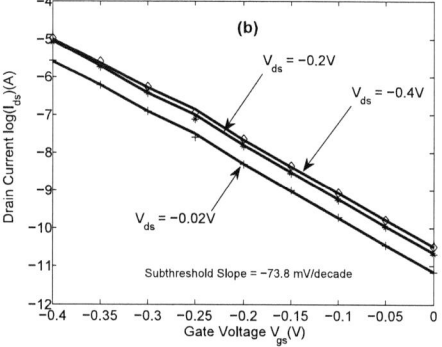

Fig. 4. Model and simulation result of subthreshold current characteristics of p-channel symmetric DG MOSFET with $L = 30nm$, $t_{ox} = 1nm$, $t_{si} = 12nm$ in (a) linear scale (b) semi-logarithmic scale. Simulation results represented by symbols, model results represented by solid lines

TABLE I
COMPARISON OF THRESHOLD VOLTAGE VALUES WITH THE SIMULATION RESULTS

Channel Length L	Body thicknes t_{si}	Model v_1	Simulation v_2	% Error $\left\lvert\frac{v_1-v_2}{v_2}\right\rvert \times 100$
22nm	10nm	−0.4087V	−0.4067V	0.2%
30nm	12nm	−0.4039V	−0.4030V	0.09%
38nm	15nm	−0.4045V	−0.4064V	0.19%
45nm	22nm	−0.3836V	−0.4014V	1.78%

Subthreshold current characteristics is obtained for the device with $L = 30nm$, $t_{si} = 12nm$, $W = 1\mu m$ at different V_{ds}. Fig 4 shows the subthreshold current characteristics in linear and semi-logarithmic scale. The model and simulation plots are found in good agreement. The extracted subthreshold slope from semi-logarithmic plot is found as $-73.8mV/decade$.

V. CONCLUSIONS

In this paper, threshold voltage model and subthreshold current model for short-channel lightly-doped p-channel symmetric DG MOSFET has been presented. The proposed threshold voltage model is suitable for sub-30-nm p-channel symmetric DG MOSFETs. Subthreshold current model has been developed considering the threshold voltage roll-off and surface roughness scattering effect. Best fit between subthreshold current model and simulation results have been obtained by introducing subthreshold-factor parameter. The proposed threshold voltage and subthreshold current model have been verified by comparing the results with simulation performed in Atlas. To derive analytical drain current expression for p-channel DG MOSFET, the strong inversion region of the device has to be modeled.

REFERENCES

[1] B. Razavi, "Design of analog CMOS integrated circuits," McGraw-Hill, 2001.

[2] H.S.P. Wong, "Beyond the conventional transistor," IBM Journal of Research and Development, vol. 46, no. 2/3, pp. 133–168, Mar 2002.

[3] N. Bhat, "Nanoelectronics Era: Novel Device Technologies Enabling Systems on Chips," Journal of the Indian Institute of Science, vol. 87, no. 1, pp. 61–74, Jan-Mar 2007.

[4] Y. Taur, "Analytical solution to a double-gate MOSFET with undoped body," IEEE Electron Device Letters, vol. 21, no. 5, pp. 245–247, May 2000.

[5] K. Kim, and J.G. Fossum, "Double-Gate CMOS: symmetrical -versus asymmetrical-gate devices," IEEE Transactions on Electron Devices, vol. 48, no. 2, pp. 294–299, Feb 2001.

[6] H. Lu, and Y. Taur, "An analytic potential model for symmetric and asymmetric DG MOSFETs," IEEE Transactions on Electron Devices, vol. 53, no. 5, pp. 1161–1168, May 2006.

[7] H. Lu, B. Yu, and Y. Taur, "A unified charge model for symmetric double-gate and surrounding-gate MOSFETs," Solid-State Electronics, vol. 52, no. 1, pp. 67–72, Jan 2008.

[8] A. Tsormpatzoglou, C.A. Dimitriadis, R. Clerc, Q. Rafhay, G. Pananakakis, and G. Ghibaudo, "Semi-analytical modeling of short-channel effcts in Si and Ge symmretrical Double-Gate MOSFETs," IEEE Transactions on Electron Devices, vol. 54, no. 8, pp. 1943–1952, Aug 2007.

[9] A. Tsormpatzoglou, C.A. Dimitriadis, R. Clerc, G. Pananakakis, and G. Ghibaudo, "Threshold voltage model for short-channel undoped symmetrical Double-Gate MOSFETs," IEEE Transactions on Electron Devices, vol. 55, no. 9, pp. 2512–2516, Sep 2008.

[10] A. Tsormpatzoglou, D. Tassis, C.A. Dimitriadis, G. Ghibaudo, G. Pananakakis, and N. Collaert, "Analytical modeling for current-voltage characteristics of undoped or lightly doped symmetric double-gate

MOSFETs," Microelectronic Engineering, vol. 87, no. 9, pp. 1764–1768, Nov 2010.

[11] K. Papathanasiou, C. Theodorou, A. Tsormpatzoglou, D. Tassis, C.A. Dimitriadis, M. Bucher and G. Ghibaudo, "Symmetrical unified compact model of short-channel double-gate MOSFETs," Solid-State Electronics, vol. 69, pp. 55–61, Mar 2012.

[12] J. Moers, S. Trellenkamp, A. Hart, M. Goryll, S. Mantl, P. Kordos and H. Luth "Vertical p-channel double-gate mosfets," 33rd Conference on European Solid-State Device Research, pp. 143–146, Sep 2003.

[13] J. Song, Y. Yuan, B. Yu, W. Xiong, and Y. Taur, "Compact modeling of experimental n- and p-channel FinFets," IEEE Transactions on Electron Devices, vol. 57, no. 6, pp. 1369–1374, Jun 2010.

[14] V. Kumari, A. Illango, M. Saxena, and M. Gupta, "Charge-based modeling of channel material-engineered p-type double-gate mosfets," IEEE 2nd International Conference on Emerging Electronics, pp. 1–4, Dec 2014.

[15] J. Song, "Compact modeling of experimental n- and p-channel FinFets," PhD dissertation, UCSD, San Diego, CA, 2010.

[16] Y. Taur, and T.H. Ning, "Fundamentals of modern vlsi devices," Cambridge University Press, 2009.

[17] B. Yu, H. Lu, L. Minijian, and Y. Taur, "Explicit continuous models for double-gate and surrounding gate mosfets," IEEE Transactions on Electron Devices, vol. 54, no. 10, pp. 2715–2722, Oct 2007.

[18] C. Lombardi, S. Manzini, A. Saporito, and M. Vanzi, "A physically based mobility model for numerical simulation of nonplanar devices," IEEE Transactions on Computer-Aided Design, vol. 7, no. 11, pp. 1164–1171, Nov 1988.

978-1-5386-1357-3/17 $31.00 © 2017 IEEE

Digital Video Stabilization- Review with a Perspective of Real Time Implementation

Mohd. Ahmed
Research Scholar, AISECT University

Abstract – **Installed Onboard cameras considering cases of off road unmanned navigating ground vehicles experience severe jitter and vibration. This leads to the prerequisite that the video images acquired from these platforms need to be heavily preprocessed to eliminate the jitter induced variations before human analysis. Digital Video stabilization system is the process of using electronic processing to control the image stability. That is, only software algorithms are used rather than hardware components such as motion sensors, actuators or floating lenses to compensate the disturbances. This makes digital stabilization more portable and cost effective among other methods. Digital stabilization can be used for real time and offline applications if the algorithms are optimized. This literature discusses the state of the art in the field of DVS with an implementation aspect of its use in challenging environment of unmanned ground vehicles where due to the dynamic nature of the vehicle, vibrations and oscillations are affect the camera resulting in a shaky and unstable video feed.**

Keywords—Video Stabilization; DVS; Motion Estimation; Motion Correction; Feature Extraction

I. INTRODUCTION

Off road navigating ground vehicles are a premier area of research with their applications ranging from both civil to military uses. The cameras onboard these type of vehicles experience severe jitter and vibration as a result of which the video images obtained from these platforms need to be subjected to heavy preprocessing so that the jitter induced effects can be completely or partially eliminated before human analysis. The foremost objective is to first detect the jitter and then to eliminate its effect on the obtained video. This in turn is composed of two subtasks, the first of which is to develop a reliable method in order to detect real-time the areas affected by jitters. Secondly, there is a need to develop a strategy to interpolate the images, without sacrificing important details (dismount targets). Different methods have been developed for different applications where stabilization is required. Optical video stabilization, Mechanical video stabilization, and Digital video stabilization are the three currently available methods that have been discussed in this literature. Each of these three methods has different motion estimation, motion correction and an associated block performing Image correction. These blocks in turn can be considered as the building blocks of a general video stabilization. Motion estimation is the process in which global motions in the frames of the video are obtained. On the other hand, motion correction is the process where intentional motions are extracted from obtained global

motions which are composed of intentional and unintentional motions. And, consequently, image correction is the process where stabilized video is produced using the estimated unintentional motions.

II. DIGITAL VIDEO STABILIZATION

Digital stabilization systems use completely electronic processing to control the image stability. That is, only software algorithms are used rather than hardware components such as motion sensors, actuators or floating lenses to compensate the disturbances. That's why digital stabilization is considered as more portable and cost effective among other listed methods. In digital stabilization, global motions in between frames are obtained by taking two adjacent frames of the video and performing a series of operations over them. Digital stabilization causes some distortions over the stabilized video. Since interpolation is utilized to correct the frames, sharp edges and high frequency details of the frames are lost. Furthermore, movements result in some content being lost in frames. In addition to visual degradations, computation cost is another weakness of digital stabilization. But digital stabilization can be used for real time applications if the algorithms are optimized. On the other hand, while other stabilization methods can be used for real time applications only, digital stabilization can be used for both real time and off-line applications. This is an advantage of digital stabilization.

Digital Stabilization can be considered as a combination of three main steps.

    a) Motion estimation
    b) Motion correction
    c) Image correction

These three steps can be thought as three independent steps and shown successively in the following figure

Fig 1: Video Stabilization Process

A. Motion Estimation:

For digital stabilization, motion estimation is the most time consuming part among all other processes. There are various types of digital motion estimation algorithms which have different types of theoretical backgrounds. But, all digital algorithms have considerable computational costs because of exhaustive image processing.

The motion Estimation is of five types in general:

1) Gradient techniques [1,2,3]
2) Pixel Recursive Techniques [4]
3) Block Matching Techniques [5,6]
4) Frequency Based Techniques [7]
5) Global Motion Estimation

All of them rely on a hypothesis that the image intensity I on the position is constant in the time period between two frames Δt and changes only due to displacement. Therefore

$$I(\vec{r}, t) = I(\vec{r} - \vec{d}, t - \Delta t) \qquad (1)$$

The displacement frame difference (DFD) is defined as

$$\text{DFD } (\vec{r}, t, \vec{d}) = I(\vec{r}, t) - I(\vec{r} - \vec{d}, t - \Delta t) \qquad (2)$$

A.1 Gradient Techniques:

For the image sequence analysis applications, gradient techniques are developed. Gradient techniques solve the optical flow constrain equation.

$$\vec{v} \cdot \vec{\nabla} I(\vec{r}, t) + \frac{\partial I(r,t)}{\partial t} = 0 \qquad (3)$$

for finding a motion vector ~v on the position ~r. For solving that, some additional constrains must be introduced. For example, a Horn-Schunck method [8] minimizes the square of the optical flow gradient magnitude

$$\left(\frac{\partial v_x}{\partial x}\right)^2 + \left(\frac{\partial v_x}{\partial y}\right)^2 \text{ and } \left(\frac{\partial v_y}{\partial x}\right)^2 + \left(\frac{\partial v_y}{\partial y}\right)^2 \qquad (4)$$

The main drawback of all the methods belonging to this group is a significant prediction error on moving objects boundaries caused by smoothness constraints and the fact that it they exhibit good accuracy only when dealing with dense motion fields.

A.2 :Pixel recursive Techniques

Pixel recursive although principally is a subset of gradient technique, in spite of that it is generally documented as separate group owing to the fact that its contribution to the field has been crucial. This group of recursively gradient techniques of prediction error minimization is known as Displaced Frame Difference (DFD) (29). An example of pixel recursive method is the Netravali-Robbins method [10], which iteratively updates the displacement vector according to the formula as per a given DFD

$$\vec{d}^{k+1} = \vec{d}^k - \in DFD(\vec{r}, t, \vec{d}^k) . \nabla_{\vec{r}} I(\vec{r} - \vec{d}, t - \Delta t) \qquad (5)$$

A.3: Block matching techniques

Block Matching Techniques work primarily on the minimization of a disparity measure. In other words, the functional procedure of block matching in current picture is matched with the block of previous picture and hence motion estimation is done. Block-matching technique minimizes a disparity between macroblocks in two frames

$$d = \arg\min_{\vec{d} \in S} \sum_{\vec{r} \in S} || I(\vec{r}, t) - I(\vec{r} - \vec{d}, t - \Delta t) \qquad (6)$$

It uses different cost functions ‖x‖ and search algorithms, which have been discussed further in [4]. Motion estimation through block-matching is an easy to implement mechanism and gives relatively good results, resulting in it being widely used in a video coding

A.4: Frequency Based Techniques

This group of motion estimation techniques relies heavily on the relationship between transforms from image domain to the frequency domain (e.g. Fourier or Gabor transform) and measures a correlation factor with different phase shift (which corresponds to a translation in the image domain)[7]

A.5 Global Motion Estimation

Global motion estimation [9,10] is the process of calculating the Global Motion Vectors (GMV) by estimating transformation between two adjacent frames of a video sequence. To calculate Global motion, featured base motion estimation is used in calculating the 2-D motion as a motion model. Local and Global features are the two broad classifications in which features can be grouped. Local features like points, edges, corners, faces, textures, and colors are generally tolerance to occlusion, but show relatively lower degrees of tolerance to noise[10].

B. Motion Correction:

The results of motion estimation part are the global motions between consecutive frames. Once motions are estimated, motion correction part distinguishes intentional and unintentional motions between each other. Since intentional movements such as panning have to be kept within the video, frames are stabilized using only unintentional motions. Like motion estimation, there are various algorithms used for motion correction too. Kalman filtering [11 - 16], fuzzy filtering [17 - 22] and lowpass filtering [23] are the most popular and widely used algorithms. In this review, all of these algorithms are examined. In addition to these algorithms, because of its basic implementation and suitable structure, moving average filtering [24] has also been reviewed.

B.1 Kalman Filtering

In video stabilization, Kalman filter is used to estimate global intentional movements of the camera from the estimated absolute frame positions. Since Kalman filters produce intentional motions, jitter on the video is obtained by subtracting the output of Kalman filter from the estimated absolute frame positions. As a result, stabilization is performed with respect to these obtained jitter estimation. Following 7 and 8 equations represent the state transition and observation equations of Kalman filter respectively for constant velocity model

$$\begin{bmatrix} x_t \\ m_t \end{bmatrix} = \begin{bmatrix} 1 & T \\ 0 & 1 \end{bmatrix} \begin{bmatrix} x_{t-1} \\ m_{t-1} \end{bmatrix} + [w]$$

$$[z_t] = [1 \quad 0]\begin{bmatrix} x_t \\ m_t \end{bmatrix} + [v] \qquad (7)$$

where x_t, m_t and z_t are estimated absolute frame position, velocity and measured absolute frame position of the frame respectively,

$$A = \begin{bmatrix} 1 & T \\ 0 & 1 \end{bmatrix}, H = [1 \quad 0] \qquad (8)$$

with a given time interval T as the interval between successive frames in the video. Since video stabilization aims to remove the jitter in x and y directions, above equations are applied both of measured absolute frame positions in x and y directions individually. The characteristics of Kalman filter can be adjusted by changing R and Q values. R value is the covariance of measurement noise and Q value is the covariance of process noise. Consequently, higher R values result to have more smooth estimations, whereas, higher Q values results to have estimations which are closer to the noisy measurements.

B.2 Fuzzy Filtering

Fuzzy filtering is a kind of filtering that uses fuzzy logic which classifies itself as a problem solving control system methodology. Fuzzy logic concept contains the following three main phases; fuzzification, fuzzy engine and defuzzification. Following figure illustrates the fuzzy concept schematically.

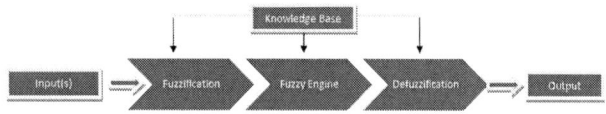

Figure 2: Fuzzy Correction System

Lets define a discrete time system to understand fuzzy filtering approach;

$$x_t = f(x_{t-1}) + w_t \qquad (9)$$

where f is a function that uses previous state to obtain the present state, w is zero mean white noise of the process and t represents the time dependency. Assume that states x_t are not measured directly. Instead, we can measure states z_t and there is a following relation between x_t and z_t

$$z_t = h(x_t) + v_t \qquad (10)$$

where h is a function and v is zero mean white measurement noise. To estimate states x_t with fuzzy filtering approach, another set of equation is defined using the recursive predictor-corrector architecture which is used commonly.

$$\hat{x}_t = f(\hat{x}_{t-1}) + g\ (z_t, \hat{x}_{t-1}) \qquad (11)$$

We can use above equation to estimate the jitter between video frames in a video sequence with fuzzy filtering approach. estimator structure. If there is a video sequence taken from a platform moving with constant velocity, we can estimate the states (which correspond to the exact frame positions) using the following 12 and 13 equations;

$$\hat{x}_t = \hat{x}_{t-1} + T\hat{v}_{t-1}$$
$$\hat{x}_t = \hat{x}_t + g(z_t, \hat{x}_t) \qquad (12,13)$$

where T is update period and v is rate of change estimation of frame motion speed;

$$\hat{v}_t = \frac{\hat{x}_t - \hat{x}_{t-1}}{T} \qquad (14)$$

Function g in Equation 12 is correction function of the system and can be defined by fuzzy logic.

B.3 Low Pass Filtering

In principal, a Lowpass filter passes low frequencies while attenuating high frequencies keeping in mind a predetermined cut off frequency. In video stabilization, desired motions like panning exhibit low frequency characteristics relative to unintentional and undesired motions. Therefore intentional motions can be extracted from the whole motion by lowpass filtering. Filter is applied to absolute frame position like Kalman and fuzzy filtering. The important point for lowpass filtering is to determine the cut off frequency. If cut off frequency is selected accurately considering the characteristics of the system, lowpass filter gives a reasonable performance on differentiation of the jitter. For lowpass filters, filter length (filter order) is another parameter that affects the stabilization performance.

B.4 Moving Average Filtering

One of the most predominant used filters is the Moving average filter, primarily because it is amongst the more easier ones to understand and use. Moving average filter remains an optimal solution to eliminate random noise while maintaining a sharp step response, alongside its simple design. Moving average filter is a kind of filter that replaces each value in a series with the average of its neighbourhood. Following equation realizes moving average filter

$$x_t = \frac{x_{t-N} + \cdots x_t + \cdots + x_{t+M}}{M + N + 1} \qquad (15)$$

where x_t is the state at time t, N representing number of previous neigbouring states and M is the number of future neighbouring states. Moving average filter takes the estimated absolute frame positions as the inputs and smooths them to produce intentional motions. Then, jitter is obtained by subtracting the intentional motions from the estimated absolute frame positions. Following formula shows the calculation of intentional motions.

$$xx_t = \frac{x_{t-N} + x_{t-N+1} + \cdots x_t}{N+1} \qquad (16)$$

where x_t are the estimated absolute frame positions with respect to reference image, xx_t are the intentional motions of the camera and N is the filter length. Since video stabilization aims to remove the jitter in x and y directions, above equation is applied to both directions individually.

C. Image Correction

Image correction is the finishing step in video stabilization. Implementation of this stage of image correction changes in accordance with the utilized video stabilization methods. That is, if we are using mechanical or optical video stabilization, image correction is done by motors and mechanical structures. However, in case of digital video stabilization, image correction is achieved through software methods resulting in digital video stabilization being the most cost effective amongst all documented methods. Digital image correction part takes the frames of the video and aligns them by shifting and rotating with respect to the output of motion correction part. There are various interpolation techniques each of which has different accuracy and computational cost. Nearest neigbourhood, bilinear, and bicubic interpolation are most commonly used interpolation techniques in the literature. Even if bicubic interpolation has the best accuracy, it has considerable computational load. On the other hand, nearest neighbourhood technique is the fastest algorithm. But it has not enough accuracy

III. LITERATURE SURVEY-PAST AND PRESENT

As already discussed, the task of video stabilization can be divided into three premier tasks of estimating the motion, leading up to correction and then finally end up correcting the Image. Hence different works in the field can be classified on the basis of the different techniques used for implementing or improving each of the three involved steps. Works have also been classified on the basis of their implementation scenario viz. Real Time implementation, Offline Implementation, application in Unmanned Vehicles etc.

A. Motion Estimation approaches:

The first issue depends on the strategy selection of motion estimation, which is crucial for fast solution and real-time application as our work primarily concentrates on the real time applications. Liang [25] in his literature used a feature based motion estimation mechanism. Lane lines and the road vanishing point were considered as global features and were extracted from the input images. Broggi [26], extracted the horizontal edges and used the obtained histogram as the specific feature by taking into account the effectiveness of automotive application. In [27], the authors applied Diamond search (DS) to get the estimation of the motion vector, which was found to have a performance to the extent of full search (FS). The advantage of DS is that it immensely reduces the complexity of the computation of the sum of absolute difference (SAD) [28] between the referenced frame and the frame under consideration. Hooke Jeeves (HJ) pattern search algorithm [29] was made use of in order to fasten up the search. Shen [30] explained a digital video stabilization which differed from the previous work in the sense that PCA-SIFT (Principal Component Analysis- Scale Invariant Feature Transform) method was used for the purpose of extracting features. In [31], Harris point detection was used with improvements through grid sampling to select evenly distributed points. The points were initially matched by feature window matching and further validated by distance criterion to get global motion matrix. Then particle filter was employed to obtain smooth scanning motion vector from original motion curves. In [32], Corner points were extracted using Good Features to Track corner detection algorithm and the extracted points were then utilised to compute the optical flow between adjacent frames. From that, the points detected from optical flow were then used to estimate the motion parameters using an affine transform model. In [33], the authors discussed a video stabilization technique based on Gray coded bit plane (GCBP) matching was discussed for translational motion analysis. This technique performed fast motion estimation using GCBP of image sequences which worked in reducing the computational load. Another unique realtime approach was discussed and presented in [34]. This research was concerning a learning based camera motion characterization scheme for alleviation of the video stabilization problem. The authors proposed a characterization scheme representing the compressed domain block motion vectors utilizing polar angles and magnitude histograms. Distinguishing features from the obtained histograms were extracted and fed to a supervised learning based hierarchical classifier that aided in recognizing the six camera motion patterns. In [35] a gray based algorithm was suggested which can compensate the translation and rotation motion of the image. However it added extra complexity since there is no need to consider the rotation motion in most of the cases of offroad vehicles apart from certain airborne applications. In this, a given window was used to let the regional field images pass. Then the odd and even field images were projected onto x and y axes to obtain the odd and even field regional gray projection curves at the same time. Once regional gray projections were obtained, the motion or mainly the transition between two consecutive odd field images was estimated using the minimum absolute difference. C.Wang [37] worked out a digital image stabilization mechanism dependent on feature point tracking and using Kanade Lucas Tomasi tracker. The authors in this research made use of feature points to estimate the trajectory between two subsequent image frames and consequently leading to motion estimation. Another work of relevance was done in [38], using Speeded Up Robust Features (SURF) descriptor. SURF is considered above the Scale Invariant Feature Transform (SIFT) approach [7] because it is computationally much lesser extensive. It involves interest point description. Interest points are based on Hessian calculations and it has been observed that Hessian using approaches have been found to be more stable and repeatable than those making use of corner detection. A vastly different approach was used in [40]. This study proposed a unique digital video stabilisation scheme based on modeling of motion imaging (MI). The biggest advantage of the method was that modeling of MI eliminated the speed motion as a result of a moving car, which was ignored in other models such as rotation as well as translation model, and estimated movement parameters of the background in video sequences captured from cameras mounted on moving cars.

B. Motion Correction methods

Kalman filter and similar controllers like PID and MVI are the prevalent motion compensating methods. However, it's difficult in these methods to preserve the panning motion faithfully with minimum time lag. Henceforth, to solve the problem, [27] proposed a modified Kalman filter which can variably adapt the parameters according to the chosen scene of camera motion. In order to solve the above problem, the authors proposed an altered Kalman filter which showed good effect in the case of motion exhibiting both panning and. The system was implemented as three one order Kalman filters with single sensor that were applied on each dimension of PGMV, the accumulated global vector, GMV. Inter frame GMV was estimated by the least square estimate through at least three pairs of motion parameters obtained by the combined DS and Hooke Jeeves motion estimation method which have been used in other methods and have been discussed before in this literature. In [30,31] particle filter was detailed to reduce the effect of jitter with low error variance, which is crucial in video stabilization. The usage of particle filter has its advantages in the sense that it has characteristics of good convergence, smoothness and robustness. Like [27], [39] also followed the Kalman filter approach but modified it for strictly for realtime applications. The proposed Kalman filter and mosaic algorithm enabled the development of a somewhat practical real time video stabilizer that produced steady video but also tried retaining the full resolution of the actual video.

C. Classification on the basis of applications:

Many works in Video Stabilization have dealt with Real-time video stabilization for unmanned aerial vehicles requiring different approaches considering lack of computational resources and for them to be applied on board UAV systems in real time [27]. However, Most methods have revolved around finding the 2D motion model to estimate the global motion path. Then the trajectory is low pass filtered to eliminate the high frequency jitter component. The low frequency parameters are then warped onto frames. This framework has been found to be really effective for scenes with very little dynamic movement which is very well to the case of UAV aerial videos. However, it is difficult to achieve instantaneous stabilization, as required for real-time processing only predicting the global motion trajectory using a given window of frames. For this problem, in [30] a video stabilization algorithm was proposed using a circular block to search and match key places. The computed affine transform was finally smoothened by numerical method of polynomial fitting and prediction method (PFPM). In Vazquez and Chang [41], a smoothening method made use of the Kanade Lucas tracker which helped in the detection of interest points. The undesired motion was compensated by making adjustments for additional rotation and displacements that the vibrations have caused. This approach was found to be able to achieve a stabilizing speed in the range of 20fps to 28fps for images with resolution in order of 320x240 pixels The implemented system was stated to be running on MAC using a 2.16GHz

Processor clock and a three frame delay. Wang [42], afterwards proposed a video stabilization method specifically designed for UAVs. In this literature, he proposed a 3 step mechanism using a FAST corner detector to locate the feature points existing in frames. In the second step, after matching, the key points were then used for estimation of affine transform to seperate false matches. In the third and final step, motion estimation was performed based on the particular model and the compensation for induced vibration was conducted on the basis of spline smoothening. [36] also used an actual video obtained from UAV as input and then SIFT points were extracted and matched for following frames. Another similar work using architecture for Harris corner detection has been presented in [43] and [44]. Another recent hardware directed solution was in [45] where the authors use a texture analysis-based strategy.

IV. CONCLUSION

Video stabilization has been a topic of active research for a while now and a number of new methodologies have been proposed bringing about changes in the mechanisms of Motion estimation, extraction of feature vectors, Better block matching techniques or better Motion correction methods with improvements to the existing motion correction filters like Kalman Filter. However, still a tradeoff exists between their implementations considering the implementation aspects as most of the methods are computationally intensive thereby limiting their use in realtime applications which is the active area of research nowadays. Applications in challenging environments like unmanned vehicle, UAV's require strict realtime applications as the payload cannot bear the computation cost of intensive processing and will fail to perform the operations in realtime which is the prime requirement of such environments.

References

Papers that have not been published, have been cited as "unpublished"

[1] Masatoshi Hino, Nozomu Hamada, "Gradient-based motion vector estimation using variable block shape" 11th European Signal Processing Conference, 2002, IEEE

[2] Toshiaki Kondo, Pramuk Boonsieng, Waree Kongprawechnon, "Improved gradient-based methods for motion estimation in image sequences", SICE Annual Conference, 2008, IEEE

[3] F. Dufaux, F. Moscheni. Motion estimation techniques for digital TV: a review and a new contribution. Proceedings of the IEEE, 83(6):858–876, 1995.

[4] J. Huang , S. Liu ; M.H. Hayes, R.M. Mersereau, "A multi-frame pel-recursive algorithm for varying frame-to-frame displacement estimation," IEEE International Conference on Acoustics, Speech, and Signal Processing, 1992. ICASSP-92., 1992.

[5] Shan Zhu; Kai-Kuang Ma, "A new diamond search algorithm for fast block-matching motion estimation," IEEE Transactions on Image Processing Year: 2000, Volume: 9, Issue: 2, Pages: 287 - 290.

[6] A. V. Paramkusam, "Efficient motion estimation algorithm on the layers," Electronics Letters, Year: 2017, Volume: 53, Issue: 7, Pages: 467 - 469,.

[7] Meng-Chun Lin ; Lan-Rong Dung, "Two-step windowing technique for wide range motion estimation," . IEEE Asia Pacific Conference on Circuits and Systems, 2008.

[8] Osama A. Omer, "Region-based Horn-Schunck optical flow estimation" Japan-Egypt Conference on Electronics, Communications and Computers(JEC-ECC),2012

[9] Yangke Liu, De Jifu, Bo Li, Qizhi Xu, "Region Real-Time Global Motion Vectors Estimation Based on Phase Correlation and Gray Projection Algorithm" 2nd International Congress on Image and Signal Processing, 2009. CISP '09.

[10] Yang Zhang, Yuquan Leng, Xu He, "A Fast Video Stabilization Algorithm with Unexpected Motion Prediction Strategy" 2015 IEEE International Conference on Advanced Intelligent Mechatronics (AIM) , 2015. Busan, Korea

[11] S. Ertürk, "Real-time digital image stabilization using Kalman filters", Real-Time Imaging, vol. 8, no. 4, pp. 317–328, 2002

[12] E. Yaman and S. Ertürk, "Image stabilization by Kalman filtering using a constant velocity camera model with adaptive process noise", International Conference on Electrical and Electronics Engineering, ELECO2001, Bursa, vol., pp.152-157, 2001

[13] G. Welch, G. Bishop, "An Introduction to the Kalman Filter", Transactions of the ASME - Journal of Basic Engineering, 82 (Series D), pp. 35-45

[14] S. Ertürk, "Image sequence stabilisation based on Kalman filtering of frame positions", Electronics Letters, 37, (20), pp. 1217-1219, 2001

[15] O. Kwon, J. Shin, J. K. Paik, "Video Stabilization Using Kalman Filter and Phase Correlation Matching", ICIAR 2005, pp. 141-148

[16] R. E. Kalman, "A New Approach to Linear Filtering and Prediction Problems," Transaction of the ASME - Journal of Basic Engineering, pp. 35-45, March 1960

[17] M. K. Güllü, S. Ertürk, "Image Sequence Stabilization using Membership Selective Fuzzy Filtering", Lecture Notes in Computer Science, Springer Verlag, 2869, pp. 497-504, 2003

[18] N. R. Hughes, G. N. Roberts, G. R. Wilson, "Application of Fuzzy Signal Processing to Three Dimensional Vision", 5th International Conference on FACTORY 2000, conference publication no: 435, April 2-4, 1997

[19] M. K. Güllü, E. Yaman and S. Ertürk, "Image sequence stabilisation using fuzzy adaptive Kalman filtering", Electronics Letters, 39, (5), pp. 429-431, 2003

[20] M. K. Güllü, S. Ertürk, "Fuzzy Image Sequence Stabilisation", Electronics Letters, 39, (16), pp. 1170-1172, 2003

[21] N. Kyriakoulis and A. Gasteratos, "A Recursive Fuzzy System for Efficient Digital Image Stabilization", Advances in Fuzzy Systems, vol. 2008, article id: 920615

[22] M. K. Güllü and S. Ertürk, "Membership function adaptive fuzzy filter for image sequence stabilization," IEEE Transactions on Consumer Electronics, vol. 50, no. 1, pp. 1-7, 2004

[23] S. Ertürk and T. J. Dennis, "Image sequence stabilisation based on DFT filtering", IEE-Proc., Vis. Image Signal Process., 147, (2), pp. 95-102, 2000

[24] J. A. Ramirez, E. Rodriguez, J. C. Echeverria, "Detrending fluctuation analysis based on moving average filtering", Physica, A 354, pp. 199-219, 2005

[25] Y.M. Liang, H.R. Tyan, S.L. Chang, H.Y.M. Liao, and S.W. Chen:"Video stabilization for a camcorder mounted on a moving vehicle," IEEE T Veh Technol, 53, (6), pp. 1636-1648, 2004

[26] A. Broggi, P. Grisleri, T. Graf, and M. Meinecke: 'A software video stabilization system for automotive oriented applications' 'Book A software video stabilization system for automotive oriented applications' (2005, edn.), pp. 2760-2764

[27] Jifei Song, Xiaohong Ma "A Novel Real-Time Digital Video Stabilization Algorithm Based on the Improved Diamond Search and

Modified Kalman Filter", IEEE 7th International Conference on Awareness Science and Technology (iCAST), Sept. 2015

[28] V. Filippo, C. Alfio, M. Massimo, G. Messina, "Digital image stabilization by adaptive block motion vectors filtering," Consumer Electronics, IEEE Transaction on, vol. 48, no. 3, pp. 796–800, Aug., 2002

[29] R. Hooke, T Jeeves, "Direct search solutions of numerical and statistical problems," Association for Computing Machinery, vol. 8, no. 2, pp. 212–229, 1961

[30] Y. Shen, P. Guturu, T. Damarla, B. P. Buckles and K. R. Namuduri, Video Stabilization Using Principal Component Analysis and Scale Invariant Feature Transform in Particle Filter Framework, IEEE Transactions on Consumer Electronics, Vol. 55, No. 3, pp. 1714-1721, 2009

[31] Juanjuan Zhu, Cheng Li, Jinli Xu, "Digital Image Stabilization for Cameras on Moving Platform" International Conference on Intelligent Information Hiding and Multimedia Signal Processing, 2015

[32] Lakshya Kejriwala, Indu Singh, "A Hybrid filtering approach of Digital Video Stabilization for UAV using Kalman and Low Pass filter"Pg 359-366, Procedia 6th International Conference On Advances In Computing & Communications, ICACC, 2016, Cochin, India, Elsevier

[33] L. Garlin Delphina, VPS Naidu, "Implementation and Validation of Video Stabilization using Simulink" National Aerospace Laboratories, Published in Researchgate

[34] Manish Okade , Gaurav Patel , Prabir Kumar Biswas, "Robust Learning-Based Camera Motion Characterization Scheme With Applications to Video Stabilization" IEEE Transactions on Circuits and Systems for Video Technology (Volume: 26, Issue: 3, March 2016) Page(s): 453 - 466

[35] W. Yang, Z. Zhang, Y. Zhang, et al. A real-time gray projection algorithm for electronic image stabilization. Proc. of the International Symposium on Photoelectronic Detection and Imaging: Advances in Infrared Imaging and Applications, 2011

[36] Ahlem Walha, Ali Wali, Adel M. Alimi, "Video Stabilization for Aerial Video Surveillance" AASRI Conference on Intelligent Systems and Control, AASRI Procedia, Pg 72-77, Elsevier, 2013

[37] Chuntao Wang, Jin-Hyung Kim, Keun-Yung Byun, Jiangqun Ni, and Sung-Jea Ko, Robust Digital Image Stabilization Using the Kalman Filter, IEEE Transactions on Consumer Electronics, Vol. 55, No. 1, pp 6-14, February 2009

[38] Tahiyah Nou Shene, K. Sridharan, N. Sudha, "Real-Time SURF-Based Video Stabilization System for an FPGA-Driven Mobile Robot" IEEE Transactions On Industrial Electronics, 2017 UNPUBLISHED

[39] Jing Dong, Haibo Liu, "Video Stabilization for Strict Real-time Applications" IEEE Transactions on Circuits and Systems for Video Technology,2017,UNPUBLISHED

[40] Jialin Yu, Ke Xiang, Xuanyin Wang, Songxiao Cao, Yang Zhang, "Video stabilisation based on modelling of motion imaging" IET Image Processing (Volume: 10, Issue: 3, 3 2016) Page(s): 177   188

[41] Vazquez M, Chang C (2009) Real-time video smoothing for small rc helicopters. In: Systems, Man and Cybernetics, 2009. SMC 2009. IEEE International Conference on, IEEE, pp 4019–4024

[42] Wang Y, Hou Z, Leman K, Chang R (2011) Real-time video stabilization for unmanned aerial vehicles. In: MVA, pp 336–339

[43] P. Hsuao, C. Lu, and L. Fu, "Multilayered image processing for multiscale Harris corner detection in digital," IEEE Trans. Ind. Electron.,vol. 57, no. 5, pp. 1799–1805, 2010

[44] T. Tsai, C. Fang, and H. Chuang, "Design and implementation of efficient video stabilization engine using maximum a posteriori estimation and motion smoothing approach," IEEE Trans. Circuits Syst. Video Technol., vol. 22, no. 6, pp. 817–830, 2012.

[45] I. Ozsarac and I. Ulusoy, "Real time FPGA implementation of full search video stabilization method," in Proc. 2012 20th IEEE Signal Proc. Commun. App. Conf., 2012, pp. 1–4.

2017 IEEE International Symposium on Nanoelectronic and Information Systems

Gate metal work function engineering for the improvement of electrostatic behaviour of doped tunnel field effect transistor

Deepak Soni, Dheeraj Sharma, Shivendra Yadav, Mohd. Aslam, Dharmendra Singh Yadav, and Neeraj Sharma

Nanoelectronics and VLSI Laboratory ,Electronics and Communication Engineering Discipline
PDPM- Indian Institute of Information Technology, Jabalpur, 482005, India. and
Department of Computer Engineering
Ramrao Adik Institute of Technology Nerul, Navi Mumbai, 400706, India.

Email: : deepaksoni@iiitdmj.ac.in, dheeraj@iiitdmj.ac.in , shivendra1307@gmail.com , mohd.aslam22d@gmail.com ,
tech.dharmendra26@gmail.com and neeraj16ks@gmail.com

Abstract— This paper reports a new configuration of tunnel field-effect transistor (TFET) for improving the current drivability of device, reduced threshold voltage, suppress ambipolar behaviour and better high-frequency response of the device. For this, a P+-I-N+ type structure has been considered, and then gate electrode is divided into three parts which is placed over the channel region. Low work function metal is applied at the source/channel junction side to increase the steepness in the band at source/channel interface, which increases electron tunnelling rate, that results in higher ON-state current and reduction in threshold voltage. At the same time, low work function metal electrode is applied at the drain/channel junction side, which is helpful for the suppression of ambipolar behaviour of the device. The simulated results show excellent improvement in ON-state current, lowering of the threshold voltage, suppression in ambipolar behaviour and drastic improvement in analog/RF parameters of the device. To show the impact of proposed modification and to analyse the device behaviour, parameters such as transfer characteristics and energy band diagram are considered as DC figure of merits, whereas transconductance (gm), output onductance (gds), gate-to-drain capacitance (Cgd)), cut off frequency (fT) and gain bandwidth product (GBP) are used as RF performance parameters. All the simulations have been performed by Silvaco ATLAS simulator.

Keywords—Band to band tunneling; Hetero gate oxide; Hetero junction; cut off frequency; Transconductance.

I. INTRODUCTION

In recent decades, the metal oxide semiconductor field effect transistor (MOSFET) has been scaled down aggressively to meet the demands of the market for providing electronic equipment of, low cost, high speed, and better Analog/RF performance. This continuous downscaling has brought fundamental physical limitations, such as sub-threshold slope ($SS \geq 60mV/dec$) at room temperature, short-channel effect, and leakage currents [1]-[3]. These are very critical problems which are associated with power crisis [3]. Since MOSFET is based on drift-diffusion phenomenon, the use of different materials, gate engineering, and other techniques does not further lower the existing limit of subthreshold slope (60 mV/dec). Thus, the research is being carried out to find new

alternative devices. Tunnel field-effect transistor (TFET) is the most suitable device for the ultra low power digital and analog circuits [4]-[8]. It offers the very low leakage current (1015 A/m order) [3], [9]-[11], steeper odulation between ON and OFF state, low subthreshold slope (<60 mV/dec) [3], [12]-[13], scalability to 10 nm without degrading leakage current due to the band to band tunnelling nature of carrier transport [17]. Therefore, it provides better immunity towards the short channel effects (SCEs), low power dissipation (static and dynamic) [14]-[15]. However, with all these potentials, it faces the serious concerns of ambipolar conduction [16] and low ON-state current [17]-[18]. Variation in doping concentration is helpful techniques for suppressing ambipolar behaviour and improving ON-state current, but the selection of lower drain doping is crucial for increasing the lateral tunneling distance at the drain/channel interface which results into suppressed ambipolar behaviour [19], [20]. Whereas, higher source doping is helpful for creating steeper source/channel tunneling junction which results in the higher band to band tunneling generation rate that leads to high ON-state current. Although, silicon cannot be doped over the 1×1020 cm-3, since, increasing the doping level above this does not increase the carrier concentration [21]. In this concern, our proposal presented in this paper is a helpful approach to creating steeper band profile at source/channel interface.

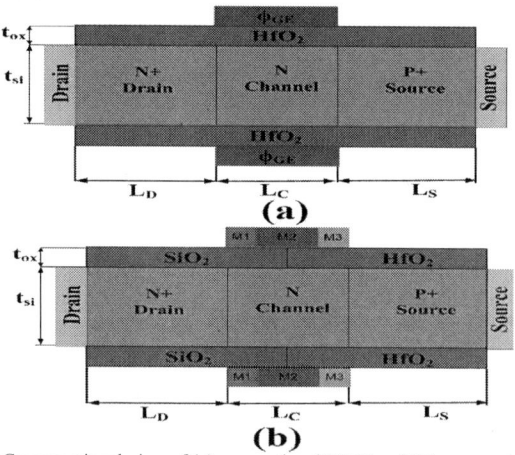

Fig.1. Cross sectional view of (a) conventional TFET and (b) proposed TFET.

978-1-5386-1357-3/17 $31.00 © 2017 IEEE

Parameters	Unit	Conventional	Proposed
Source doping	cm-3	10^{20}	10^{20}
Drain doping	cm-3	10^{20}	10^{20}
channel doping	cm-3	10^{17}	10^{17}
Drian length (L_G)	nm	100	100
Source length (L_S)	nm	100	100
Channel length (L_C)	nm	50	50
Silicon thickness (t_{si})	nm	10	10
Gate electrode workfunction (φ_{GE})	eV	4.53	-
Gate electrode (M1) workfunction (φ_1)	eV	-	3.9
Gate electrode (M2) workfunction (φ_2)	eV	-	4.53
Gate electrode (M3) workfunction (φ_3)	eV	-	3.9
Oxide thickness (t_{ox})	nm	2	2

To resolve the above-mentioned issues, in this article gate engineered doped TFET has been resented. In the proposed device the gate electrode is divided into three parts, namely, tunneling gate (M1), gate (M2), and gate (M3), with their work, functions $\varphi 1$, $\varphi 2$, and $\varphi 3$, respectively were ($\varphi 1 = \varphi 3 < \varphi 2$). Low work-function $\varphi 1$ (low $\varphi 1 < \varphi 2$) is used at source side to create abruptness and improving e-tunneling rate at the source/channel junction, which results in to higher drain current in on state, improved high frequency response and better linearity as compare to conventional doped TFET, whereas low work function ($\varphi 3 < \varphi 2$) at drain side is applied for the suppression of ambipolar current and improving RF performance. Along with this, low-k dielectric material (SiO2) is applied at the drain side which is also helpful for improving ambipolar behaviour and high-frequency figure of merits. Further paper is arranged as follows: section 2 presents the device architecture and simulation parameters, section 3 DC and Analog/RF performance of both devices in comparative manner. Finally, the conclusion is summarized in section 4.

II. SIMULATION SETUP AND DEVICE DESIGN

Fig.1 (a) Conventional DG-TFET (b) Hetero-Junction DG-TFET (proposed device).

The cross-sectional view of the devices is shown in Fig.1 (a-b). Fig. 1 (a) shows the conventional TFET and Fig. 1 (b) shows proposed device which is similar as conventional except the gate contact is divided into three parts which are having different work function $\varphi 1$, $\varphi 2$ and $\varphi 3$ for gate electrode M1, M2 and M3 espectively and low-k dielectric(Sio2) material is applied at he drain side. The other physical parameters of the device used for simulation are listed in Table 1. The 2D-ATLAS, Silvaco [22] has been used for simulation. Since, the TFET follows band-to-band tunneling mechanism so given that Non-local BTBT model is used. Besides this, a band gap narrowing model with some

other physical models; like, concentration dependent, SRH recombination model and Fermi-Dirac statistics are also used.

III. DC CHARACTERISTICS & RF PARAMETERS

Fig. 2 (a) represents the electric field of the devices in the thermal state and it can observe that electric field is higher at the source/channel interface and lower at drain/channel interface in case of proposed device, which is the effect of proposed modification at the source/drain side of the gate electrode. The Higher electric field is helpful for achieving higher tunneling rate which turns in to higher ON-state current at the same time lower electric field at drain/channel junction is helpful for preventing tunneling of charges in the ambipolar state. In addition to this low work function metal at the source, side pushes the band downward, reduces the barrier width and creates abruptness at the source/channel junction as shown Fig. 2 (b).

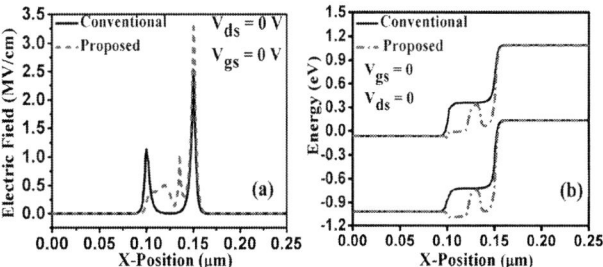

Fig.2 (a) Electric field and (b) Energy Band Diagram under Thermal Equilibrium.

Further, in ON-state of the device as Vgs increases it pushes band downward in the source region and increases tunneling area and abruptness so electron tunnel from valance band of the source to the conduction band of the channel region. As depicted in Fig. 3 (a). Due to higher tunneling rate of electrons at the source channel junction electron current density is higher for the proposed device which is illustrated in Fig. 3 (b). Furthermore, due to higher electric filed and reduced barrier width at the source/channel junction higher electron tunneling rate is achieved by proposed device as a result higher ON-state current is delivered along with this threshold voltage calculated by constant current method [23] and it is found proposed device has 250 mV less threshold voltage as compare to conventional device, which is helpful to scale down the power supply as depicted in Fig.4 (a). Moreover, the output characteristic of the device shown in Fig. 4 (b) and it can infer that drain current is higher for the proposed device as we vary Vds drain current increases and saturates at 0.75 V; therefore, device is useful for designing low power circuit application.

Fig. 5 (a) shows the behavior of transconductance (gm) at with Vgs it is expressed as $g_m = \dfrac{\partial I_{ds}}{\partial V_{gs}}$ The mutual conductance commonly term as transconductance is defined as the partial differentiation of the transfer characteristics. gm of the device shows the sensitivity (change in drain current)of the device against the variation in gate voltage. For the proposed device transconductance is higher as compare to the conventional device so it can state that proposed device has higher sensitivity. Higher transconductance is helpful for improving the DC performance as well as a high-frequency performance of the device. Along with this, Fig. 5 (b) shows the output transconductance (gds) of the device it is reciprocal of output resistance and can express as $gds = change\ in\ Ids\ /\ change\ in\ Vds$.

In a case of proposed device gds is higher, initially, for the lower value of drain voltage gds increases and after a peak value it starts to fall due to saturation in drain current. Furthermore, the most significant parameter which decides the radio-frequency (RF) performance of the device is the gate to drain capacitance. (Cds) is lower for the proposed device due to proposed modification at the drain side (low work function metal gate and low-k dielectric) low work function metal increases the depletion width at the drain/channel junction and low-dielectric decouple the terminals, therefore, resultant gate to drain capacitance decreases with increase in Vgs as illustrated in Fig. 6 (a).

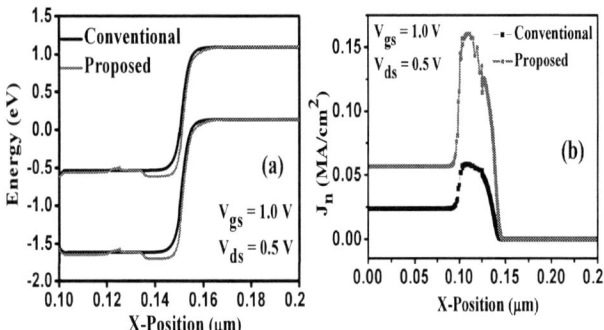

Fig.3 (a) Energy Band Diagram (b) electron current density under ON-State.

Cut-off frequency (fT) is the important parameters for the analysis of high frequency responses. fT can be defined as the operating frequency of device at which short circuit current gain decreases up to unity. Expression for cut-off frequency (fT) can be expressed as:

$$f_T = \dfrac{g_m}{2\pi(C_{gd} + C_{gs})}$$

Fig. 6 (b) shows the variation in cut-off frequency (fT) as a function of V gs. It can observe that for lower gate voltages fT increases due to increment in gm, if we further increase the

gate voltage fT increases up to a peak and then start to fall. This decrement in fT is due to the mutual influence, increased cgd and the reduction in transconductance (gm). The transconductance reduces with increase in voltage as a result of charge carrier mobility degradation. gm is high for the

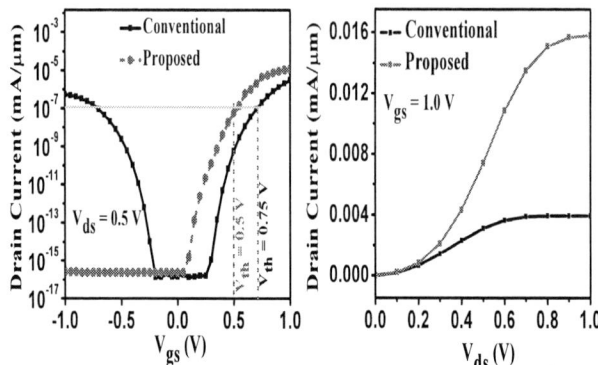

Fig.4 (a) Transfer Characteristics I$_{ds}$ - V$_{gs}$ and output characteristics

Proposed device due to consideration of low work function metal at the source side and cgd is lower for proposed device due to low-k dielectric and low work function metal gate at the drain side. Thus, collectively both parameters improve the cut-off-frequency of proposed device. Gain bandwidth product (GBP) is also one of the mostsignificant parameters for evaluating RF performance of the device, and it is a trade-off between two parameters which is gain and bandwidth. GBP is computed as GBP = (gm)/ ($20\pi cgd$) with the help of TCAD simulations. It observed thatGBP for proposed device is 2.66 times greater than conventional as shown in Fig. 7. The reason for the improvement in GBP is same which is explained earlier for fT , because of both the dependent parameters (GBP and fT) have the same relation with same variable parameters, i.e., with gm and cgd.

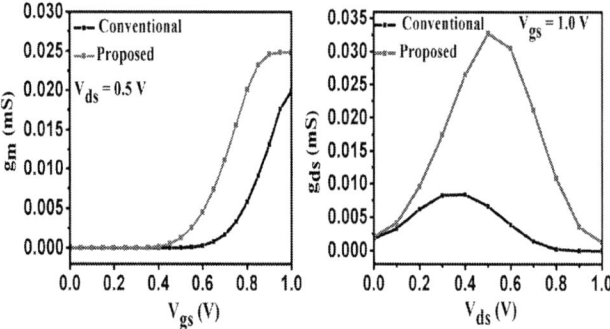

Fig.5 Variation in (a) gm and (b) gds with Vgs

978-1-5386-1357-3/17 $31.00 © 2017 IEEE   192

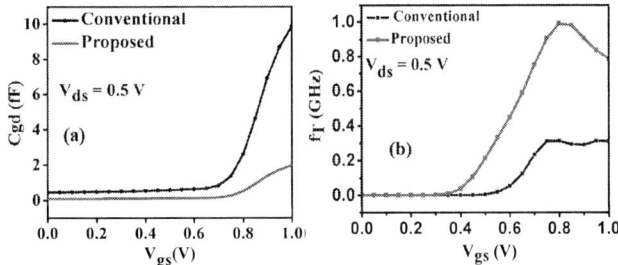

Fig.6 Variation in (a) Cgd and (b) fT with Vgs .

$$f_T = \frac{g_m}{2\pi(C_{gd} + C_{gs})} \quad \dots\dots\dots\dots (3)$$

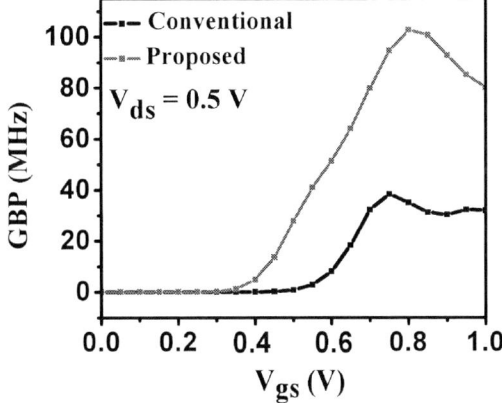

Fig. 7. Variation in GBP with Vgs.

IV. CONCLUSION

In this given paper, we have proposed a metal work function based approach for enhancing the ON-state current with the reduction in threshold voltage. Where low work function metal is used at the source side is helpful for generating higher electric field and abruptness at source/channel junction it turns in to higher ON-state current and low threshold voltage. Further, a modification which is low work function metal along with low-K dielectric material at the drain side widening the drain/channel junction barrier and prevent to charge carrier for tunneling. This technique suppresses the ambipolar current up to off state current in the ambipolar state. Apart from this proposed modification at the drain side is helpful for improving the RF performance of the device. Therefore, proposeddevice seems to be useful for implementing the low power circuit design.

Acknowledgements

☐
☐The authors would like to thank the Science and Engineering Research Board, Department of Science and Technology, Government of India☐ (established through an act of parliament) for providing the financial support to carry

out this work. As this work has been implemented under the project Implementation of Sigma Delta Modulator Using Nanowire
Electrically Doped Hetero Material Tunnel Field Effect Transistor (TFET) for Ultra Low Power Applications which is funded by this board.

References

[1] The international technology roadmap for semiconductors, http://www.itrs2.net.

[2] S. Bangsaruntip, G.M. Cohen, A. Majumdar, J.W. Sleight, Universality of short-channel effects in undoped-body silicon nanowire MOSFETs, IEEEElectron Device Lett. 31 (9) (Sept. 2010) 903-905.

[3] K.K. Young, Short-channel effect in fully-depleted SOI MOSFETs, IEEE Trans. Electron Devices 36 (2) (Feb. 1989) 399-402.

[4] J.P. Colinge, FinFETs and Other Multi-transistors, Springer, 2008.

[5] K. Boucart, A.M. Ionescu, Double gate tunnel FET with high-k gate dielectric, IEEE Trans. Electron Devices 54 (7) (Jul. 2007) 1725-1733.

[6] A.M. Ionescu, H. Riel, Tunnel field-effect transistors as energy efficient electronic switches, Nature 479 (7373) (Nov. 2011) 329-337.

[7] E.H. Toh, G.H. Wang, G.S. Samudra, Y.C. Yeo, Device physics and design of double-gate tunneling field-effect transistor by silicon film thicknessoptimization, Appl. Phys. Lett. 90 (26) (Jun. 2007) 2635071-2635073.

[8] V. Vijayvargiya, S.K. Vishvakarma, Effect of drain doping profile on double gate tunnel field effect transistor and its influence on device RF performance, IEEE Trans. Electron Devices 13 (5) (Sep. 2014) 974-980.

[9] A. Revelant, P. Palestri, P. Osgnach, L. Selmi, Calibrated multi-subband Monte Carlo modeling of tunnel-FETs in silicon and III-V channel materials, SolidState Electron. 88 (Oct. 2013) 54-60.

[10] Wang H., Chang S., Hu Y., ET AL.: 'A novel barrier controlled tunnel FET', IEEE Electron Device Lett., 2014, 35, (7), pp. 798–800

[11] Lee M.J., Choi W.Y.: 'Effects of device geometry onhetero-gate-dielectric tunneling field-effect transistors', IEEEElectron Device Lett., 2012, 33, (10), pp. 1459–1461

[12] Hraziia A.V., Amara A., Anghel C.: 'An analysis on the ambipolarcurrent in Si double-gate tunnel FETs', Solid-State Electron., 2012, 70, pp. 67–72

[13] Abdi D.B., Jagadesh Kumar M.: 'Controlling ambipolar current in tunneling FETs using overlapping gate-on-drain', IEEE J. Electron Devices Soc., 2014, 2, (6), pp. 187–190

[14] Verhulst A.S., Vandenberghe W.G., Maex K., ET AL.: 'Tunnelfield-effect transistor without gate-drain overlap', Appl. Phys. Lett.,2007, 91, (5), pp. 053102–053103

[15] Ryusuke Nishitani, Hiroshi Iwasaki, Yusuke Mizokawa1 and Shogo Nakamura,: 'An XPS Analysis of Thermally Grown Oxide Film on GaP', Japanese Journal of Applied Physics., 1978, 17, (2), pp. 321

[16] Mookerjea S., Krishnan R., Datta S., ET AL.: 'On enhanced miller capacitance effect in interband tunnel transistors', IEEE Electron Dev. Lett., 2009, 30, (10), pp. 1102–1104

[17] Silvaco Int., Santa Clara, CA, USA: 2014, 'ATLAS DeviceSimulation Software'.

[18] H. J. M. Veendrick,"Short-circuit dissipation of static CMOS circuitry and its impact on the design of buffer circuits," *IEEE J. Solid-State Circuits* , vol. 19, no. 4, pp. 468-473, Aug 1984.

[19] M. S. Kim, H. Liu, X. Li, S. Datta, and V. Narayanan, "A SteepSlope Tunnel FET Based SAR Analog-to-Digital Converter," *IEEE Trans. Electron Devices*, vol. 61, no. 11, pp. 3661-3667, Nov. 2014

[20] H. Schmid, M. T. Bjrk, J. Knoch, S. Karg, H. Riel, and W. Riess, "Doping Limits of Grown in situ Doped Silicon Nanowires Using Phosphine," *Nano lett.*, vol. 57, no. 4, pp. 820-826, Apr. 2009.

[21] J. Madan and R. Chaujar, "Interfacial charge analysis of heterogeneous gate dielectric-gate all around-tunnel FET for improved device

978-1-5386-1357-3/17 $31.00 © 2017 IEEE

reliability," *IEEE Trans. Device Mater.*,vol. 16, no. 2, pp. 227234, Jun. 2016.

[22] ATLAS device simulation soft, *Silvaco, Santa Clara, CA, USA*, 2012.

[23] Bhagwan Ram Raad, Sukeshni Tirkey, Dheeraj Sharma, and Pravin Kondekar, "A New Design Approach of Dopingless Tunnel FET for Enhancement of Device Characteristics," *IEEE Trans. Electron Devices.*,vol. 64, no. 4, pp. 1830-1836, Apr. 2017.

A Comparative Study of GaP/SiGe Hetero Junction Double Gate Tunnel Field Effect Transistor

Dharmendra Singh Yadav, Dheeraj Sharma, Sukeshni Tirkey, Deepak Soni, Deepak G. Sharma, Shriya Bajpai and Neeraj Sharma

Nanoelectronics and VLSI Laboratory ,Electronics and Communication Engineering Discipline

PDPM- Indian Institute of Information Technology, Jabalpur, 482005, India. and

Department of Computer Engineering

Ramrao Adik Institute of Technology Nerul, Navi Mumbai, 400706, India.

Email: tech.dharmendra26@gmail.com ,dheeraj@iiitdmj.ac.in, sukeshni@iiitdmj.ac.in , deepaksoni09@gmail.com,

deepak.sharma@iiitdmj.ac.in, shriya.bajpai@iiitdmj.ac.in and neeraj16ks@gmail.com

Abstract— **This manuscript presents a detailed study of Hetero Junction Dual Gate Tunnel Field Effect Transistor (HJ-DG-TFET) for improving the drain current and degrading the ambipolar behavior. The proposed device employs hetero-material for device analysis. Drain current of the device increases because of narrowing of tunneling barrier width at the source-channel junction with the application of low band gap material at the source/channel junction. Hence, the proposed device shows superior performance over conventional DG-TFET in terms of sub threshold slope, drain current, resistant to parasitic capacitance and high frequency parameters. Further, the optimization of channel length and drain voltage is also performed to ensure the reliability of the device. Hence, the overall performance of the HJ-DG-TFET is examined for better efficiency of the device.**

Keywords—Band to band tunneling; Hetero gate oxide; Hetero junction; Gallium Phosphide; Silicon-Germanium.

I. INTRODUCTION

The scaling of MOSFET technology has reached its optimum level because of its inability to surpass the limit of sub threshold swing of 60 mV/decade and large leakage current. To overcome this limitation, a new device of Tunnel Field Effect Transistor (TFET) was proposed. This is mainly based on the tunneling phenomenon of charge carriers. This gives the device ability to surpass the limitation of sub threshold slope which can be approached to less than 60 mv/decade in this device [1-4]. TFET has emerged as a promising candidate to replace conventional MOSFET for low power applications because of its better adaptation in DC and RF domain [5-7]. The ON-state current is obtained from the tunneling of electron from valence band of source to conduction band of channel unlike the thermionic emission of charge carriers in MOSFET. However, the lower amount of ON-state current [6] and ambipolar conduction [8] are still the issues affecting the performance of the device. Therefore, various methods has been utilized in order to overcome the problem of ON-state current and ambipolarity of the device. Different semiconductor material is utilized in place of conventional Silicon to enhance the performance of the device. Hence, in this manuscript, we have proposed such a Hetero-Junction based Dual Gate TFET (HJ-DG-TFET). The proposed device is composed of two different material of Silicon-Germanium (SiGe) and Gallium Phosphide (GaP). Wider band gap material of GaP is incorporated at the drain

and channel region whereas lower band gap material of SiGe is utilized at the source region. Tunneling probability of holes decreases as the tunneling barrier width increases due to the employment of high band gap material (III-V group semiconductors compounds) rather than using silicon [9]. On the other hand, incorporation of lower band gap material there is narrowing of tunneling barrier width. This increases the tunneling probability of electrons to tunnel through the junction. Thus, ON-state current of the device increases while suppressing the ambipolarity of the device [10-14]. The process of thermal oxidation of GaP, an oxide is formed by which further the formation of $GaPO_4$ can be done [15].

The employment of hetero material increases the ON-state current leading to improved RF performance. Therefore, we have presented a comparative study of conventional DG-TFET and HJ-DG-TFET to evaluate its DC and RF characteristics. The RF parameters considered are Gate-to-Drain Capacitance (C_{gd}), Transconductance (g_m), Cut-Off Frequency (f_T) and Gain Bandwidth Product (GBP).

This paper is divided into different sections as follows: Section II gives the structural details of the simulation setup and device design. Section III demonstrates the device characteristic and RF performances of conventional DGTFET and HJ-DG-TFET. Section IV gives the optimization of channel length (L_G) and the drain voltage (V_{ds}). Section V summarizes the important conclusion drawn from the study

II. SIMULATION SETUP AND DEVICE DESIGN

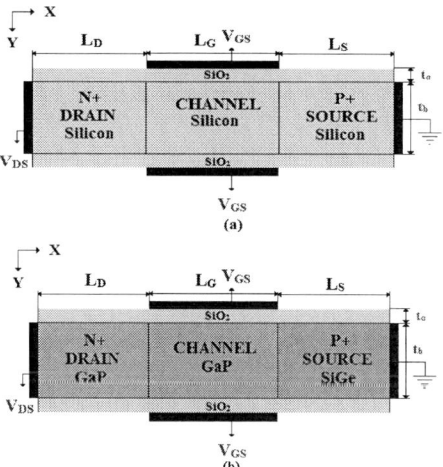

Fig.1 (a) Conventional DG-TFET (b) Hetero-Junction DG-TFET (proposed device).

Fig. 1a – 1b depicts the 2-D cross-sectional view of different TFET considered in this paper. Different architecture of Conventional DG-TFET and Hetero Junction Dual Gate TFET (HJ-DG-TFET) are represented in the Fig. 1a and Fig. 1b respectively.

The dual gate TFET can be considered basically as a p-i-n diode as depicted in the Fig. 1a – 1b. The parameters of the device considered are presented in the Table 1.

Table I

Parameters of the Device	Symbol	Values
Source doping concentration	N_P	1×10^{20} cm^{-3}
Channel doping concentration	N_I	1×10^{17} cm^{-3}
Drain doping concentration	N_D	5×10^{18} cm^{-3}
Channel Length	L_G	50 nm
Gate Oxide Thickness	t_{ox}	3 nm
Body Thickness	t_b	10 nm

Different models are incorporated in the simulation so as to observe the physical effect in the device. The foremost model incorporated is Band-to-band tunneling (BTBT) model as TFET is based on the tunneling phenomenon. Narrowing of band gap takes place due to Band Gap Narrowing (BGN) model. To get physical effect of tunneling, NON-LOCAL BTBT model is considered. Wentzel-Kramer-Brillouin (WKB) method is used for numerical calculations. Some simulation models like Auger recombination, Fermi-Dirac statistics, concentration-dependent mobility, field-dependent mobility and Shockley Read Hall (SRH) recombination model are included. The simulations have been carried out by 2D-ATLAS, Silvaco TCAD (Version 5.19.20.R) [17].

III. DC Characteristics & RF Parameters

In this section, the study of DC as well as High Frequency (RF) characteristics of the proposed device is presented. The study is carried out under Thermal equilibrium and ON-state.

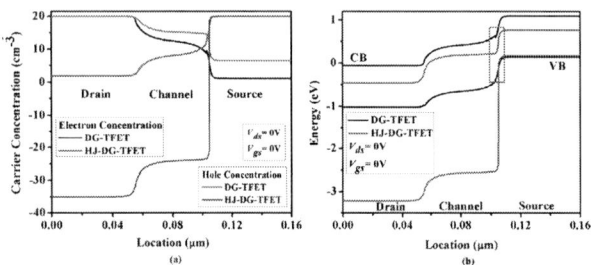

Fig.2 (a) Carrier Concentration (b) Energy Band Diagram under Thermal Equilibrium.

At the thermal equilibrium (V_{gs}= 0V), as no biasing is applied so the concentration of electrons and holes have similar concentration in the channel region for both of these devices. However, the proposed device shows earlier accumulation of charge carriers at the source-channel junction. This can be observed from the Fig. 2a. Therefore, the

tunneling width at source-channel junction gets narrower while that at drain-channel junction gets wider as observed from the Fig. 2b. However, as there is no biasing, there is no driving capability of the current so no drain current is produced in this case.

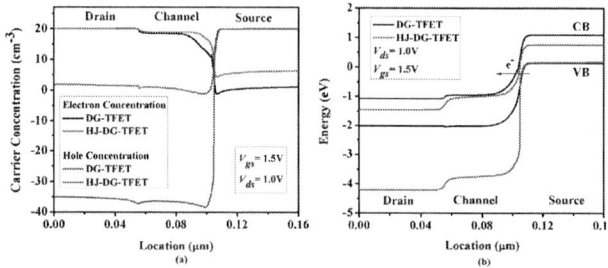

Fig.3 (a) Carrier Concentration (b) Energy Band Diagram under ON-State.

Under ON-state condition of the device, the applied biasing reduces the tunneling barrier width as depicted from the Fig. 3b. Thus, the electron gets tunnel into the channel region increasing the concentration of electron in the channel region in comparison to the concentration of holes. This phenomenon can be noticed in the Fig. 3a. It is observed that HJ-DG-TFET has highest electron concentration in the channel region than conventional TFET. Hence, tunneling probability of electron increases in the source-channel region and hence, accumulation of charge carriers is increased.

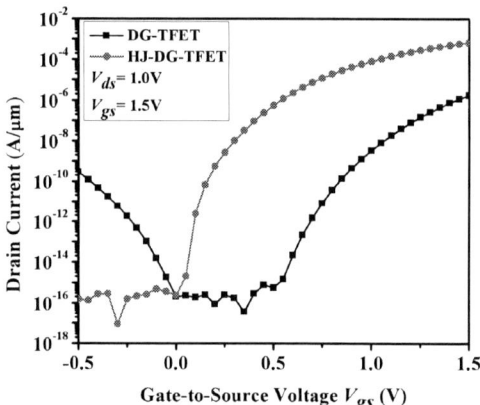

Fig.4 Transfer Characteristics I_{ds} - V_{gs} of DG-TFET and HJ-DG-TFET.

The analysis of drain current (I_{ds}) as a function of gate voltage (V_{gs}) is presented in the Fig.4. The introduction of hetero material in the device shows enhanced ON-state current in comparison to the conventional DG-TFET. The reason behind this is lower band gap material employed at the source region. This increases the rate of tunneling of electron from valence band of source to conduction band of channel. The proposed device of HJ-DG-TFET also shows suppression of ambipolar current as depicted in Fig. 4. This is because of higher band gap material at drain-channel region causes widening of band gap at drain-channel junction. Thus, tunneling of holes from conduction band of drain to valence band of channel becomes impossible and hence reduced ambipolar current is observed

978-1-5386-1357-3/17 $31.00 © 2017 IEEE

in HJ-DG-TFET (Fig. 4). The parameters obtained from the analysis shown in the Table II:

Table II

Device	ON-state current (A/μm)	Threshold Voltage (V)	Subthreshold Swing (mv/decade)
DG-TFET	1.7×10^{-6}	1.21	42.3
HJ-DG-TFET	6.6×10^{-4}	0.4	16.2

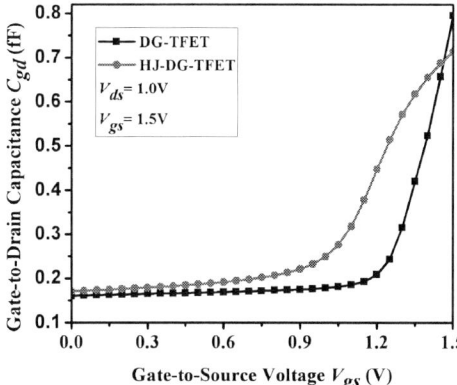

Fig.5 Gate-to-Drain Capacitance (C_{gd}) of DG-TFET and HJ-DG-TFET.

The performance of the device degrades with the presence of parasitic capacitance. Parasitic capacitance consists of Gate-to-drain capacitance (C_{gd}), Gate-to-source capacitance (C_{gs}), and Gate-to-gate capacitance (C_{gg}), out of which C_{gd} has greater effect on device performance. It is expressed as:

$$C_{gd} = \frac{\partial Q_g}{\partial V_d} \dots\dots\dots\dots\dots (1)$$

From the analysis, it is observed that C_{gd} of the proposed HJ-DG-TFET is less than conventional DG-TFET. The main reason behind this is the reduction of Density of State (DOS) in the channel region.

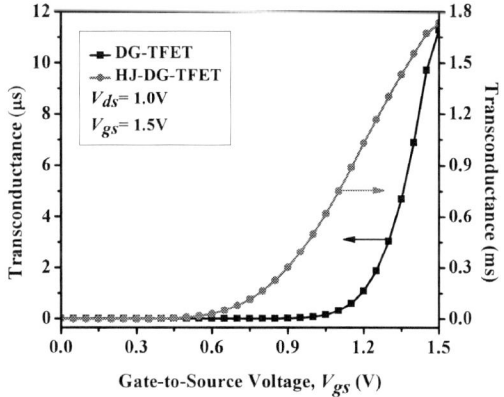

Fig.6 Transconductance (g_m) of DG-TFET and HJ-DG-TFET.

Transconductance (g_m) is defined as the ratio of change in output current to change in input voltage. Thus, g_m

is proportional to drain current. Since there is an increase in drain current of the device (Fig. 4), so transconductance (g_m) of the device also increases. Therefore, transconductance of HJ-DG-TFET is significantly larger than conventional DG-TFET. Hence, the proposed device generates considerably large current with small applied bias voltage. This results in the better sensitivity of the device. Fig. 6 represents the transconductance (g_m) as a function of gate voltage (V_{gs}).

$$g_m = \frac{\partial I_{ds}}{\partial V_{gs}} \dots\dots\dots\dots\dots (2)$$

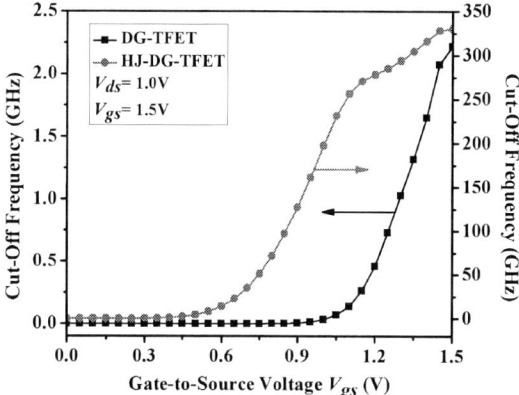

Fig.7 Cut-Off Frequency (f_T) of aDG-TFET and HJ-DG-TFET.

The frequency at which current gain becomes unity is called Cut-Off frequency (f_T). The Cut-off Frequency is dependent on transconductance (g_m) and summation of C_{gd} and C_{gs}. Increasing ON-state current leads to increase in g_m and decrement in the value of C_{gd}. This leads to increase cut-off frequency of the device. The HJ-DG-TFET shows superior value of f_T in comparison with conventional DG-TFET as depicted through Fig. 7.

$$f_T = \frac{g_m}{2\pi(C_{gd} + C_{gs})} \dots\dots\dots\dots\dots (3)$$

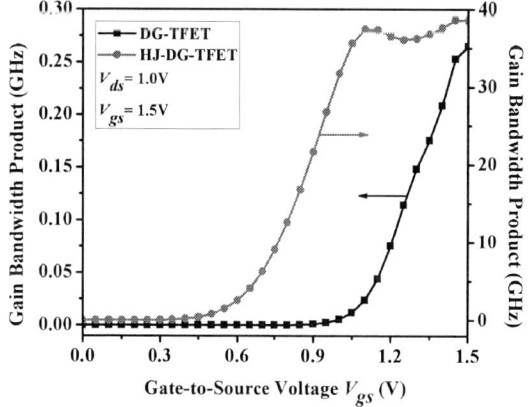

Fig 8 Gain Bandwidth Product (f_A) of DG-TFET and HJ-DG-TFET.

Gain Bandwidth Product (GBP) is also proportional to transconductance (g_m) and gate-to-drain capacitance (C_{gd}). The GBP also shows similar behavior of increased bandwidth in comparison to the bandwidth of conventional DG-TFET. GBP of HJ-DG-TFET is relatively higher than conventional DG-TFET as observed from the Fig. 8.

$$f_A = \frac{g_m}{20\pi C_{gd}} \quad \text{.............................. (4)}$$

IV. OPTIMIZATION OF CHANNEL LENGTH (L_G) & DRAIN VOLTAGE (V_{DS})

A. Optimization of Channel Length (L_G)

Table III

DC - RF Parameters	HJ-DG-TFET (L_G)			
	40nm	50nm	60nm	70nm
I_{ON} (A/µm)	6.78×10^{-4}	6.6×10^{-4}	6.41×10^{-4}	6.25×10^{-4}
I_{Off}(A/µm)	1.5×10^{-16}	2.4×10^{-16}	3.3×10^{-16}	4.8×10^{-16}
I_{Amb}(A/µm)	2.79×10^{-16}	1.5×10^{-16}	1.58×10^{-16}	1.51×10^{-16}
C_{gd}(fF)	0.568	0.714	0.852	0.983
g_m (ms)	1.8	1.73	1.66	1.6
f_T(GHz)	416	330	273	232
GBP(GHz)	50.6	38.6	33.4	30.6

The results obtained from optimization of channel length (L_G) are depicted in the Table III. The foremost aim of this optimization is to overcome the fabrication difficulties while fabricating channel at the nanometer regime. Channel length of 40nm, 50nm 60nm and 70nm are taken into consideration for this study.

Table III shows minor variation in the drain current of the device on downscaling of the channel length. This is due to dependence of drain current on the resistance of source-channel barrier. Downscaling of the channel length leads to change in the resistance of channel region. Thus, drain current does not get affect by change in channel length because of its dependency on the tunneling barrier resistance instead of channel resistance. Similarly, there is less variation in the ambipolarity behavior of the device. As the length of channel increases, there is an increase in the inversion layer formed in the channel region. Hence, gate-to-drain capacitance (C_{gd}) of the device decreases with decrement in the channel length of the device (Table III). The transconductance (g_m) is dependent on the drain current and drain voltage of the device. Since drain current of the proposed device shows minor variations with downscaling of the channel length. Therefore, transonductance also shows minor variations towards downscaling of channel length (Table III). Cut-off frequency (f_T) depends upon the transconductance (g_m) and combination of gate-to-source capacitance (C_{gs}) and gate-to-drain capacitance (C_{gd}). Transconductance of the device very minute variation with change in channel length while C_{gd} decreases

with downscaling of channel length. This results in increment in the value of Cut-off frequency (f_T) with downscaling of the channel length as observed from the Table III. Similarly, Gain Bandwidth Product (GBP) has similar behavior as such of cut-off frequency towards the variations in the channel length. This is because of GBP depends upon both transconductance (g_m) and gate-to-drain capacitance (C_{gd}). Therefore, the value of cut-off frequency and gain bandwidth product increases with decrease in the channel length (Table III). From this study, it is observed that at L_G= 40nm, the performance of the device enhances. But fabrication process for the device at this level may have some difficulties. However for $L_G \leq$ 50nm, there is less variation in the device performance. Therefore, a feasible length of channel L_G= 50nm is taken into consideration for further analysis.

B. Optimization of Drain Voltage (V_{ds})

This study is used to determine the RF performance of the device at low operating voltage. For the analysis, the drain voltage of 0.6V, 0.8V, 1.0V, 1.2V, and 1.4V are taken into consideration. The result obtained from the analysis are depicted in Table IV. The ON-state current of the proposed device increases with increase in the applied drain voltage. Increase in drain voltage leads to increment in the driving force for the current which increases the drain current.

Table IV

DC - RF Parameter	HJ-DG-TFET Drain Voltage (V_{ds})				
	0.6V	0.8V	1.0V	1.2V	1.4V
I_{ON} (A/µm)	3.92×10^{-4}	5.41×10^{-4}	6.6×10^{-4}	7.3×10^{-4}	7.6×10^{-4}
I_{Off} (A/µm)	1.1×10^{-16}	1.5×10^{-16}	2.4×10^{-16}	3.0×10^{-16}	5.2×10^{-16}
I_{Amb} (A/µm)	3.9×10^{-16}	1.5×10^{-16}	1.5×10^{-16}	2.1×10^{-16}	1.8×10^{-16}
C_{gd}(fF)	0.884	0.819	0.714	0.54	0.345
g_m (ms)	0.901	1.32	1.73	2.09	2.29
f_T(GHz)	146	225	330	513	778
GBP GHz)	16.2	25.5	38.6	70.2	112

However, ambipolar conduction still remains at the constant level. With increase in the drain voltage on the device, the inversion layer starts to pinch off from the source end [16]. This decreases the length of inversion layer which leads to degradation in the gate-to-drain capacitance (C_{gd}). Thus, parasitic capacitance reduces with increase in the drain voltage. Transconductance (g_m) of the device is proportional to drain current. From Table IV, it is observed that transconductance increases with increasing drain voltage. This is because of increase in the drain current with increased drain voltage. Cut-off frequency (f_T) and Gain Bandwidth Product (GBP) are proportional to Transconductance (g_m) and Gate-to-Drain Capacitance (C_{gd}). The value of f_T and GBP increases with increasing drain voltage. This is due to increase in g_m and decrease in C_{gd} with enhancement in drain voltage.

The proposed device shows better results for $V_{ds} \geq 1.0V$ observed from the optimization of drain voltage (V_{ds}). Considering the fact that there has been a lot of researches taking place for developing ultra-low power electronic devices, a reliable drain voltage of $V_{ds} = 1.0V$ is considered in this manuscript.

V. CONCLUSION

In this paper, a comparative study of conventional Dual Gate TFET (DG-TFET) and Hetero-Junction based (GaP/SiGe) Dual Gate TFET (HJ-DG-TFET) is studied with the help of simulation. The important findings obtained from study are as follows:-

i. There is an enhancement in the ON-state current of the proposed device while the ambipolar current is reduced to great extent. This is because of introduction of low band gap material at the source region and a high band gap material at the drain/channel region.

ii. Increment in the ON-state current leads to lowering of threshold voltage and reduces the sub threshold slope in comparison with the conventional DG-TFET

iii. The RF performance of the proposed device shows that it is compatible to work under such conditions.

iv. The proposed device of HJ-DG-TFET shows g_m about 150 times, f_T about 135 times and f_A about 150 times that of conventional DG-TFET.

v. Furthermore, the optimization of channel length (L_G) and drain voltage (V_{ds}) states that at $L_G = 50nm$ and $V_{ds} = 1V$, the proposed device delivers better performance than conventional DG-TFET

Hence, the proposed device shows great potential to perform in DC as well as in high frequency domain.

References

[1] The international technology roadmap for semiconductors, http://www.itrs2.net.

[2] S. Bangsaruntip, G.M. Cohen, A. Majumdar, J.W. Sleight, Universality of short-channel effects in undoped-body silicon nanowire MOSFETs, IEEEElectron Device Lett. 31 (9) (Sept. 2010) 903-905.

[3] K.K. Young, Short-channel effect in fully-depleted SOI MOSFETs, IEEE Trans. Electron Devices 36 (2) (Feb. 1989) 399-402.

[4] J.P. Colinge, FinFETs and Other Multi-transistors, Springer, 2008.

[5] K. Boucart, A.M. Ionescu, Double gate tunnel FET with high-k gate dielectric, IEEE Trans. Electron Devices 54 (7) (Jul. 2007) 1725-1733.

[6] A.M. Ionescu, H. Riel, Tunnel field-effect transistors as energy efficient electronic switches, Nature 479 (7373) (Nov. 2011) 329-337.

[7] E.H. Toh, G.H. Wang, G.S. Samudra, Y.C. Yeo, Device physics and design of double-gate tunneling field-effect transistor by silicon film thicknessoptimization, Appl. Phys. Lett. 90 (26) (Jun. 2007) 2635071-2635073.

[8] V. Vijayvargiya, S.K. Vishvakarma, Effect of drain doping profile on double gate tunnel field effect transistor and its influence on device RF performance, IEEE Trans. Electron Devices 13 (5) (Sep. 2014) 974-980.

[9] A. Revelant, P. Palestri, P. Osgnach, L. Selmi, Calibrated multi-subband Monte Carlo modeling of tunnel-FETs in silicon and III-V channel materials, SolidState Electron. 88 (Oct. 2013) 54-60.

[10] Wang H., Chang S., Hu Y., ET AL.: 'A novel barrier controlled tunnel FET', IEEE Electron Device Lett., 2014, 35, (7), pp. 798–800

[11] Lee M.J., Choi W.Y.: 'Effects of device geometry onhetero-gate-dielectric tunneling field-effect transistors', IEEEElectron Device Lett., 2012, 33, (10), pp. 1459–1461

[12] Hraziia A.V., Amara A., Anghel C.: 'An analysis on the ambipolarcurrent in Si double-gate tunnel FETs', Solid-State Electron., 2012, 70, pp. 67–72

[13] Abdi D.B., Jagadesh Kumar M.: 'Controlling ambipolar current in tunneling FETs using overlapping gate-on-drain', IEEE J. Electron Devices Soc., 2014, 2, (6), pp. 187–190

[14] Verhulst A.S., Vandenberghe W.G., Maex K., ET AL.: 'Tunnelfield-effect transistor without gate-drain overlap', Appl. Phys. Lett.,2007, 91, (5), pp. 053102–053103

[15] Ryusuke Nishitani, Hiroshi Iwasaki, Yusuke Mizokawal and Shogo Nakamura,: 'An XPS Analysis of Thermally Grown Oxide Film on GaP', Japanese Journal of Applied Physics., 1978, 17, (2), pp. 321

[16] Mookerjea S., Krishnan R., Datta S., ET AL.: 'On enhanced miller capacitance effect in interband tunnel transistors', IEEE Electron Dev. Lett., 2009, 30, (10), pp. 1102–1104

[17] Silvaco Int., Santa Clara, CA, USA: 2014, 'ATLAS DeviceSimulation Software'.

Design and Simulation of SF-FinFET and SD-FinFET and Their Performance in Analog, RF and Digital Applications

Syed Samsuz Zaman*, Pankaj Kumar*, Manash Pratim Sarma[†], Ashok Ray[‡] and Gaurav Trivedi[§]

*PG Student, Dept. of ECE, Gauhati University, Email: mr.samsuzzaman123@gmail.com

[†]Assistant Professor, Dept. of ECE, Gauhati University

[‡]Assistant Professor, Dept. of ECE, North Eastern Regional Institute of Science and Technology, Itanagar

[§]Assistant Professor, Dept. of EEE, IITG, Email: trivedi@iitg.ernet.in

Abstract—This paper presents TCAD simulation of Triple gate Step Fin FinFET (SF-FinEFT) and Step Drain FinFET(SD-FinFET) field effect transistor. The electrical characteristics of the devices are compared with conventional Fin-FET(cFinFET/FinFET). Various performance parameter like ON current I_{on}, leakage current I_{off}, $\frac{I_{on}}{I_{off}}$ ratio, sub-threshold slope, DIBL have been observed. Also to check the device functionality in Analog, RF and Digital applications a comparative study of various parameters has been done between the proposed devices and conventional FinFET. We have used strained silicon as channel material for all the devices. As the number of transistors per square area on a chip keeps on increasing, the performance metrics for a planar MOSFET degrades and hence it faces various challenges in nano-meter regime. The need for a new device technology to control the unavoidable challenges was felt hence a 3-D Multi-gate transistor have emerged. These newer 3-D devices showed better performance in the nano-meter regime. This paper analyzes parameters like the trans-conductance (g_m), drain conductance (g_d), trans-conductance generation factor, total gate capacitance (C_{gg}), cut-off frequency (f_t), gain frequency product(GFP) for analog and RF performance. For digital application noise margin and propagation delay have been extracted.

Index Terms—SF-FinFET, SD-FinFET, SCEs, Analog, RF and Digital FOMs.

I. INTRODUCTION

FIN shaped field effect transistor technology has evolved as a result of relentless increase in the scaling of conventional MOSFET. Now-a-days IC chip consist of billions of transistors and designer team aims to have technology with higher drive current, low sub-threshold slope, low drain induced barrier lowering, high packing density, and also with longer battery lifetime, high energy efficient electronics in low power and high performance operation.

As the gate length decreases drive current increases but need to reduce the gate oxide thickness to maintain gate control on channel [1]. The reduction of conventional silicon dioxide as gate dielectric thickness has reached its limit below which performance of MOSFET decreases. This is because of the increases in direct tunneling of carriers to gate, which result in increasing of static power and in circuit operation [1]. Device simulation using high-k dielectric can get rid the gate tunneling problem [1,2,3]. High-k dielectric as gate oxide on silicon decrease the performance by increasing short channel effects

(SCEs). SCEs rises due to the fringing electric field from gate to source/drain or source/drain to channel [1] and punch-through problem [4]. Using silicon on insulator technology punch-through problem can be avoided [5]. And fringing electric field problem can be reduced by using gate stack structure i.e. using silicon dioxide between channel and high-k dielectric region [1,2]. This is done by maintaining constant equivalent oxide thickness (EOT). To increase the gate control on channel charge multi-gate structure like Double-gate (DG), Triple-gate (TG), Gate-all round (GAA) MOSFET transistors are also attractive choice [6].

Strained silicon increases mobility of carriers, thus it modifies the current transport properties [7] and increases the drive current. A brief study can be found in [8]. Gate length to fin width ratio should be high for low DIBL and high I_{on} to I_{off} ratio [9]. This paper simulates the SF-FinFET, SD-FinFET and cFinFET structures and compares various performance parameters of SF-FinFET and SD-FinFET with cFinFET. This proposed SD-FinFET structure shows 1.19X improvement on I_{on} current and 1.34X improvement in trans-conductance. An improvement in total gate capacitance and cut-off frequency is observed for SF-FinFET. Improvement on noise margin and propagation delay is also observed in SF-FinFET. This paper has been organized as follows : In section II Device structure and Simulation setup are described. The device electrical Performance and also the performance on analog, RF and Digital application parameter are discussed in section III and finally the paper is concluded on section IV.

II. DEVICE STRUCTURE AND SIMULATION SET-UP

The 3-D view and doping concentration of SF-FinFET and SD-FinFET is shown in Fig. 1(a) and (b) respectively. The structure is a combination of Silicon and strain silicon on insulator technology and 3-D FET technology. Strained silicon is used in the channel region. Strained silicon increases mobility of carrier thus modify the carrier transport properties [7]. Thus increases the drive current. Silicon is used in source and drain region. Fig. 2(a) and (b) shows the schematic of Fin structure of SF-FinFET and SD-FinFET respectively. Equivalent oxide thickness (EOT) is considered at 1.1 nm to reduce the leakage current by reducing the tunneling current [1,5]. Here SiO_2

TABLE I
DIMENSION FOR SF-FINFET AND SD-FINFET

Dimension		S-FinFET	SD-FinFET
Width	W1,W3	3nm	-
	W2	6nm	-
	W4,W6	-	3nm
	W5	-	4nm
Height	H1,H3	5nm	-
	H2	20nm	-
	H4,H6	-	7nm
	H5	-	16nm

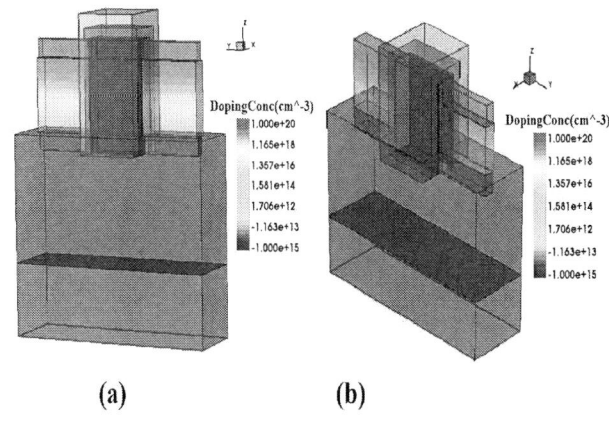

(a)   **(b)**

Fig. 1. 3D structure and doping level of (a) SF-FinFET and (b) SD-FinFET.

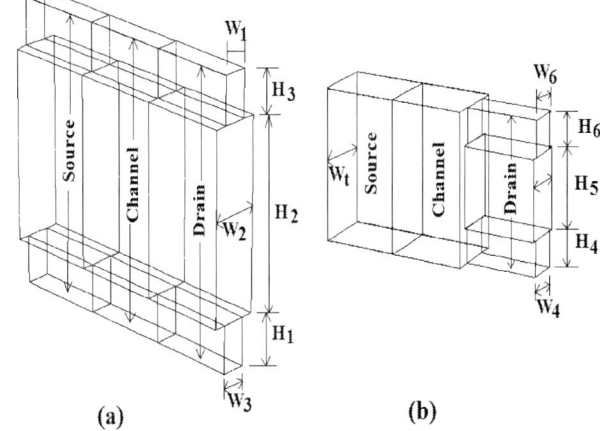

(a)   **(b)**

Fig. 2. 3D schematic view of fin of (a) SF-FinFET and (b) SD-FinFET device.

and HfO_2 are used as dielectric material. SiO_2 is used as interface oxide layer with thickness of 0.6nm [5]. SiO_2 and HfO_2 has dielectric constant of 3.9 and 22 respectively. All the devices i.e. P-channel and N-Channel SD-FinFET, SF-FinFET and FinFET are made with gate length 20nm. Channel region is undoped. Source/Drain and substrate are doped with concentration of 1.0×10^{20} cm^{-3} and 1.0×10^{15} cm^{-3} respectively. Boron active concentration and Phosphorus active concentration are used as P-type and N-type dopant. Buried oxide (BOX) width of the devices are 30nm. The gate material used here is PolySilicon. Fin height and fin width are 30nm and 6nm respectively for conventional FinFET simulated in this paper.

In SD-FinFET device heights H4 and H6 are equal and widths W4 and W6 are equal. But the width of fin in source side W_t is greater than width W5 of drain side. In SF-FinFET widths W3 and W1 same and heights H1 and H3 are same. Table I shows height and width parameter for SF-FinFET and SD-FinFET.

As, the channel width and height of FinFET is given by [10]-

$$W = 2H_{fin} + W_{fin} \qquad (1)$$

From Fig. 2 the width of channel region of SF-FinFET is-

$$W1a = 2H_1 + W_3$$
$$W2b = 2H_2 + W_2$$
$$W3c = 2H_3 + W_1$$

And fin height is H_{fin}= H1 + H2 + H3 The channel effective width is given by-

$$W_{eff} = W1u + W2b + W3c \qquad (2)$$

In order to consider the mobility degradation due to high surface scattering at semiconductor to insulator interface, Lombadri mobility model is used in our simulation. It includes scattering phenomenon like phonon scattering and column scattering. Mobility model to account for doping and velocity saturation dependency are activated. To include the high doping concentration in the device Band-gap Narrowing model is enabled. High doping concentration reduces the mobility of the carrier. So, doping dependence mobility model have been included in the mobility model to account the effect of mobility of carriers. To generate results at nano-meter regime, quantum effects have been included [11]. Quantum drift-diffusion model is used for accuracy of result and Fermi-Dirac transport model is also used in the simulation. We have activated SRH generation and recombination model to evaluate the leakage current. For minority carrier recombination Auger

recombination model is used. The current , charge and voltage at every electrode can be measured at steps of bias ramp through quasi-stationary simulation [12]. Numerical technique Newton is enables for getting the solution. Temperature during the simulation is 300K. To validate our simulation , compared the result with previous literature data.

III. RESULT

Surface Potential of the device along the channel is shown in Fig. 3. Surface Potential in the channel region should be higher in order to control the short channel effect [13]. Surface Potential of all the device is almost equal in the channel region. Electron density and electron mobility along the channel are shown in Fig. 4 and Fig. 5 respectively. Electron density and electron mobility of SD-FinFET are observed more than the other two devices. Descending order of electric density of in channel region of the structure are SD-FinFET > SF-FinFET > FinFET. Surface Potential,electron density and electron mobility are seen at V_g=1V and V_d= 1V.

Out-put characteristics is plotted in Fig. 6(a) for V_{ds}= 2V and V_{gs}= 1.5V. The transfer characteristics $I_d - V_{gs}$ in linear scale at drain voltage V_{ds}= 1V and gate voltage V_{gs}= 1V is shown Fig. 6(b). It is observed high drain current in SD-FinFET device because of the improvement in electron density

978-1-5386-1357-3/17 $31.00 © 2017 IEEE

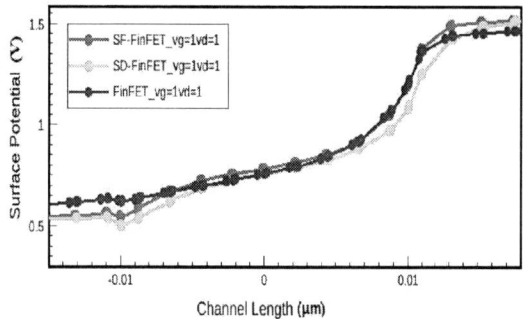

Fig. 3. Surface Potential plot of the device along the channel length.

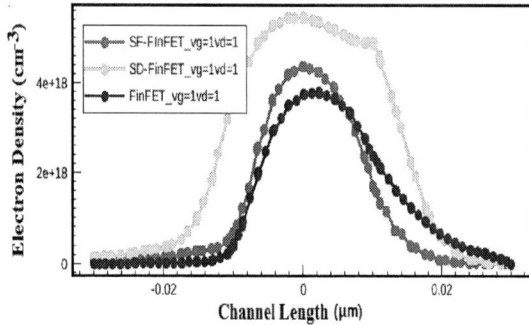

Fig. 4. (a) Electron Density plot of the devices along the channel length.

and electron mobility in the channel region. The off state leakage currents are found to be 2.56×10^{-13}A, 4.51×10^{-13}A and 4.577×10^{-13}A for SF-FinFET, SD-FinFET and FinFET respectively. I_{off} current is the current when transistor is in OFF state means when no gate voltage is applied. All the devices shows low off state leakage current. I_{on} values are found to be 25μA, 35.6μA and 30.6μA for SF-FinFET, SD-FinFET and FinFET respectively. I_{on} should be as high as possible for high switching speed device [14]. The I_{on} and I_{off} are important parameters for the circuit. It has been achieved good $\frac{I_{on}}{I_{off}}$ ratio with this work. The high $\frac{I_{on}}{I_{off}}$ ratio

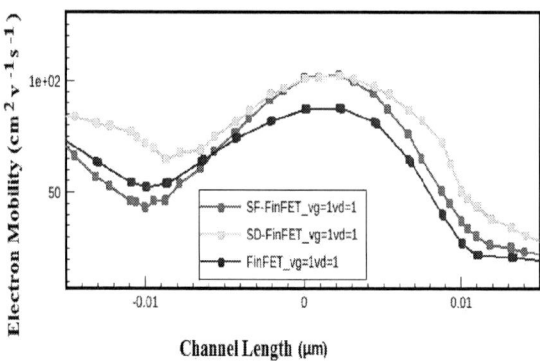

Fig. 5. Electron Mobility plot of the devices along the channel length.

is an important figure of merit(FOM) for better switching speed application [14]. Threshold Voltage is calculated using constant current method. Various method to find threshold voltage are discussed in paper [15]. Threshold Voltage of the devices are found to be 0.42V, 0.45V and 0.44V respectively for SD-FinFET, SF-FinFET and FinFET. Other important parameters are like sub-threshold slope, DIBL, I_{on}, $\frac{I_{on}}{I_{off}}$ are tabulated in Table II and Table III for the proposed N-channel and P-channel devices and compared with the conventional FinFET.

Out-put characteristics plot for P-channel device is shown in Fig. 7. For P-channel device I_{off} are -1.84×10^{-13}A, -1.182×10^{-13}A and -2.197×10^{-13}A respectively for FinFET, SF-FinFET and SD-FinFET. Sub-threshold slope and DIBL is calculated using the following formula-

$$SS = [\frac{\delta log I_d}{\delta V_{gs}}]^{-1} \qquad (3)$$

$$DIBL = \frac{\Delta V_t}{\Delta V_{ds}} = \frac{V_{t1} - V_{t2}}{V_{ds2} - V_{ds1}} \qquad (4)$$

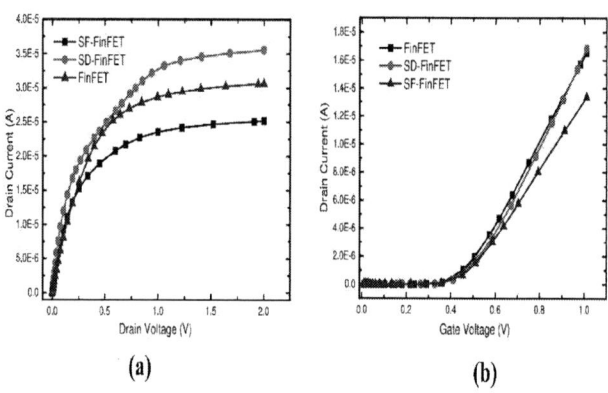

Fig. 6. (a) I_d-V_d plot and (b) I_d-V_g plot of the devices.

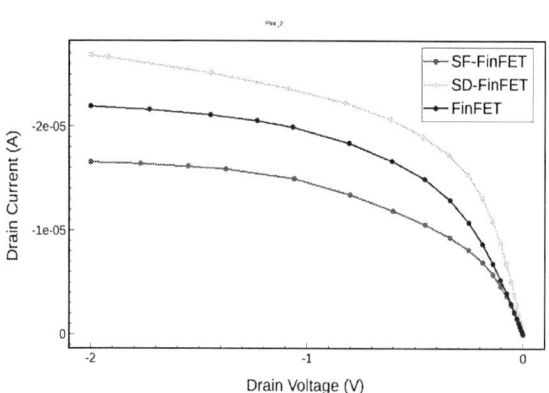

Fig. 7. I_d-V_d plot P-channel of the devices.

A. Analog Parameter

High Density integrated circuit are facing problem in the improvement of high device performance in terms of low

978-1-5386-1357-3/17 $31.00 © 2017 IEEE

TABLE II
COMPARISON OF DEVICE ELECTRICAL PARAMETER FOR N-CHANNEL

Device	$I_{on}(\mu A)$	SS($\frac{mV}{decade}$)	DIBL($\frac{mV}{V}$)	$\frac{I_{on}}{I_{off}}$(A)
FinFET	30.6	70.89	22.15	6.68×10^7
SF-FinFET	25	71.13	23.49	9.76×10^7
SD-FinFET	35.6	73.23	32.12	5.76×10^7

TABLE III
COMPARISON OF DEVICE ELECTRICAL PARAMETER FOR P-CHANNEL

Device	$I_{on}(\mu A)$	SS($\frac{mV}{decade}$)	DIBL($\frac{mV}{V}$)	$\frac{I_{on}}{I_{off}}$(A)
FinFET	-21.9	71.91	30.15	1.19×10^8
S-FinFET	-16.57	70.79	33.88	1.40×10^8
SD-FinFET	-26.89	70.36	35.83	1.224×10^8

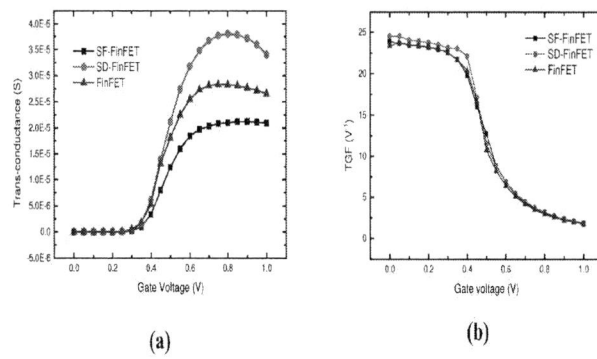

Fig. 8. (a)Trans-conductance plot (b) TGF plot of the device V_d=1V.

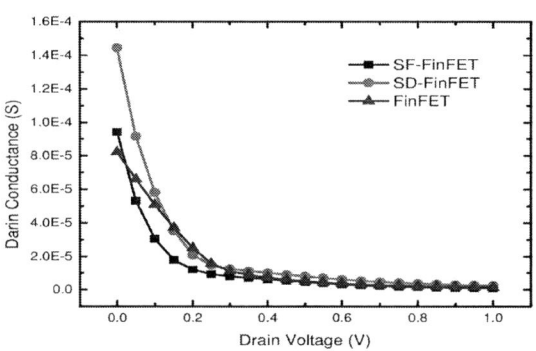

Fig. 9. Drain-conductance plot at V_g=1V w.r.t. drain voltage.

power consumption and high frequency of operation [1]. Analog circuit performance parameters measured here are trans-conductance g_m, drain conductance g_d, Trans-conductance Generation factor(TGF) and gain. These parameter are calculated at drain bias equal to 1V swiping the gate voltage from 0 to 1V.

Trans-conductance plot for SF-FinFET, SD-FinFET and cFinFET is shown in Fig. 8(a) Trans-conductance is an important parameter to evaluate gain. From the Fig. 8(a) it is observed that trans-conductance is high in SD-FinFET device compared to the other two device in the applied drain bias. This high trans-conductance is achieved due to the more change in drain current in a particular gate to source voltage and after the peak value it saturates due to the channel length modulation(CLM) effect in the saturation region. Trans-conductance efficiency(TGF) is a quality factor representing the translation of trans-conductance for a certain level of drain current. TGF value is maximum at weak inversion. TGF plot is shown in Fig. 8(b). Maximum value of TGF is achieved for SD-FinFET device in low gate voltage. High value of TGF bring forfeit in high linearity microwave system because of power consumption issues [1]. Mathematical equations for trans-conductance, Drain conductance, trans-conductance generation factor and gain are-

$$g_m = \frac{\partial I_d}{\partial V_{gs}} \tag{5}$$

$$g_d = \frac{\delta I_d}{\delta V_{ds}} \tag{6}$$

$$TGF = \frac{g_m}{I_d} \tag{7}$$

$$Gain, A_v = \frac{g_m}{g_d} \tag{8}$$

Fig. 9 shows drain conductance as a variation of drain voltage at gate bias equal to 1V. The drain conductance should be low in the saturation region i.e. drain output resistance should be high for the current to saturate. Output resistance is reciprocal of drain conductance.

B. RF Parameter

RF application parameter analyzed here are total gate capacitance C_{gg}, Cut-off frequency f_t and gain frequency product(GFP). And these parameter are plotted against gate voltage at V_d=1V.

Total gate capacitance,C_{gg} plot against gate voltage is shown in Fig. 10(a). The total gate capacitance is extracted through AC small signal analysis. During extraction of capacitance AC frequency is set to $1MHz$ and gate voltage is ramp through step size of 0.05V from 0 to 1V. SF-FinFET shows lower gate capacitance. After the device reaches the saturation region gate capacitance saturates because in this region Channel length modulation occur which maintain constant charge in the channel region. f_t is frequency when current gain is unity. Fig. 10(b) shows f_t variation w.r.t. gate voltage. f_t is mostly affected by geometrical parameter and parasitic capacitance [16]. From the plot it is observed that SD-FinFET shows lower cut-off frequency. The difference in g_m and gate capacitance gives the difference in f_t. The peak of f_t is

TABLE IV
COMPARISON OF ANALOG AND RF PARAMETER OF THE DEVICES

Parameter	cFinFET	SF-FinFET	SD-FinFET
$g_m(\mu S)$	28.3	21.2	38.0
$g_d(\mu S)$	82.2	94.2	114
TGF(V^{-1})	23.9	23.4	24.5
Gain(dB)	30.8	32.4	36.1
$C_{gg}(pF)$	1.16×10^{-2}	9.82×10^{-3}	2.33×10^{-2}
f_t(GHz)	433	378	334
GFP(GHz)	1.49×10^4	1.51×10^4	2.06×10^4

observed at that point when minimum gate capacitance and peak of trans-conductance occur.

$$f_t = \frac{g_m}{2 * \pi * (C_{gs} + C_{gd})} = \frac{g_m}{2 * \pi * C_{gg}} \quad (9)$$

$$GFP = \frac{g_m}{g_d} * f_t \quad (10)$$

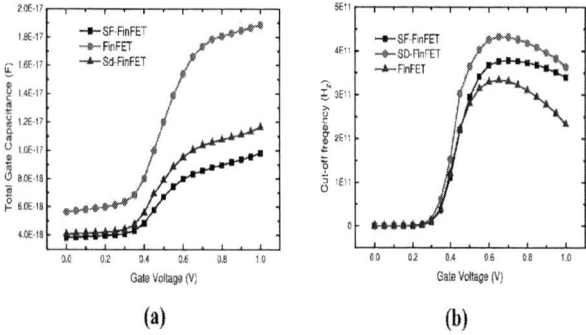

Fig. 10. (a) Total Gate Capacitance plot (b) Cut-off Frequency Plot w.r.t. gate voltage.

GFP is an important FOM for operational amplifier in high frequency amplifier application. It is shown in Fig. 11 against gate bias. As gate voltage increases from sub-threshold region it starts to increase and find an optimum point and in saturation region it decrease. From the curve we can observe that SD-FinFET gives a better result. By observing the curve circuit designer can get the optimum point between gain and speed. Maximum value of Important analog and RF parameter are shown in Table IV.

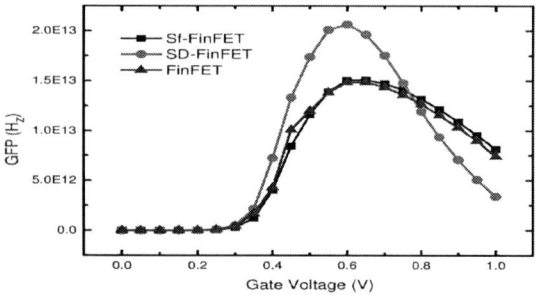

Fig. 11. (a) GFP plot at V_d=1V with a variation of gate voltage.

C. Digital Parameter

Inverter are called the nucleus in digital design. The CMOS inverter technology requires pull-up and pull-down network. The P-channel and N-channel MOS FinFET transistors schematic diagram are given in paper [17]. Fig. 12(a) shows the inverter circuit diagram using that PMOS and NMOS FinFET structure. Mixed mode simulation is done for inverter circuit. Voltage transfer characteristics curve are shown in Fig. 12(b) for for the devices. The VTC plot shows that maximum output voltage is obtained for input

TABLE V
COMPARISON OF NOISE MARGIN AND PROPAGATION DELAY FOR THE DEVICES

Parameter	cFinFET	SF-FinFET	SD-FinFET
V_{IL}(V)	0.442	0.431	0.424
V_{IH}(V)	0.543	0.516	0.543
NM_L(mV)	441.95	425.96	423.51
NM_H(mV)	456.34	483.39	486.48
TR(mV)	100.91	85.18	89.11
τ_{PHL}(ps)	2.11	1.62	4.54
τ_{PLH}(ps)	0.49	0.3	2.27
τ_P(ps)	1.3	0.96	3.41
Gain	-14.58	-15.5	-14.5

logic '0' an minimum output voltage is obtained for input logic '1'. The steepness of output logic switch is more in SF-FinFET than conventional FinFET. The descending order of steepness as observed is SF-FinFET > SD-FinFET > cFinFET. $V_{OH}, V_{OL}, V_{IH}, V_{IL}, NM_H, NM_L$ and transition region(TR) are tabulated for all the device in Table VI. for $V_{dd} = 1V$ and $V_{in} = 1V$. The result are obtained using output capacitance, C_{out} equal to 1×10^{-17} F.

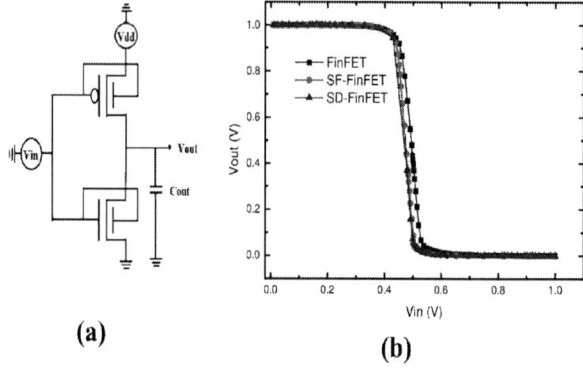

Fig. 12. (a) Inverter Circuit using FinFET device (b) VTC curve for the devices.

So, noise margin gives the tolerance of unwanted signal in digital circuit. So, the analysis of noise margin of inverter circuit is important figure to study the performance of digital application of the devices. Lowest transition is obtained for SF-FinFET.

Switching characteristics for SF-FinFET, FinFET inverter circuit are shown together in Fig. 13. To evaluate the switching characteristics mixed mode simulations are performed using a pulse of 10ps delay time, 10ps rise time, 10ps fall time, 60ps ON time and 140ps period of one cycle for 200ps time for a voltage of amplitude 1V. After performing the simulation $\tau_{PHL}, \tau_{PLH}, \tau_p$ and gain are calculated for output capacitance or load capacitance of 1×10^{-17} F and tabulated in Table V. SF-FinFET shows fast switching as compared to the other two device. Here in our simulation we consider output capacitance or load capacitance 1×10^{-17}, because all the devices simulated here has total gate capacitance range in between 1×10^{-17}F to 1×10^{-18}F. A description about load capacitance is given paper [20]. Gain can be measured taking

978-1-5386-1357-3/17 $31.00 © 2017 IEEE

the derivative of VTC curve. So, inverter gain can be given by $\frac{\partial V_{out}}{\partial V_{in}}$. Fig. 14 shows gain plot of CMOS inverter circuit.

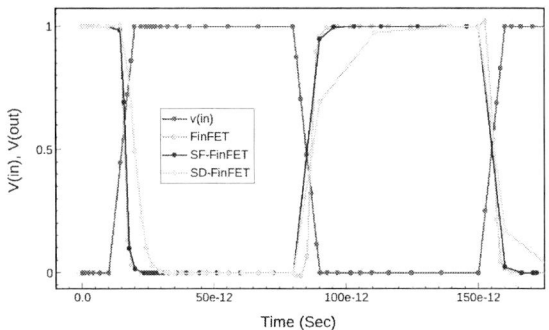

Fig. 13. Switching characteristics plot of the devices.

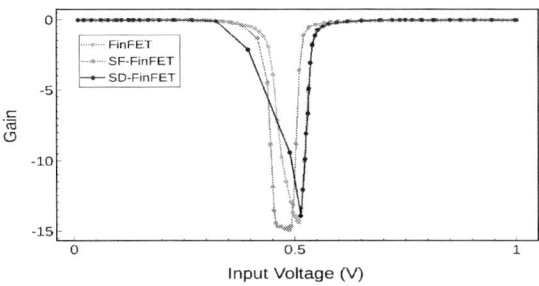

Fig. 14. Gain plot of inverter circuit.

IV. CONCLUSION

This paper bring the novelty of Step Fin FinFET(SF-FinFET) and Step Drain FinFET(SD-FinFET). The performance parameters of the new devices are analyzed and compared with conventional FinFET. A study has been made on SCEs of the devices like Sub-threshold slope(SS), Drain Induced Barrier Lowering(DIBL) and also important figure of merits(FOMs) for analog, RF and Digital application like trans-conductance, drain conductance, intrinsic gain(A_v), Cutoff frequency(f_t), NM_H, NM_L, Propagation delay etc are analyzed. We can observed 1.19X increment of drain current in SD-FinFET compared to the conventional FinFET. This improvement on I_{on} current may due the improved electron mobility and electron density profile in the channel region. We can also observe 1.34X improvement in Trans-conductance in SD-FinFET. P-channel SD-FinFET also gives higher drain current as compared to the other two P-channel devices. Lower total gate capacitance C_{gg} is observed in SF-FinFET due to the step size in the channel region. SF-FinFET shows improvement in digital performance parameter. It is found high noise margin, very low transition region. And the propagation delay for SF-FinFET and FinFET is found to be 0.96ps and 1.20ps respectively, which means SF-FinFET provide fast switching speed in digital design. The improvement in digital parameter is due to the less gate capacitance in SF-FinFET device.

V. ACKNOWLEDGEMENT

Authors are very much thankful to Center for Advanced Computing, IIT Guwahati for providing the resources to carry out this work.

REFERENCES

[1] Pradhan, K. P., S. K. Mohapatra, P. K. Sahu, and D. K. Behera.,"Impact of high-k gate dielectric on analog and RF performance of nanoscale DG-MOSFET.", *Microelectronics journal 45*, no. 2 (2014): 144-151.

[2] Kumar, M. Jagadesh, and Anurag Chaudhry.,"Two-dimensional analytical modeling of fully depleted DMG SOI MOSFET and evidence for diminished SCEs.", *IEEE Transactions on Electron Devices 51*, no. 4 (2004): 569-574.

[3] Maszara, W.P.; Lin, M.-R.,"FinFETs Technology and circuit design challenges",*in Solid-State Device Research Conference (ESSDERC), 2013 Proceedings of the European*,vol., no., pp.3-8, 16-20 Sept. 2013.

[4] Veeraraghavan, Surya, and Jerry G. Fossum.,"Short-channel effects in SOI MOSFETs.", *IEEE Transactions on Electron Devices 36*, no. 3 (1989): 522-528.

[5] Narendar, Vadthiya, and R. A. Mishra.,"Analytical modeling and simulation of multigate FinFET devices and the impact of high-k dielectrics on short channel effects (SCEs).",*Superlattices and Microstructures 85* (2015): 357-369.

[6] Pandey, Archana, Swati Raycha, Satish Maheshwaram, Sanjeev K. Manhas, Sudeb Dasgupta, Ashok K. Saxena, and Bulusu Anand.,"Effect of load capacitance and input transition time on FinFET inverter capacitances.", *IEEE Transactions on Electron Devices 61*, no. 1 (2014): 30-36.

[7] Rim, Kern, Judy L. Hoyt, and James F. Gibbons.,"Fabrication and analysis of deep submicron strained-Si n-MOSFET's.", *IEEE Transactions on Electron Devices 47*, no. 7 (2000): 1406-1415.

[8] Maiti, Chinmay K., Swapan Chattopadhyay, and L. K. Bera.,"Strained-Si heterostructure field effect devices.", *CRC Press*, 2007.

[9] Kumar, Abhishek, and Sruti Suvadarsini Singh.,"Optimizing FinFET parameters for minimizing short channel effects.", *In Communication and Signal Processing (ICCSP), 2016 International Conference on*, pp. 1448-1451. IEEE, 2016.

[10] G. Pei, J. Kedzierski, P. Oldiges, M. Ieong, and E. C. C. Kan,"FinFET design considerations based on 3-D simulation and analytical modeling.", in *IEEE Trans. Electron Devices* , vol. 49, no. 8, pp. 1411-1419, Aug. 2002.

[11] Sarkar, Angsuman, Aloke Kumar Das, Swapnadip De, and Chandan Kumar Sarkar.,"Effect of gate engineering in double-gate MOSFETs for analog/RF applications", *Microelectronics Journal 43,*, no. 11 (2012): 873-882.

[12] K.P. Pradhan, Priyanka, P.K. Sahu,"Temperature dependency of double material gate oxide (DMGO) symmetric dual-k spacer (SDS) wavy FinFET",*Superlattices and Microstructures, 2015* .,vol., no., pp.355-361, 21 Dec. 2015.

[13] Ramkrishna, B. S., B. Jena, S. Dash, and G. P. Mishra., "Investigation of electrostatic performance for a conical surrounding gate MOSFET with linearly modulated work-function.", in *Superlattices and Microstructures*, 101 (2017): 152-159.

[14] K.P. Pradhan, Priyanka, Mallikarjunarao, P.K. Sahu,"Exploration of symmetric high-k spacer (SHS) hybrid FinFET for high performance application", *Superlattices and Microstructures*, Volume 90, February 2016, Pages 191-197.

[15] Ortiz-Conde, Adelmo, FJ Garca Snchez, Juin J. Liou, Antonio Cerdeira, Magali Estrada, and Y. Yue. "A review of recent MOSFET threshold voltage extraction methods.", *Microelectronics Reliability 42*, no. 4 (2002): 583-596.

[16] David M. Binkley, "Tradeoffs and Optimization in Analog CMOS Design", in *John Wiley Sons, Ltd.,Publication*

[17] Debjit Bhattacharya and Niraj k. Jha,"FinFETs: From Device to Architecture",*Hindawi Publishing Corporation. Advance in Electronics Volume 2014*, Article ID 365689,21 pages,;published 7 september 2014.

[18] Synopsys, TCAD Sentaurus User Guide, Version 2014.

[19] Neil H. E. Weste and David Money Harris, "CMOS VLSI Design-A Circuits and Systems Perspective", in *Pearson Education, Inc., publishing as Addison-Wesley*, Fourth Edition

[20] Baidya, Achinta, Trupti Ranjan Lenka, and Srimanta Baishya., "Mixed-mode simulation and analysis of 3D double gate junctionless nanowire transistor for CMOS circuit applications.", in *Superlattices and Microstructures* ,100 (2016): 14-23.

A Design Methodology for MOS Current Mode Logic VCO

Abir J Mondal, Alak Majumder
Department of Electronics & Communication Engineering,
NIT Arunachal Pradesh
Yupia, India
abir_jm@hotmail.com

Bidyut K Bhattacharyya
Department of Electronics & Communication Engineering,
NIT Agartala
Agartala, India
bkbhatta1@yahoo.com

Abstract— A new model of low power MOS current mode logic (MCML) voltage controlled oscillator (VCO) is proposed. The proposed MCML VCO interpolates nine MCML inverters in ring structure with additional transistors biased using control voltage, V_{c1}, to control the current and frequency of the ring oscillator so made. Each inverter in the ring is made of two different current sources, where the one being biased using a constant voltage source (V_c) pumps current constantly and the other pumps current depending on arrival data. The current mode logic (CML) reduces the power consumption and the CML inverter with extra transistor improves the gain of the VCO by increasing the slope of the output frequency, f_{VCO} vs. V_{c1}. The frequency can be tuned between 292 MHz and 1.54 GHz by varying V_{c1} from 0.6 V to 1.4 V for V_c=0.6 V. An average current of 497 μA (895 μW) was consumed by these nine MCML inverters to generate signals with a frequency of 1.07 GHz from a 1.8 V supply.

Keywords—MCML; CML inverter; current source; VCO; phase error

I. INTRODUCTION

Two major obstacles in the production of LSIs operating at gigahertz order frequencies are the excessive design margin that must be considered to accommodate deviations in circuit delay, clock skew and the power dissipation resulting from the gates being operated at a single frequency. Although the performance of CMOS circuits improves with scaling, conventional CMOS circuits cannot simultaneously satisfy the speed and power requirements. Conventional CMOS circuits generate significant supply noise. MOS current mode logic (MCML) has emerged as a logic style that can achieve the much needed high speeds while consuming less power at high frequencies [1]. However, the performance of conventional MCML circuits is found to be affected by supply noise and current. It is observed that with the increase in switching frequencies the current increases almost linearly. Therefore, building conventional MCML circuits are getting even complex with the advancement in silicon technologies. The greatest challenge in MOS current mode signalling is designing an efficient circuit which has almost a constant current beyond certain frequency and also least sensitivity to supply noise.

Hassan et. al [2] portrayed the advantages of a conventional MCML inverter with CMOS circuit. In addition, a MCML ring oscillator (VCO) is designed and analysed in order to verify the accuracy of the proposed MCML inverter in 180 nm technology CMOS process. A temperature compensated circuit [3] operating at 1 MHz and 3.3 V supply voltage is described and analysed using 350 nm technology CMOS process. Reference [4] describes the operation of a de-skew buffer operating at 1.8 V supply and designed using 180 nm CMOS process. A supply sensitive structure with resistive biasing and operating at 2.5 GHz is designed and analysed in 180 nm CMOS process [5]. Puigdemont et. al [6] portrayed a variable length VCO designed using 90 nm Xilinx FPGA. Wang [7] introduces current reuse technique to improve the loop gain using single ended delay cells, but reduces the frequency of operation significantly. A gated current starved circuit operating at 25 MHz is designed and analysed using 5 V supply and 500 nm CMOS process [8]. Zhang et. al [9] portrayed a 1 MHz temperature compensated RO operating at 3 V and designed using 350 nm CMOS process.

Conventional VCOs require current starved inverters to generate oscillation frequency and even additional circuitries are required to compensate for the current drive loss due to temperature and supply voltage. In order to operate at gigahertz frequencies MCML VCOs are also described and analysed as an alternative to conventional CMOS VCO. It is noted that in spite of the use of above techniques, the power dissipation remains a significant factor. Further, it is important to mention that no detail methodology exists to mathematically verify the MCML inverter operation and its subsequent use in VCO. Even though the high-speed application of MCML circuits is well accepted, Fig. [1] portrays the problem associated with such MCML circuits while operated at higher switching frequencies.

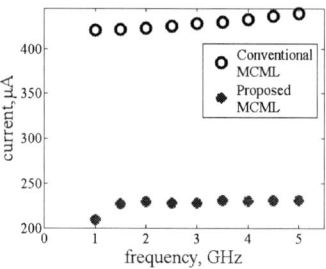

Fig. 1 Plot of current vs frequency

To reduce the consumption of current with increasing switching frequencies, this work attempts to present a low power MCML inverter providing an improved current versus frequency plot. At first, mathematical relations are generated to understand the operation of MCML inverter and its subsequent implementation. Thereafter, a MCML VCO is designed with the proposed MCML inverter and analysed using TSMC 180 nm technology CMOS process. In addition, a methodology to determine phase error is described to indicate what fraction of clock edge will be valid if a 0.1 V, 100 MHz signal is superimposed on supply voltage, Vdd. The effect of supply noise on oscillation frequency is also observed and the ways to reduce it are also analysed.

II. MCML INVERTER

Fig. 2 illustrates the proposed MCML inverter circuit with two current sources, I_1 and I_2 meant for generating the inverter output. One of the current sources (I_2 or M_2) biased using a constant voltage, V_c pump current continuously, while the other current source (I_1 or M_1) pump current depending on the arrival of data at the input of the gate of M_1. When the input of M_1 is high, current is driven by M_1, otherwise there is no current flow through M_1 with low input. The gate M_1, being the signal, controls the current swing which in turn changes the voltage of V_{out}. Eq. (1) displays the voltage at V_{out}, also denoted by VH, when the gate of M_1 is on,

$$R_1 \times (I_1 + I_2) = VH = V_{out} \tag{1}$$

Similarly, when the gate, M_1 turns off, there is no current flow through M_1, which effectively means $I_1=0$ and the voltage at V_{out}, now denoted by VL, can be written using (2),

$$R_1 \times I_2 = VL = V_{out} \tag{2}$$

Using the above two equations and knowing the required VL and VH one can approximate the currents I_1 and I_2 for the circuit shown in Fig. 2. Consequently, the values of I_1 and I_2 will help in determining the corresponding transistor sizes for M_1 and M_2. The current source I_2 is set with a NMOS M_2 with its gate driven by a dc to get a constant current provider.

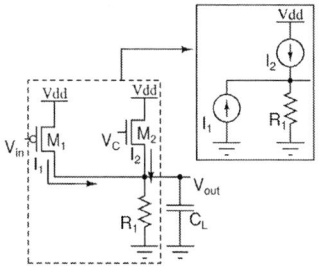

Fig. 2 Schematic of proposed MCML inverter

R_1 in the above equations controls the current drive capability of M_2 or, in other words, defines the lower limit of V_{out}. The current source I_1 is set using a PMOS M_1 along with its gate driven by a pulse switching at data rate. M_1 pumps current to increase the V_{out}, thus defining the upper limit, VH.

III. IMPLEMENTATION OF MCML VCO

Fig. 3 shows the implementation of VCO operating at 1.07 GHz with an interpolating ring structure. To obtain controlled oscillator voltage, the delay at each stage needs to experience necessary variation. Each stage here resembles the MCML inverter (shown in Fig. 2) connected with additional transistors (M_{19}-M_{27}) biased using V_{c1}. These transistors control the amount of current available to discharge the capacitive load of each stage, thereby controlling the frequency. Simultaneously, the transistors M_{19}-M_{27} control the discharge current. Fig. 4 portrays the frequency performance, f_{VCO} of the MCML VCO as a function of V_{c1} for different V_c. The VCO receives the V_{c1} and generates a clock with its frequency tuned by the magnitude of the V_{c1}. The tuning characteristics of the VCO should be as linear as possible. It is apparent from Fig. 4 that for V_{c1} between 1.8 V and 1.3 V, the slope of the measured frequency, i.e. the VCO gain, is almost constant for different V_c. The slope of output frequency becomes linear beyond 1.3 V thus providing a frequency tuning range (FTR) between 292 MHz and 1.54 GHz.

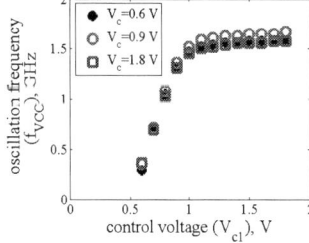

Fig. 4 Plot of oscillation frequency versus control voltage

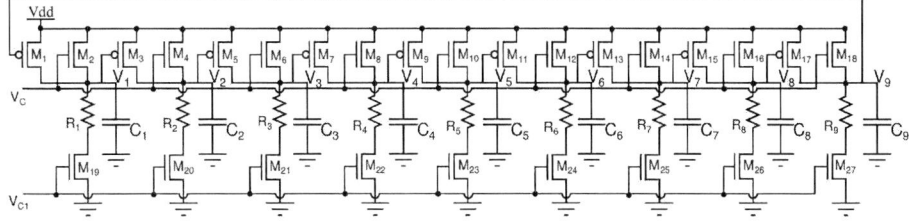

Fig. 3 Schematic of proposed MCML VCO

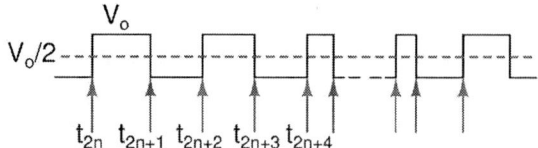

Fig. 5 Phase error calculation methodology for the proposed MCML VCO

A. Methodology and understanding of phase error in MCML VCO

Fig.5 demonstrates the methodology developed to understand the phase error from the output signal of a clock. At the outset, considering the output of the clock circuit a reference line is drawn following the midpoint of the voltage swings, thus cutting the clock's rising and falling edges at various points. As evident in the figure, for rising edges it cuts at t_{2n} (n=0,1,2...) and for falling edges at t_{2n+1} (n=0,1,2...). The difference, (3) gives the clock pulse width.

$$\Delta t_{2n} = t_{2n+1} - t_{2n} \qquad (3)$$

If Δt_{2n} is constant, for every n the phase error is zero. If Δt_{2n} varies as a function of time then there are phase errors from clock edges to clock edges. Further, dividing Δt_{2n} by the fundamental angular frequency (ω_{2n}) of the clock signal at a particular point will give rise to the phase error. The actual period of the clock signal at any given point is given by (4), where T_{2n} is a running variable.

$$T_{2n} = t_{2n+2} - t_{2n} \qquad (4)$$

Therefore (5) is calculated for every section to understand the phase error of the clock output. For large number of sample sizes, the average and standard deviation of this phase error are also calculated. In Fig. 6 the measured phase error distribution is shown by injecting 100 MHz noise in V_{c1} at 10 mV (Fig. 3) fluctuations in voltage. The DC voltage of V_{C1} at that time was 0.8 V, which is about 1.25% of Vdd in magnitude. It is interesting to see that the probability distribution obeys the Central Limit Theorem and it is Gaussian for large sample sizes, even though there are 5 degree phase shifts. For instance, a clock frequency of 3 GHz corresponds to a clock period of 333 psec. The 5 degree phase error will correspond to 29 psec offset from one clock edge to other, which is about 9%.

$$\phi_{2n} = \Delta t_{2n} \times \frac{2\pi}{T_{2n}} \qquad (5)$$

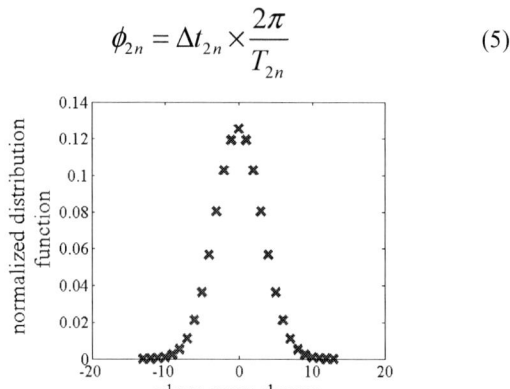

Fig. 6 Simulated phase error distribution for V_{c1}=0.8 V

IV. SIMULATION RESULTS

The circuits were designed and simulated using TSMC 180 nm technology CMOS process and a supply voltage of 1.8 V. Fig. 7a portrays the plot of oscillation frequency versus V_{c1} for different Vdd. The simulation results show that for V_{c1} between 1.8 V and 1.3 V, slope of f_{VCO} remains constant. The FTR is 1.28 GHz appearing between 353 MHz and 1.63 GHz at a 1.8 V. The plot of Fig. 7b indicates that the measured current increases linearly for different V_{DD} as V_{c1} and f_{VCO} varies. In order to compare the proposed MCML-VCO circuit with the one developed by Hassan et al [2], the two circuits are simulated using TSMC 180 nm technology in Lt spice. Fig. 8 shows the f_{VCO} as function of the controlled voltage for both the two circuits while operated at V_{DD}=1.8 V. Evidently, the MCML VCO [2] results in a linear slope for a small V_{c1} variation compared to the proposed MCML VCO at 1.8 V. Further, the summary of performance data in Table 1 clearly depicts that the power consumption of the proposed MCML VCO is much lower than that of MCML VCO [2] with comparable FTR.

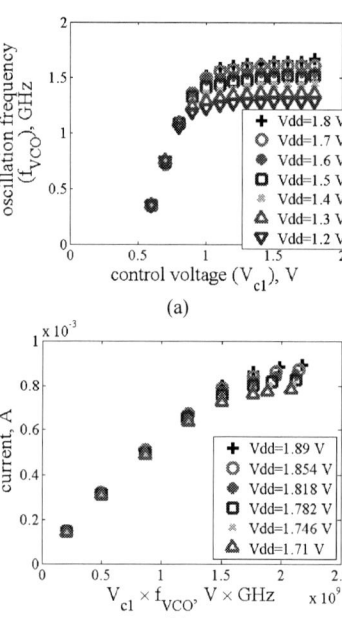

Fig. 7 Plot of a) simulated frequency and b) current versus control voltage with Vdd variations

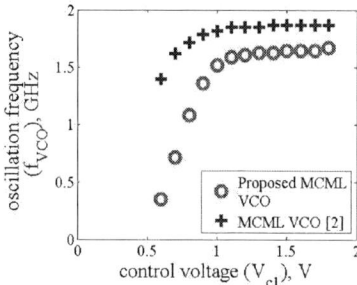

Fig. 8 Plot of simulated f_{VCO} with V_{c1} for the proposed and conventional VCOs

TABLE I COMPARISON WITH CONVENTIONAL MCML VCOs

Ref.	Units	[2]	[7]	This work
Tech.	μm	0.18	0.18	0.18
FTR	Hz	1.4 G-1.85 G	12.6 M-48 M	292 M-1.54 G
V_{DD}	V	1.8	1.2	1.8
V_{cl}	V	0.8	1.2	0.8
f_{VCO}	Hz	1.7 G	47M	1.07G
P_{DC}	Watt	6.84 m	1.8m	895 μ

V. CONCLUSIONS

In this paper, a new low power MCML VCO is proposed. Eliminating the use of an external reference frequency, the proposed MCML VCO achieves low power and wide frequency tuning range using CML inverter and extra transistor. The proposed MCML VCO being independent of V_{dd} helps in high speed high performance CPU. Because, initially when the CPU starts clocking, the large instantaneous current drawn by the device generates voltage drop in V_{DD}. In real life this causes functional failure. In the proposed method the isolated power supply V_{cl}, which draws very small current with low voltage fluctuations, controls the clock frequency of VCO. This frequency can be tuned from 292 MHz to 1.54 GHz by varying the V_{cl} from 0.6 V to 1.8 V while consuming a current of between 145 μA and 859 μA.

References

[1] M. Yamashina and H. Yamada, "An MOS current mode logic (MCML) circuit for low-power sub-GHz processors," IEICE Trans. Electron., vol. E75-C, no. 10, pp. 1181-1187, Oct. 1992.

[2] H. Hasan, M. Anis and M. Elmasry, "MOS current mode circuits: analysis, design and variability," IEEE Trans. VLSI., vol. 13, no. 8, pp. 885-898, Aug. 2005.

[3] K. N. Leung, C. H. Lo, P. K. T. Mok, Y. Y. Mai, W. Y. Leung and M. J. Chan, "Temperature-compensated CMOS ring oscillator for power-management circuits," Electronic Letters, vol. 43, no. 15, pp. 786-787, July 2007.

[4] Y. G. Chen, H. W. Tsao and C. S. Hwang, "A fst locking all-digital deskew buffer with dut-cycle correction," IEEE Trans. VLSI, vol. 21, no. 2, pp. 270-280, Feb. 2012.

[5] X. Gui and M. M. Green, "Design of CML ring oscillators with low supply sensitivity," IEEE Trans. VLSI, vol. 60, no. 7, pp. 1753-1763, Apr. 2013.

[6] J. P. Puigdemont, F. Moll and A. Calomarde, vAll digital simple clock synthesis through a glitch free variable length ring oscillator," IEEE Trans. Circuits and Systems II: Express Brief, vol. 61, no. 2, pp. 90-94, Jan. 2014.M.

[7] S. F. Wang, "Low voltage, full swing, voltage controlled oscillator with symmetrical even-phase outputs based on single-ended delay cells," IEEE Trans. VLSI, vol. 23, no. 9, pp. 1801-1807, Sept. 2015.

[8] S. Zhang, A. Li, Y. Han, L. Jie, X. Han and R. C. C Cheung, "Temperature compensation technique for ring oscillators with tail current, " Electronic Letters, vol. 52, no. 13, pp. 1108-1110, June 2016.

2017 IEEE International Symposium on Nanoelectronic and Information Systems

A 90nm Novel MUX-Dual Latch Design Approach for Gigascale Serializer Application

Monalisa Das[†,‡], Alak Majumder[ǁ,‡], Abir J Mondal[¥,‡] and Bidyut K Bhattacharyya[†,♯]

‡VLSI Design Laboratory,
Department of Electronics & Communication Engineering,
National Institute of Technology, Arunachal Pradesh,
Yupia, Dist.– Papumpare, Arunachal Pradesh 791112, India.
♯Department of Electronics & Communication Engineering,
National Institute of Technology, Agartala 799046, India.
†*monalisadasece@gmail.com*, ǁ*majumder.alak@gmail.com*, ¥*abir_jm@hotmail.com*, †*bkbhatta1@yahoo.com*

Abstract— **As the high speed electronic systems are in the midway of getting shifted from conventional parallel data transmission to new high data rate serial link, design of an unit cell (i.e. MUX) for the serializer interface has become an area of interest. In this paper, we have demonstrated current mode logic based novel 2:1 multiplexer featuring dual latch to be steered by either CLK or CLK$_{BAR}$. The new design approach is simulated for 90nm CMOS technology using Cadence Virtuoso platform at a power supply of 1Volt with 10GHz switching frequency. The main noted point of this circuit is that it generates a massive output swing of 955mV (95.5% of the power supply) and outsmarted the prior arts by offering an average power and delay of 61.02µW and 32.93ps respectively. The robustness of the architecture is tested in different process corners with 'no skew' and '5% process skew' through 200 runs of Monte-Carlo. The energy/bit, RMS Jitter, PP Jitter and BER of the proposed design read a value of 1.74fJ, 4.41ps, 15.01ps and <10^{-11} respectively.**

Keywords— *Current Mode Logic (CML), Multiplexer, Latch, MUX-Latch, High frequency, Data rate*

I. INTRODUCTION

The trend to create a faster communicating environment, has forced the communication researchers to switch from parallel to serial ways of data transmission through channels. During the era of such serial communication, it is urged to achieve a data rate in gigascale range. Besides that, the designer takes care of a circuit arrangement to withstand all the limitations related to faster transmission, such that quality of the signal is maintained with a higher level of accuracy. The main component focused today by the researchers for faster transmission and reception is the Serializer-Deserializer (SerDes). This component plays an important role of converting a parallel data set to serial data and vice-versa for transmission and reception of signals. Such SerDes system generally consist of MUX and DEMUX circuits respectively arranged in a tree like fashion.

Although it is mentioned in [1] that CMOS logic is the most common to design any circuitry, it limits the circuit performance at higher data rate due to power supply noise and larger delay. Thus, the evolution of Current Mode Logic (CML) serves better in this regard, as it is based on differential pair concept, a better supplement for noise reduction. The

advantages of CML over voltage mode (VM) logic in terms of performance improvement, specific features and structural merit has paved the use of CML for designing continuous time analog circuits [2]. Performance improvement of CML over VM includes high speed, low power consumption at higher frequencies, high signal dynamic range, low crosstalk and low switching noise. However the CML has its own disadvantages like reduced voltage headroom, more chip area and large power consumption compared to CMOS logic [3].

It has been mentioned in [4] that, the reduced swing of CML output helps in delay minimization to produce output at a faster rate. But practical realization of such circuits in serializers, leads to a major drawback due to cascaded MUX-tree arrangement, which may lead to the loss of peak-to-peak swing of the signal at the final output. Though there's one way to increase the swing by using a high resistive load at the pull up (which is found to be almost >25 KΩ); but practically it is an absurd method as the fabrication of high value resistors is tedious as well as consumes larger area. Thus, one way to overcome this is to use latch circuits, which holds and keeps track of the output to obtain an improved swing compared to the conventional one.

In this paper, a novel approach of designing a MUX- Dual Latch circuit is unveiled to achieve large output swing. The rest of the paper is organized as follows: Section II discusses a brief review on the prior works of MUX and MUX-Latch circuits for SerDes applications. This is followed by the explanation of the proposed dual Latch based CML MUX circuit and its performance analysis in Section III and Section IV respectively. A comparison chart is presented in Section V and finally, Section VI concludes the work.

II. PRIOR RELATED WORK

A. Existing CML based MUX Circuits

Different techniques are implemented to overcome the limitations of the conventional approach so as to obtain a high speed MUX for SerDes applications. Alioto et.al [5] proposed a BJT based CML model to design high speed MUX with the frequencies ranging from 6 to 20 GHz. Like the conventional one, this model also uses a pull-up resistor. However, in [6] an

978-1-5386-1357-3/17 $31.00 © 2017 IEEE

active PMOS load is used as a replacement of this pull-up resistor to obtain a data rate of 10 Gbps and lesser power consumption. Another approach is presented by Tondo et.al [7], using a combination of both CML and CMOS logic to design a high speed 2:1 MUX and DEMUX with a data rate of 40 Gbps. But the requirement of conversion blocks to design such combination circuits increases its complexity. The most effective method to increase the output swing as well as to obtain a faster data rate is presented in [8], which uses passive components, such as inductors to form a transformer coupling (TC) technique to see better output swing at a speed of 40 Gbps. But the physical design of inductors is itself a challenging task. Besides improving speed, it is also necessary to maintain signal quality, which may be served using a latch circuit embedded between the differential outputs of each MUX available in Serializer design by giving penalty in transistor count to some extent. Thus a new CML based MUX-Latch is to be designed to attain faster transmission and reception in serial links with better signal quality.

B. Existing CML based MUX-Latch Circuits

As mentioned in [9], latch is basically a storage element which can also be used in circuits to perform 'HOLD' operation so as to obtain almost full voltage swing at the output of the CML based MUX circuit. The main issue in conventional MUX is the logic 0 level, which is somewhat solved by the use of MUX-Latch concept [10]. It is also worth mentioning that, a latch samples the data in the one half period of CLK and holds the same in the other half period. A Conventional CML MUX-Latch circuit is shown in figure 1.

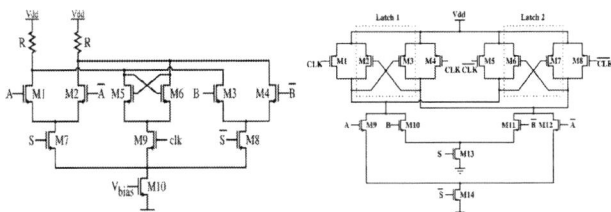

Fig. 1. Conventional MUX-Latch   Fig. 2. Proposed Mux-Dual Latch

Another new MUX-Latch circuit is designed with a latency combined approach and is described in [11]. This architecture utilizes a CMOS latch to hold the output swing. But one common problem in all the MUX-Latch circuits so far is that the low logic levels of both the differential outputs are not same. This is because the latch works only for the half of the time period of CLK thereby helping only single end of the differential outputs to get proper logic 0. This creates serious signal integrity problems when such outputs are being used to drive next stages of cascaded MUX-tree circuit (i.e. Serializer). So, in this work an effort is made to address this problem with a Mux-Dual Latch circuit to observe better performance in terms of speed, area, power and jitter.

III. PROPOSED CML BASED MUX-LATCH

An overall limitation recorded from the previous circuits sets the necessity to design a circuit that mitigates the problem of unbalanced swing as well as maintains faster data rate with

better signal quality. As it can be inferred that a single latch in MUX cannot serve the purpose of balancing the output swing, the concept of dual latch is incorporated. Such dual latches can be connected in series as well in parallel to each other in the circuit. But series latches produce more delayed output than the parallel one.

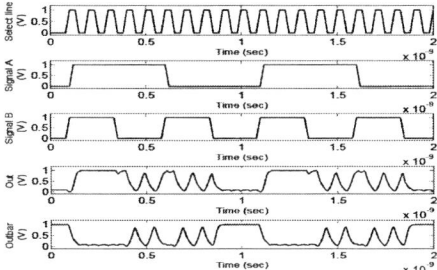

Fig. 3. Transient of 2:1 CML MUX-Latch using dual PMOS latch

The proposed MUX using parallel dual latch circuit is shown in figure 2. The PMOS based latching part is controlled by clock signal (CLK) which is kept same as the select line (S) of MUX, designed using NMOS transistors. The circuit is arranged in such a way that both the latch remain active for CLK = 0 or 1. The working of this circuit possess two modes of operation: Logic mode and Latch mode. If CLK=0, the input A will hit the output, whereas B will reach output only when CLK=1. It first follows the operation of basic Conventional MUX of providing complemented outputs, which can be termed as the Logic mode. This logic is hold by both the latch circuits irrespective of what is available in CLK to maintain the desired level. It is worth mentioning that the Latch mode works simultaneously with the Logic mode. The major highlight in this proposed circuit is that the logic 0 level approaches almost the ground (GND) level, which was not the case in prior arts. The transient waveform of this circuit considering A=0 and B=1 is shown in figure 3, where CLK is running at a frequency of 10 GHz and the circuit measures an output signal swing ranging from 0.045Volt to 1Volt.

IV. SIMULATION RESULTS

The proposed CML based 2:1 MUX-Dual Latch circuit is designed and simulated for 90nm CMOS technology using CADENCE Virtuoso Platform at a power supply of 1Volt and CLK switching frequency of 10GHz with a rise time and fall time of 20ps each. To prove its robustness, we have performed different analyses as discussed below:

A. DC and AC analysis

The DC operating points of the proposed circuit are given in Table I. The lower and higher noise margin reads a value of 75.31mV and 69.94mV.

TABLE I. DC OPERATING POINTS

Parameters	Operating Points
V_{OH}	1V
V_{IL}	171.41 mV
V_{IH}	930.06 mV
V_{OL}	96.1 mV
NM_L ($-V_{IL} - V_{OL}$)	75.31 mV
NM_H (=$V_{OH} - V_{IH}$)	69.94 mV

The AC analysis of the proposed MUX- Dual Latch provides a Unity Gain Frequency (UGF) and Bandwidth (BW) of 17.14 GHz and 23.13 GHz respectively.

B. Delay Analysis

The implementation of two Latches in the proposed circuit offers the possibility of larger delay, which reads 32.93ps while driving a load of 0.5fF. Variation of delay with power supply at different loads is plotted in figure 4.

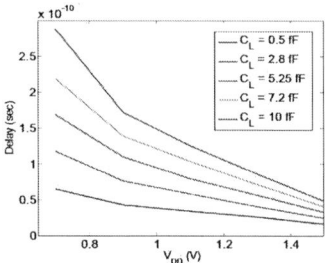

Fig. 4. Delay analysis vs Supply Voltage at different loads

C. Power Analysis

The total average power of the proposed circuit is found to be 61.02µW, as displayed in table II. To analyze its performance at different load conditions, figure 5(a) is plotted to show the reduction of average power with scaling down of power supply. Keeping the input signals A and B fixed at 0 and 1 respectively the average power variation with increasing CLK frequency is observed and displayed in figure 5(b). The linear fit line for this plot gives a dynamic and static power of 4µW/GHz and 40µW respectively.

TABLE II. DETAILS OF AVERAGE POWER AT 10GHz

Input Combinations		Average Power (µW)
A	**B**	
0	0	29.015
0	1	88.748
1	0	88.748
1	1	37.569
Total Average Power (µW)		61.02

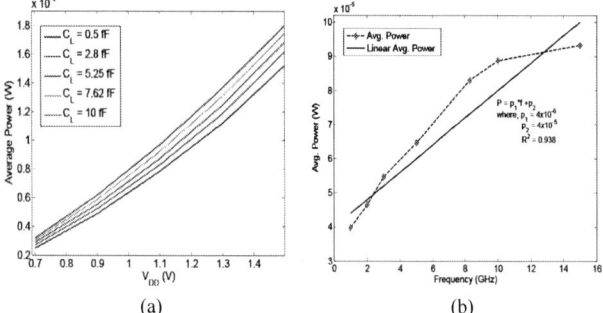

(a)   (b)

Fig. 5. (a) Average Power Vs V_{DD} at different load conditions (b) Average Power Variation with increasing Frequency

D. Noise Analysis

The output noise and phase noise as a function of increasing frequency is plotted in figure 6, which reads the output noise and phase noise to be -159.42 dB and -151.59 dBc/Hz respectively at 10MHz offset.

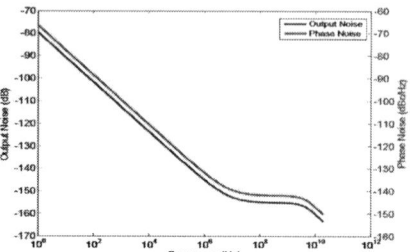

Fig. 6. Noise analysis of the Proposed circuit

E. Performance at Higher Frequencies

Performance of the proposed 2:1 MUX-Dual Latch circuit at much higher frequencies is checked and plotted in figure 7. It can be inferred that as the frequency gets increased beyond 10 GHz, the output peak-to-peak swing is detected to get decreased and beyond 33 GHz, it gets reduced drastically.

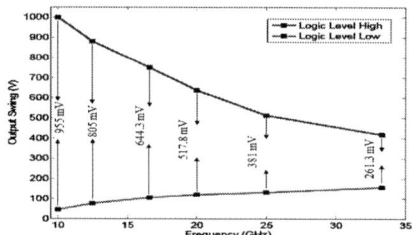

Fig. 7. Output swing variation with increasing Frequency

F. Process Corner Analysis

To test the robustness of the proposed circuit, it is run under extreme process corners like Nominal, Fast-Fast (FF), Fast-Slow (FS), Slow-Fast (SF), and Slow-Slow (SS). The transient at different corners is shown in figure 8, which infers that the output swing at nominal is measured to be from 45 mV (low logic level) to 1000mV (high logic level). For all the corners the logic high level is fixed, whereas the logic low level is shifted to 88 mV, 47 mV, 210 mV, and 101 mV for FF, FS, SF and SS respectively.

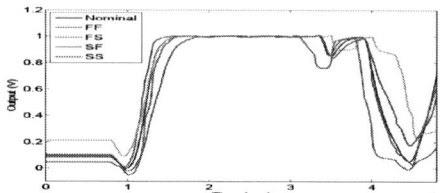

Fig. 8. Output Transient Response at Extreme Corners

G. Monte Carlo Simulation

Monte Carlo Analysis is performed to check the circuit's reliability at extreme conditions. Figure 9 depicts the histogram analysis for three different parameters: delay, power and PDP. After 200 runs, the mean and standard deviation of delay (Refer figure 9(a)) are found to be 33.04ps and 3.11ps respectively. So, it can be inferred that for 3σ statistical variation, the delay ranges from 23.7ps to 42.3ps. Similarly, from figure 9 (b) and 9(c), the average power and PDP values are detected to be ranging from 53.5µW to 67.42µW and 1.4fJ to 2.58fJ respectively. Table III shows the variation in average

978-1-5386-1357-3/17 $31.00 © 2017 IEEE   212

TABLE III. PROCESS VARIATION OF THE PROPOSED CIRCUIT

Process	No skew			5% skew					
	Avg. Power (μW)	Delay (ps)	PDP (fJ)	Avg. Power (μW)		Delay (ps)		PDP (fJ)	
				Mean	Standard Deviation	Mean	Standard Deviation	Mean	Standard Deviation
Nominal	61.02	32.93	2.01	60.02	5.32	32.64	1.82	1.95	0.12
FF	74.29	27.12	2.01	72.97	6.50	26.85	1.48	1.95	0.13
FS	53.84	33.88	1.82	52.81	5.27	33.68	1.97	1.78	0.11
SF	58.17	31.46	1.83	57.54	4.65	31.32	2.59	1.79	0.03
SS	45	43.68	1.96	44.41	4.05	43.35	3.35	1.92	0.10

power, delay and PDP (Power-Delay-Product) for no skew and 5% skew conditions at different corners. 5% skewed analysis is done for different parameters (V_{DD}, length (L), and rise time, fall time) to determine circuits tolerance to extreme variations. It is observed that at different conditions the mean and standard deviation values remain almost same, which proves its reliability for better performance.

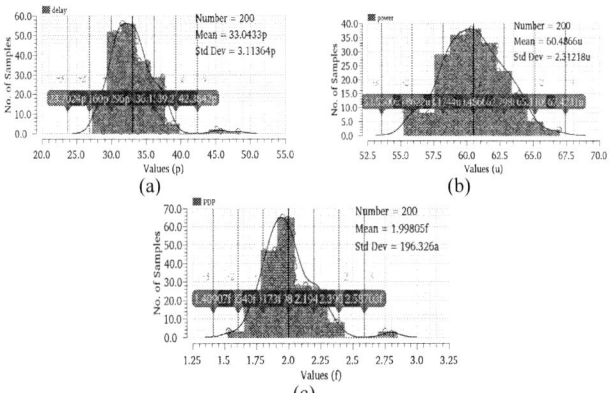

Fig. 9. Histogram of proposed MUX-Dual Latch circuit for (a) Delay (b) Average Power and (c) PDP

H. Eye Diagram of Proposed MUX-Latch

Eye diagram of the output waveform at different frequencies are generated for 200 runs of Monte Carlo Simulation and is shown in figure 10. It can be realized from the figure that with increasing clock frequency the eye generated from the output signal gets deteriorated. Table IV summarizes different parameters related to eye measurement at different frequencies like 5GHz and 10GHz, while offering a data rate of 10Gbps and 20Gbps respectively. The eye width and height are seen to maintain an inversely proportional relation with frequency.

Bit Error Rate (BER) of the proposed 2:1 MUX-Latch circuit is calculated using the principle of Bathtub curve and it reads $< 10^{-11}$ thereby referring this circuit to be a better option for Serializer design.

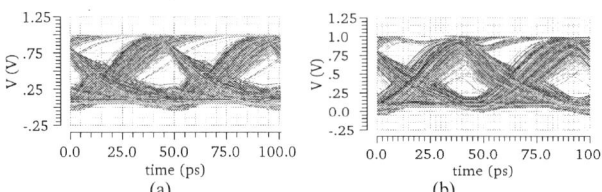

Fig. 10. Generated Eye diagram at (a) 5 GHz and (b) 10 GHz

TABLE IV. EYE PARAMETERS

Parameters	Data Rate (Gbps)	
	10	20
P-P Jitter (ps)	10.42	15.01
RMS Jitter (ps)	2.92	4.41
Level 0 mean (mV)	132.36	169.43
Level 0 std. dev (mV)	104.75	135.91
Level 1 mean (mV)	935.8	855.76
Level 1 std. dev (mV)	112.97	143.68
Eye Amplitude (mV)	760.62	725.46
Eye Height (mV)	141.9	119.29
Eye Width (ps)	47.11	45.1
Eye S/N	3.68	3.59
Eye Rise Time (ps)	19.32	16.65
Eye Fall Time (ps)	21.27	12.88
BER	$<10^{-11}$	$<10^{-11}$

V. COMPARISON WITH PRIOR ART

The performance comparison of the proposed 2:1 MUX-Dual Latch circuit with other existing works is shown in Table V. Different works in this table used passive components to increase the speed, but such designs limit its use due large area and power consumption. Hence, the proposed architecture is found to outsmart the prior arts in terms of performance metrics. The use of only 14 transistors in proposed circuit, makes the design quite simple.

From Table V, it can be clearly conferred that the proposed circuit will play a better role than existing CML circuits for serial link design. The energy dissipated per bit is calculated for 2ns transient and is found to be as small as 1.74fJ/bit only.

VI. CONCLUSION

A process variation aware current mode design of MUX-Dual Latch circuit is unearthed in this article to address the issues present in prior arts. The circuit arrangement doesn't only offer lesser energy dissipation, but also is capable of sampling data with respect to the clock frequency up to 33GHz (corresponding to 66Gbit/s) by maintaining a decent output swing. The superiority of this proposed architecture has led its way to be a useful supplement in Serializer design of high speed on-chip serial links and network routing. Also, due to its low power consumption, it may become a potential candidate to be used as communication link in future FPGAs.

Acknowledgment

We would like to thank MEITY, Government of India for the financial assistance under SMDP-C2SD Project to carry out this work.

TABLE V.   COMARISON WITH PRIOR ARTS

Parameters	This Work	Ref. [15]	Ref. [10]		Ref. [14]	Ref. [9]	Ref. [8]	Ref. [13]	Ref. [12]
Technology	90nm CMOS	90nm CMOS	90nm CMOS		90nm CMOS	65nm CMOS	65nm CMOS	120nm CMOS	250nm Bi-CMOS
Function	2:1	4:2	4:1	8:1	2:1	16:1	2:1	2:1	2:1
Features	CML MUX-Latch using dual Latch concept	CML	CML Mux-Latch		TC CML	CML	Transformer coupled technique	TC CML	CML/ECL Latch topology
Supply (V)	1	1.2	0.5		0.7	1.2	0.7	1.5	2.5
Transistor Count	14		10+passive components	17+passive components	8+passive components	--	8+passive components	10+passive components	7+passive components
Swing (mV)	904	866	--	--	50-100	761	175	100	150
CLK Freq. (GHz)	10	--	6	6	--	20	11	20	--
Data Rate (Gbps)	20	16	6	12	60	40	40	40	14
Power (μW)	61.02	72149	170000	308000	100000	1.6×10^6	7700	10000	146700
Delay (ps)	32.93	17.6	--	--	--	--	--	--	--
PDP (fJ)	2.01	1269	--	--	--	--	--	--	--
RMS Jitter (ps)	4.41	--	4.39/7.47	2.26/4.66		--	1.8	--	--
P-P Jitter (ps)	15.01	7.235	32/40	28/30		--	9.78	--	18
BER	$<10^{-11}$	--	$<10^{-12}$	$<10^{-12}$	--	$<10^{-11}$	--	--	--
Energy/bit (fJ)	1.74	--	--	--	--	--	--	--	--

References

[1] Anna Pena Martinez, "Design of MOS Current-Mode Logic Standard Cells, " *M. Tech Thesis, Microelectronics System Laboratory (LSM), EPFL*, 2007.

[2] Abdullah Al Owahid, "Design of 3.33GHz CML Processor Datapath," *M. Tech Thesis, Auburn University*, 2012.

[3] Rogers, John, Plett, Calvin, Dai, Foster, "Integrated Circuit Design for High-Speed Frequency Synthesis," *Chapter 5, Artech House Publication*, 2006.

[4] H. Hassan, M. Anis and M. Elmasry, "MOS current mode circuits: analysis, design, and variability," in *IEEE Transactions on Very Large Scale Integration (VLSI) Systems*, vol. 13, no. 8, pp. 885-898, Aug. 2005.

[5] M. Alioto and G. Palumbo, "Modeling and optimized design of current mode MUX/XOR and D flip-flop," in *IEEE Transactions on Circuits and Systems II: Analog and Digital Signal Processing*, vol. 47, no. 5, pp. 452-461, May 2000.

[6] A. Tanabe *et al.*, "0.18-μm CMOS 10-Gb/s multiplexer/demultiplexer ICs using current mode logic with tolerance to threshold voltage fluctuation," in *IEEE Journal of Solid-State Circuits*, vol. 36, no. 6, pp. 988-996, Jun 2001.

[7] D. F. Tondo and R. R. Lopez, "A low-power, high-speed CMOS/CML 16:1 serializer," *2009 Argentine School of Micro-Nanoelectronics, Technology and Applications*, San Carlos de Bariloche, 2009, pp. 81-86.

[8] F. T. Chen, J. M. Wu and M. C. F. Chang, "40-Gb/s 0.7-V 2:1 MUX and 1:2 DEMUX with Transformer-Coupled Technique for SerDes Interface," in *IEEE Transactions on Circuits and Systems I: Regular Papers*, vol. 62, no. 4, pp. 1042-1051, April 2015.

[9] N. Nedovic *et al.*, "A 3 Watt 39.8–44.6 Gb/s Dual-Mode SFI5.2 SerDes Chip Set in 65 nm CMOS," in *IEEE Journal of Solid-State Circuits*, vol. 45, no. 10, pp. 2016-2029, Oct. 2010.

[10] W. Y. Tsai, C. T. Chiu, J. M. Wu, S. S. H. Hsu and Y. S. Hsu, "A Novel Low Gate-Count Pipeline Topology With Multiplexer-Flip-Flops for Serial Link," in *IEEE Transactions on Circuits and Systems I: Regular Papers*, vol. 59, no. 11, pp. 2600-2610, Nov. 2012.

[11] Travis William Lovitt, "Single Stage Latency Combined Multiplexer and Latch Circuit", United States Patent, Patent No. – US 9350355B1, May, 2016.

[12] O. Schrape, M. Appel, F. Winkler and M. Krstić, "Low-power design methodology for CML and ECL circuits," *2014 24th International Workshop on Power and Timing Modeling, Optimization and Simulation (PATMOS)*, Palma de Mallorca, 2014, pp. 1-5.

[13] D. Kehrer, H. D. Wohlmuth, H. Knapp, M. Wurzer and A. L. Scholtz, "40-Gb/s 2:1 multiplexer and 1:2 demultiplexer in 120-nm standard CMOS," in *IEEE Journal of Solid-State Circuits*, vol. 38, no. 11, pp. 1830-1837, Nov. 2003.

[14] D. Kehrer and H. D. Wohlmuth, "A 60-Gb/s 0.7-V 10-mW monolithic transformer-coupled 2:1 multiplexer in 90 nm CMOS," *IEEE Compound Semiconductor Integrated Circuit Symposium, 2004.*, 2004, pp. 105-108.

[15] I. Jang, Y. Lee, S. Kim and J. Kim, "Power-Performance Tradeoff Analysis of CML-Based High-Speed Transmitter Designs Using Circuit-Level Optimization," in *IEEE Transactions on Circuits and Systems I: Regular Papers*, vol. 63, no. 4, pp. 540-550, April 2016.

978-1-5386-1357-3/17 $31.00 © 2017 IEEE

Leakage Reduction in DT8T SRAM Cell using Body Biasing Technique

Rajani Suthar, Kirti S. Pande, Murty N.S.
Department of Electronics and Communication Engineering
Amrita University
Bengaluru, India
rajanisuthar911@gmail.com, kirtispande@yahoo.co.in and ns_murty@blr.amrita.edu

Abstract—**During recent years SRAM has become the topic of crucial research due to increased demand in mobile applications. For low power applications, leakage reduction is very important. This brings the incentive to design low leakage SRAM cells. In this work, a Dynamic Threshold 8T (DT8T) SRAM cell is proposed to elevate the performance in terms of maintaining stability and leakage reduction by using variable threshold voltage transistors. During write and hold mode, drivability of transistors is improved with forward body biasing. For reducing the leakage current, reverse body biasing is used during read mode of operation. The proposed circuit is designed in 45 nm CMOS technology and is simulated using Cadence Virtuoso platform. At supply voltage of 0.5 V, read leakage currents through feedback transistors are observed as 0.261 pA and 0.198 pA when body is reverse biased at 0.6 V and these are reduced by approximately 84% compared to D8T SRAM cell. This proposed SRAM cell draws 7.31 nA leakage current in standby mode which is 24% lesser compared to conventional 6T SRAM cell.**

Keywords—SRAM; Subthreshold Leakage current; Body biasing; Dynamic Threshold.

I. INTRODUCTION

Rapid advancement of CMOS technology has provided many advantages like faster and denser integration and low operating power for chip designing. To achieve the above mentioned benefits, CMOS devices have been scaled down dramatically [1]. Transistor scaling improves the integration density but also leads to increase in leakage current. For long term operation, efficient power management is required in portable, wireless and biomedical applications.

With domination of VLSI devices in electronics industry, on-chip memory is essential for high throughput, low latency microprocessors and system-on-chip products. Static Random Access Memories (SRAMs) are used as cache memory in modern embedded system applications because they are faster compared to DRAM (Dynamic Random Access Memory). For battery operated devices the power consumption is crucial [2]. It is found that SRAM occupies 90% die area in System of Chip (SoC) [3]. Therefore, SRAM is dominant participator in the total power consumption of the system. Scaling of supply voltage has helped in reducing the dynamic power consumption. Subthreshold leakage current increases, when threshold voltage is scaled down. With the scaling of MOS technology, the main challenge is to reduce leakage power in SRAM cell.

The traditional 6T SRAM cell structure is presented in Fig. 1. This cell comprises of two cross-coupled CMOS inverters for storing data and two access transistors for data transfer. The sizing of access transistors calls for a trade-off between the write ability during write operation and the data retention in read mode [4]. In subthreshold region, a memory circuit faces more challenges because of the high standby power and instability. The main goal at lower power supply is that SRAM cell should perform write, hold and read operations conveniently and achieve power savings as well.

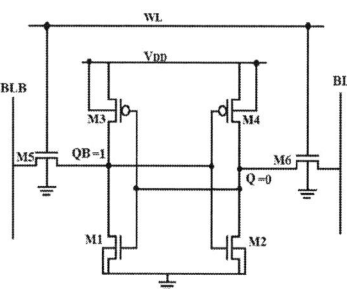

Fig. 1. Traditional 6T SRAM cell

This paper proposes a Dynamic Threshold 8T (DT8T) SRAM cell suitable for subthreshold operations and improved cell performance in terms of stability, speed and leakage current. To check out the effectiveness of the DT8T cell, the proposed cell is compared with the conventional 6T cell and existing D8T cell [5].

The rest of the paper is organized as follows. Section II covers literature survey of different SRAM cells. Section III presents the background of D8T SRAM cell. The proposed DT8T SRAM cell is explained in Section IV. Section V analyses the simulation results of the proposed DT8T SRAM and gives the comparison with the conventional 6T SRAM cell and D8T SRAM cell [5]. Finally, concluding remarks are presented in Section VI.

II. LITRATURE SURVEY

In ultra-low voltage operations, process variations increase the failures in SRAM [6]. Various SRAM cell architectures are designed for reliable subthreshold applications [7] and [8]. These bit cells with varying number of transistors enhance the read stability at the price of increased leakage current and additional hardware. Among the different leakage currents in

978-1-5386-1357-3/17 $31.00 © 2017 IEEE

the nanometer CMOS, the subthreshold leakage current is the most dominant. Various techniques as multi threshold voltage (Vth) techniques, body bias control and stack effect-based methods are used for leakage control. A 10T SRAM cell based on stack effect is introduced for better stability and low leakage [9]. An Adaptive Voltage Level (AVL) circuit is added to 7T and 8T SRAM cells for minimizing the subthreshold leakage current at the cost of extra circuitry [10].

To avoid the stability problem in SRAM cell, Dynamic Threshold technique is introduced for low power applications. A single ended 11T SRAM cell based on Schmitt trigger inverter configuration which uses dual-threshold CMOS technology is presented in [11]. This cell consumes low power during the hold operation and exhibits high read static noise margin (RSNM) and hold static noise margin (HSNM). Dynamic Threshold MOS based 6T (DTM6T) cell is used to avoid the disturbances in subthreshold region [12]. With dynamic threshold transistors, cell performance is improved in terms of read noise margin (RSNM) and write noise margin (WSNM). To manage leakage, speed and stability, a body bias controller is designed which is added to the existing 6T SRAM cell [13]. This circuit controls the value of threshold voltage.

For maintaining stability with reduction of leakage in the cell, many cells are designed based on the idea of breaking the feedback loop of back to back inverters present in the traditional 6T SRAM cell. A 7T SRAM cell is proposed for the same [14]. A write bitline balancing circuit is added for power reduction in the designed 7T cell. For subthreshold operation, a Differential 8T (D8T) SRAM has been designed [5]. This cell splits the feedback structure of two consecutive inverters with the utilization of two additional transistors. When the D8T cell is compared to the existing 6T and 8T SRAM cell, it is found that with improvement in read stability and write ability, cell leakage increases in standby mode. To reduce this standby leakage a Modified Differential 8T (MD8T) cell is created [15]. With the same structure as D8T cell, this uses additional transistors named as sleep transistors for leakage current reduction. Due to extra transistors the speed reduces, and this is compensated using forward body biasing for feedback-breaking transistors.

III. EXISTING D8T SRAM CELL

Fig. 2. illustrates the structure of the existing differential 8T (D8T) SRAM cell which utilizes differential sensing scheme.

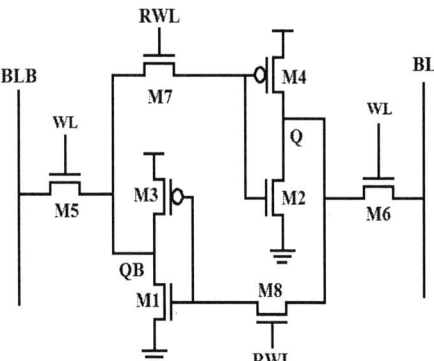

Fig. 2. Existing D8T SRAM cell

To eliminate read disturbance, two additional transistors M7 and M8 are used. These transistors are set between the access nodes (Q and QB) and storage nodes (Q2 and QB2) to break the feedback connection between the inverters during read operation. Wordline (WL) controls the pass transistors M5 and M6. Another signal named read wordline (RWL) controls M7 and M8 transistors.

The voltage levels of RWL and WL are properly controlled for different modes of operation. These values are depicted in Table I.

TABLE I. VOLTAGE LEVELS DURING WRITE, HOLD AND READ OPERATIONS

Signal	Write	Hold	Read
WL	V_{DD}	0	V_{DD}
RWL	V_{DD}	V_{DD}	0

For writing a data into cell, both RWL and WL are kept at V_{DD} and opposite values are given at bitlines. During hold mode WL is disabled and RWL is enabled to hold the data stored in cell. Both write and hold modes in D8T SRAM cell are like in the customary 6T SRAM cell. When a data is read, WL signal is high at V_{DD} and RWL is low. With low RWL signal, M7 and M8 transistors are turned off. This isolates the storage nodes from read operation path. The raise of ΔV at node QB does not influence the node voltage of QB2 and thus cell stability is maintained during read operation. To improve the write ability, access transistors are kept approximately three times wider than pull down transistors. Pull-up devices are minimally sized for less area. With wider access transistors higher write SNM is achieved.

This D8T SRAM cell improves the cell performance with increased read stability and writing ability but having disadvantage of increase in leakage current during read and hold modes. In order to further improve the cell performance of existing D8T cell, a modified DT8T SRAM cell is proposed as discussed in Section IV.

IV. PROPOSED DYNAMIC THRESHOLD 8T (DT8T) SRAMCELL DESIGN

The circuit diagram of the proposed DT8T cell is shown in Fig. 3. The proposed DT8T cell is having structure identical to D8T cell [5]. The only modification required is the connecting of substrate of all the transistors (except M7 and M8) to the storage nodes. This routing is done to achieve dynamic body biasing dependent on the stored data. M7 and M8 transistors are used to isolate the feedback loop while reading the data. This action improves the read stability but leads to speed reduction in write operation. During read mode, leakage through these off transistors M7 and M8 may degrade the cell performance.

For data retention, dynamic body biasing is provided for transistors by connecting their substrates to the Q or QB nodes. In the proposed cell, substrates of M1, M3 and M5 are connected to Q while substrates of M2, M4 and M6 are

978-1-5386-1357-3/17 $31.00 © 2017 IEEE

connected to QB. By changing the body biasing, threshold voltage of a transistor can be reduced or increased [16]. Variable body bias voltage (V_{BB}) for the M7 and M8 transistors is applied in two ways:

- Forward body biasing during write and hold mode for better drivability of the cell.

- Reverse body biasing during read mode which reduces leakage current through feedback transistor.

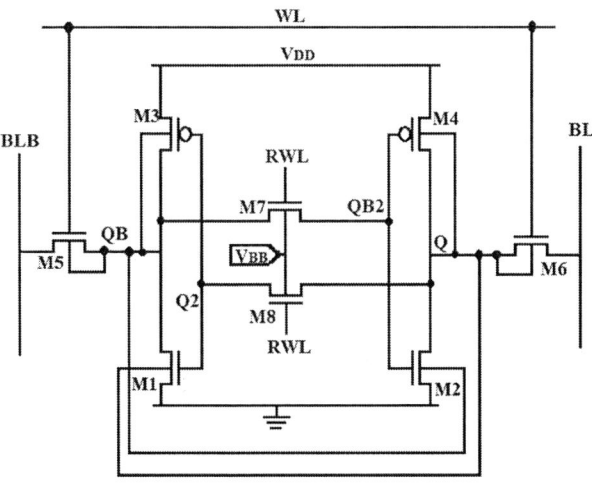

Fig. 3. Proposed DT8T SRAM cell

A. Cell performance

For writing data, both the WL and the RWL are kept at V_{DD}. When a data is to be written into the cell, the desired value is applied to the bit line BL. In order to write data "1", the BL and BLB are kept at V_{DD} and 0 respectively and for writing data '0', we assert BL as 0 and BLB as V_{DD}. With activation of WL, M5 and M6 provide a path for the charge transfer. As M7 is on, QB2 has a slower change towards V_{DD}. This slow transition is caused by the small current flowing through M7. This is why QB2 does not obtain its final voltage value. In D8T cell, the QB2 reaches upto 347.3 mV. With the use of dynamic body biasing in the proposed DT8T cell, this QB2 value increases to 409.1 mV.

Further improvement in write mode is obtained by applying forward body biasing for M7 and M8. Applying forward body biasing reduces the threshold voltage of these devices. This increases the current flow in write mode and hence speed is increased. During hold operation RWL signal is held at V_{DD}, which activates the latch structure of the cross-coupled inverters. When WL is low, M5 and M6 cut out the cell from BL and BLB. The current drawn in this state is stated as standby leakage current. For read mode, WL is high and RWL is low. M7 and M8 transistors are turned off, thus separating the storage nodes. During read operation, bit lines BL and BLB are precharged to V_{DD}. In the bit cell, M7 and M8 are off. This causes the flow of subthreshold leakage current through them.

This current disturbs the values of different nodes. To reduce this leakage current, reverse body biasing is applied for M7 and M8. Due to this, threshold voltage of these transistors increases and the leakage current decreases.

B. Transistor sizing

The proposed DT8T cell has same cross coupled inverter structure of 6T SRAM which helps to store the information. The transistors are sized optimally in order to have better write ability considering cell area limitation. Table II shows the comparative study of different sizing values used for all the transistors in traditional 6T cell, D8T cell [5] and the proposed DT8T cell. For better performance, the devicesM3, M4, M1 and M2 and M7 and M8 are minimally sized as 120 nm. Wider access transistors are allowed in order to have high write margin.

TABLE II. SIZING COMPARISON OF 6T, D8T AND PROPOSED DT8T SRAM CELLS

Cell Type	Width of Transistors (nm)							
	M1	M2	M3	M4	M5	M6	M7	M8
Conventional 6T cell	360	360	120	120	240	240	-	-
D8T cell [5]	120	120	120	120	300	300	120	120
Proposed DT8T cell	120	120	120	120	240	240	120	120

V. SIMULATION RESULTS

In this section, the simulation results of the proposed DT8T SRAM cell are discussed. The cell is implemented using Cadence Virtuoso Analog Design Environment in 45 nm technology. We have also implemented the conventional 6T cell and existing D8T cell [5] to compare the results with the proposed cell. Fig. 4 illustrates the waveforms obtained by performing transient analysis on the proposed DT8T SRAM cell during write, hold and read modes at the supply voltage of 0.5 V. The body bias voltage for M7 and M8 is controlled by signal V_{BB}. This body biasing is done at 0.5 V forward bias for write and hold margin keeping supply voltage limitation. For read operation, reverse body biasing is varied from 0 V to 0.6 V. The reverse bias voltage of 0.6 V is chosen according to breakdown voltage for NMOS transistor.

A. Write operation

For writing data into the cell WL and RWL are activated. If data 0 (Q node) has to be written in the cell then this value 0 is given at BL and opposite value is written at BLB. With BL at 0 V and BLB at V_{DD}, data 0 is written at Q node and data QB node is written as 1. When no body bias is given to the M7 and M8 transistors, it is observed that the voltage at QB2 is 409.1 mV, which is just 82% of the final voltage. To increase the voltage at QB2, forward body biased is applied for M7 and M8 transistors. When the forward body bias is used for M7 and M8, their threshold voltages decreases which results in increase

978-1-5386-1357-3/17 $31.00 © 2017 IEEE   217

in drivability of these transistors. Hence, the QB2 voltage is pulled to higher value. The node voltage of QB2 reaches about 94.6% of the supply voltage value with forward body bias of 0.5 V.

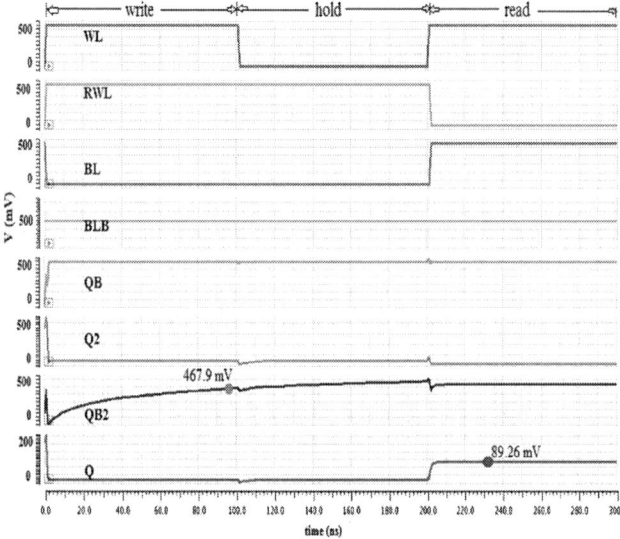

Fig. 4. Transient response of proposed DT8T SRAM cell

B. Read operation

For read operation, WL is at V_{DD} and RWL is at 0. While reading the data 0, stored in the cell, bitlines are precharged at V_{DD}. During read mode, node Q is at 190.14 mV instead of 0 V due to the leakage current through M7 and M8. This node voltage at Q is reduced to 89.26 mV in the proposed cell using reverse body biasing for off transistors M7 and M8. Thus the data is retained in the cell without any disturbance.

During write and hold mode, forward body bias of 0.5 V is applied while reverse body bias of 0.6 V is provided for read operation. When data is written in D8T cell, QB2 node voltage is at 412.5 mV and is increased to 467.9 mV in DT8T cell. Table III compares the different node voltages in D8T SRAM cell and DT8T SRAM cell with variable body bias of M7 and M8 for different modes of operations.

TABLE III. NODE VOLTAGES WITH VARIABLE BIAS VOLTAGE IN THE PROPOSED DT8T CELL

Node	Node voltages of D8T Cell [5] (mV)			Node voltages Proposed DT8T Cell (mV)		
	Write	*Hold*	*Read*	*Write*	*Hold*	*Read*
QB	500	500	500	500	500	500
Q2	0.0028	1.31	21.42	0.0018	0.86	11.38
QB2	412.5	423.7	400.9	467.9	473.2	459.1
Q	0.0012	1.44	190.14	0.0011	0.53	89.26

Due to reverse body bias during read (when RWL is 0 V), leakage currents in M7 and M8 decrease rapidly in both D8T

cell and the proposed DT8T cell. The comparison of these leakage currents in read mode is shown in Fig. 5.

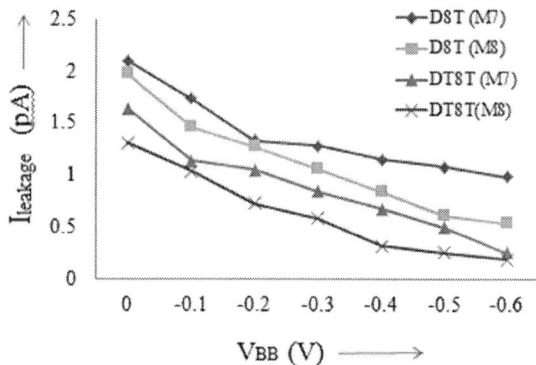

Fig. 5. Comparison of leakage current during read mode

C. Hold operation

During hold mode, WL is deactivated and RWL is activated. The circuit is in idle mode when no operation takes place but the standby leakage current flows through the devices. Due to this leakage, the active currents are wasted in the cell. As a result, the total power consumption is increased. In order to reduce this leakage in the cell, an estimation of leakage current is necessary.

For a conventional 6T SRAM cell, in standby mode the subthreshold leakage current flows due to M6, M1 and M4 (when data stored at Q node is 0). For D8T SRAM cell and the proposed DT8T SRAM cell these leakage paths are similar, that means same transistors M6, M1 and M4 are responsible for leakage current flow in the cells. Leakage currents of M7 and M8 are added in total standby leakage calculation.

The leakage current comparison for conventional 6T SRAM cell, existing D8T SRAM cell [5] and the proposed DT8T cell is shown in Fig. 6. The proposed DT8T cell reduces the leakage current and improves the cell performance compared to existing cells.

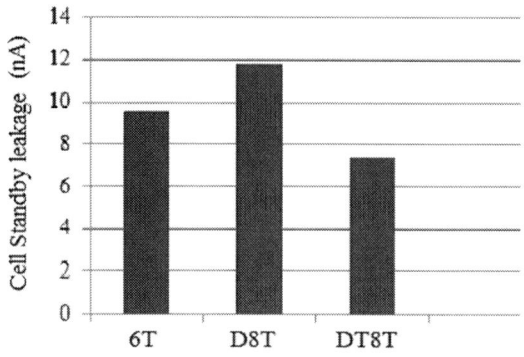

Fig. 6. Comparison of standby leakage current

VI. CONCLUSION

The proposed Dynamic Threshold 8T (DT8T) SRAM cell is implemented in 45 nm technology and simulated using Cadence Virtuoso simulator. The cell is provided with the supply voltage of 0.5 V and operated in subthreshold region. When reverse body biasing body is applied at 0.6 V the subthreshold currents through M7 and M8 are found as 0.261 pA and 0.198 pA respectively. These leakage currents are reduced by approximately 84% using reverse body biasing technique. During hold mode, subthreshold leakage currents flows through off transistors in conventional 6T SRAM cell, D8T SRAM cell [5] and the proposed DT8T SRAM cell, which are 9.58 nA, 11.83 nA and 7.31 nA respectively. Overall standby leakage of the proposed DT8T cell is 24% lesser compared to existing 6T SRAM cell at supply voltage of 0.5 V.

As future work, an array of the proposed DT8T SRAM cell can be designed to validate the functionality.

REFERENCES

[1] Jan Rabaey, "Low Power Design Essentials", Springer 2007.

[2] B.R. Upadhyay and T. S. B. Sudarshan, "Low Power Predictive Placement Cache Scheme for Embedded System," International Conference on Embedded Systems, pp.250-254, 2014.

[3] International Technology Roadmap for Semiconductors (ITRS) 2011.

[4] Jawar Singh, Saraju P.Mohanty and Dhiraj K. Pradhan, "Robust SRAM Cell Designs and Analysis", Springer Science 2013.

[5] R. Saeidi, M. Sharifkhani and K. Hajsadeghi, "A Subthreshold Symmetric SRAM Cell With High Read Stability," IEEE Transactions on Circuits and Systems II: Express Briefs, vol. 61, no. 1, pp. 26-30, Jan. 2014.

[6] A. J. Bhavnagarwala, Xinghai Tang and J. D. Meindl, "The impact of intrinsic device fluctuations on CMOS SRAM cell stability," IEEE Journal of Solid-State Circuits, vol. 36, no. 4, pp. 658-665, Apr 2001.

[7] B. H. Calhoun and A. P. Chandrakasan, "A 256-kb 65-nm Sub-threshold SRAM Design for Ultra-Low-Voltage Operation," IEEE Journal of Solid-State Circuits, vol. 42, no. 3, pp. 680-688, March 2007.

[8] S. Pal and S. Arif, "A Single Ended Write Double Ended Read Decoupled 8T SRAM Cell with Improved Read Stability and Writability," International Conference on Computer Communication and Informatics, pp. 865-870, Jan.2015.

[9] P. S. Grace1 and N. M. Sivamangai, "Design of 10T SRAM Cell for High SNM and Low Power," IEEE International Conference on Devices, Circuits and Systems, pp.281-285, 2016.

[10] M. Yadav, S. Akashe and Y. Goswami, "Analysis of Leakage Reduction Technique on Different SRAM Cells," International Journal of Engineering Trends and Technology, Vol. 2,pp.78-83, 2011.

[11] D. Sreenivasan, D. Purushothaman, K. S. Pande and N. S. Murty, "Dual-threshold single-ended Schmitt-Trigger based SRAM cell," 2016 IEEE International Conference on Computational Intelligence and Computing Research, pp. 1-4, 2016.

[12] S. Pal and A. Islam, "Device bias technique to improve design metrics of 6T SRAM cell for subthreshold operation," International Conference on Signal Processing and Integrated Networks, pp. 865-870, 2015.

[13] R. Lorenzo and S. Chaudhury, "Body Biasing Scheme to Control Leakage, Speed and Stability in SRAM Cell Design," International Conference on Computing, Communication and Sensor Network, pp. 11-15,2014.

[14] S. Akashe, A. Srivastava and S. Sharma, "Calculation of Power Consumption in 7 Transistor SRAM Cell using Cadence Tool," International Journal of Advances in Engineering & Technology, pp. 189-194, Sept 2011.

[15] P. Sreelakshmi, K. S. Pande and N. S. Murty, "SRAM cell with improved stability and reduced leakage current for subthreshold region of operation," 2015 IEEE International Conference on Computational Intelligence and Computing Research, 2015, pp. 1-5.

[16] A. Verma, A. Mishra, A. Singh and A. Agrawal, "Effect of Threshold Voltage on Various CMOS Performance Parameter," International Journal of Engineering Research and Applications, ISSN : 2248-9622, vol. 4, pp.21-28, April 2014.

2017 IEEE International Symposium on Nanoelectronic and Information Systems

A Single-Ended Read Decoupled 9T SRAM Cell for Low Power Applications

S. R. Mansore
Electronics and Communication Engg. Dept.
Ujjain Engineering College,
Ujjain, India
sr_mansore@rediffmail.com

R. S. Gamad and D. K. Mishra
Electronics and Instrumentation Engineering, Dept.,
Shri G. S. Institute of Technology and Science
Indore, India
rsgamad@gmail.com, mishrad_k@hotmail.com

Abstract— **A single-ended nine transistor (9T) Static Random Memory (SRAM) cell is presented in this paper which improves read stability and write ability. The cell employs separate access transistors for read and write operations to eliminate the conflicting design requirement on access transistors. The cell employs feedback loop cut off scheme along with power supply interuption scheme to enhance the write ability. Simulation is done on 65nm standard CMOS technology on cadence. Simulation results show that read SNM (RSNM) and write SNM (WSNM) of the proposed cell are 2.77x and 1.12x larger respectively, than those of conventional 6T cell at 0.4V. Proposed cell consumes 0.98x lesser leakage power than the conventional 6Tcell.**

Keywords- Leakage power, low power, RSNM, SRAM, WSNM.

I. INTRODUCTION

Battery operated portable devices such as smart phones, laptops and biomedical instruments require low power consumption in order to prolong their battery life. SRAM is an important block of these devices. As per ITRS statement, a substantial part of the chip area is occupied by the SRAM block [1], [2]. SRAM is the primary consumer of the power of the total chip power because of its large area occupancy. Thus, total chip power consumption is determined by the power consumed by the SRAM. Hence, design of SRAM with minimum power consumption is desirable. Low power consumption can be achieved by operating the SRAM at the low power supply. Also, to improve integration density, transistor dimension is continuously shrinking to submicron technology. However, at scaled technology and reduced supply voltage, SRAM stability is degraded dramatically due to process, voltage and temperature (PVT) variation [3], [4]. Hence, it is necessary to design an SRAM with enhanced stability. Fig. 1 shows the schematic of the conventional 6T cell. This cell has a simple structure but suffers from read disturb and conflicting read versus write design requirement on the access transistor. Let us assume that cell stores bit '1' (Q = 1). During a read operation, voltage division action between transistors M3 and M6 causes rise in voltage at node QB. If, this voltage bump is equal to switching threshold voltage of the inverter M1-M2, bit-flip can take place. This problem is referred to as read disturb. Therefore, strong pull-down transistor and a weak access transistor

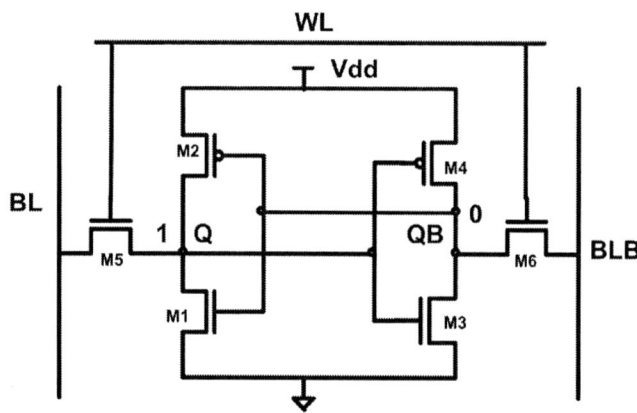

Figure 1. Conventional 6T SRAM cell

are required for the successful read. However, strong access and weak pullup transistors are required for successful write operation. Thus, the conventional 6T cell suffers from read versus write design conflict [5], [6].

Various cell configurations are found in the literature to overcome the problems of the conventional 6T cell. Designs proposed in [7]- [10] use separate read and write ports to eliminate the problem of read versus write design conflict. Cells in [11] and [12] offer enhanced RSNM but require negative bit line scheme to perform a successful write. In this work, we present a read decoupled 9T bit cell. The proposed 9T SRAM cell employs separate transistors for read and write operations to eliminate the read versus write design conflict. The rest of the paper is organized as follows. Section II describe the operation of the proposed cell. In section III simulation results and discussion is presented. Section IV gives the conclusion

II. PROPOSED SINGLE ENDED 9T CELL

The schematic of proposed 9T cell is depicted in Fig. 2. Truth Table of the proposed cell is given in Table I. Transistors M2- M4 form left inverter while the right inverter is formed by transistors M5- M6. Transistors M1 is used to make transistor M2 weaker during write '0' operation, while transistor M3 is turned OFF to perform successful write '1' operation. M7 is write access transistor controlled by Write Word Line (WWL) for writing data

978-1-5386-1357-3/17 $31.00 © 2017 IEEE
220

from bit line BL in to the storage node Q. Transistors M8 and M9 form a separate read buffer for read operation.

In the hold mode, WWL and Read word line (RWL) are forced to logic '0' while WL and VSS are forced to VDD. The cell retains the stored bit as long as power supply is ON.

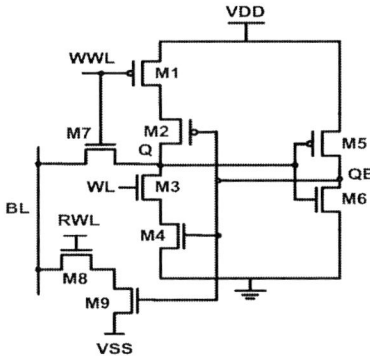

Figure 2. Proposed 9T SRAM cell

TABLE I. TRUTH TABLE OF THE PROPOSED 9T SRAM CELL

Operation	WWL	RWL	WL	VSS	BL
Write '0'	'1'	'0'	'1'	'1'	'0'
Write '1'	'1'	'0'	'0'	'1'	'1'
Read	'0'	'1'	'1'	'0'	'1'
Hold	'0'	'0'	'1'	'1'	'1'

To read the content of the cell, bit line BL is precharged to VDD and RWL is enabled to turn ON transistoe M8. Control signal WL is raised to VDD while VSS and WWL are pulled down to ground. Now, if '0' is stored in the cell (Q = '0', QB = '1') then, M9 turns ON and BL is discharged to ground through M8 and M9. If '1' is stored in the cell (Q= '1', QB='0') then, M9 is OFF and BL remains at its precharged value.

The write operation is performed by raising WWL, WL and to VDD and lowering RWL to ground. For writing a '0' in the cell, BL is loaded with data '0'. High WWL turns ON M7 to discharge node Q to the ground. To write a '1' in the cell WWL is forced to VDD while WL and RWL are pulled to ground. Low WL disables the transistor M3 which disconnects the storage node Q from ground during write '1' mode. Now, data '1' is applied to BL which causes current to flow from BL to node Q through M7. As a result, the voltage at node Q is raised which switches ON M6. Consequently, QB is pulled down to ground through M6 and finally '1' is written into the cell. After successful write operation, the cell goes into hold mode by raising WL to VDD and lowering WWL to ground.

III. SIMULATION RESULTS AND DISCUSSION

In this section, parameters such as RSNM, WSNM and power consumption of the proposed 9T and conventional 6T bit cell are simulated and compared. Simulation is done in 65 nm standard CMOS technology on Cadence. For comparison, the authors also designed and simulated the conventional 6T cell. For the conventional 6T cell, width of pull-down, access and pull-up transistors is chosen as 300nm, 200nm and 150nm, respectively.

Width of pull-down, write access and pull-up transistors in the proposed cell is equal to 300nm, 300nm, and 150nm, respectively. The transistor width in the read buffer cells is set to 200nm while other transistors width in the cell is equal to 150nm. Channel length of all transistors is 65nm.

A. Read Stability

Read stability is measured in terms of RSNM. Fig. 3 illustrates the voltage transfer characteristics (VTCs) of the proposed 9T and the conventional 6T cells during read operation at VDD = 400mV. RSNM is measured as the side length of the largest square embedded inside the butterfly curves formed by the read VTCs [13]. In the proposed design, read current does not flow through the storage node and therefore voltage at '0' storing node does not rise above the ground), hence, proposed 9T cell exhibits RSNM equal to that of HSNM. From the Fig. 3, it is noticed that the proposed cell offers 2.77x larger RSNM than that of the conventional 6T cell at Vdd = 400mV. Fig.4 shows the RSNM as a function of the supply voltage for proposed 9T and the conventional 6T cell.

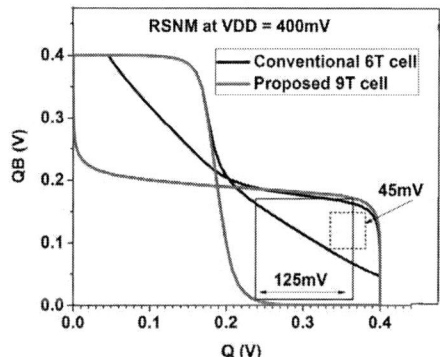

Figure 3. Read butterfly curves

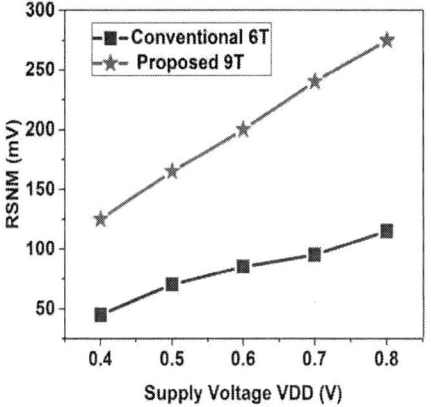

Figure 4. RSNM versus Supply Voltage

B. Write Ability

Write ability is expressed in terms of WSNM. During write operation, OFF transistors M1 and M3 disconnect the storage node Q from VDD and ground. Therefore, the cell achieves enhanced write ability compared to the 6T cell. Fig. 5 shows the WSNM versus VDD curves for proposed and the conventional 6T cells. From Fig. 5, it is noticed that 9T cell offers 1.12x larger WSNM as compared to the conventional 6T cell at 0.4V.

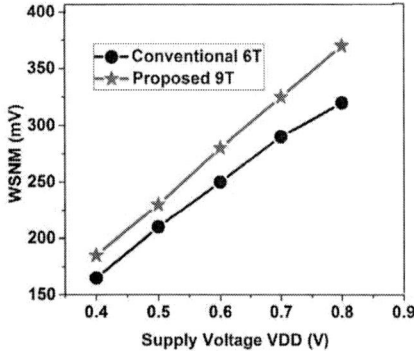

Figure 5.   WSNM versus Power Supply

C. Hold Stability

Hold SNM measures the cell stability when the cell is in standby mode. During hold state, 9T cell behaves like the conventional 6T cell therefore, HSNM of the proposed cell is almost equal to that of the conventional 6T cell.

D. Read and Write delay

For differential read SRAM such as conventional 6T cell, read delay is defined as the duration between WL activation and the instant when 50mv difference is reached between BL and BLB [14], [15]. For a single- ended SRAM cell, read delay is defined as the time elapsed between the instant when RWL is asserted to the instant when bit line discharged by 50 mV from its initial precharged level [7]. Table II compares the read and write delay of proposed and conventional 6T cells at VDD = 400mV. Proposed design shows penalty in the read delay due to higher BL capacitance.

Write '0' delay is defined as the duration between the instant when the word line is activated to the instant when the the voltage at '1' storing node drops to 10% of its initial high level. Similarly, write '1' delay is the time elapsed between the instant when the word line is activated to the instant when the voltage at '0' storing node rises to 90% of VDD[14], [15]. Due to single ended write operation, proposed cell shows larger write delay than that of the conventional 6T as illustrated in Table II.

E. Read /Write power

During read operation, bit line BL is discharged only if '0' is stored in the cell. BL remains at its precharged value if a '1' is stored. However, for differential read cells, one of the bit lines is always discharged. Therefore, for equal probability of '0' and '1', BL (for single ended read) is discharged only for half of the times as compared to differential read. Hence, on an average, read power of the single ended cell is lesser than that of differential read cells. Table II shows the power consumption during read operation. This power consumption is the average of read '0' and read '1' power consumption. From the Table II, it can be observed that the read power consumption of the proposed 9T cell is lesser than that of the conventional 6T cell.

TABLE II.   DELAY AND POWER CONSUMPTION FOR READ AND WRITE OPERATION AT VDD=0.4V

Operation	Read		Write	
	Delay (ns)	Power (nW)	Delay (ns)	Power (nW)
Conventional 6T	380	31.59	7.2	30
Proposed 9T	604	15.45	7.9	12.18

Unlike the conventional 6T cell, proposed design employs single bit line for the write operation. Therefore, low power is consumed in charging and discharging of single bit line as compared to differential bit lines. Also, proposed single ended cell needs half the number of write drivers as compared to differential write cell [16]. As a result, write power dissipation is also reduced as compared to that of differential write operation. Table II records the write power consumed (average power consumed during write '0' and write '1' operations) by the proposed 9T cell. It is observed that the proposed cell consumes lesser power during write operation.

F. Leakage Power

A major portion of an SRAM remains idle for most of the times. However, leakage current does flow through cells of the idle portion. This causes dissipation of leakage power by the SRAM cell [17]. Among various leakage currents, the sub-threshold leakage is the dominating component. In the proposed cell, left inverter consists of four transistors in series. Therefore, stacking effect results in reduced leakage current. Table III compares the leakage power for proposed design and conventional 6T cell at different voltages. It is observed that for Q = '0' (Q = '1') our cell dissipates 0.57x (0.98x) lesser leakage power compared to conventional 6Tcell at 0.4V.

TABLE III.   LEAKAGE POWER COMPARISON AT VDD= 0.4V

VDD (V)	Conventional 6T cell (pW)	Proposed 9T cell (pW)	
		For Q = '0'	For Q = '1'
0.4	7.45	7.32	4.19
0.5	12.48	12.29	6.86
0.6	19.8	19.65	10.65
0.7	30.26	30.15	15.95
0.8	45.0	45.0	23.0

IV. CONCLUSION

A 9T SRAM cell with enhanced write ability has been presented in this paper. Use of separate read and write ports eliminates the read versus write design conflict. The cell also eliminates the read disturb problem by employing separate read buffer. The proposed cell offers 1.12x larger WSNM and 2.77x larger RSNM compared to that of the conventional 6T cell at 0.4V. The cell consumes lower switching power due to use of single bit line. Our cell dissipates 0.98x lesser leakage power than that of conventional 6T cell at 0.4V.

REFERENCES

[1] S.R. Sridhara et al. "Microwatt embedded processor platform for medical system-on-chip applications, " IEEE J. solid state circuits, vol. 46,no.4, pp. 721-730, Apr. 2011

[2] "International Technology Roadmap for Semiconductors" Semiconductor Industry Association(SIA), San Francisco, CA, available at: http//www.itrs.net/Links/2013chapters/2013 Executive Summary.pdf, 2013.

[3] S. Pal and A. Islam, "Variant tolerant differential 8T SRAM cell for ultralow power applications, "IEEE trans. computer aided design of integrated circuits and systems, vol. 35, no. 4, pp. 549 -558, Apr. 2016.

[4] A. Bhavnagarwala, X Tang and J. Meindl, " The impact of intrinsic device fluctuation on CMOS SRAM Cell stability," IEEE J. solid state circuits, vol. 36, no. 4, pp. 658--665, Apr. 2001.

[5] Sung-Mo Kang and Yusuf Leblebici, CMOS digital integrated circuits analysis and design, 3rd ed. New Delhi, TMH, 2003.

[6] S. Ataei, J. E. stine and M. R. Gauthaus "A 64 kb differential single-port 12T SRAM design with a bit-interleaving scheme for low voltage operation in 32nm SOI CMOS," 34th intr. Conf on compuer design (ICCD) 2016, pp. 499-506.

[7] G. Pasandi and S. M. Fakhraie, "A New Sub-Threshold 7T SRAM Cell Design with capability of Bit-Interleaving in 90 nm CMOS," 978-1-4673-5634-3/13/$31.00 ©2013 IEEE

[8] C. B. Kushwaha and S. K. Vishwakarma, "A single-ended with dynamic feedback control 8T sub threshold SRAM cell," IEEE trans. very large scale integration (VLSI) systems," vol. 24, no.1, pp. 373-377, Jan. 2016.

[9] M. F. Chang, S. W. Chang, P.W. Chou and W. C. Wu, "A 130 mV SRAM with expanded write and read margins for subthreshold applications," IEEE J. Solid state circuits, vol. 46, no. 2, pp. 520-529, Feb. 2011.

[10] L. Wen, Z. Duan, YI Li and X. Zeng, "Analysis of read disturb-free 9T SRAM cell with bit-interleaving capability," Microelectronics Journal, vol. 45, pp. 815-824, 2014.

[11] Ming-Hsien Tu et al., "A single -ended disturb-free 9T sub-threshold SRAM with cross-point data-aware write word-line structure, negative bit-line and adaptive read operation timing tracing, "IEEE J. Solid state circuits, vol. 47, no. 6, pp. 1469-1482, June 2012.

[12] C. Y. Lu et al., "A 0.325 V, 600-kHz, 40-nm 72-kb 9T subthreshold SRAM with aligned boosted write word line and negative write bit line write-assist," IEEE Trans. Very Large Scale Integr. (VLSI) Syst., vol. 23, no. 5, pp. 958–962, May 2015

[13] E. Seevinck, F. J. List and J. Lohstroh, "Static noise margin analysis of MOS SRAM cells, "IEEE J. Solid -State Circuits, vol.22, no. 5, pp. 748-754, Oct. 1987.

[14] J. P. Kulkarni, K. Kim, and K. Roy.," A 160 mV robust Schmitt trigger based sub-threshold SRAM," IEEE J. Solid state circuit circuits, vol.42, no.10, pp. 2303-2313 Oct 2007.

[15] S. Ahmad et al., "Single ended Schmitt-trigger based robust low power SRAM cell," IEEE trans. very large scale integration (VLSI) system, vol. 24, no.8, pp. 2634-2642, Aug. 2016.

[16] G. Pasandi, and S. M. Fakhraie, "A 256-kb 9T near-threshold SRAM with 1k cells per bitline and enhanced write and read operations," IEEE trans IEEE Trans. Very Large Scale Integr. (VLSI) Syst, vol. 23, no.11, pp. 2438-2446, Nov 2015.

[17] S. Akashe, S. Bhushan and S. Sharma, "Modeling and simulation of high level leakage power reduction techniques for 7T SRAM cell design," Journal of Microelectronics, Electronic Components and Materials, vol. 42, no. 2, 83-87, 2012.

Current Profile Generated by Gating Logic Reduces Power Supply Noise of Integrated CPU Chip

Alak Majumder[‡] and Pritam Bhattacharjee[†]

VLSI Design Laboratory,
Department of Electronics & Communication Engineering,
National Institute of Technology, Arunachal Pradesh,
Yupia, Dist.– Papumpare, Arunachal Pradesh 791112, India.
[‡]*majumder.alak@gmail.com*, [†]*pritam@ieee.org*

Abstract— **With the continuous advent of CMOS, process technologies is extending threat to the noise immune capability of CMOS circuits and the power consumed by them. In present day scenario, though there are a lot of techniques that exist for power reduction, the study of power–supply noise (PSN) based on those techniques is almost unattended in literature. Modern clock gating is one of the best techniques to reduce dynamic and static power dissipation by curbing down the switching activity of the operating clock as well as blocking the direct path between the power lines during logic transition. Therefore, in this paper, we have incorporated gating logic to offer solution to PSN occurrence in CMOS circuits by controlling di/dt, which is generated by the linear current ramp of present day high performance CPU. It is witnessed that, the gated architectures generate very less di/dt with respect to their non-gated counterpart, resulting a noted amount of reduction in PSN.**

Keywords—power supply noise; simultaneous switching noise; ground bound noise; clock gating; LECTOR

I. INTRODUCTION

When a processor system is turned ON, the silicon chip running with a clock frequency close to 3-4 GHz, comes out of sleep mode and the logic gates inside the chip start working and draw current simultaneously. This leads to a large current drawn by the device, thereby producing a large voltage drop on the power supply due to $L\frac{di}{dt}$, where L is the inductance of the main supply connecting to the silicon device to the first level decoupling capacitance (which is available basically inside the package) and $\frac{di}{dt}$ is the rate of change of current pumped by the chip. It causes the transistors inside the chip to run at lower speed, eventually leading to the functional failure of the CPU (central processing unit) [1]. The present day high performance processors are packaged with Organic Land Grid Array (OLGA) with a thickness of around 0.03–0.04 inch having an inductance of about 24pH, which is the primary cause of $L\frac{di}{dt}$ [2, 3].

With the advancement of process technologies, circuit design using CMOS is facing major challenge as switching noise is appearing to be a prominent constraint [4]. This switching noise happens to be in scene, when logic swing in digital gates allows more number of charge carriers to pass through the substrate of a MOS device. Digital integrated circuits (IC)s, built in virtue of static CMOS, are generally quite immune to noise as it has comparably high noise margin. But, with the scale down of technology, the power supply and threshold voltage get lowered, which increases the probability of switching noise occurrence. In order to keep a track on this noise commonly referred as power–supply noise (PSN), one has to note the maximum of dynamic current flow due to power supply (V_{DD}), as it is directly proportional to the number of charge carriers flowing through the device [5]. There are many digital techniques reported in literature to reduce power–supply switching noise [5-7]. But most of these design styles are based on current–mode logic (CML). As CML designs are power hungry, employing them in reduction of PSN may not look a healthy option. However, the sole target has been to restraint the inside components of an IC from having the maximum switching activity so that the current peaks become minimum. The conventional method to reduce these current peaks is to install a series of buffers in the signal paths, viz. introduction of deliberate skew in the signal paths [8]. However, the maximum switching activity offered in a digital IC is by the operating clock. Thereby, it has been also reported in [9] to cleave the clock to multiple sub–clocks with correlative skew so that the peaks of the dynamic current flow gets configured to a reduced amount by a factor of two. But in both the cases, the major constraint occurring is the breaching of timing criterions that are required to be maintained in an IC.

The traditional approach to control the switching activity of operating clock has been clock gating (CG), which came into practice with the emergence of Intel Pentium IV [10]. This technique, in general, is incorporated inside the IC to reduce the dynamic power consumption and it does not create any timing criticality in the normal operation. Thereby, in this paper, we accentuate clock gating to be a designated technique to optimize PSN.

The organization of rest of the paper is as follows: Section II enunciates the different sources of noise emerging in integrated circuits and elucidates the prime contributors of PSN. In Section III, the fundamental description of clock gating and its types are discussed. A comparative analysis of those CG types to minimize the PSN is presented in Section IV. In section V, we conclude our paper.

978-1-5386-1357-3/17 $31.00 © 2017 IEEE

II. Sources of Noise in Integrated Circuits

There are abundant sources of noise in integrated circuits [11, 12] which may be broadly classified to internal noise sources and external noise sources. Among the internal noise sources, leakage noise affects the most, whereas, for the case of external noise source, the major contributors are power–supply noise, input noise and ground–bound noise.

The occurrence of these noise sources are elaborated as below:

A. Leakage Noise

The current leak through a MOS transistor exponentially increases with its threshold voltage, which is always decreasing with the reduction of V_{DD} during scale down of process technology. Therefore, there is the possibility of charge loss in a circuit. Because of this sub-threshold leakage current flow which can degrade the output logic make it highly prone to noise.

B. Power–Supply Noise

In integrated circuit packages, there exists parasitic resistance and inductance in the power lines, which are connected directly to the package pins. As this package pins are exterior to the package, they are highly prone to external noise. During a chip to chip communication, a driver gate, contaminated with DC noise, can degrade the logic voltage level at the input of the receiver [12]. Basically, the power–supply noise (PSN) is depicted as the voltage instability occurring against the power lines of the power distribution network (PDN) in the integrated circuit. Whenever inside the chip, a numerous number of logic gates are switching concurrently, a notable proportion of current will be drawn from the V_{DD}. This current flow down the parasitic resistances and inductances within the power line, leads to both dynamic and static voltage instability. Thereby, it arise a high

probability of noise generation which is explicitly depicted through figure 1. Actually, the range of voltage that is available for a switching circuit is V_{DD}–V_{gnd} indicated as *voltage headroom* [13]. But due to noise implantation, the voltage headroom gets reduced to V_{DD}–V_{gnd}–($\S V_p$+$\S V_g$) where $\S V_p$ and $\S V_g$ corresponds to the power–supply noise voltage, given as:

$$\S V_p = i(t)\sum_{k=1}^{n} R_{p_k} + \frac{di(t)}{dt}\sum_{l=1}^{n} L_{p_l} \qquad (1)$$

where L_p and R_p are the parasitic inductance and resistance respectively at the power supply ends of PDN and 'n' denotes the number of inductances and resistances contained within the PDN. The first term of equation (1) refers to the voltage drop across the parasitic resistances which is solely dependent on the immensity of the current $i(t)$ during a transient time. However, the second term in that equation depicts the voltage drop across the parasitic inductance and is analogous to the rate of change in $i(t)$ and termed as $L\dfrac{di}{dt}$ noise [1]. Therefore, the reduction of power–ground noise directly depends on the amount of $i(t)$ and $\frac{di(t)}{dt}$.

C. Input Noise

The noise occurring at the input of a logic gate is referred to as input noise. This noise generally occurs when two adjacent wires approaching to the input of a logic gate, get capacitive coupled with each other and let happen cross–talk. The input noise become prominent with gradual scale down of process technology, where the interconnect shrinks in lateral dimensions keeping the vertical dimensions almost constant [12].

D. Ground–Bound Noise

Ground–bound noise occurs at a situation when the output of a device switches from high to low, causing voltage change at the other terminals of the device. Basically, it is a voltage oscillation within the ground pin of an IC package and the ground pin of the die on which it is fabricated. Therefore, the scenario of this noise comes in picture only in a densely packed IC, if there is no low resistive logic gate connected to the ground. Due to this, there lies a chance of current flow through the ground pin to develop a voltage drop, which creates an increase in the device threshold voltage level and voltage level of a static output. This phenomenon is also referred as simultaneous switching noise (SSN) [14]. Also, the faster change of voltage per unit time by the circuit topology used in the IC, worsens the SSN. Generally, in the IC datasheet, the maximum change of voltage per unit time is indicated as Voltage Output Low Pulse (*Volp*). Therefore, in order to avoid ground bounce noise, output peak voltage, which is directly proportional to the amount of $i(t)$, should always be less than that of *Volp* limit.

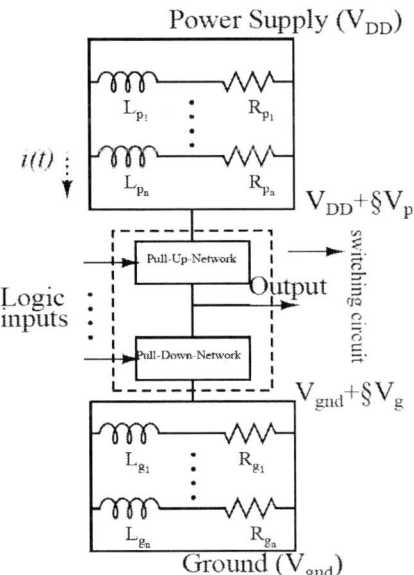

Fig. 1. Power–ground noise appearing due to parasitic impedances.

From the above exploration of noise sources in ICs, it is quite comprehensible that PSN manifest the significant amount of noise injection in the CPU chips. Though, the other noise sources contaminate the working voltage level of CPU chips, it is comparably negligible to that affected by PSN. Therefore, we concentrated in reduction of PSN.

III. CLOCK GATING IN INTEGRATED CIRCUITS

The clock gating technique is contemplated to be an intelligent proposition to reduce power dissipation through the sequential elements of integrated circuits as shown in figure 2.

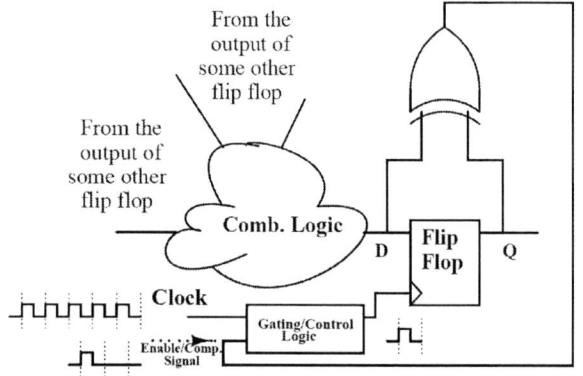

Fig. 2. Clock gating concept in integrated circuits.

With clock gating, the triggering rate of clock signal is reduced to a measurable extent as the continuous triggering of clock is blocked by a control logic which allows the clock to trigger only during the instantaneous change in the inputs. So, by implementing CG, the switching activity of clock is brought down to a very nominal value. The recent circuit–level clock gating styles are reported in [15] and [16] amongst which Double–Gated CG (DG–CG), NC²MOS–CG and LECTOR–based CG (LB–CG) are quite popular unlike the primitive gating styles such as AND/OR gate–based CG reported in [17] which uses an Enable signal to control the propagation of clock signal to the gated clock. The NC²MOS–CG, DG–CG and LB–CG use a comparison signal (based on the present data and the previous data output) to generate the gated clock. Even if, the present data continuously switch, the gated clock will be different from the original clock signal. The gated clock will only switch when the comparison signal is logically high. In fact, both DG–CG and LB–CG is reported in [18-20] to engender gated clock for driving the complex component architectures of ICs. Implementation of these clock gating techniques lead to cutback the peaks of current drawn from V_{DD}, thereby decreasing dynamic current flow to reduce the dynamic power dissipation within IC. But, none of these gating techniques have been highlighted to control the $\frac{di(t)}{dt}$ (a salient component of PSN). As switching activity and $\frac{di(t)}{dt}$ can be brought to a controllable limit incorporating CG, reduction of PSN is feasible by them. Hence, CG may be employed as an alternative technique to reduce the PSN in digital ICs.

IV. EFFECT OF CLOCK GATING ON NOISE OCCURRENCES

It has been predicted in the previous section that clock gating can be an apt option to reduce PSN as they are

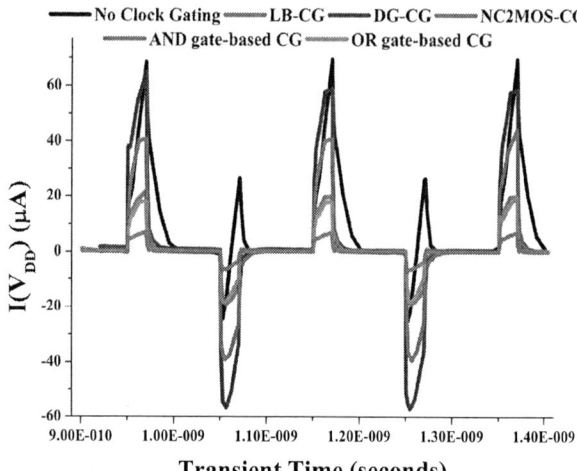

Fig. 3. Average peaks of current drawn from V_{DD} by CG techniques.

competent to manage the current ramps due to clocking activities in ICs. However, the rudimentary CG techniques viz. AND/OR gate–based CG are not that much suitable for PSN reduction as one of the inputs to its gating logic is the 'Enable' which is an open signal and extensively prone to input noise. As the PSN contamination is mostly dependent on $i(t)$, the amount of instantaneous current $i(t)$ drawn from V_{DD} by different clock gating techniques are to be measured. In order to do so, the CG techniques are implemented on MASTER–SLAVE configured D–Flip–Flop under the environment of 90nm Predictive Technology Model (PTM) [21] with a V_{DD} of 1.1 Volt at 5GHz clock. In figure 3, the transient response of $i(t)$ is projected for the CG techniques along with their no–gating correspondent. It is observed that NC²MOS–CG has pumped the least $i(t)$ peak that has been drawn from V_{DD} and it is about 90.4% less in amount than that for the case of no clock gating. In this frame of reference, OR gate–based CG and LB–CG are at a thin line in comparison to NC²MOS–CG, as they pump about 73.89% and 71.33% less $i(t)$ peak respectively than that pumped by their no–gating peer, whereas, the AND gate–based CG and DG–CG pump about 41.47% and 15.37% less $i(t)$ peak respectively than their no–gating peer. This analysis is also carried out in 65nm PTM [22], where the peak of $i(t)$ drawn from V_{DD} for the case of NC²MOS–CG, LB–CG, DG–CG, AND gate–based CG and OR gate–based CG is 86.74%, 75.12%, 27.49%, 49.91% and

TABLE I.   POWER SUPPLY NOISE BY DIFFERENT CG TECHNIQUES IN 90NM PTM FOR V_{DD} OF 1.1VOLT AT 5GHz CLOCK

Clock Gating Techniques	Avg. $i(t)$ (µA)	Avg. $\frac{di(t)}{dt}$ (A/s)	PSN (µV)	Circuit Overhead
OR gate–based CG	9.57	12.612	0.287	06
AND gate–based CG	10.244	0.40048	0.307	06
DG–CG	17.369	25.868	0.521	30
LB–CG	9.895	21.167	0.297	20
NC²MOS–CG	9.726	15.75	0.292	13
No clock gating	28.439	117.54	0.855	00

978-1-5386-1357-3/17 $31.00 © 2017 IEEE

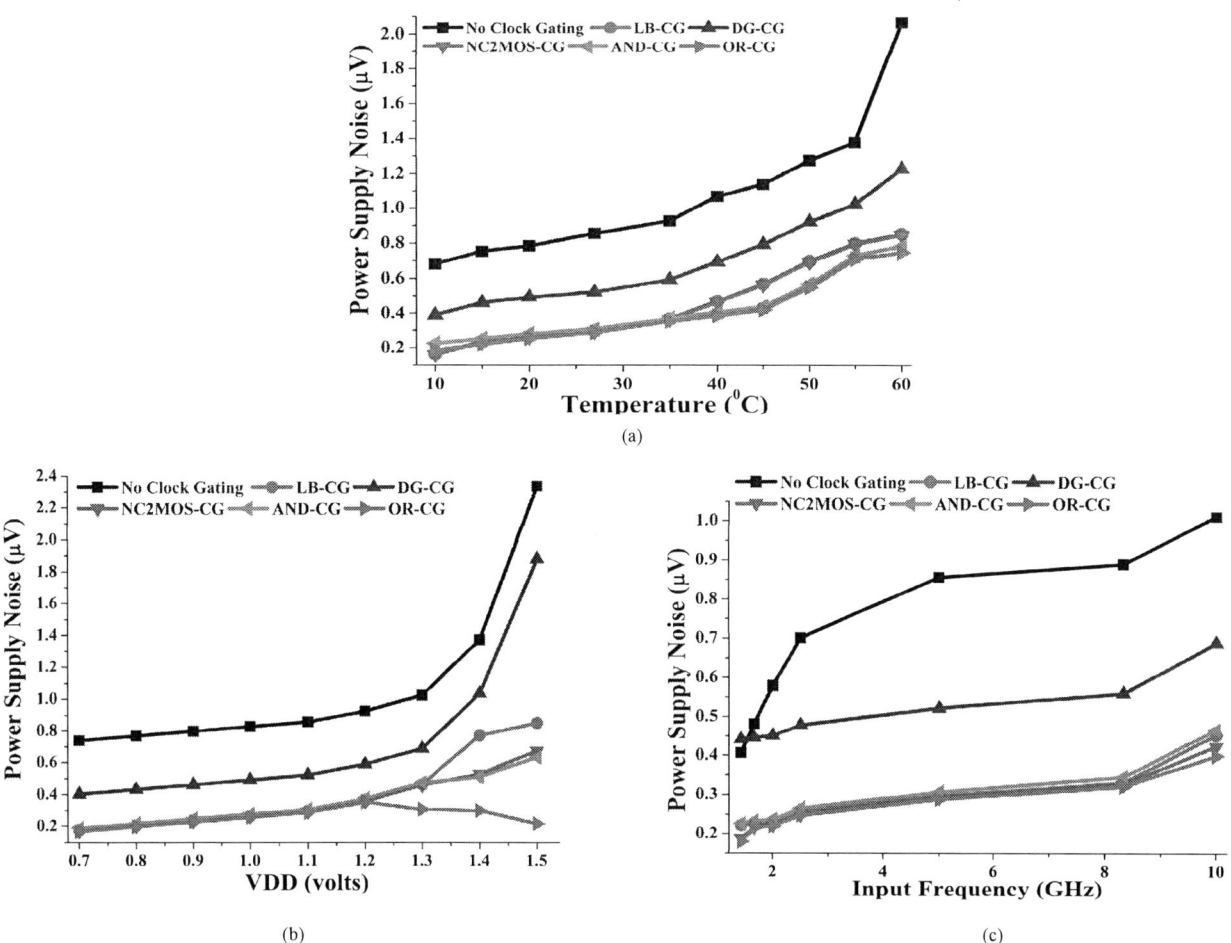

Fig. 4. Change in Power Supply Noise with the incorporation of CG techniques tested as the function of (a) temperature; (b) power supply (V_{DD}); (c) input frequency.

75.41% respectively less than that of no clock gating. But as both AND/OR gate–based CG are having the input noise issue, we have preferred to look into the PSN reduction possibility through DG–CG, LB–CG and NC²MOS–CG. The LB–CG, NC²MOS–CG and DG–CG have the ability to cutback the values of §V_p and §V_g (for a particular value of 'L_p' and 'R_p') leading to mitigation of PSN. However, in LB–CG, as the gating is done only at the MASTER end, it restrict the static current (also due to the incorporation of LECTOR in the gating logic) and dynamic current flow simultaneously. Therefore, the average $\frac{di(t)}{dt}$ across V_{DD} in LB–CG individually happens to be 18.17% and 81.99% less in comparison to that in DG–CG and no–gating peer as depicted through Table–I. In fact, the circuit overhead of LB–CG is 33.33% less than that of DG–CG. However, in this context, the average $\frac{di(t)}{dt}$ of NC²MOS–CG is 25.59%, 39.11% and 86.60% less in contrast to that in LB–CG, DG–CG and no–gating respectively, also giving an advantage of 35% and 56.66% in terms of circuit overhead with respect to LB–CG and DG–CG. This infers that, the best scope of PSN prohibition is possible using

NC²MOS–CG and the second best by LB–CG. Even more, in order to check the reliability of CG techniques for PSN proscription, we have estimated the probable amount of PSN in 90nm PTM considering the values of L_p and R_p to be 24pH and 0.03Ω respectively as per the recent standard of OLGA reported in [1]. Thereby, it is observed that NC²MOS–CG, LB–CG and DG–CG releases 65.84%, 65.26% and 39.06% less PSN than their no–gating counterpart.

The release of PSN by CG techniques along with their no–gating peer is tested in different temperature zone shown in figure 4(a) and it is observed that the PSN release is comparably high at higher temperatures. Even more, the PSN generation by CG techniques plus the PSN behavior by no clock gating is also checked for different power supply (V_{DD}) and input frequency shown in figure 4(b) and 4(c) respectively. Hence, it is visible that increase in V_{DD} and input frequency has increased the release of PSN for almost all CG techniques as well as for no clock gating. However, the magnitude of increase is not quite high. In fact, the magnitude of increase by the CG techniques is extremely low in comparison to that in the case of no clock gating, compelling the CG techniques to be viable option for power supply noise reduction even at varied input conditions.

V. CONCLUSION

In this paper, we have suggested the incorporation of clock gating to reduce power supply noise which is one of the major issues appearing in modern day low–power integrated circuits. The clock gating techniques successfully reduce the current ramps within silicon chips while switching from their OFF to ON state, instigating less chance of PSN contamination in comparison to the no–gating peer. It has been also observed that all the emphasized clock gating techniques are capable of reducing PSN even when operated in lower technology nodes like 65nm PTM.

Acknowledgment

We are thankful to Ministry of Electronics & Information Technology, Government of India, for providing us the financial grant under Visvesvaraya PhD Scheme & SMDP–C2SD project.

References

[1] Laskar, Nivedita, et al. "A New Current Profile Determination Methodology Incorporating Gating Logic to Minimize the Noise of CPU Chip by 40%." *Journal of Circuits, Systems and Computers* (2017): 1850049.

[2] Intel. Pinned packaging. In: 2000 Packaging Data book; Chapter 13. pp. 01-26.

[3] Bhattacharyya, B. K., & Baral, D. (2013). Method to simulate rise time of current drawn by a microprocessor. *IEEE Transactions on Components, Packaging and Manufacturing Technology*, 3(10), 1731-1736.

[4] Aragones, Xavier, Jose Luis Gonzalez, and Antonio Rubio. *Analysis and solutions for switching noise coupling in mixed-signal ICs*. Springer Science & Business Media, 2013.

[5] Allstot, David J., et al. "Folded source-coupled logic vs. CMOS static logic for low-noise mixed-signal ICs." *IEEE Transactions on Circuits and Systems I: Fundamental Theory and Applications* 40.9 (1993): 553-563.

[6] Ng, Hiok-Tiaq, and David J. Allstot. "CMOS current steering logic for low-voltage mixed-signal integrated circuits." *IEEE Transactions on Very Large Scale Integration (VLSI) Systems* 5.3 (1997): 301-308.

[7] Albuquerque, E., J. Fernandes, and M. Silva. "NMOS current-balanced logic." *Electronics Letters* 32.11 (1996): 997-998.

[8] Heydari, Payam, and Massoud Pedram. "Ground bounce in digital VLSI circuits." *IEEE Transactions on Very Large Scale Integration (VLSI) Systems* 11.2 (2003): 180-193.

[9] Vittal, Ashok, et al. "Clock skew optimization for ground bounce control." *Proceedings of the 1996 IEEE/ACM international conference on Computer-aided design*. IEEE Computer Society, 1997.

[10] Bentley, Bob. "Validating the Intel® Pentium® 4 microprocessor." *Dependable Systems and Networks, 2001. DSN 2001. International Conference on*. IEEE, 2001.

[11] Castro, Javier, et al. "A switching noise vision of the optimization techniques for low-power synthesis." *Circuit Theory and Design, 2007. ECCTD 2007. 18th European Conference on*. IEEE, 2007.

[12] Ding, L., & Mazumder, P. (2004). "On circuit techniques to improve noise immunity of CMOS dynamic logic." *IEEE Transactions on Very Large Scale Integration (VLSI) Systems*, 12(9), 910-925.

[13] Salman, Emre. "Switching Noise and Timing and Characteristics in Nanoscale Integrated Circuits." ProQuest, 2009.

[14] Akl, Charbel J., Rafic A. Ayoubi, and Magdy A. Bayoumi. "An effective staggered-phase damping technique for suppressing power-gating resonance noise during mode transition." *2009 10th International Symposium on Quality Electronic Design*. IEEE, 2009.

[15] Strollo, A. G. M., E. Napoli, and D. De Caro. "Low-power flip-flops with reliable clock gating." *Microelectronics journal* 32.1 (2001): 21-28.

[16] P. Bhattacharjee, A. Majumder, T. D. Das, "A 90 nm Leakage Control Transistor Based Clock Gating for Low Power Flip Flop Applications", *IEEE 59th International Midwest Symposium on Circuits and Systems (MWSCAS)*, pp. 381-384, 2016 ©IEEE.

[17] Shinde, Jitesh, and S. S. Salankar. "Clock gating—A power optimizing technique for VLSI circuits." *India Conference (INDICON), 2011 Annual IEEE*. IEEE, 2011.

[18] Bhattacharjee, P., & Majumder, A. (2016, December). "LECTOR Based Gated Clock Approach to Design Low Power FSM for Serial Adder." In *Nanoelectronic and Information Systems (iNIS), 2016 IEEE International Symposium on* (pp. 250-254). IEEE.

[19] P. Bhattacharjee, B. Nath and A. Majumder. "LECTOR Based Clock Gating for Low Power Multi–Stage Flip Flop Application." In *16th International Conference on Electronics, Information and Communication (ICEIC)* (pp. 106-109). ©IEIE 2017.

[20] Bhattacharjee, Pritam, Alak Majumder, and Bipasha Nath. "A 23.52 μW/0.7 V Multi-stage Flip-flop Architecture Steered by a LECTOR-based Gated Clock." *IEIE Transactions on Smart Processing & Computing* 6.3 (2017): 220-227.

[21] http://ptm.asu.edu/modelcard/2006/90nm_bulk.pm

[22] http://ptm.asu.edu/modelcard/2006/65nm_bulk.pm

2017 IEEE International Symposium on Nanoelectronic and Information Systems

Binary Counter Based Gated Clock Tree for Integrated CPU Chip

Bipasha Nath[t], Alak Majumder[¥]
VLSI Design Lab, Department of ECE
National Institute of Technology Arunachal Pradesh
Yupia, India - 791112
[t]nathbipasha.dmg@gmail.com, [¥]majumder.alak@gmail.com

Abstract— **The conventional clock tree, made of some buffers, is prone to large current with the increase in switching clock frequency. Due to the high current peak at the clock edge, the power and ground noise (PGN / PSN) of a CPU chip increases significantly to engender a crucial effect on the circuit performance metrics such as the delay and power dissipation of the entire chip. This current peak also gets increased when numerous signals, driven by the neighboring sources, switch simultaneously. To eliminate the above stated issue, we have come up with a new approach of designing clock tree consisting of a binary counter and an enable signal generator with reset logic. The generated enable signals control the switching of different gates of the clock tree at different time, thereby helps to yield a new current profile by reducing both the current peak and current ramp, the main culprit of power and ground noise. This work of Binary Counter based Gated Clock Tree Circuit (GCTC) is simulated for 90nm CMOS technology using CADENCE Virtuoso platform at a power supply of 1.2Volt and 5 GHz operating frequency. It is observed that the proposed GCTC outsmart the conventional method by mitigating average power, average current, current ramp and PSN by 1.88%, 9.97%, 91.9% and 83.43% respectively.**

Keywords— Binary Asynchronous Counter, Enable Signal Generator, Lector AND, Binary Clock Tree

I. INTRODUCTION

Over the last four decades the integrated circuit industry has evolved in a tremendous pace. This success has been driven by the scaling of device sizes leading to higher and higher integration capability, which have enabled more functionality and higher performance. The impressive evolution of modern high performance microprocessors have resulted in chips with over a billion transistors as well as multi-GHz clock frequencies[1]. Clock signals are important in synchronous circuits to synchronize different data signals arriving from different parts of the integrated circuit, such that the correct data is available for computation. Modern clock distribution design continues to face challenges in spite of significant advances in the last decade. Some challenges are still to be eliminated: the first is to support higher clock frequencies based on the strong correlation between frequency and chip performance and secondly, as the technology scaling allows higher level of integration and larger die size leading to

higher clock loading and larger distances, the clock network needs to be traversed. The final challenge is that the technology scaling leads to an increase in on-die variations that may degrade clock performance if not properly addressed.

Although a lot of research efforts have been made to minimize the total power consumption caused by the clock tree, almost no attention has been paid to the minimization of the current peak and ramp (di/dt) caused by the clock tree. The most common strategy for clock distribution is to insert a large number of buffers along the paths from clock source to destination, forming a buffered-tree structure [2]. Each transition of the clock signal changes the state of each capacitive node within the clock tree in contrast to the switching activity of any combinational block, where the change of logic states is dependent on the logic function. Therefore, the clock tree remains one of the major sources of power dissipation in modern day VLSI circuits and hence, we have come up with a process of designing this clock distribution network to address the said issue.

The paper is organised as follows: the second section defines the existing problem in clock tree circuit. In section 3, the proposed clock tree circuit is described in details. This is followed by the simulation results and different analyses done to compare its performance with conventional circuit. Finally, section 5 concludes the work.

II. PROBLEM STATEMENT

When the modern CPUs / a Silicon chip, running at a frequency close to 3-5GHz, wakes up from sleep mode, the functional logic blocks inside it start performing computation and hence, draw significant current simultaneously. This large current produces a substantial voltage drop on the power supply due to Ldi/dt, where L is the inductance of the main supply connected to the silicon device to the first level decoupling capacitance inside the package and di /dt is the current ramp inside the chip. This voltage drop subjects the chip to run at lower speed, which guides to the functional failure of the CPU [3, 4]. The high performance modern processors are packaged with Organic Land Grid Array (OLGA) with a thickness of around 0.03–0.04 inch having an inductance of about 24pH, which is the primary cause of Ldi/dt [5]. However, large current leads to huge IR drop, which also contributes to a large power supply noise.

We are thankful to MEITY, Government of India for providing us financial support through SMDP-C2SD Project.

978-1-5386-1357-3/17 $31.00 © 2017 IEEE   229

The major part of the current flows during the switching of the clock, as the present conventional clock trees are consisting of some buffers. Existing solutions to minimize the power supply noise are listed below.

(i) Use of assigning polarities to clock buffers and determining buffer sizes to fully exploit the effects of buffer sizing together with polarity assignment satisfying the clock skew constraint [6].

(ii) To schedule the clock arrival times of flip-flops (FFs) in order to disperse the peak current [7].

(iii) Formulated the clock arrival time scheduling problem as a 0-1 integer linear program [8].

(iv) To assign positive polarity onto a half of clock buffers and negative polarity onto the remaining half of the clock buffers [9].

None of the above approaches could solve the power supply noise issue, as they lead to timing criticality. Hence, there is a need to redesign the clock tree to generate a new current profile such that the ramp of current may be reduced.

III. PROPOSED GATED CLOCK TREE

To eliminate the loopholes of the clock distribution tree, we have come up with altogether different approach of distributing clock by incorporating a control unit which consists of a Binary Counter, an enable signal generator with reset logic and a switch based on AND gate, which are depicted in the subsequent sections.

A. Enable Signal Generator (ESG)

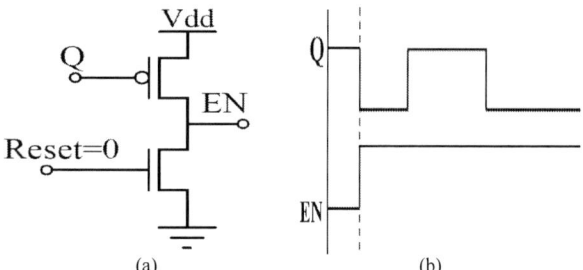

(a)   (b)

Fig. 1. Enable Signal Generator (a) Schematic (b) Ideal Transient

At the beginning of operation, there is no charge at common drain terminal and thus initially 'EN' must be equals to '0'. For normal operation, NMOS is kept in OFF state by connecting Reset pin to GND. Now when Q = 1, PMOS is OFF and therefore, no flow of charge through PMOS takes place and EN remains at '0'. Now, whenever Q changes from '1' to '0', PMOS gets turned ON to charge EN to '1' and EN retains that value forever as there is no discharge path present until Reset pin is connected to logic 1. Thus, signal 'EN' doesn't depend on the signal Q for rest of the time interval. It means that the first (1-0) transition in 'Q' changes the EN signal from '0' to '1'. Thus. different enable signals (EN) have been generated from different Q signals to control the binary clock-tree circuit.

B. Lector AND based Switch

The logic AND may be considered as a switch when one of the inputs is made to work as a control variable (EN) such that the input will be travelled to the output only when EN goes high. Though the conventional AND can serve the purpose, we have introduced LECTOR based AND to reduce the current peak, a major culprit of PSN.

The basic way to reduce leakage current is the effective stacking of transistors in the path from supply voltage to ground. This is based on the observation which says that 'a state with more than one transistor OFF in a path from supply voltage to ground is far less leaky than a state with only one transistor OFF in any supply to ground path' [10]. In our method, we have used two leakage control transistors (M3 and M4). These M3 and M4 are introduced between the nodes N1 and N2 of the pull-up and pull-down logic of the NAND gate. The drain nodes of the transistors M3 and M4 are connected together to form the output node of the NAND. The source nodes of the transistors are connected to nodes N1 and N2 of pull-up and pull-down logic respectively. The switching of transistors are controlled by the voltage potentials at nodes N1 and N2 respectively. This wiring configuration ensures that one of the transistors (M3 or M4) is always near its cutoff region, irrespective of the input applied to the NAND gate. A Lector based inverter is cascaded to make it AND gate as shown in the Fig. 2.

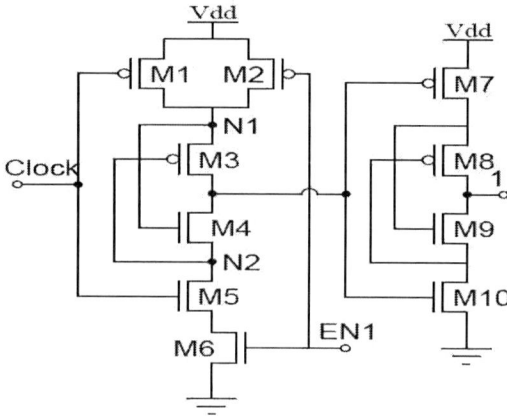

Fig.2. LECTOR based AND Gate

C. Architecture and Working principle

The figure 3(a) displays the proposed method of designing a clock distribution circuit, which consists of mainly two parts namely Control Circuit and Clock Tree Circuit (CTC). Control circuit consists of a binary counter and ESG with reset logic. where the outputs of a counter (Q1, Q2, Q4, Q5) and an OR gate (Q3) are used to feed the ESG circuit to generate some enable signals (EN1, EN2, EN3, EN4 and EN5) as shown in figure 3(b). Whenever reset goes high all the enables will come down to logic zero and remain there until the reset switches to low. On the other hand, the new clock tree unit is unearthed to deal with the distribution of clock through some enable controlled switches, made of LECTOR AND gate.

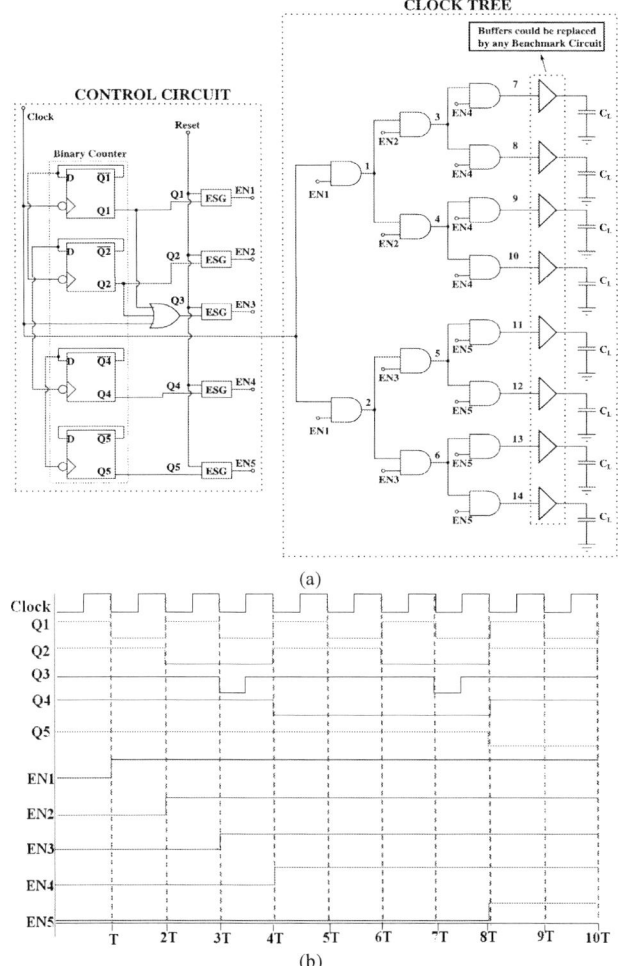

(a)

(b)

Fig. 3. Proposed binary counter based gated clock tree circuit (a) schematic (b) ideal transient

The binary counter employed here is asynchronous by design and made of a negative edge triggered D-FF presented in [11]. While generating these EN signals, the time of first ever transition from 'high to low' of Q signal has to be taken care of, as because the enable signals will switch from 'low to high' as soon as the ESG circuit detects that first transition of Q from 'high to low'. The enable signals will remain at this high level until the reset pin is connected to logic 1. The signals Q1, Q2, Q3, Q4 and Q5 are generated such that first ever 'high to low' transition for each of them occur at clock time T, 2T, 3T, 4T and 8T respectively and thus produces EN1, EN2, EN3, EN4 and EN5 as displayed in figure 3b. Let us consider, (1, 2), (3, 4), (5, 6), (7, 8, 9, 10) and (11, 12, 13, 14,) are the AND gates which are controlled by signals EN1, EN2, EN3, EN4 and EN5 respectively. These enable signals will turn the respective AND gates ON at T, 2T, 3T, 4T and 8T time respectively and will float them in ON state for the rest of the time interval until the reset pin is connected to logic 1. The enable signals may be generated to switch from 'low to

high' at other time positions different than the times mentioned here (T, 2T, 3T, 4T and 8T) depending on the requirement. This methodology helps the total current to ramp slowly before it reaches the peak value (when all gates are ON) thereby reducing the di/dt, the leading scorer of the power supply noise and simultaneous switching noise. To observe the current profile using the proposed binary gated clock tree circuit (GCTC), we have feed several buffer circuits, which may be replaced by different benchmark circuits or different logic blocks inside a core processor.

IV. PERFORMANCE ANALYSIS

The proposed gated clock tree is designed and simulated on Cadence Virtuoso platform using 90nm CMOS technology at 1.2 Volt supply voltage. The transient analysis of the proposed circuit is shown in the figure 4.

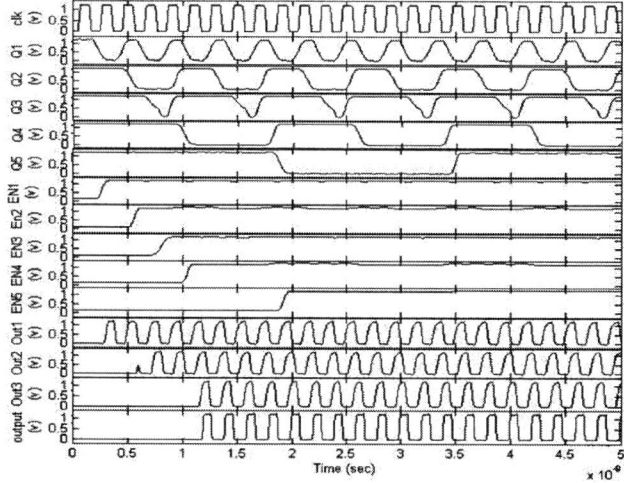

Fig. 4. Transient Analysis of Proposed Gated Clock Tree Circuit

The clock frequency is varied up to a maximum of 5 GHz for the smooth functioning of the proposed circuit. The current profile of Conventional CTC and Proposed GCTC is measured and displayed in figure 5, from which we can note that the current peak of conventional one reads a value of 366µA, whereas the GCTC measures as small as 277µA only.

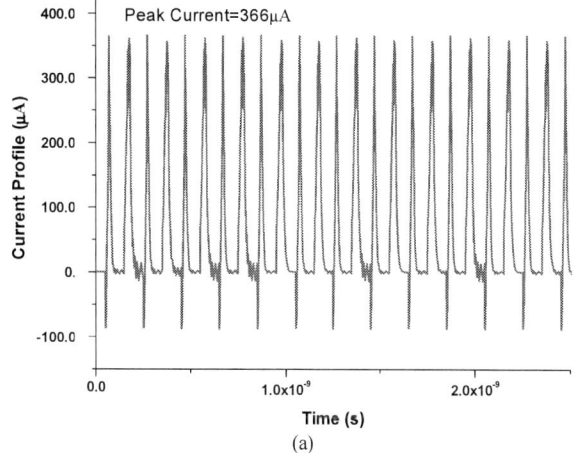

(a)

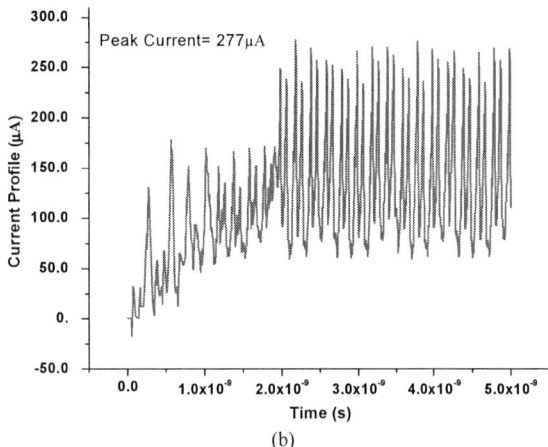

(b)

Fig. 5. Generated current profile (a) Conventional CTC (b) GCTC

It is also seen from figure 5(b) that the rate of change of current is proposed GCTC is less as compared to the conventional CTC. As the new current profile ramps slowly to reach the peak value, it steers a lesser di/dt thereby leading to lower PSN.

The performance metrics of the proposed and conventional CTC are displayed at table 1. The on-chip PSN is calculated considering the package resistance (R) and package inductance (L) to be 0.03Ω and 24pH respectively [12]. We have observed 1.88% savings in average power and 9.79% lesser current in the proposed gated CTC with respect to its conventional counterpart. As per our proposal, the proposed circuit has explicitly outsmarted the conventional one in current ramp (di/dt) and total power supply noise (PSN). It is witnessed that the proposed gated CTC has got a huge reduction of 91.9% and 83.43% in terms of current ramp (di/dt) and power supply noise (PSN) respectively.

TABLE 1: COMPARISON OF PERFORMANCE METRICS

Parameters	Proposed Circuit	Conventional Circuit
Technology (nm)	90	90
Supply Voltage (Volt)	1.2	1.2
Operating Frequency (GHz)	5	5
Average Power (μW)	157	160
Output Noise Voltage (aV/Hz$^{1/2}$)	60.287	60.058
I_{RMS} (μA)	129	143
Current Ramp (di/dt) , (A/μs)	0.124	1.54
PSN (μV)	6.84	41.3

Performance parameters like RMS value of the power supply current, current ramp (di/dt) and PSN are also investigated as a function of clock frequency and supply voltage in three different temperatures such as 0°C, 27°C and 90°C. All these analyses are plotted from figure 6 to figure 8, where from it is witnessed that the said parameters get increased with the increment of supply voltage and frequency. However, all the parametric values including PSN are found to be much lesser in proposed Gated Clock Tree Circuit against its conventional approach while varying temperature, supply voltage and frequency.

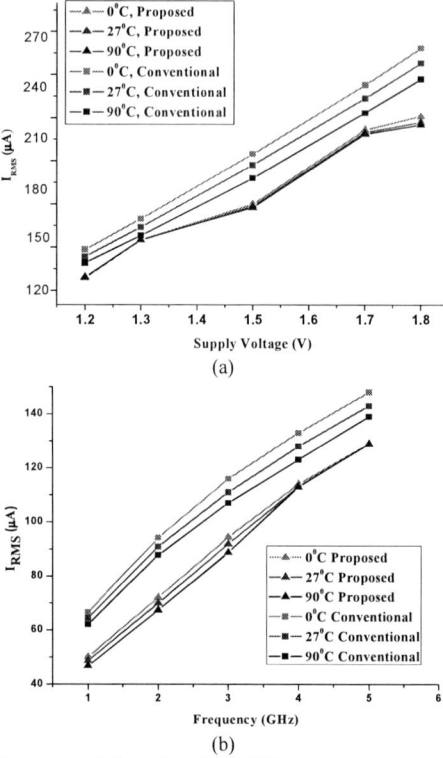

Fig. 6. RMS current vs (a) supply voltage (b) frequency

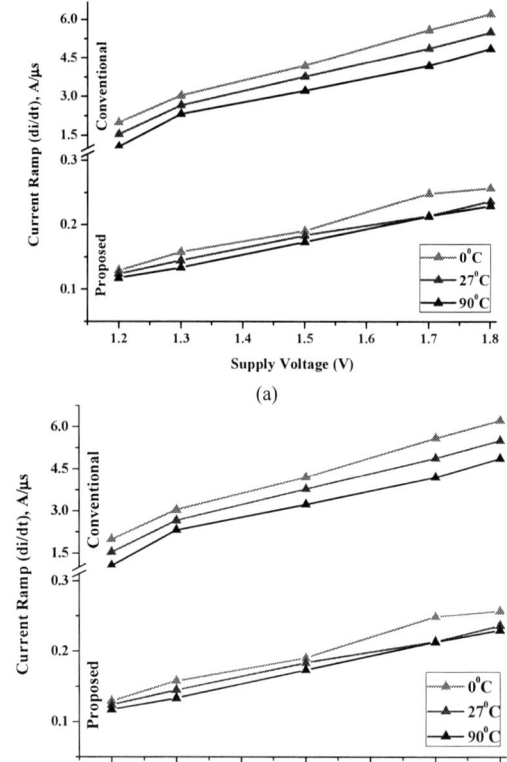

Fig. 7. Current profile vs (a) supply voltage (b) frequency

978-1-5386-1357-3/17 $31.00 © 2017 IEEE

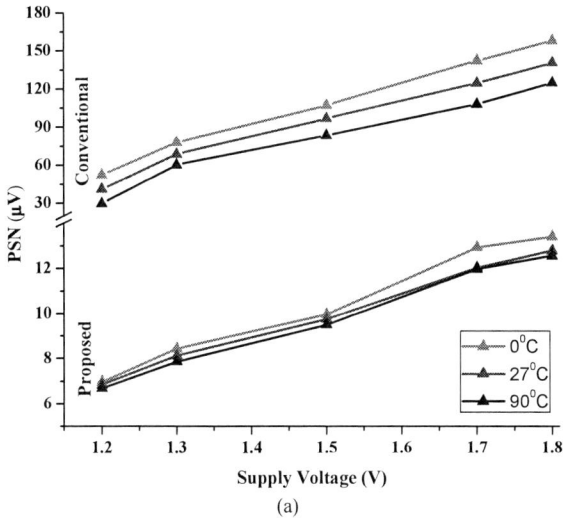

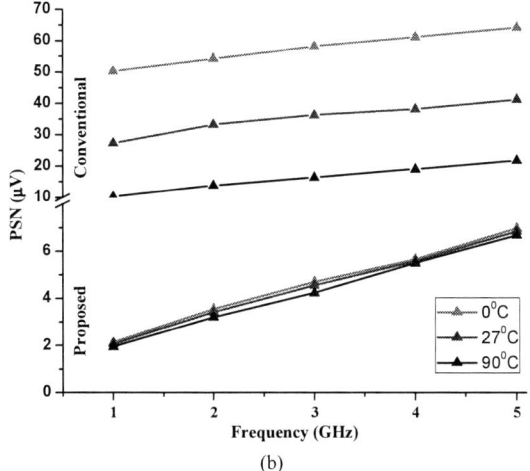

Fig. 8. Power supply noise vs (a) supply voltage (b) frequency

V. CONCLUSION

In this article, a new arrangement of clock tree circuit is presented where different parts of the circuit are gated at different time using a binary asynchronous counter and an enable signal generator with reset logic. The different enables generated are employed at different parts of the CTC to allow them to flow current at different time. The GCTC is designed up to 3rd stage, output of which is used to steer some buffer circuits to generate a current profile. This new current that ramps very slowly to reach the peak current thereby reducing the power supply noise. From the several analyses done, it is observed that the proposed PVT aware GCTC has outplayed the conventional approach by reducing di/dt and PSN by 91.9% and 83.43% respectively, which are some great number. There remains several scopes to deal with this work in near future. Some of them are jotted down below:

1. Buffer circuits could be replaced by some benchmark circuit to prove its effectiveness.
2. Final physical design and post-layout analysis are to be done for proving its worth.
3. Testing of proposed GCTC in a core architecture of Nehalem or Broadwell-E processors.

References

[1] Martin Hansson, "Low-Power Clocking and Circuit Techniques for Leakage and Process Variation Compensation", Dissertation, Linköping university of technology, 2008.

[2] Yow-Tyng Nieh, Shih-Hsu Huang, Sheng-Yu Hsu, "Minimizing Peak Current via Opposite-Phase Clock Tree", 42nd Design Automation Conference, 2015, Proceedings.

[3] Bhowmik, Suman, et al. "Reduction of Noise Using Continuously Changing Variable Clock and Clock Gating for IC Chips." IEEE Transactions on Components, Packaging and Manufacturing Technology, 2016

[4] N. Laskar, S. Debnath, A. Majumder and B.K. Bhattacharyya. "A new cuurent profile determination methodology incorporating gating logic to minimize the noise of CPU chip by 40%", Journal of Circuits, Systems, and Computers ©World Scientific Publishing Company, 2017.

[5] Intel. Pinned packaging. In: 2000 Packaging Data book; Chapter 13. pp. 01-26.

[6] H Jang, D. Joo, T. Kim, "Buffer Sizing and Polarity Assignment in Clock Tree Synthesis for Power/Ground Noise Minimization", IEEE trans.on CAD of Integrated Circuits & Systems, Vol.30, No.1, Jan 2011.

[7] L. Benini, P. Vuillod, A. Bogliolo, and G. D. Micheli, "Clock skew optimization for peak current eduction," J. VLSI Signal Process, vol.16, nos. 2–3, pp. 117–130, 1997.

[8] A. Vittal, H. Ha, F. Brewer, and M. M. Sadowska, "Clock skew optimization for ground bounce control," in Proc. IEEE/ACM Int. Conf.Comput.-Aided Design, 1996, pp. 395–399.

[9] Y.-T. Nieh, S.-H. Huang, and S.-Y. Hsu, "Minimizing peak current via opposite-phase clock tree," in Proc. IEEE/ACM Des. Autom. Conf., 2005, pp. 182–185.

[10] N. Hanchate, and N. Ranganathan,, "LECTOR: A Technique for Leakage Reduction in CMOS Circuits", IEEE Transactions on Very Large Scale Integration (VLSI) Systems, Vol. 12, No. 2, February 2004.

[11] B. Nath, A Majumder, M. Das, A.J. Mondol, P. Chakraburty, B.K. Bhattacharyya,"Voltage Keeper Based 28.27 µW New Frequency Divider Circuit in 90nm Technology for Gigascale Serdes Application", IEEE VLSI Circuit & System Letter, Vol: 3, Issue:2, June 2017.

[12] K. Tasreen, "On-chip power supply noise: scaling, suppression and detection", PhD Thesis, University of Waterloo, Ontario, 2012.

2017 IEEE International Symposium on Nanoelectronic and Information Systems

Design and Analysis of Schmitt Trigger Based 10T SRAM in 32 nm Technology

Amit Singh Rajput
Jiwaji University
Gwalior, India
0915.ec@gmail.com

Manisha Pattanaik
ABV-IIITM
Gwalior, India
manishapattanaik@iiitm.ac.in

R.K. Tiwari
Jiwaji University
Gwalior, India
phy05@rediffmail.com

Abstract— **Low stability and higher process parameter variations are the critical issues in SRAM cell in nanometer technology. In this paper, we propose a Schmitt trigger based differential SRAM bit cell to solve scaled technology issue. Schmitt triggers based SRAM cell provides higher read stability and higher process variation tolerance because of its internal positive feedback. This paper proposes a cell with which utilize read-write decouple technique and Schmitt trigger technique to enhance cell acceptability at low technology. To estimate the effect of process parameter variation on stability Monte Carlo simulation are performed using 32 nm predictive-technology-model (PTM) with 30% threshold voltage variations. Proposed cell achieves 1.56 higher read static noise margin compared to the traditional ST-1 and ST-2 cell respectively. In this cell, Sensitivity to process parameter variation is reduced by 5.9× and 2.5× as compared to ST-1 and ST-2 cell. The results indicate that the proposed cell may be a better substitution of other Schmitt trigger based SRAM cell at scaled technology node.**

Keywords—process parameter variation, Schmitt trigger, SRAM, static noise margin, threshold voltage variation

I. Introduction

Embedded caches are expected to occupy 90% of the total die area of a system-on-a-chip (SoC)[1] because of its high data processing and data storage. A memory cell, which is the fundamental unit of SRAM memory is scaled down with each technology node to increase cell density. High-density cell uses the smallest devices size in technology, making it more vulnerable to process variations. One technique available in the literature to make process invariant SRAM cell is to replace data storage latch (back to back connected inverter) with a Schmitt trigger inverter. Schmitt trigger inverter is much more stable to external noise compared to conventional inverter because of its variable switching threshold voltage. Static Noise Margin (SNM) is a parameter to quantify memory cell stability against noise. Kulkarni et al.[1] Proposed a Schmitt trigger based SRAM cell (ST-1) with enhancing read stability (SNM). However, ST-1 suffers from the read write conflict and read disturbance problem[2]. A modified version of Schmitt trigger-based SRAM cell (ST-2) is given in Ref. [3]. This ST-2 cell affords higher read SNM and improved write-ability along with higher process tolerant ability. However, this circuit takes an undesirable amount of bit line power consumption (because four transistors are attached to bit line unlike two in ST-1) and peripheral circuit area (because it has the two-word line). Moreover, a low value of I_{on}/I_{off} current ratio in ST-2 cell limit number of the cell

connected to a single bit line. Ahmad et al. proposed [4] a Schmitt trigger based single-ended 11T SRAM cell, which provides higher read-write stability and lower power consumption because of the single-ended design (One-half of the active power for BL switching). However, it occupies more area and access time penalty (because of single ended operation) as compared to the differential cell.

This paper introduces a new Schmitt trigger based differential 10T SRAM cell design. The proposed cell gives higher read stability because of jointly implementation of ST technique and Read write decoupled technique. Furthermore, it also provides lower process variation sensitivity.

The rest of the paper is organized as follows. Schmitt trigger based SRAM cells (ST-1 and ST-2) are explained in Section II. Proposed Schmitt trigger based 10T SRAM cell and its read, write and hold operation are analyzed in Section III. Read stability and process variations affect are compared and analyzed in Section IV, and last Section V concludes the paper.

II. Schmitt Trigger Based SRAM Cell

In the Schmitt trigger based SRAM cell, conventional cross coupled inverter pair are replaced from Schmitt trigger inverters. Fig. 1. shows Two possible structures of Schmitt trigger inverters. Conventional Schmitt trigger (ST) inverter requires six transistors to design a cell as shown in Fig. 1(a). To implement a cell with conventional ST inverter require 14 transistors, which would give large area penalty. ST inverter is shown in Fig. 1(b) remove area penalty because it takes five transistors to implement an ST inverter. To understand the working Fig. 1(b). Consider a case, when V_{in} is the transition from "0" to VDD, voltage V_a rises to a to a positive value because transistor M1 and M3 form a voltage divider network thus gate to source voltage(VGS) of M2 become negative. This negative V_{GS} increases switching threshold of the transistor M2. Therefore, inverter preserves V_{out} until V_{in} is less than switching threshold voltage of the inverter.

A. ST-1

Fig. 2. shows the Schematic diagram of the ST-1 memory cell. In this cell transistor M3, M1, M7 and M8 forms first ST inverter(Inv-1) while Transistor M6, M4, M9 and M10 makes second ST inverter (Inv-2). Access transistor M2 and M5 are used to transfer data from BL and BLB to Q and QB

978-1-5386-1357-3/17 $31.00 © 2017 IEEE

234

respectively on the activation of write word line (WWL) signal

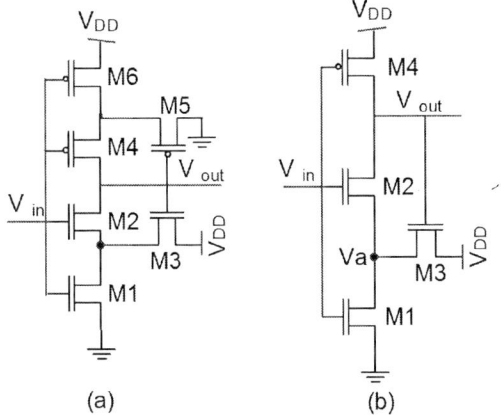

Fig. 2. Schmitt triggers inverter

During the read operation (assume QB="0" and Q=VDD); BL is precharged to VDD, and WWL is at VDD. Transistor M5, M4, and M9 of inv-2 forms a voltage divider network. Therefore the voltage of node QB is raised above the ground potential. If this voltage is higher than switching threshold voltage of Inv-1, data stored in Q may be flipped resulting a read failure operation. In Inv-1 a voltage divider network forms by transistor M7 and M8 increase voltage V_a. This positive V_a improves switching threshold of Inv-1; thus Inv-1 preserve data "1".

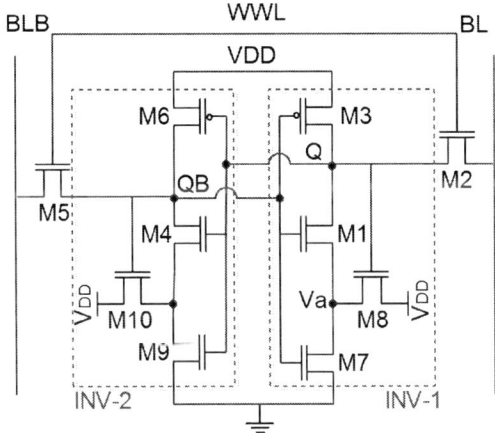

Fig. 2. Schmitt triggers based 10T SRAM cell (ST-1) [3]

ST-1 has two-bit lines and one-word line same as conventional 6T SRAM cell. Therefore it can be used in place of 6T without any change in memory architecture. ST-1 SRAM cell provides better read stability, and improved process tolerance capabilities compare to the conventional 6T cell. It can operate reliably at sub threshold voltage whereas traditional SRAM fails to operate.

B. ST-2

Fig. 3. shows a more process-tolerant version of the St-1 memory cell.It has two-bit lines BL and BLB, two-word line (WWL and RWL) also two write access transistors M2 and M5. During the write operation, WWL and RWL are connected to VDD while during reading time RWL is connected to VDD whereas WWL is connected to "0". The main difference between ST-1 and ST-2 is that in the case of ST-2 feedback is provided by RWL signal whereas in the case of ST-1 feedback is provided by internal storage node Q and QB.

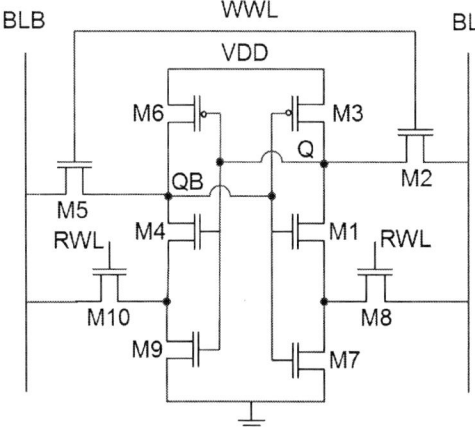

Fig. 3. Schmitt trigger based 10T SRAM bit cell (ST-2)

In the case of ST-1 feedback mechanism is available until the high voltage is available at internal nodes. The feedback mechanism lost when voltage starts to fall. Whereas in the case of ST-2 feedback mechanism is provided by separate RWL signal, so it is strong.

III. PROPOSED SRAM CELL

Fig. 4. shows the proposed cell with minimum feature size (W=L=32nm) transistor. The upper sub circuit consists of cross-coupled inverters (M1-M3 and M4-M6) to store single bit information. Moreover, write access transistors (M5 and M2) used to transfer data from two-bit lines (BL and BLB) to the storage node(Q and QB) on the activation of write word line (WWL) signal. The lower part of the cell is made of the read-access transistor (M7, M8 and M9, M10) to read data from Q and QB when read write (RW) line asserted high. Transistor M7 and M9 are used to decrease leakage current in hold mode

A. Read operation

Before start read operation BL and BLB are pre charged to VDD. In the second step, RW transitions to VDD while WWL is maintained at "0". If a "1" is stored at node Q, BL is discharged through the transistors M7 and M8.

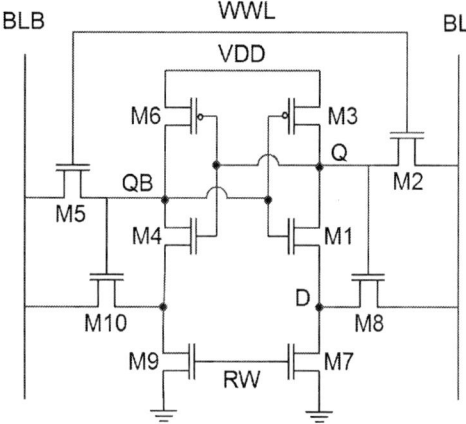

Fig. 4. Proposed Schmitt trigger based 10T SRAM cell

Alternatively, if a "0" is stored at Node Q, BL is maintained at VDD.In the read operation, when Q is high; it turns on the transistor M8. The intermediate node D rises to a positive voltage. The switching threshold voltage of inverter M1-M3 is increased because of the negative gate to source voltage of transistor M1. The increased threshold voltage keeps Q high until QB is not more than to switching threshold voltage of inverter M1-M3.

In the proposed cell, positive feedback is provided by the internal nodes Q and QB like as ST-1 cell. In the proposed cell feedback mechanism is available until applied node voltage is less than switching threshold voltage of the inverter. Once the storage nodes start transitioning from high state to low state, the feedback mechanism is lost.

B. Write operation

Before a write operation, the BL is discharged to GND while BLB is charged to VDD to force a "0" onto Node Q. To start the write operation, the WWL transitions to VDD while the RW is maintained at "0". The strength of both inverters is reduced because inverter M1-M3 and M4-M6 are disconnected from ground potential. Therefore cell data flip easily unlike to 6T cell.

C. Hold operation

BL and BLB are precharged to VDD while WWL and RW are connected to GND. Leakage current is decreased due to stacking effect of M7 and M9 transistors.

IV. CELL PERFORMANCE AND DISCUSSION

This part of the paper gives read stability and inter-die process analysis comparison of the proposed cell with ST-1, ST-2.

A. Read Stability

SNM is a parameter used to quantify cell robustness against noise.SNM is equivalent to the highest value of DC noise voltage that can be tolerated by the cell before changing

states[6]. SNM is calculated graphically as the length of a side of the largest square that can be fixed within the smallest lobes of the butterfly curve[7]. The circuit used to estimate stability is shown in Fig. 5. In the first step both bit lines BL and BLB are precharged to VDD, then RW and WWL is biased to VDD and GND respectively .In the second phase, N1 is swept from "0" to VDD while measuring node voltage Q. In the next step, N2 is swept from "0" to VDD while measuring the voltage at node QB in the same way. At last measured node voltages, Q and mirrored QB are plotted to generate ''butterfly curve''[8]. We used 32nm PTM model with following technology parameters for simulation: VDD =0.9V, vth0n = 0.501, vth0p = -0.452 , and T=25°c as used by [13].

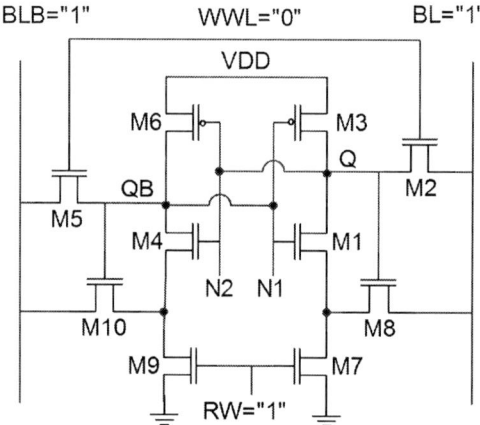

Fig. 5. Test circuit to evaluate read SNM of proposed 10T SRAM cell

The read SNM of the proposed 10T, ST-1 and ST-2 are plotted in Fig. 6. It is clear from Fig. 6. that proposed cell provides highest read SNM among the all SRAM cell discussed. Proposed cell gives 1.8x and 1.6x higher RSNM as compared to ST-1 and ST-2 at 0.9V.The SRAM cell having RSNM of 25% of VDD is considered to be fairly good for read-stability[4]. Proposed cell gives 38%, read SNM value for supply voltage 0.9V, which is reasonably acceptable.

B. Process Variations

This part of the paper presents inter-die process analysis of the proposed cell. Memory fabrications involve many processing steps, imperfection in these steps is the cause of parameter (intra-die and inter-die) variation[9].Transistor threshold voltage variation is critical parameter among all parameter variations [10][11].Two statistical device models are used to estimate parametric variation effect: corner models and Monte Carlo models[12]. Corner method is useful in studying the worst-case analysis, however; Monte Carlo simulation gives statics parameter like standard deviation mean(σ) and mean (μ), so it is used here. Three sigma variation of 30% Vth with independent Gaussian distributions is applied here.

SNM in this paper is estimated with 5000 MC run [1] to get higher accuracy. Variability (σ/μ) is a parameter used to measure process tolerant capability of memory cell[14]. Fig. 7. shows the comparative analysis of SNM variability

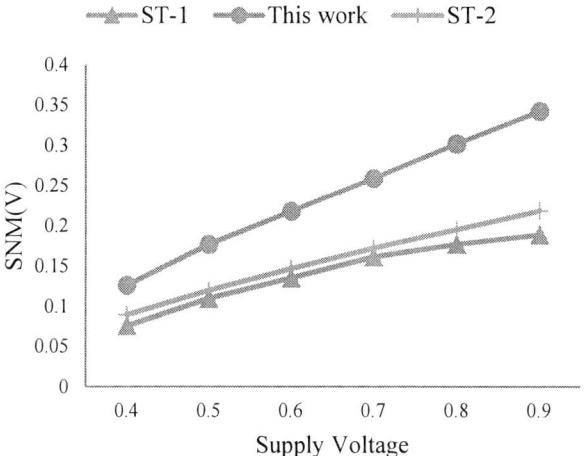

Fig. 6. Read SNM at different voltage.

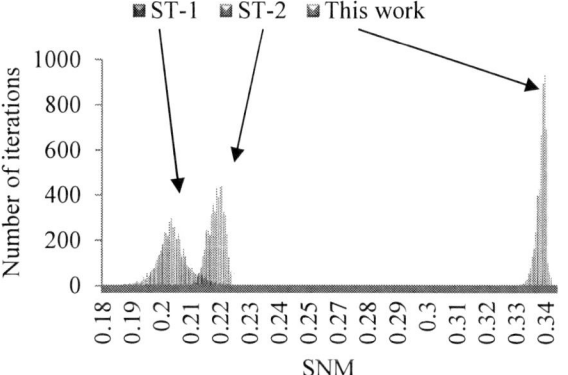

Fig. 7. SNM variation distribution at 0. 9 V and 25°C

It is clear from the Fig. 7. Proposed design shows 5.9 ×, 2.5× less sensitivity to SNM distribution at 0.9V. Detail comparisons of ST-1and ST-2 proposed cells are shown in Table-1.

TABLE I. Comparison of SRAM bit-cells (0.9V, 25°C)

	ST-1	**ST-2**	**This Work**
Reading scheme	Differential	Differential	Differential
Writing scheme	Differential	Differential	Differential
Read disturb free	No	No	Yes
Read margin	1.8×	1.5×	1×
#NMOS in read path	3	2	2
SNM variability	5.9×	2.5×	1×
#WL	1	2	2
#BL	2	2	2

V. CONCLUSION

This paper presents Static notice margin analysis of a new Schmitt trigger based SRAM cell. Proposed cell gives 1.8x and 1.6x higher RSNM as compared to ST-1 and ST-2.

The proposed design presents 5.9× and 2.5× less sensitivity to SNM distribution at 0.9V compared to ST-1 and ST-2 SRAM cell respectively. Simulation result indicates the proposed cell may be a better replacement of ST-1 and ST-2 at nanometer technology. As Future work, we plan to give leakage current analysis of proposed cell.

ACKNOWLEDGMENT

Authors are thankful to their respective organization for their support.

REFERENCES

[1] J. P. Kulkarni, K. Kim, and K. Roy, "A 160 mV Robust Schmitt Trigger Based Subthreshold SRAM," *IEEE J. Solid-State Circuits*, vol. 42, no. 10, pp. 2303–2313, 2007.

[2] C. Lo and S. Huang, "P-P-N Based 10T SRAM Cell for Low-Leakage and Resilient Subthreshold Operation," *IEEE Int. Solid-State Circuits Conf.*, vol. 46, no. 3, pp. 695–704, 2011.

[3] J. P. Kulkarni and K. Roy, "Ultralow-Voltage Process-Variation-Tolerant Schmitt-Trigger-Based SRAM Design," *IEEE Trans. Very Large Scale Integr. Syst.*, vol. 20, no. 2, pp. 319–332, 2012.

[4] S. Ahmad, M. K. Gupta, N. Alam, and M. Hasan, "Single-Ended Schmitt-Trigger-Based Robust Low-Power SRAM Cell," *IEEE Trans. Very Large Scale Integr. Syst.*, pp. 1–9, 2016.

[5] J. Boley, J. Wang, and B. H. Calhoun, "Analyzing Sub-Threshold Bitcell Topologies and the Effects of Assist Methods on SRAM V MIN," *J. Low Power Electron. Appl.*, no. 2, pp. 143–154, 2012.

[6] B. H. Calhoun and A. P. Chandrakasan, "Static Noise Margin Variation for Sub-threshold SRAM in 65-nm CMOS," *IEEE J. Solid-State Circuits*, vol. 41, no. 7, pp. 1673–1679, 2006.

[7] E. Seevinck, "Static-Noise Margin Analysis of MOS SRAM Cells," *IEEE J. Solid-State Circuits*, vol. 22, no. 5, pp. 748–754, 1987.

[8] A. Islam and M. Hasan, "A technique to mitigate impact of process , voltage and temperature variations on design metrics of SRAM Cell," *Microelectron. Reliab.*, vol. 52, no. 2, pp. 405–411, 2012.

[9] K. A. Bowman, S. G. Duvall, and J. D. Meindl, "Impact of Die-to-Die and Within-Die Parameter Fluctuations on the Maximum Clock Frequency Distribution for Gigascale Integration," *IEEE J. Solid-State Circuits*, vol. 37, no. 2, pp. 183–190, 2002.

[10] S. R. Sarangi, B. Greskamp, R. Teodorescu, J. Nakano, A. Tiwari, and J. Torrellas, "VARIUS : A Model of Process Variation and Resulting Timing Errors for Microarchitects," *IEEE Trans. Semicond. Manuf.*, pp. 1–25, 2007.

[11] S. Pal and A. Islam, "Variation Tolerant Differential 8T SRAM Cell for Ultralow Power Applications," *IEEE Trans. Comput. Des. Integr. Circuits Syst.*, 2015.

[12] K. Qian, "Variability Modeling and Statistical Parameter Extraction for CMOS Devices," University of California at Berkeley, 2015.

[13] G. K. Reddy, K. Jainwal, J. Singh, and S. P. Mohanty, "Process Variation Tolerant 9T SRAM Bitcell Design," in *Thirteenth International Symposium on Quality Electronic Design (ISQED)*, 2012, pp. 493–497.

[14] S. Pal, A. Bhattacharya, and A. Islam, "Comparative Study of CMOS- and FinFET-based 10T SRAM Cell in Subthreshold regime," in *2014 IEEE International Conference on Advanced Communications, Control and Computing Technologies*, 2014, pp. 507–511.

Gap in pagination due to unavailable paper.

Pages 238-242

2017 IEEE International Symposium on Nanoelectronic and Information Systems

A Power, Thermal and Reliability-Aware Network-on-Chip

Ashish Sharma[1], Yogendra Gupta[2], Sonal Yadav[1,3], Lava Bhargava[2], Manoj Singh Gaur[1], Vijay Laxmi[1]

[1] Dept. of CSE, Malaviya National Institute of Technology, Jaipur, India
ashishsharma.fitt@gmail.com, (gaurms, vlaxmi)@mnit.ac.in

[2]Dept. of ECE, Malaviya National Institute of Technology, Jaipur, India
yogen.571@gmail.com, lavab@mnit.ac.in

[3]Dept. of CCE, Manipal University, Jaipur, India
sonaldv4@gmail.com

Abstract—**Scaling of transistor towards the nano-scale era is continue with every process generation as per Moore's law. The smaller feature size of Complementary Metal-Oxide Semiconductor (CMOS) technology increases the volume of on-chip transistors. Unicore processor's chip design approaches to the end of the line due to the power wall. To accommodate increasing transistor count within power wall, open a new era of computing is called "many-core" architecture. The proliferation of on-chip cores has lead to gain in performance and throughput of chip Multi-Processor (CMP). However, it shifts the paradigm from computational to communication-centric design. To scale many core communication, Network-on-Chip (NoC) emerges as a solution for many core System-on-Chip (SoC).**

1. Introduction

Rapid shrinking of oxide thickness in CMOS scaling increased the gate-oxide electric field, and make transistor transistors susceptible to wearout. Transistor wearout effects such as Hot Carrier Injection (HCI), Time Dependent Dielectric Breakdown (TDDB), Negative Bias Temperature Instability (NBTI), Electromigration (EM), Stress Migration (SM) and Thermal Cycles (TC) started to affect device performance within the lifetime of a circuit. It raised the importance of addressing the sustainability of on-chip communication architectures. Wearout due to aging-induced performance degradation in the NoC can create a serious bottleneck, undermining system-level efficiency and reliability. The manycore with NoC architectures have experienced thermal and power inconsistencies that eventually affects the reliability of NoC. Therefore, it increases the susceptibility of the incorrect functionality of hardware, hence decrease the reliability of a CMP. These reliability issues, motivating the researchers to consider the long-term durability in design approaches to improves the lifetime of devices. The International Technology Roadmap for Semiconductors (ITRS) recent studies show that a 10-fold decrease of transistor wear-rate will be needed in the next ten years to maintain current design lifetimes [1]. Any failure mechanism leads to generate a permanent fault; hence the system is no longer

useful. The wearout of an individual processing unit (core) in CMP may not damage the whole system, but a single fault in the NoC may lead to complete system failure due to protocol-level deadlock and disjoint connectivity of components. The aging effect deviates the transistor parameters from design specifications and degrades the circuit performance. The change in path delay due to aging creates the timing uncertainty in the circuit. Sometimes this uncertainty not only in path delay but also lead to operation failure after a period. Therefore, these issues motivate the researchers to consider the reliability analysis and long-term durability in NoC design approaches. The power and thermal induced effect decrease the reliability of NoC. Therefore, we need an extensive analysis and design for enhancing the reliability of NoC. The primary objective of this tutorial is as follows:

- **Introduction to NoC**: In this section, we briefly discuss the NoC parameters and router micro-architecture and its components like Virtual Channel Buffers (VCB), Routing Algorithms (RA), Virtual Channel Allocation (VCA), Switch Allocator (SA) and Crossbar with existing mechanisms.

- **Low-Power Architecture and techniques for NoC**: NoC has a significant impact on the power, area, and performance of many-core architectures. The contribution of NoC in the total power budget of a CMP is approximately 30 to 40% [2], and the router buffer consumes most of it. Therefore, the designers need to work on designing of low power on-chip communication architecture by reducing the power consumption of buffers. This section address the available low-power architecture and technique for NoCs.

    1) HVT router to increase the reliability of NoCs [3]: As technology downscaling the power is a critical parameter for design. The leakage is a dominating component in the power budget. The low threshold voltage (LVT) cell has fast switching but consumes more dynamic and leakage power. To reduce the power consumption, we used high threshold voltage (HVT) cell routers,

978-1-5386-1357-3/17 $31.00 © 2017 IEEE   243

which work on lower frequency and consumes very less leakage power compare to LVT. To optimize router power, we have replaced power-hungry LVT based buffer through HVT based low-power buffer in NoC router. Further for more power optimization, we design HVT transistor based NoC router.

2) Hybrid Buffer based Router: State-of-the-art NoC router buffers is made by SRAM, which is neither power nor area efficient. The FIFO buffers consume a significant amount of NoC router power and area. STT-MRAM has near zero leakage power and very high density as compared to the SRAM memory. Other features that make STT-MRAM fascinating are high endurance, high integration density, non-volatility, higher reliability, scalability. Therefore, by using STT-MRAM in the Network-on-Chip buffer [4] design has significant merits since with the same area budget NoC router can incorporate more virtual channel buffers as compared to SRAM because of higher package density of STT-MRAM. Power gating is an effective technique to reduce static power consumption of NoC Architecture [5], but it could potentially incur a substantial performance penalty. We could design coarse-grained power gated NoC router microarchitecture with hybrid buffers using both SRAM and STT-MRAM (Spin Transfer Torque Magnetic RAM). We have used hybrid buffers to mitigate the area and performance overhead associated with power gating technique.

3) Routing-logic based port enabled clock gating for low-power NoC router [6]: We have introduced the routing logic port enabled clock gating technique at input channel buffers. In this technique, the primary clock gating circuit is added to each input channels. The clock and wired enabled signal are input to a latch. The latch output and the clock signal are input to a AND gate that will generate the gated signal for input channel buffer. The routing logic selected the routing path and enabled the corresponding wired enabled signal out of the five different directional ports. Only Logic 1 input channel buffer contains the active clock and other are disabled at the same time. The disable input channel does not provide any activities at that time that save the buffers power.

4) Low Power Multiple NoCs: Unlike Single-NoC(S-NoC), M-NoCs [7] not only partition the physical link width but also split the respective number of VCs between multiple physical links. Rather than initiating few parallel transmission with complex VC allocation logic, M-NoCs favour more the number of simple, independent and parallel data flows. The physical link width of S-NoC is divided by the number of multiple NoC networks to decide the link width of a single one out of multiple networks. In the proportion of link width, the flit width is determined for individual NoC. Smaller flit width of M-NoCs not only decreases router's crossbar size in a quadratic way [8] but also reduces the size of flit buffers. Thus, the total occupied area and static power of M-NoC's router are smaller than the equivalent S-NoC router. Real integrated circuit implementations of M-NoCs in last one and a half decade are 1) MIT's RAW architecture (four-NoCs, 2002) [9] was the first multiple NoC architecture, subsequently 2) TRIPS chip (two-NoCs, 2007) [10], 3) Tilera chips (five-NoCs, 2007) and 4) Adaptive Epiphany (three-NoCs, 2009) [11] which has recently gained popularity due to kilo-core architecture (Epiphany-V, 2016).

- **Thermal-Aware NoCs**: The dynamic power consumption produces the heat in a circuit. This heat decapitation produces thermal capacitance (J/K), temperature difference (K), thermal resistance (K/W) and heat transfer rate (W) is used for thermal modeling. Thermal profile depends on the floorplan and power of NoC components. In this section, we discuss the thermal profile creation and hotspot detection mechanism for NoCs. A node/router is called a hotspot node when its temperature is higher than other nodes. A hotspot depends on the NoC parameters, traffic type and load.

- **Reliability Estimation for NoCs**: The reliability estimation of NoC is done using modified REST tool [12] for including new SM and TC lifetime failure models. This tool uses the Monte Carlo method to estimate the MTTF of a SoC based on the chip characteristics (floorplanning, voltage, temperature). The REST takes at least two inputs to calculate the SoC reliability, floorplan (area information) and temperature traces of each component. In this section, we address the power and thermal induced effect on the reliability of NoCs.

- **HiPER-NIRGAM Framework Live Demonstration** [13], [14], [15]: In this part, we address the performance, power, thermal and reliability estimation of NoCs. HiPER-NIRGAM is a framework that supports the NoC parameters and technology parameters based power, thermal and reliability estimation for NoC and its components.

- **Pre-Silicon NBTI Aging-Aware Design of Reliable NoC Router**: The rapid shrinking of CMOS design margins in deep submicron technology has made

978-1-5386-1357-3/17 $31.00 © 2017 IEEE   244

ageing mechanism such as Negative Bias Temperature Instability (NBTI), a prime concern in NoC design. In this paper, we propose a novel pre-silicon, NBTI stress aware circuit-to-system level solution for a reliable NoC router. To achieve reliability, we develop a NoC NBTI aging-aware timing analysis framework based on the real workload stress. This section also addresses the most NBTI stressed path and cell in a set of potential critical path (PCP) of NoC router micro-architecture. The NBTI mitigation algorithm used a minimum number of multi V_{dd} Cells.

References

[1] ITRS. Itrs 2009 report on reliability. Available at. http://www.itrs2.net/itrs-reports.html.

[2] R. Marculescu, U. Y. Ogras, L. S. Peh, N. E. Jerger, and Y. Hoskote, "Outstanding research problems in noc design: System, microarchitecture, and circuit perspectives," *IEEE Transactions on Computer-Aided Design of Integrated Circuits and Systems*, vol. 28, no. 1, pp. 3–21, Jan 2009.

[3] R. Ansar, P. Upadhyay, M. Singhal, A. Sharma, and M. S. Gaur, "Characterizing impacts of multi-vt routers on power and reliability of network-on-chip," in *Contemporary Computing (IC3), 2015 Eighth International Conference on*. IEEE, 2015, pp. 476–480.

[4] H. Jang, B. S. An, N. Kulkarni, K. H. Yum, and E. J. Kim, "A hybrid buffer design with stt-mram for on-chip interconnects," in *2012 IEEE/ACM Sixth International Symposium on Networks-on-Chip*, May 2012, pp. 193–200.

[5] J. Zhan, J. Ouyang, F. Ge, J. Zhao, and Y. Xie, "Hybrid drowsy sram and stt-ram buffer designs for dark-silicon-aware noc," *IEEE Transactions on Very Large Scale Integration (VLSI) Systems*, vol. 24, no. 10, pp. 3041–3054, Oct 2016.

[6] A. Sharma, M. S. Ansar, Ruby Gaur, L. Bhargava, and V. Laxmi, "Reducing fifo buffer power using architectural alternatives at rtl," in *VLSI Design and Test (VDAT), 2016 20th International Symposium on*. IEEE, 2016, pp. 1–2.

[7] Y. J. Yoon, N. Concer, M. Petracca, and L. P. Carloni, "Virtual channels and multiple physical networks: Two alternatives to improve noc performance," *IEEE Transactions on Computer-Aided Design of Integrated Circuits and Systems*, vol. 32, no. 12, pp. 1906–1919, Dec 2013.

[8] S. Yadav, V. Laxmi, and M. S. Gaur, "A power efficient dual link mesh noc architecture to support nonuniform traffic arbitration at routing logic," in *Proceedings of the 2016 29th International Conference on VLSI Design and 2016 15th International Conference on Embedded Systems (VLSID)*, ser. VLSID '16. Washington, DC, USA: IEEE Computer Society, 2016, pp. 69–74. [Online]. Available: http://dx.doi.org/10.1109/VLSID.2016.104

[9] M. B. Taylor, J. Kim, J. Miller, D. Wentzlaff, F. Ghodrat, B. Greenwald, H. Hoffman, P. Johnson, J.-W. Lee, W. Lee, A. Ma, A. Saraf, M. Seneski, N. Shnidman, V. Strumpen, M. Frank, S. Amarasinghe, and A. Agarwal, "The raw microprocessor: a computational fabric for software circuits and general-purpose programs," *IEEE Micro*, vol. 22, no. 2, pp. 25–35, Mar 2002.

[10] P. Gratz, C. Kim, K. Sankaralingam, H. Hanson, P. Shivakumar, S. W. Keckler, and D. Burger, "On-chip interconnection networks of the trips chip," *IEEE Micro*, vol. 27, no. 5, pp. 41–50, Sept 2007.

[11] A. Varghese, B. Edwards, G. Mitra, and A. P. Rendell, "Programming the adapteva epiphany 64-core network-on-chip coprocessor," in *2014 IEEE International Parallel Distributed Processing Symposium Workshops*, May 2014, pp. 984–992.

[12] A. Y. Yamamoto and C. Ababei, "Unified reliability estimation and management of noc based chip multiprocessors," *Microprocess. Microsyst.*, vol. 38, no. 1, pp. 53–63, Feb. 2014. [Online]. Available: http://dx.doi.org/10.1016/j.micpro.2013.11.009

[13] A. Sharma, M. S. Gaur, L. Bhargava, V. Laxmi, and M. Zwolinski, "Hiper-nirgam: A tool chain based framework for modeling thermal-aware reliability estimation in 2d mesh nocs," in *DATE 2015 University Booth*, 2015.

[14] A. Sharma, P. Upadhyay, R. Ansar, V. Laxmi, L. Bhargava, M. S. Gaur, and M. Zwolinski, "A framework for thermal aware reliability estimation in 2d noc," in *VLSI Design and Test (VDAT), 2015 19th International Symposium on*. IEEE, 2015, pp. 1–6.

[15] M. Jaipur and University of Southampton. NIRGAM NoC Simulator. Available at. http://nirgam.ecs.soton.ac.uk/People.php.

Microprocessor Based Physical Unclonable Function

Sudeendra kumar K, Sauvagya Sahoo, Abhishek Mahapatra, Ayas Kanta Swain, K.K.Mahapatra
kumar.sudeendra@gmail.com, sauvagya.nitrkl@gmail.com, kmaha2@gmail.com
National Institute of Technology, Rourkela

Abstract- **Research on Physical Unclonable Functions (PUF) is well established topic in the field of hardware security. PUF is useful in many security applications like IC metering, IP protection and cryptographic key generation. The PUF circuits proposed in the past are dedicated circuits which are extra overhead in terms of area and power. Utilizing the existing circuit structures like microprocessor, power rails, etc to design PUF can be seen in recent literature. In this paper, we propose a PUF topology based on microprocessor and CRP generation method. We present the interim result in terms of hamming distance to prove sufficient randomness in path delays in the hardware multiplier of OpenMSP430 microprocessor which can be exploited to design the PUF. The simulation and statistical analysis technique is also discussed.**

Keywords: **Physical Unclonable function, Microprocessor, Hardware security.**

I. INTRODUCTION

Security has become one of the prime concerns in modern chip design. The major security problems are: - reverse engineering, side-channel attacks, hardware Trojans, IP violations and counterfeiting [1]. Counterfeiting is a major threat, which affects both revenue and reputation of the chip vendor. The chip vendors categorize semiconductor applications as: - Automotive, Consumer electronics, Communication and networking, Biomedical and Industrial control. With the ubiquitous networking happening due to IoT based products and services, the above mentioned applications will undergo significant re-engineering. In the process of this re-engineering, the growth and revenue of chip makers will mainly depend on the success of their products and services in IoT domain. The core tenets in the IoT design are: - agility, scalability, cost and security. The known challenges in an IoT ecosystem are: - Identification or authentication for addressing an IoT node, choosing a right connectivity technique, maintaining data compliance across network and security.

Security is of prime importance, because it is a part of challenge and also it is one of the core tenets in the IoT design. In the list of security issues, the problem of identification and authentication of an IoT node in the network of millions of nodes is very important. The conventional cryptography based authentication techniques may not support millions of nodes and vulnerable to different varieties of side channel attacks. Physical Unclonable Function (PUF) is a promising security primitive, which is widely investigated to solve the security problems in hardware security and IoT domain [2]. PUF circuits are used in

hardware metering schemes against IC counterfeiting, chip authentication and in designing IP protection techniques [3] [4]. PUF produces a unique response for an input challenge. The input challenge and corresponding response obtained from PUF circuit is called challenge-response pair (CRP). The large number of CRP's are collected from each chip produced. CRP's are unique to a given chip, which can be used for identification of chip and also further can be used in the identification of IoT node, in which chip is used. The unique CRP for a chip is derived from the process variation that occurs during chip fabrication. With complete knowledge of PUF circuit, it is impossible to manufacture an identical circuit with same CRP's. The uniqueness in CRP's for every chip produced comes from process variation, which is hidden and distinct. PUF circuits are helpful in solving the security issues in hardware security and cryptography [2].

The circuits which pick up maximum process variation are selected to design the dedicated PUF circuits. Ring oscillator (RO) based PUF circuits are discussed widely in the literature [2]. Arbiters, SRAM cells and flip-flops based PUF circuits are found in literature [2] [5]. A majority of PUF circuits are dedicated structures, which occupy reasonably good amount of area on silicon and affect the power budget. And also, integration of PUF circuits into chip needs extra effort in terms of placement and routing. Some PUF implementations are proposed in the past using on-chip modules which are primarily not designed to use as PUF [6]. Typical examples are microprocessor PUF in [7]. Microprocessor is used as PUF in a separate mode called PUF-mode. In the similar way, there are several PUF implementations use SRAM cells, Scan or Design for Testability (DFT) structures [6], power rails [8], clock networks [9] etc. In this paper, we focus on: -

- Survey and comparative analysis of all PUF implementations which use existing on-chip structures like processor, memory, DFT etc.
- Analysis of different microprocessor based PUF implementations.
- Novel technique to design a PUF using microprocessors.

The organization of the paper is as follows: - Section II discusses background and motivation to design intrinsic PUF using available on-chip components or sub-modules. Section III introduces the openMSP430 microprocessor. Section IV presents a novel technique to use microprocessor as PUF. Finally section V concludes the paper.

978-1-5386-1357-3/17 $31.00 © 2017 IEEE

II. BACKGROUND

PUF circuits found in research literature are implemented need dedicated structures which tax the silicon area and power consumption. In design effort, special care is required in placement and routing of dedicated PUF structures. Dedicated PUF structures may also increase number of I/O pins, which is a crucial resource. So using the existing modules, which are primarily required for functionality or testing of the chip can be used as PUF circuits. This separate class of PUF circuits has advantage in terms of silicon area and power consumption. Most of the PUF circuits are based on delay. Ring Oscillator PUF (RO-PUF) [5] is a common delay PUF produces CRP's by comparing the frequency of two ring oscillators. The widely discussed PUF structure after RO-PUF is Arbiter PUF. Arbiter PUF derives its randomness from the delay of two paths in cascaded switches, which generates unique CRPs [2] [5]. Both RO-PUF and Arbiter PUFs are dedicated structures. The PUF structures which are capable of producing large number of CRPs are known as Strong PUFs. The PUF with small number of CRPs are called Weak PUFs [2].

The quality of PUF is decided by security properties: - uniqueness, uniformity and reliability. Uniqueness is an ability of PUF to generate a unique response in a specific chip among the group of chips of same type for same challenge. Uniqueness is measured using hamming distance. Inter-chip variation of PUF response is measured using uniqueness. An ideal value of uniqueness is 50%. Reliability of PUF is ability to generate the same response, when same stimulus is applied repeatedly. The ideal value of reliability is 100%. Reliability of PUF CRP's is affected by temperature, supply voltages and aging. Uniformity of the PUF is an estimation of the proportion of 1's and 0's in the response of PUF. The ideal value of uniformity is 50%. Uniformity is calculated using percentage of the hamming weight of the response. The details on calculation of security properties can be found in [2] [5].

Motivation to Build PUF based on microprocessor: - Microprocessor is a versatile module in modern day System on Chip (SoC) devices. When microprocessor is not used as PUF circuit, it can be used several other purposes like any other processor. The delays in microprocessor can be exploited to design an efficient PUF circuit. There are several sections of microprocessor in which delays are sensitive to process variation and randomness in delays can be used to generate the quality CRPs. Few examples of such sensitive sections are: - memory to processor interfacing, ALU of the processor (mainly combinational logic), MAC unit (Multiply and Accumulate unit) and memory management. Delays in microprocessor sections are sensitive to intra-die variations, which is helpful in generating the quality CRPs. The microprocessor is a large circuit and strong PUF can be designed.

Comparative Analysis of Different PUF Implementations using on-chip structures: - Several PUF implementations using on-chip structures can found in research literature. This class of PUF implementations can be classified as: -

- Microprocessor based PUF implementations [7] [10],
- Test structures based PUF implementations [6] [11],
- Memory/SRAM based PUF's [12] [13],
- Other delay based PUF implementations [14].

Microprocessor PUF: - A.Maiti et al, in [7], propose microprocessor PUF, which accepts assembly program as challenge and produce response based on the delay in the data path or control path. The delay value is captured by over-clocking the microprocessor operation. The instructions are characterized as input challenge to PUF, fail when clock frequency is increased. Based on the pass count between the passing frequency and failing frequency, response is generated. This microprocessor PUF is verified on 32-bit LEON3 processor using the SPARC instruction set on Xilinx Spartan 3E FPGA. The uniqueness of this PUF is of acceptable quality and very good reliability. J.Kong et al propose a PUF, by adding the arbiter circuits at the output of ALU of processor [10]. The randomness in ALU and arbiter circuit is used to generate the CRPs. Authors of [10], also proposes an algorithm which leverages aging phenomenon in chips to improve the inter-chip and intra-chip variation, which leads to better quality of CRPs. The proposed PUF and algorithm is verified in 45 nm technology.

Test structures based PUF: - B. Niewenhuis proposed PUF [15] based on the power-up states of scan chains. Scan chains are inserted into chip to make the post fabrication tests easy and cost effective. The randomness required for PUF is derived from the scan chain power up states. This PUF implementation is verified in 65nm CMOS process. Y.Zheng in [11], propose a PUF by exploiting the path delay variation of scan flip-flops to generate the CRPs. The scan chains which are spread across the chip provided the large pool of scan paths to create large number of CRPs. Circuit simulation results for 1000 chips show high uniqueness of 49% at room temperature and this PUF is sensitive to environmental variations like temperature etc. PUF is also validated on FPGA and results at room temperature are presented. Y.Zheng in [6] proposes an extension to [6] called DScanPUF. DScanPUF is based on delay measurement structure consisting of PLL and multiple clock delay lines to measure the delays in scan path. Based on the delays, CRPs are generated. DScanPUF is validated on 31 FPGA chips and results show good security properties and circuit simulations are performed at 45nm CMOS process. F.Saqib in [16], presents the PUF design based on the embedded test structure called REBEL. REBEL is similar to logic analyser to perform the analysis of temporal behavior of signals from emerging paths of the logic core. REBEL based PUF is called as HELP (Hardware Embedded deLay PUF), which extracts the randomness from the path delays of the logic core. The security properties of HELP are evaluated across temperature and supply voltage variations.

Memory/SRAM based PUF: - PUF's are designed using randomness derived from the inner node voltages of storage circuits like SRAM and flip-flop. RESP (Retrofitted Embedded SRAM PUF) [13] utilizes the voltage scaling induced access failures in SRAM array to generate the challenge-response pairs. RESP extracts the randomness from write access failures under scaled supply voltage from a set of SRAM cells. The access failures occur in only few cells, due to device level process variation. After writing initial values into the SRAM cell at the scaled supply, the content of SRAM cells is read out to create the CRPs of that chip. Initial values and voltage levels will act like challenges, which make large number of challenges that can be fed into SRAM to generate unique responses for each chip produced. The MECCA (Memory Cell based Chip Authentication) PUF proposed in [12] leverages the randomness from the read/write access failures based on the word line duty cycle of the SRAM cells. The large number of challenge response pairs can be generated using word line controllability. The MECCA PUF contains a SRAM array and programmable delay generator. The input challenge decides the number of SRAM cells and selected cells are written with logic '0' or logic '1' into the cells with standard write duty cycle. The word line duty cycle is reduced using programmable delay circuit. The write operation is performed with reduced duty cycle and finally values are read out from the cells to create the response equal to number of cells. MECCA PUF is simulated for 1000 chips with 10% inter-die variations. The simulation results show the high uniqueness with 49.9% and good reproducibility.

Other delay based PUF: - Suzuki proposes delay based PUF called Glitch PUF [14]. Glitch PUF exploits glitches from delay variations between the gates and pulse propagation of each gate. The Glitch PUF behaves differently on each individual chip due to process variation. A random challenge is fed into the logic core and glitch wave forms are acquired and converted to response bits. Any reasonable size digital logic core can be used as Glitch PUF. In [14], AES (Advanced Encryption Standard) is used to generate the glitches. The security properties of glitch PUF are comparable with other standard PUF circuits.

The comparison of all the intrinsic PUF's discussed above is analysed in Table I. The observations are: -
- Except SRAM PUF, all other PUF topologies are based on delay.
- Accurate delay measurement and conversion of delay into response bits is crucial.
- The PVT performance of delay circuit or voltage regulator (RESP PUF) is crucial for the performance of the PUF.
- In most of the intrinsic PUFs, the CRP collection is complex. The exhaustive characterization of both PUF and additional circuit added to handle challenge creation is required before enrolment.
- The quality of PUF and number of CRPs depends upon the size of the circuit. The higher the size of the circuit, more CRPs can be generated in delay PUF.
- Generally, SRAM based PUF, are considered as weak PUF. The voltage and temperature variations will affect the SRAM PUF CRP reliability.
- The quality of test structure based PUF depends upon the number of scan chains in the design. Careful PUF

characterization is required before enrolment of CRPs, to get reliable CRPs. With less number of scan chains (in smaller designs) the PUF will suffer from low intra-die and inter-die variations.

Based on the above observations, size of circuit is important to get the quality PUF with large number of CRP. An 8/16-bit microprocessors makes a good candidate for designing PUF. Microprocessor based PUF implementations are not explored well. The two processor based PUF implementations methods are discussed below: -
- In microprocessor PUF described in [7], delay variability in data path and control path is used to generate the PUF response. Depending on the instruction running on the processor, particular set of data path and control path are activated. The paths will have variable delay for every chip produced. The path delay depends upon the instruction and its operands being executed. Increase in clock frequency leads to setup and hold violations. This will make the instruction fail. Each instruction with different operands will fail at different frequencies. The point of failure is called frequency failure point (FFP). Before reaching the point of failure, instruction will fail partially at different frequency points. Between the full successful execution of instruction and complete failure point, the instruction will fail partially, when over-clocked. Setup and hold a failure due to over-clocking occurs through a region and instruction does not fail at one point. The instruction is executed for 'n' number of times at different frequencies and number of successful executions of instruction is recorded in the region. The number of successful executions is called pass count (pc). Based on pc, the response is generated.
- An important PUF topology is Glitch PUF in [14]. Glitch PUF is carefully designed using extensive simulation studies on delay. The numerous SDF files (Standard Delay Format) are created with small variations in delay parameters, which reflect the randomness in the process variations for simulations. Based on the accurate simulations, occurrence of glitches is determined and challenges are designed accordingly. Glitches are captured and converted into response bits.

The concepts used in microprocessor PUF in [7] and glitch PUF [14] can be further extended to build better PUF circuits. An exhaustive path delay analysis of the complete microprocessor through statistical static timing analysis (SSTA) and designing PUF circuit based on the path delays is not yet explored. A complete timing analysis from fetching the instruction and operands from memory to final execution and its modelling is important in designing PUF. In this paper, we perform the SSTA on hardware multiplier of OpenMSP 430 microprocessor and using the path delays of the multiplier, new PUF is proposed.

978-1-5386-1357-3/17 $31.00 © 2017 IEEE

TABLE I

COMPARISON OF DIFFERENT INTRINSIC PHYSICAL UNCLONABLE FUNCTIONS

Intrinsic PUF Implementation	Validation	Need for additional circuit	Complexity in operation	Security Properties	Challenges with this technique
Microprocessor PUF [7]	FPGA	On board precise high speed pulse generator	Characterising the each instruction of microprocessor for different levels of over-clocking makes the input challenge design difficult.	Each instruction will have separate security values. Reliability is better than uniqueness. Reliability =98% and Uniqueness =38%	Choosing a right instruction to use as input is a challenge. Calibration is required for response collection.
Processor PUF [10]	Circuit Simulation (ASIC)	Arbiter at the output of ALU of the processor.	This paper does not mention the suitable method for CRP collection.	Uniqueness = 43% and Reliability = 98%. Properties are validated on circuit simulations based on Hamming Distance (HD).	Choosing the aging based input vectors for challenge to PUF is challenge. Post silicon characterization is also complex.
Scan PUF [11]	FPGA and Circuit Simulation (ASIC)	NA	Relatively simple operation.	Comparable with standard PUF implementations.	Low Inter-die and Intra die variations.
DScan PUF [6]	FPGA and Circuit simulation	On-chip precise programmable PLL/Clock generator required	CRP collection is complex. Responses are first stored in memory and read out. Challenge design is based on PLL characterization.	Uniqueness is high (49%) and Reliability is good for voltage fluctuations and temperature.	Challenge design is not clear. Location of scan path is part of challenge. Careful challenge design based on timing analysis data is followed.
Hardware Embedded Delay PUF (HELP) [16]	FPGA	On-chip precise programmable PLL/Clock generator required	CRP collection is complex. CRP quality will vary on the performance of Data Collection Engine (DCE). The DCE characterization is required prior CRP collection.	Hamming distance is presented in paper. Security properties are not explicitly discussed.	Sufficiently large logic core is required for better PUF or strong PUF implementation
RESP (Retrofitted Embedded SRAM PUF) [13]	ASIC	Programmable reference generator and voltage regulator	Input challenges depend upon the precision of voltage generator and regulator.	Uniqueness =49% for standard voltage and uniqueness reduces when operating voltage is decreased.	Low Inter-die and Intra die variations. Using only voltage as challenge will lead to this problem. Resolution of voltage generator is crucial.
MECCA (Memory Cell based Chip Authentication) PUF [12]	ASIC	Programmable delay generator	PVT analysis of delay generator is crucial	Uniqueness is 49%. Reliability at normal operating conditions is 87%.	Voltage variations lead to unreliable CRPs.
Glitch PUF [14]	FPGA	Glitch generator Delay circuit and Memory.	Careful design and analysis of glitch acquisition module is crucial.	Hamming distance is presented in paper. Security properties are not explicitly discussed	Low inter-chip and intra chip variations. Jitter corrections and shape judgements. Size of logic core should be reasonably large.

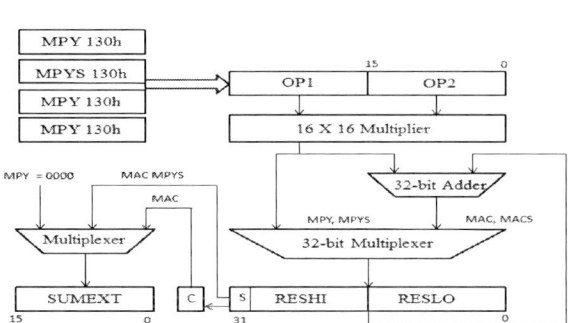

Fig. 1.OpenMSP430 Microprocessor Hardware Multiplier

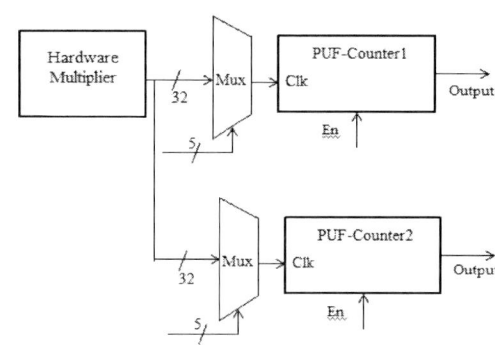

Fig. 2.Proposed Microprocessor PUF

978-1-5386-1357-3/17 $31.00 © 2017 IEEE

III. OPENMSP430 MICROPROCESSOR

The openMSP430 is instruction cloned design which is compatible with Texas Instruments MSP 430 microcontroller. It is Von Neumann architecture with single address space for both instruction and data. The complete design and gcc tools (compiler and debugger) are available in open source [17]. Design is silicon proven multiple times and it is suitable for both FPGA and ASIC implementation. The microprocessor core is fully compatible to connect memory cores and RAM/ROM of any size up to 32Kb. The components of microprocessor are: - ALU, serial debug interface, Memory backbone for memory management, clock module, SFR, Watchdog and 16X16 hardware multiplier. All features of original MSP 430 are implemented in hardware multiplier of OpenMSP430. We use hardware multiplier to build the PUF circuit.

Hardware multiplier: - Hardware multiplier is a peripheral and does not interfere with CPU activities. The multiplier registers supports unsigned and signed multiplication, unsigned and signed multiply accumulate. The block diagram of hardware multiplier is shown in figure 1. Hardware multiplier has got two 16-bit operand registers: - OP1 and OP2 and three register to store result: - RESLO, RESHI and SUMEXT. RESLO stores lower word and RESHI stores the high word of the result and SUMEXT stores information about the result. The operand OP1 has four addresses, used to select the different modes of multiplication. Writing the first operand to address will select the type of multiply operation. Writing the second operand into appropriate register initiates the multiplication. Multiplication is performed on the values stored in OP1 and OP2 and result is stored in RESLO, RESHI and SUMEXT. Repeated multiplication operations can be performed without reloading OP1, if the value in OP1 is used for successive multiplications. More details on hardware multiplier can be found in datasheet of MSP 430 [18].

IV. PROPOSED MICROPROCESSOR PUF

The block diagram of proposed microprocessor PUF is shown in figure 2. The proposed PUF uses the hardware multiplier, two 32-bit multiplexers, two 32-bit PUF-counters. The output of hardware multiplier block consists of 32-bits, which is connected to multiplexer. The multiplexer chooses the any one output of hardware multiplier block of microprocessor to drive the PUF-counter. The output of hardware multiplier is connected to the PUF-counter. The size of PUF-counter is 32-bits. The central portion of the PUF-counters output from 9th bit to 24th bit is stored in general purpose general purpose register of the microprocessor and taken out at the output ports. The 32-bit response is PUF response is generated out of two PUF counters.

Challenge Design and Response collection in the proposed PUF: - The each multiplexer has got 5 select lines to choose the output of hardware multiplier. The input to the select lines makes a part of the challenge to PUF circuit. The other portion of the challenge is assembly program which triggers the different paths of the hardware multiplier. The response generated from the counter are stored in general purpose registers of the microprocessor for further processing. The development of assembly program as a challenge to PUF is crucial for CRP generation. The randomness in the process variation reflected in interconnect delays in data path of hardware multiplier will be different for every chip fabricated. The Multiply-Accumulate instruction mode (MPY and MPYS mode) has got the inherent feedback in the functionality. Using MPY mode, data can be looped several times. This feature is used effectively in challenge design to PUF.

Implementation and Results: - The openmsp430 design from opencores is modified to operate in two modes: - normal mode and PUF mode. The 10 select lines and 'En" pin are added to the input/output pins of the microprocessor design. The hardware multiplier output section is modified to add the multiplexer and PUF-counter. The additional circuits are active only during PUF mode and in normal mode circuits are detached from the output of hardware multiplier.

Statistical STA of hardware multiplier: - Static Timing Analysis (STA) is deterministic and analysis is based on fixed delays for all timing arcs in the design. Statistical modeling of variations of all parameters like cell timing models, interconnect parasitics is required to perform Statistical Static Timing Analysis (SSTA). This means that timing models should be described in terms of mean and standard deviation for both global and local parameters. The interconnect parasitics described in terms of mean and standard deviation is used in delay calculations. Every delay is represented by a mean and standard deviation. SSTA process combines the delays of timing arcs to calculate the path delay, which is also in terms of mean and standard deviation. SSTA maps the standard deviation with respect to independent parameters (both process and interconnect related) to calculate the overall standard deviation of path delay. The slack is also obtained as statistical variable with its nominal value and standard deviation. Timing windows for noise and crosstalk are also modelled statistically and added to SSTA. Based on the path slack distribution, SSTA calculate and report the all relevant statistical parameters of slack for every path of interest [19].

A complete openMSP430 microprocessor is simulated using assembly programs in Cadence NcSim. Synopsys Design Compiler is used to synthesize the microprocessor using TSMC65nm library. Synopsys PrimeTime is used for STA. Cadence Encounter is used for place and route the design.

The Hamming distance analysis is performed on the bits generated for a given challenge from the output of hardware multiplier, to ensure the randomness in the output of hardware multiplier. The procedure is based on the statistical analysis of path delays: -

- The VCD (Value Change Dump) files for gate level netlist of the design are captured for different challenges (assembly programs) during gate level simulations.

978-1-5386-1357-3/17 $31.00 © 2017 IEEE

- Post layout parasitics (SPEF) files are extracted from place and route. Both VCD and SPEF files are used in Synopsys PrimeTime for STA.
- Design is synthesized at 50MHz and STA is performed. Further, SSTA is performed and mean and standard deviation of delays in different paths of hardware multiplier is studied. Based on this, effect of global and local variations on delays is analysed. Based on the statistical data, the timing constraints file (SDC and SDF) used in synthesis are changed accordingly to reflect the actual delays of chips after fabrication. The timing constraints are varied in SDC and SDF between -3σ to +3σ from the standard values. In this experiment, 64 different SDC files are created and different gate level netlist is created with each SDC. The STA analysis is performed on each gate level netlist to calculate path delays in hardware multiplier.
- Path delay values are collected from Primetime is processed to convert them into bits. The delay value of each path is compared with standard value (50 MHz delay values) to generate the bits. The hamming distance performed for all 64 instances (64 different SDC used for synthesis) with standard 50 MHz SDC. The average hamming distance is 42%.

REFERENCES

[1] M. Tehranipoor and C. Wang, "*Introduction to Hardware Security and Trust*", Newyork, NY, USA; Springer-2011.

[2] C.Herder et al, "Physical Unclonable Functions and Applications: A tutorial", Proceedings of the IEEE, Volume: 102, Issue: 8, Aug. 2014.

[3] Abhishek Basak, et al, "Security Assurance for System-on-Chip Designs With Untrusted IPs", IEEE Trans. Information Forensics and Security, 1515-1528, June-2017.

[4] X. Wang, Y. Zheng, A. Basak, and S. Bhunia, "IIPS: Infrastructure IP for Secure SoC Design," IEEE Transaction on Computers, 2014.

[5] C.H.Chang et al, "A Retrospective and a Look Forward: Fifteen Years of Physical Unclonable Function Advancement", IEEE Circuits and Systems Magazine, Volume 17, Issue-3, 2017.

[6] Y.Zheng, et al, "DScanPUF: A Delay-Based Physical Unclonable Function Built Into Scan Chain", : IEEE Transactions on Very Large Scale Integration (VLSI) Systems, Volume: 24, Issue. 3, March 2016.

[7] Abhranil Maiti, Patrick Schaumont, A novel microprocessor-intrinsic Physical Unclonable Function, 22nd International Conference on Field Programmable Logic and Applications (FPL), 2012.

[8] Helinski, "A Physical Unclonable Function Derived from the Power Distribution System of an Integrated Circuit", Ph. D dissertation submitted to University of New Mexico, December-2010.

[9] Y. Yao et al, "ClockPUF: Physical Unclonable Functions based on clock networks", Design, Automation & Test in Europe Conference & Exhibition (DATE), 2013.

[10] J.Kong, "Processor-Based Strong Physical Unclonable Functions With Aging-Based Response Tuning" IEEE Transactions on Emerging Topics in Computing, Volume: 2, Issue: 1, March 2014.

[11] Yu Zheng et al, "ScanPUF: Robust ultralow-overhead PUF using scan chain" 18th Asia and South Pacific Design Automation Conference (ASP-DAC), 2013.

[12] Aswin R Krishna et al, "MECCA: a robust low-overhead PUF using embedded memory array" 13th International conference on Cryptographic hardware and embedded systems (CHES-2011).

This confirms the randomness in the hardware multiplier can be used to design PUF and output of hardware multiplier can be used to drive the PUF-counter to get the more stable response.

V. CONCLUSION AND FUTURE WORK

In this paper, we present the microprocessor based intrinsic PUF. We have discussed intrinsic PUF topologies and comparative analysis is also presented. The inherent path delays in the large combinational circuits in the microprocessor can be exploited to design PUF. Statistical STA of Path delays in hardware multiplier in openmsp430 microprocessor used in this experiment prove that sufficient random variation in path delays exists. The interim results are presented in this paper and our future work is: -

- Analyse the microprocessor (openmsp430) completely through statistical STA for designing better quality PUF circuit.
- Characterization of instruction set for better challenge design
- FPGA implementation and large scale CRP collection.
- Analysis of PUF quality using security metrics like uniqueness, uniformity and reliability.

[13] Yu Zheng et al, "RESP: A robust Physical Unclonable Function retrofitted into embedded SRAM array" 50th ACM/EDAC/IEEE Design Automation Conference, 2013.

[14] D.Suzuki, "The glitch PUF: a new delay-PUF architecture exploiting glitch shapes", 12th International conference on Cryptographic Hardware and Embedded Systems (CHES'10), 2010.

[15] B. Niewenhuis et al, "SCAN-PUF: A low overhead Physically Unclonable Function from scan chain power-up states", International Test Conference, 2013.

[16] J.Aarestad et al, "HELP: A Hardware-Embedded Delay PUF", IEEE Design & Test, Volume: 30, Issue: 2, April 2013.

[17] www.opencores.org.

[18] http://www.ti.com/lit/ds/symlink/msp430g2253.pdf.

[19] J.Bhasker, Static Timing Analysis, http://www.springer.com/in/book/9780387938196 Springer-2010.

On-Chip RO-Sensor for Recycled IC Detection

Sauvagya Ranjan Sahoo, Sudeendra K, A. Mahapatra, A.K. Swain, K.K.Mahapatra

National Institute of Technology, Rourkela

email:{sauvagya.nitrkl, kumar.sudeendra, kmaha2}@gmail.com

Abstract— The presence of recycled IC in the supply chain impacts the reliability of electronics systems used in critical applications. This paper presents a ring oscillator (RO)-based sensor to detect the recycled ICs. Although, ICs are used for a short duration, the proposed sensor is able to detect it. In this paper the modified RO shows more sensitive to negative bias temperature stability (NBTI) aging mechanism. Simulations are carried out using 90 nm CMOS technology to validate efficiency of proposed sensor for detection of recycled IC. Further, the RO with more number of cascaded inverters can detect the ICs used for a few days.

Keywords— Ring Oscillator (RO); Recycling; Aging; process variation (PV).

I. INTRODUCTION

In the modern times, the demand for reliability and security in chip design is increasing. Reliability and Security are the new tenets added to the chip design along with the traditional ones like power, performance and area. Further, globalization of semiconductor supply chain [1] from design to end-user induces some security issues. Vulnerable points in supply chain may lead to infiltration of counterfeit components by an adversary. Counterfeit parts damage both reputation and revenue of the chip maker. Further, a counterfeit IC affects the reliability and performance of the system in which they are used. It will be further catastrophic when used in critical applications.

A counterfeit IC [2] may be an unauthorized copy of original manufacturer or a part which is over produced in foundry, cloned design, and defective, out of specification, tampered, recycled or used IC. Recycled IC's are generally extracted from obsolete PCB's and packed as new IC's. Recycled IC's will enter the supply chain more easily than other types of counterfeit parts. The recycled or used ICs make for 80%-90% of all the reported counterfeiting incidents [3]. Further, as reported in [4], the number of recycled IC in supply chain increases with time.

A recycled or used IC generally recovered from a obsolete system or circuit boards and then sold as a genuine component. During recycling process [5], IC is removed from the PCB generally at very high temperature and then undergo processes like cleaning, repackaging and remarking is done to make it look like as a fresh/new IC. So a used/recycled IC possesses various defects like

➤ Electrical defects like resistive open/short
➤ Aging [6] related issues like performance degradation, out of specification behavior, etc.

The countermeasure against using recycled IC is required to prevent catastrophes in critical equipment used in defense, aerospace and medical etc., in order to avoid mission failure. It is essential to develop a novel on-chip anti-counterfeiting scheme which is capable of detection of recycled IC even it is used for short period of time. Further, the added on-chip mechanism must be cost effective.

The rest of this paper is organized as follows, a brief discussion related to conventional recycled IC detection approach and impact of aging is given in section II. The architecture and recycled IC detection approach of the proposed modified RO sensor is briefed in section III.

Simulation results are presented in section IV and finally we concluded in section V.

II. PRELIMINARIES

When an IC is being used for a prolonged period then its performance degrades due to aging mechanism and this type of degradation is irreversible. So for detecting a recycled IC in the supply chain it is essential to understand the impact of aging upon IC.

Prior work to monitor the overproduction of IC coming out from the foundry includes metering techniques or assignment of unique ID to individual IC using PUF. Although hardware metering technique [7] prevents overproduction of IC but it is not a solution for recycled IC detection. A complete survey on different types of counterfeit IC detection is given in [2]. There are primarily three different counterfeit detection schemes: physical, electrical and aging based fingerprints. As aging is continuous and irreversible phenomenon, hence aging based approach is advantageous to use in detection of recycled IC. Although physical and electrical testing can be used for counterfeit IC detection but these techniques are more complex and time consuming. As aging causes permanent degradation in threshold voltage of MOSFET [6], so several threshold voltage dependent parameters like drain current, propagation delay etc. is used in several proposals [8-10, 13] to detect the recycled IC. In path-delay based fingerprint [8], the delay variation due to aging is used to detect the recycled IC. Although this approach does not consume additional power and area but it requires fingerprints from the golden IC to detect the recycled IC. A support vector machine (SVM) based recycled IC detection is proposed in [9], which requires measurements from golden IC to train SVM. Similarly dynamic current variation [1] in the symmetrical path is used to detect the recycled IC. In all the above techniques the test engineer requires measurement from golden IC comes out of the foundry. Instead of relying on golden IC, sensor based approach [10, 13] is used to detect the recycled IC.

A sensor based approach for recycled IC detection is proposed in [10]. It uses a RO-based on-chip sensor to predict the usage time of the recycled IC. The primary aging mechanisms like NBTI, HCI (Hot Carrier Injection) causes continuous degradation in threshold voltage of MOSFET and the rate of degradation increases depending upon the usage time. An on-chip RO based NBTI monitoring technique is proposed in [11, 12]. Due to the impact of NBTI, distributed ROs in the fresh IC and recycled IC possess difference in oscillation frequency and the difference increases with increase in usage time. This approach is used to design the RO-based sensor [10], which consists two RO i.e. reference RO and a stressed RO. The stressed RO is subjected continuous NBTI by driving it into non-oscillating mode and the reference RO is cutoff from supply voltage to avoid any aging due to NBTI. During authentication mode the oscillation frequency of both the RO is compared to predict the usage time of recycled IC/ chip under test (CUT). A control module generates necessary control signal to increase the aging of stressed RO and enables both the RO for frequency comparison during authentication mode. Larger frequency difference between both the RO shows higher usage time for CUT. Although, recycled IC detection is more efficient by this approach but it can detect only the ICs used for a long period of time (for a period of few months). A new RO-based sensor is proposed in [13], which exploits the aging more efficiently to detect the ICs used only for few hour/day called as NBTI aware RO-sensor. The basic difference between both the architecture (in [10] and [13]) is the way of adding more NBTI stress into the RO as shown in Fig.1.

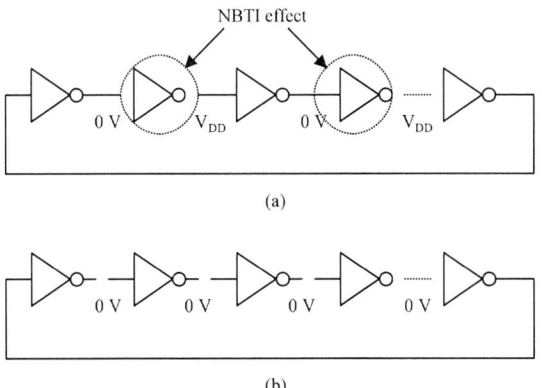

(a)

(b)

Fig. 1. (a) NBTI effect on RO [10] (b) all inverters are under NBTI stress [13]

In the case of RO-based sensor [10] (Fig.1 (a)) half of the PMOS experience NBTI whereas in [13], all the PMOS experience NBTI. As a result the degradation in oscillation frequency is higher which led to detection of ICs used for a short duration of time. A detail analysis of NBTI impact on oscillation frequency of RO is given in the section III. Further, in order to detect the recycled IC which are used only for a short duration of time ROs are designed with large number of cascaded inverters [13].

(A) Aging Impact on IC

Aging is the primary cause for permanent degradation in the performance of IC over time. Once an IC is being used continuously on the board several aging sources affects its performance. Out of several aging sources [6] like NBTI, HCI, Electromigration etc., NBTI and HCI are the primary aging mechanism which causes significant degradation in the threshold voltage (V_t) of MOSFET. The dependency of V_t on aging is discussed as follows:-

1) NBTI

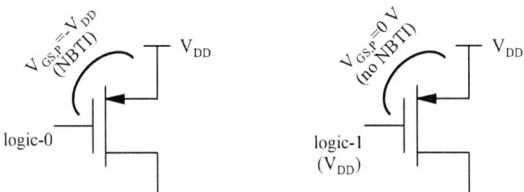

Fig. 2. NBTI effect on PMOS

NBTI causes degradation in the threshold voltage of PMOS when a continuous negative bias is applied across its gate to source terminal. The NBTI effect on PMOS is shown in Fig. 2. The negative bias ($V_{GS,P}=-V_{DD}$) causes few Si-H bonds to break as a result generates traps in the Si/SiO$_2$

interface. These traps led to shift in threshold voltage. After the removal of negative bias PMOS recovered its threshold voltage partially. The change in V_t due to NBTI [6] is:-

$$\Delta V_{t,stress} = A_{NBTI}t_{ox}\sqrt{C_{ox}(V_{DD}-V_t)}e^{\left(\frac{V_{DD}-V_t}{t_{ox}E_0}-\frac{E_\alpha}{kT}\right)}t_{stress}^{0.25}$$

(1)

Where A_{NBTI} is a constant proportional to aging rate, t_{ox} is the oxide thickness, C_{ox} is gate capacitance per unit area E_0 and E_α are device dependent parameter, k is Boltzmann constant and t_{stress} is the duration of stress/aging.

From (1) when a PMOS experience continuous negative bias for larger duration (t_{stress}), the shift in its threshold voltage is higher results in overall performance degradation of CUT.

2) HCI

HCI occurs mainly in the NMOS devices by the energetic carriers generated due to logic switching or AC stress at its gate terminal. These energetic carriers creates trap in the gate dielectric led to non-recoverable shift in V_t of NMOS. The mathematical modelling of shift in V_t [6,14], given as follows

$$\Delta V_{t,HCI} = A_{HCI}\alpha fe^{\frac{V_{DD}-V_t}{t_{ox}E}}t_{stress}^{0.5}$$

(2)

Where A_{HCI} is a constant depends upon aging rate, α is the activity factor, f is the frequency and E is a constant equals to 0.8 V/nm. The shift in V_t mainly depends upon switching activity at the gate terminal of NMOS and the duration of applied stress.

So both NBTI and HCI cause degradation in the threshold voltage of MOSFET and the rate of degradation increases with increase in stress duration.

III. PROPOSED RO SENSOR

(A) Architecture

The architecture of the proposed RO sensor is shown in Fig. 3. The basic difference between the proposed architecture and conventional N-CDIR [13] sensor is the way the RO section undergoes aging. The proposed architecture consists of two RO i.e. RO with higher stress $(RO)_{HS}$ and RO with lower stress $(RO)_{LS}$. The function of $(RO)_{HS}$ and $(RO)_{LS}$ is similar to stressed RO and reference RO respectively. The supply voltage section for both the RO is designed in such a way that during stress mode both the RO undergoes different aging and in the authentication mode both the RO oscillates at same supply voltage. The frequency measurement section consist a counter (CNTR) and TIMER block which is similar to the conventional N- CDIR sensor [13].

(B) Operating Mode

A decoder is used to generate four different control signals for each mode of operation. The four different operating modes are represented in Table 1.

TABLE 1 MODES OF OPERATION

Mode [M1M0]	SLP	$(EN)_{HS}$	$(EN)_{LS}$	RO_SEL	Function	
00	0	0	X	X	Sleep mode for both the RO	
01	1	0	0	X	Stress mode for both $(RO)_{HS}$ and $(RO)_{LS}$	
10	1	0	1	0	Authentication mode	Measure the frequency of $(RO)_{LS}$
11	1	1	0	1		Measure the frequency of $(RO)_{HS}$

1) Test Mode:

In the manufacturing or test mode (M1M0=00), both the ROs are driven into sleep mode i.e. cut-off from supply. Because in the test mode, CUT is subjected to a higher supply voltage/Temperature environment which led to additional aging. To prevent degradation due to aging, decoder enables the control signal SLP by assigning logic-0. For SLP=0, both the ROs are cut-off from supply voltage as a result experience no aging. The SLP remains at logic-1 in remaining operating modes to drive appropriate supply voltage into the RO section.

2) Stress mode:

The stress mode for both the RO is enabled for M1M0=01. The decoder generates necessary control signal (as shown in Table 1) to drive both the RO into non-oscillating mode and to inserts different amount of NBTI stress.

As SLP is at logic-1, it drives different supply voltage to the RO section i.e. $V_{DD}-V_{t,n}$ (where $V_{t,n}$ is the threshold voltage of NMOS) to the $(RO)_{LS}$ through NMOS (T_{N1}) and V_{DD} to $(RO)_{HS}$ through PMOS (T_P). These two different supply voltages led to different amount of NBTI stress [6]. As a result both the RO exhibit different rate of degradation in oscillation frequency.

3) Authentication mode:

In the authentication mode (M1M0=10 or 11) the decoder generates the necessary control signals to measure the oscillation frequency (f_{osc}) of both the RO. In this mode both the RO oscillates at a supply voltage of $V_{DD}-V_{t,n}$. For M1M0=10, the decoder drives logic-0 to RO_SEL and logic-1 to $(EN)_{LS}$ to drive the $(RO)_{LS}$ into oscillating mode. $(RO)_{LS}$ starts oscillating at a scaling voltage of $V_{DD}-V_{t,n}$ through T_{N1} as SLP is at logic-1. Further, the f_{osc} of $(RO)_{HS}$ is measured in the mode M1M0=11. The decoder drives the $(RO)_{LS}$ into non-oscillation mode and $(RO)_{HS}$ into oscillation mode. The $(EN)_{HS}$ is driven to logic-1 to drive a scaling supply voltage of $V_{DD}-V_{t,n}$ through T_{N2}. The decoder drives logic-1 to RO_SEL to measure the f_{osc} of $(RO)_{HS}$.

The above different modes of operation clarifies that both the ROs are remain in non-oscillation mode throughout the life time and driven into oscillation mode only during authentication mode. Out of primary aging mechanism NBTI effect is more pronounced in the non-oscillating mode [14] results in degradation in f_{osc} over time.

(C) NBTI effect on RO

The RO in our proposed architecture is designed by using conventional CMOS inverter. The effect of NBTI on both the RO in non-oscillating mode is shown in Fig. 4. In the stress mode, $(RO)_{LS}$ is driven by a supply voltage scaling of $V_{DD}-V_{t,n}$ and $(RO)_{HS}$ is driven by a supply voltage of V_{DD}. As shown in Fig. 4, alternate logic 0 and 1 (V_{DD}) appears at the gate terminal of PMOS in the cascaded chain of inverter. The logic-0 at the input of the PMOS devices (in the cascaded inverter chain) results in a negative bias voltage across gate to source ($V_{GS,P}$). Due to supply voltage scaling the PMOS in the $(RO)_{LS}$ experience less negative bias ($V_{GS,P} = V_{t,n} - V_{DD}$) than the PMOS in the $(RO)_{HS}$, which experience a bias of $V_{GS,P}= -V_{DD}$.

Due to different amount of negative bias, degradation in threshold voltage of PMOS in $(RO)_{HS}$ is higher than $(RO)_{LS}$. As a result both the RO experience different rate of degradation in f_{osc}. Further, with increase in stress duration the difference in oscillation frequency of both the RO increases which helps in detecting a recycled IC through f_{osc} comparison.

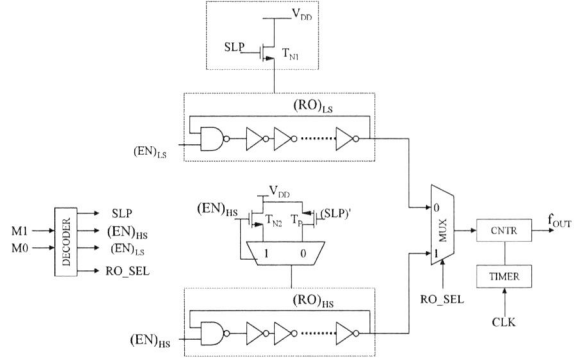

Fig. 3. Proposed Modified RO Sensor

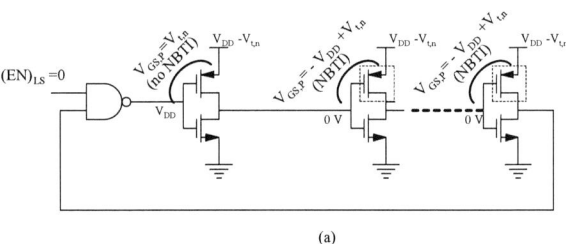

(a)

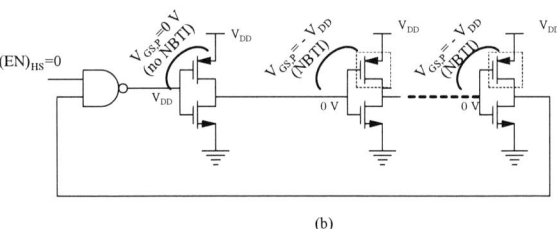

(b)

Fig. 4. NBTI effect on (a) $(RO)_{LS}$ (b) $(RO)_{HS}$

(D) Registration and Authentication

The recycled IC detection procedure is divided into two parts, registration or fingerprint generation from a new sample of IC and authentication of recycled/used IC. Initially, the fingerprint is collected for the new IC by measuring the frequency difference between both the RO i.e. $F_{DIF} = F_{LS} - F_{HS}$. Where F_{LS} and F_{HS} is the oscillating frequency of $(RO)_{LS}$ and $(RO)_{HS}$ respectively. A set of frequency across samples of fresh IC is collected to measure the spread in F_{DIF}. Although both the RO oscillates at same supply voltage i.e. $V_{DD}-V_{t,n}$, the magnitude of F_{DIF} is approximately zero, but the manufacturing PV [14] causes slight difference is oscillation frequency. Further, this difference may be positive or negative. This variation in F_{DIF} across all samples is used to create the fingerprint for CUT. During the authentication phase if the F_{DIF} for the CUT is out OF the range of fresh IC sample then CUT is treated as a recycled IC otherwise it is assumed to be a fresh IC.

The spread in F_{DIF} for a group of fresh IC ($F_{dif,0}$) and aged IC ($F_{dif,T}$:after T-days) is shown in Fig. 5. If both the spreads are far apart then a recycled IC can be easily detected. The overlapping area between the spread in Fig. 5 indicates the unpredictable region i.e. in this region a recycled IC may be assumed as a fresh IC. Further, with rise in spread the number of samples of undetected recycled IC increases.

978-1-5386-1357-3/17 $31.00 © 2017 IEEE

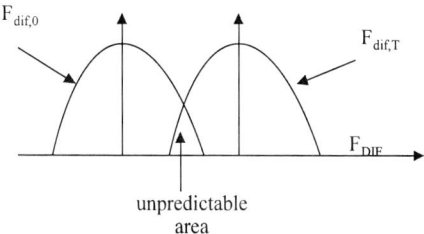

Fig. 5. Distribution of F_{DIF} for fresh IC and used IC

IV. SIMULATION RESULTS & DISCUSSION

The proposed modified RO sensor is implemented in Virtuoso environment, 90 nm CMOS technology library from UMC is used for simulation purpose. RelXpert simulator in virtuoso environment is used to measure the degradation in oscillation frequency of RO due to aging. Monte Carlo simulation with 300 iterations is carried out using statistical transistor model provided by foundry. The reason for using Monte Carlo run to measure the impact of process variation on the oscillation frequency of RO prior to chip fabrication. The f_{osc} from the fresh IC and aged IC is collected over multiple instances to predict whether the test IC is a recycled one or not. The simulation is carried out at a nominal supply voltage of 1V and 27^0C.

The simulation result for the proposed RO based sensor is shown in Fig. 6(a-c). The result shows variation in frequency difference for both fresh IC and recycled IC (which undergoes continuous aging). The F_{DIF} is measured at T=0D i.e. fresh sample of IC and the variation across 300 instances is shown in Fig. 6. The legend T=0D shows the variation in F_{DIF}. Ideally F_{DIF} should be zero but manufacturing PV causes each RO to oscillate at unique frequency which led to either positive or negative value of F_{DIF} (T= 0D in Fig. 6 (a-c)). As shown in Fig. 6 (a) the fresh IC has a spread from -10 to 10 MHz. Similarly F_{DIF} is measured after a stress interval of 6 and 15-days (D). For our analysis we have considered two aging instances (6D and 15D). In order to measure the F_{DIF} at different aging intervals, first both the ROs are undergoes continuous aging for a period of (6D and 15D) to extract the aged SPICE netlist. Finally, Monte Carlo simulation is carried out to measure the oscillation frequency of both the RO ($(RO)_{HS}$ and $(RO)_{LS}$) across multiple instances in order to calculate the spread of F_{DIF} after T=6D and 15D.

Fig. 6 (a) shows the spread in F_{DIF} after T=0, 6 and 15 days for the RO designed with 15 stages of cascaded inverter. As shown in the figure with increase in aging duration the magnitude of F_{DIF} (frequency difference between the RO) increases. The frequency difference bar chart shows overlap region between the two distribution of F_{DIF} i.e. $(F_{DIF})_{T=0}$ and $(F_{DIF})_{T=6D}$. This overlap region indicates a dilemma i.e. it is difficult to predict whether the CUT is a fresh or recycled one. Further, at T=15D there is no overlapping between the spread of $(F_{DIF})_{T=0}$ and $(F_{DIF})_{T=15D}$. The above simulation result confirms that at higher aging interval, a recycled IC can be easily detected without any error.

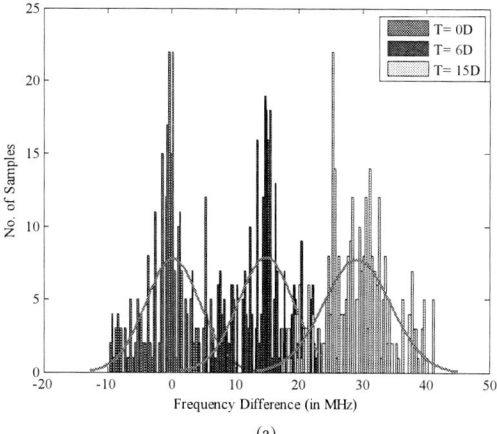

(a)

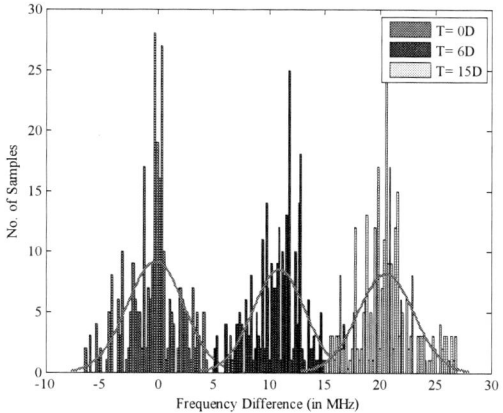

(b)

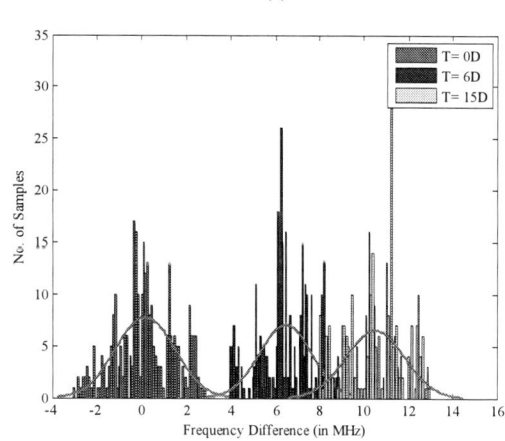

(c)

Fig. 6. Distribution of F_{DIF} (a) RO with 15 stage (b) RO with 25 stage (c) RO with 51 stage

Further, both the ROs are designed with more number of cascaded inverters i.e. 25 and 51 stages to measure the spread in F_{DIF}. The spread at different aging instances (6D and 15D) is shown in Fig. 6(b and c). The simulation result shows lower spread in F_{DIF} with increase in number of inverter stages. This reduction in spread is due to inverse relationship of spread in F_{DIF} with number of inverter in RO [13].

978-1-5386-1357-3/17 $31.00 © 2017 IEEE   255

The spread in F_{DIF} is also lowered in all the considered aging instances i.e. $(F_{DIF})_{T=6D}$ and $(F_{DIF})_{T=15D}$. The simulation result shows that the spread for 51 stages of RO (Fig. 6(c)) is much narrower than both 15th and 25th stages. This narrower spread eliminates the overlapping region even at lower aging instance. As shown in Fig. 6(b), the overlapped area in the case of 25-stages RO between $(F_{DIF})_{T=0}$ and $(F_{DIF})_{T=6D}$ is less than that of 15-stages of RO. So the probability of incorrect prediction reduces significantly, whereas by using 51 stages RO there is no misprediction or unpredicted area for the aging instance of 6D (Fig. 6(c)).

Table 2 represents the number of samples in the overlap region between the proposed RO sensor and conventional RO sensor in [10]. The proposed RO sensor shows lower number of samples in the overlapped region than the conventional RO for equal aging duration. Further, with increase in number of stages less number of samples is found in the overlap region due to reduction in spread of F_{DIF}.

TABLE 2 MISPREDICTION COMPARISON

No. of stages in RO	Aging duration $<T>$	No. of samples in overlap region $<$in %$>$	
		RO sensor [10]	Proposed RO sensor
15	6D	11.27	7.25
25		6.26	1.38
51		1.25	0
15	15D	2.92	1.81
25		0.88	0
51		0	0

V. CONCLUSION

The proposed modified RO-sensor detects the recycled ICs used for a few days. The different amount of NBTI stress on RO results in generating higher frequency difference in a recycled IC than the fresh IC. In the proposed circuit, RO with more number of cascaded inverters increases the efficiency and the proposed sensor is able to detect an IC used only for a few days.

REFERENCES

[1] Yu Zheng, A. Basak and S. Bhunia, "CACI: Dynamic current analysis towards robust recycled chip identification," *51st ACM/EDAC/IEEE Design Automation Conference (DAC)*, pp. 1-6, 2014.

[2] U. Guin, K. Huang, D. DiMase, J. M. Carulli, M. Tehranipoor and Y. Makris, "Counterfeit Integrated Circuits: A Rising Threat in the Global Semiconductor Supply Chain," in *Proceedings of the IEEE*, vol. 102, no. 8, pp. 1207-1228, 2014.

[3] Businessweek. (2008). *Dangerous Fakes*, New York, NY, USA [Online]. Available: http://www.businessweek.com/magazine/content/ 08_41/b4103034193886.htm.

[4] (2010). Bureau of Industry and Security, U.S. Department of Commence. *Defense Industrial Base Assessment: Counterfeit Electronics*[Online].
Available: http://www.bis.doc.gov/defenseindustrialbaseprograms/ osies/defmarketresearchrpts/final_counterfeit_electronics_report.pdf

[5] B. Hughitt, "Counterfeit electronic parts," in *Proc. NEPP Electron. Technol. Workshop*, 2010.

[6] A. Tiwari and J. Torrellas, "Facelift: Hiding and slowing down aging in multicores," in *Microarchitecture, 41st IEEE/ACM International Symposium*, pp. 129-140, 2008.

[7] F. Koushanfar. *Hardware Metering: A Survey* [Online]. Available: http://aceslab.org/sites/default/files/05-fk-metering.pdf, 2011.

[8] X. Zhang, K. Xiao and M. Tehranipoor, "Path-delay fingerprinting for identification of recovered ICs," *IEEE International Symposium on Defect and Fault Tolerance in VLSI and Nanotechnology Systems (DFT)*, pp. 13-18, 2012.

[9] K. Huang, J. M. Carulli and Y. Makris, "Parametric counterfeit IC detection via Support Vector Machines," *IEEE International Symposium on Defect and Fault Tolerance in VLSI and Nanotechnology Systems (DFT)*, pp. 7-12, 2012.

[10] X. Zhang and M. Tehranipoor, "Design of On-Chip Lightweight Sensors for Effective Detection of Recycled ICs," in *IEEE Transactions on Very Large Scale Integration (VLSI) Systems*, vol. 22, no. 5, pp. 1016-1029, 2014.

[11] J. Keane, X. Wang, D. Persaud and C. H. Kim, "An All-In-One Silicon Odometer for Separately Monitoring HCI, BTI, and TDDB," in IEEE Journal of Solid-State Circuits, vol. 45, no. 4, pp. 817-829, 2010.

[12] J. Keane, W. Zhang and C. H. Kim, "An Array-Based Odometer System for Statistically Significant Circuit Aging Characterization," in *IEEE Journal of Solid-State Circuits*, vol. 46, no. 10, pp. 2374-2385, 2011.

[13] U. Guin, D. Forte and M. Tehranipoor, "Design of Accurate Low-Cost On-Chip Structures for Protecting Integrated Circuits Against Recycling," in *IEEE Transactions on Very Large Scale Integration (VLSI) Systems*, vol. 24, no. 4, pp. 1233-1246, 2016.

[14] S. R. Sahoo, S. Kumar, K. Mahapatra and A. Swain, "A Novel Aging Tolerant RO-PUF for Low Power Application," *IEEE International Symposium on Nanoelectronic and Information Systems (iNIS)*, pp. 187-192, 2016.

2017 IEEE International Symposium on Nanoelectronic and Information Systems

MSM: Performance Enhancing Area and Congestion Aware Network-on-Chip Architecture

Tuhin Subhra Das, Prasun Ghosal

Indian Institute of Engineering Science and Technology,
Shibpur, Howrah 711103, WB, India
Email: {tuhin, p_ghosal}@it.iiests.ac.in

Abstract—Network-on-chip (NoC) has already proven its efficiency in removing communication bottlenecks during designing future portable high-end computing devices with many core architectures. Improving performance in many core architecture without degrading area and power requirements is a real challenge to a chip designer. Proposed work primarily focuses on designing an efficient NoC architecture with an interesting router-PE and router-router connection that finally leads to lower packet latency. Moreover, proposed architecture increases the saturation point allowing packets more flexibility in quasi-minimal routing in congested situation without major degradation in its performance. Simulation studies over 64 processing elements show **40%** and **7%** improvements compared to state-of-the-arts (namely CMesh-X2 and X-mesh of same sizes) in random traffic by the proposed topology. Moreover, it achieves **120%** and **18%** lower latency in transpose traffic under congestion aware adaptive routing claiming a very negligible (∼ 1.8 %) area overhead.

Index Terms—Network-on-Chip, NoC Topology, Performance, Area and Congestion Aware NoC architecture.

I. INTRODUCTION

Continuous down scaling of CMOS technology and its diminishing performance gain at deep submicron regime has forced semiconductor industry towards integration of hundreds and more numbers of components on a single die that may includes arrays of processor and co-processor, memory block, CPU, DSP, or others functional logic blocks. Renounced chip manufacturer like Intel's Teraflop has designed 80-core processor prototype forming a 10×8 [1] PEs (processing elements) array in 65 nm CMOS technology. Tilera's designer team has devised another 2D mesh SoC (System-on-Chip) system with 64 cores. According to ITRS 2011 executive report [2] number of these integrated processing elements (PEs) within a SoC will be in order of thousand by 2020. However, performance and scalability of this many core based SoC system is restricted due to the limitation shared buses bandwidth. Here, NoC (Network-on-Chip) meets this higher bandwidth demands after imitating the packet switch based computer networking system in chip communication scenario and thereby making system scalable in large scale dimension. In NOC, PEs are normally connected to one or more dedicated routers, assigned for the communication job. Current NoC based systems mostly devised on a 2D-Mesh [3] topology since it matches very well to the planar grid structure making it feasible to implement on chip. However, a common pitfall of 2D mesh observed in large scale dimension is its longer packet delay due to large hop counts. CMesh [4] an alternate of

conventional 2D-Mesh reduces these hop counts and number of required switches quarter to the Mesh leveraging more space for the computing fabrics. But CMesh saturates at lower traffic workload compared to generic Mesh due to its narrow channel bandwidth [5]. Performance of this CMesh further improved in CMesh-X2 [5], introducing a second parallel network to the existing CMesh architecture. HLMGS another variant of CMesh is proposed in [6] offering hierarchical connection to make it more flexible in large scale dimension but its performance is also limited by low saturation workload when low area comes into consideration that connects four PE to a router (HLMGS-4). Wang and Bandi in [7] proposed another low area based X-Network with a new PE-router connections while reducing packet delay 18-50% compared conventional architecture. In proposed work, we further improve these performances in term of latency and saturation points with a negligible area overhead compared to conventional Mesh architecture.

This paper illustrate an idea over congestion and area aware novel NoC architecture. Section I states about available state-of-the-art area aware NoC architectures. Formation of proposed topological structure and details of routing are described in section II. Section III evaluates performance in terms of latency, cost of additional area and calibrates experimental results with existing related state-of-the-art works. Section IV concludes this paper mentioning areas to investigate further and scope of improvement.

II. PROPOSED WORK

A. MSM (Mesh-Star-Mesh) topology

The idea of proposed topology is motivated from the area efficient, low latency based CMesh and its two improve variant viz. HLMGS and X-network. Figures 1(c) and 2(b) are example of proposed MSM architecture connected in 4×4 and a 6×6 PEs array using 13 and 25 number of switches respectively. Router organization in proposed architecture is formed after considering two distinguishable Mesh connection (i.e. level-1 Mesh, level-2 Mesh) interconnected through diagonal links in star connection. Here, Mesh connection at level-1 (connects all even positioned routers) requires total N1 number of routers, equals to $(\sqrt{N}/2 + 1)^2$ (for all PEs $N \in$ is a square and even number). Mesh at level-2 (connects all odd position routers) requires N2 numbers of router, i.e. equal to $(\sqrt{N}/2)^2$. Here, star links act as bridging between

978-1-5386-1357-3/17 $31.00 © 2017 IEEE   257

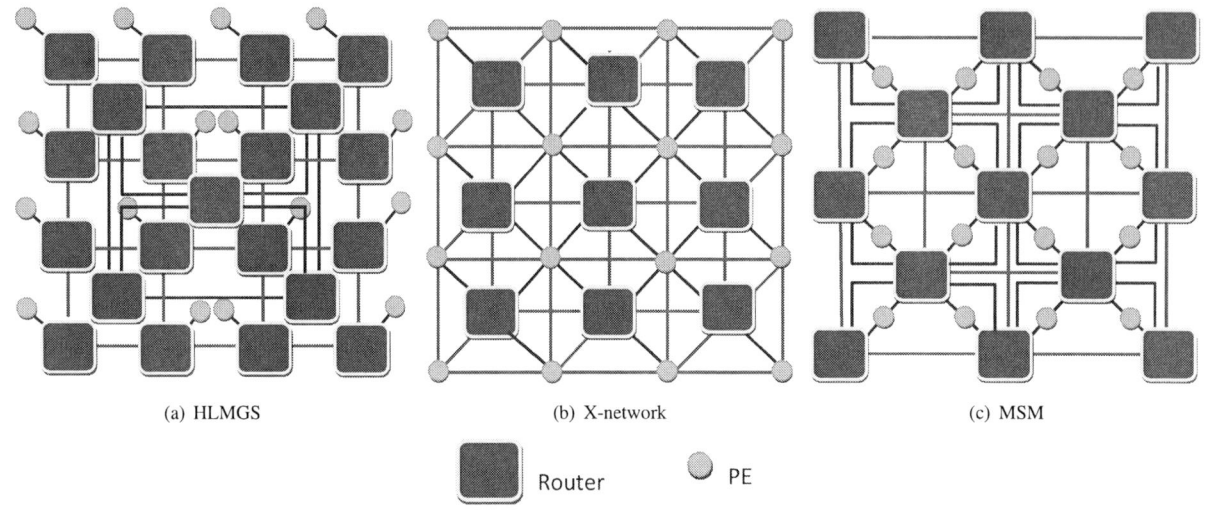

(a) HLMGS   (b) X-network   (c) MSM

    ■ Router   ● PE

Fig. 1: Area and congestion aware NoC topologies of 4 x 4 PE array.

two mesh connection and thereby allowing packets to switch from one Mesh to another level Mesh connection in congested situation. Each router in proposed (except at the corner and boundary) network connected to four PES locally, while each of these PE unit is exclusively connected to two routers. Each PE having exact two ports (Port1 and Port2) connect to two neighboring routers. One port is connected to level-1 mesh router while another is connected to a level-2 mesh router. As each PE is connected to more than one router, we require a controller unit. This controller unit is capable of selecting an output port after inspecting its sink PE id. All PEs connect to in proposed network are classified into two types. PEs located either in even row and even column position or in odd row and odd column position belongs to one class s-type (similar-type). While PEs located either in even row and odd column position or in odd row and even column position belongs to o-type (opposite-type). Port1 and Port2 of s-type connects to south-east (SE) and north-west (NW) ward router whereas Port1 and Port2 of o-type PE connects to north-east (NE) and south-west (SW) ward routers respectively.

1) Assignment of Router's id: Routers in a 6×6 PE array (figure 2(a)) having nodes id {R0, R1, R2, R3, R4, R5, R6, R7, R8, R9, R10, R11, R12, R13, R14, R15} are assigned to level-1 Mesh while routers having node id {R16, R17, R18, R19, R20, R21, R22} are assigned to level-2 Mesh connection. Routers node position at level-1 Mesh connection is represented by its co-ordinates $(X_{ni}Y_{ni})$ derived from its node id Rn_i as follows: $X_{ni} = (n_i/\sqrt{N1}) * 2$; $Y_{ni} = (n_i\%\sqrt{N1}) * 2$. Router's co-ordinate id at level-2 Mesh is derived as: $X_{ni} = ((n_i - N1)/\sqrt{N2}) * 2 + 1$; $Y_{ni} = ((n_i - N1)\%\sqrt{N2})) * 2 + 1$; where N1 and N2 represent total number of routers in level-1 and level-2 Mesh respectively.

B. Routing procedure

In generic Mesh connection a PE unit simply forward packet to its locally connected router. As proposed architecture connects more than one routers to a PE unit, a controller unit is required that monitor status of adjacent routers and to forwards packet based on the availability output port and its sink PE id.

1) Selection of source and sink router: (i) Selecting the nearest sorce router and nearest destination router(NSND [5] selection): In this selection scheme, router connected to source PE and nearest to the destination is selected as source router while the router connected with sink PE and nearest to the source router is chosen as candidate destination router.

(ii) Selecting the minimum load source router and nearest destination router (MLND [5] selection): In this selection process controller units are given flexibility to select a source router whichever is less congested 1 subject to that port is not congested. The router nearest to the source and connected to sink PE is selected as destination router. A details on selection of this source and sink router node are given in algorithm 1 and 2 .

2) Congestion aware adaptive XY-Star routing: After packets are injected in to the network via controller unit, it route through available XY (row and column wise) and diagonal links. XY-routing in Mesh based connection, routes packet first along X-direction until reaches to the same column where the destination router lies and then move towards Y-direction. Proposed architecture shorten this routing path length by allowing packet to move through shortcut diagonal express links in addition to these X and Y links. It has been reported in [8] that diagonal links improves system performance dramatically offering with more alternate paths in congested situation. To explore more optional path, proposed routing also allows selection of an output port in quasi minimal region (a rectangle region formed after connecting two nodes at the opposite end of the diagonal of rectangle [8]) when required port is not available due to congestion. We adopt number free flits buffer that are available in output port of neighboring router's as an local congestion index [9] to select an alternate route in proposed routing approach. Figures 7 shows much higher latency saturation point when congestion

978-1-5386-1357-3/17 $31.00 © 2017 IEEE

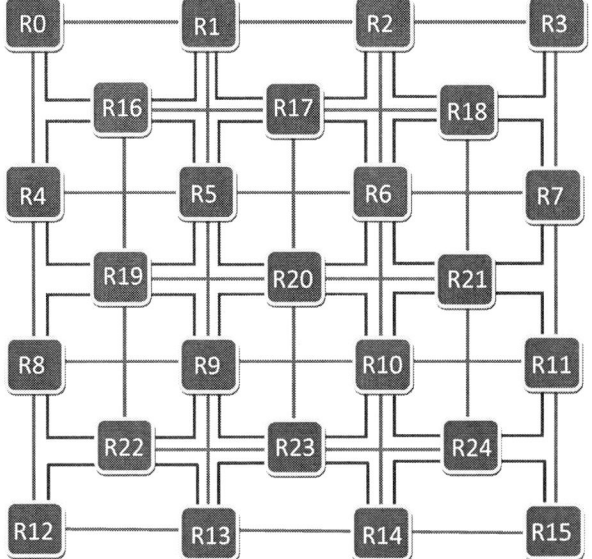

(a) Internal connections among routers in 6 x 6 PE array.

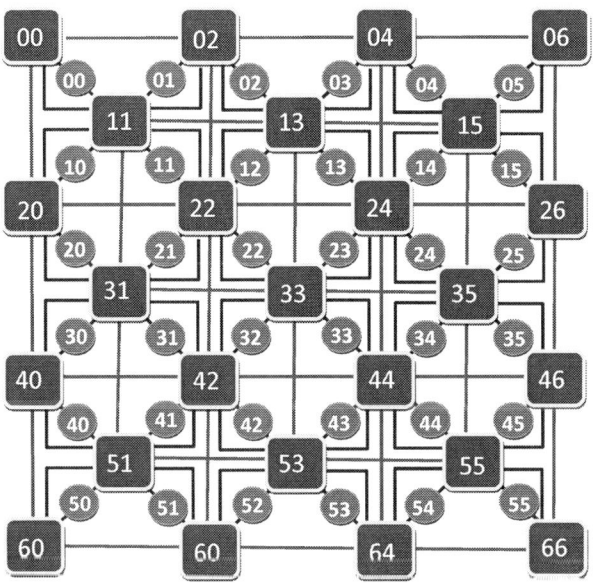

(b) Connections among routers and PEs in 6 x 6 PE array.

Fig. 2: Router-PE connectivity in proposed MSM network of 6 x 6 PE array.

aware adaptive routing is applied compared to its fixed OP (output port) selection policy (see figure 6).

3) Selection of path from source to sink router: Controller unit attached to each PE forwards packet to one of its two connected routers based on the selection procedure as defined in 1 (Here, LSE, LNW, LSW and LNE signify the local ports connect to the routers in south-est, north-west, south-west and north-east direction respectively). Each packet stored in the source router belongs one specific type (like SE, NW, SW and NE) based on the offset value between source-sink PE id as

described details in algorithm 3. Based on this packet type and its sink PE id a destination router is chosen at the source router according to the method as defined in algorithm 2. Packets follow both deterministic XY-Star routing in low congested network and congestion aware quasi-minimal adaptive routing in congested network and finally reaches to the destination router following a near optimal path. After packet received at the destination router its sink PE id is fetched to select an appropriate output port according to the method as defined in algorithm 4 and release packet from the network.

Input : Source PE p_i, sink PE p_j
Output : $\mu(i,j)$ returns output port of p_i for the sink p_j
Definition: $X_{offset} \leftarrow X_{pj} - X_{pi}$; $Y_{offset} \leftarrow Y_{pj} - Y_{pi}$;
Co-ordinate of current router (X_{ni}, Y_{ni});
Co-ordinate of destination PE (X_{pj}, Y_{pj});
Co-ordinate of destination router (X_{nj}, Y_{nj});

---

if p_i= s type **then**
  if ($X_{offset} > 0$ or $Y_{offset} > 0$) **then** $\mu(i,j) \rightarrow$ LSE
  else $\mu(i,j) \rightarrow$ LNW
  end if
else
  if ($X_{offset} > 0$ or $Y_{offset} < 0$) **then** $\mu(i,j) \rightarrow$ LSW
  else $\mu(i,j) \rightarrow$ LNE
  end if
end if

Algorithm 1: Procedure of selecting output port by the controller unit.

Input : Source router n_i, sink PE p_j
Output : $\rho(i,j)$ returns destination router id n_j for the sink p_j at n_i

---

if (sink p_j= s-type) **then**
  if (packets = SE type) **then**
    $X_{nj} \leftarrow X_{pj}$ and $Y_{nj} \leftarrow Y_{pj}$
  else// (packets = NW type)
    $X_{nj} \leftarrow X_{pj} + 1$ and $Y_{nj} \leftarrow Y_{pj} + 1$
  end if
else// (i.e. sink p_j = o-type)
  if (packets = SW type) **then**
    $X_{nj} \leftarrow X_{pj}$ and $Y_{nj} \leftarrow Y_{pj} + 1$
  else // (packets = NE type)
    $X_{nj} \leftarrow X_{pj} + 1$ and $Y_{nj} \leftarrow Y_{pj}$
  end if
end if

Algorithm 2: Procedure of selecting destination router n_j for at source router n_i for the sink p_j.

Input : Source router n_i, destination PE p_j
Output : $\eta(i, j)$ returns packet type
Definition: $X'_{offset} \leftarrow X_{pj} - X_{ni}$; Y'_{offset}
$\leftarrow Y_{pj} - Y_{ni}$;

---

if ($X'_{offset} \geq 0$ and $Y'_{offset} \geq 0$) **then** $\eta(i, j) \rightarrow$SE-type
else if ($X'_{offset} < 0$ and $Y'_{offset} < 0$) **then**
$\eta(i, j) \rightarrow$NW-type
else if ($X'_{offset} \geq 0$ and $Y'_{offset} < 0$) **then**
$\eta(i, j) \rightarrow$SW-type
else $\eta(i, j) \rightarrow$ NE-type
end if

Algorithm 3: Procedure of classifying packet types at source router.

Input : Destination router n_j, sink PE p_j
Output : $\lambda(j, j)$ returns output port of router n_j for
the sink p_j
Definition: $X''_{offset} \leftarrow X_{pj} - X_{nj}$; Y''_{offset}
$\leftarrow Y_{pj} - Y_{nj}$;

---

if ($X''_{offset} = 0$ and $Y''_{offset} = 0$) **then** $\lambda(j, j) \rightarrow$LSE
else if ($X''_{offset} < 0$ and $Y''_{offset} < 0$) **then** $\lambda(j, j) \rightarrow$LNW
else if ($X''_{offset} == 0$ and $Y''_{offset} < 0$) **then** $\lambda(j, j) \rightarrow$LSW
else $\lambda(j, j) \rightarrow$ LNE
end if

Algorithm 4: Procedure of selecting output port at destination router n_j .

C. Data packet format

We consider 32 bits wide flit data format where header reserves 8 bits for mentioning sink PE id and 8 bits for the target destination router id. Additional 4 bits sequence number is used to fix the out-of order packet delivery issue and 8 bits for mentioning sub data id. 2 bits for mentioning message data types and 2 bits for its VC id. While body and tail flit reserve 2 bits for mentioning flit data type and 2 bits for VC id and rest are used as payload data.

III. PERFORMANCE EVALUATION METHODOLOGY

A. Estimation of area overhead

The cost of additional chip area is calculated using DSENT [10] tool considering in details configuration of router specification (i.e. arbiter, port number) and the global wire (wire width and spacing) as given for 45 nm technology node. Estimated area overhead is calculated over the chip area that connects 64 PEs each of size 1.56 mm^2 (dividing 100 mm^2 chip area over the 64 Tiles). Router area and global link area are considered after connecting input and output port according to given topologies where each input port having 2VC of buffer depth size of 4 flits (flit size 32 bits). A detail summary on these area comparison results are jotted down in

Table I.

B. Evaluating framework

To evaluate performance, we conducted an experiments using popular NoC (Noxim [11]) simulator that consists of 64 PEs equivalent to an 8×8 PE array connection. We customized and model each topology according to router's ports connectivity, updating its routing logic and all others functional changes required by the concerned topologies. Two very commonly used synthetic traffics viz. random and transpose are employed in the simulation process with an increasing packet injection rate that follow an exponential distribution. Simulation initial warm up time set to 1000 cycles and then executed for period of 10,000 cycles. Two parallel FIFO buffers with capacity of 4 flits have been used considering flit width of 32 bits. In simulation process, we employed the NSND (nearest source router and nearest destination router) selection process in deterministic XY and XY-Star routing while MLND (minimum loaded source router and nearest destination router) is employed in congestion aware based quasi minimal routing approach to exploit its full potentiality.

TABLE I: Area overhead calculation (using DSENT).

Network topology	Router area (mm^2)	Global wire area (mm^2)	Total area (mm^2)	Area overhead (%)
Mesh	1.21	2.62	103.83	
CMesh-X2	1.02	2.85	103.87	0.06
X-Network	1.57	2	103.57	- 0.2
HLMGS-4	0.79	3.6	104.39	0.5
MSM	2.41	3.35	105.76	1.8

Experimental results shows that proposed architecture outperform its nearest counterpart X-mesh and C-mesh with 7% and 40 % lower packet delay when random traffics are applied. Whereas in case of transpose traffic proposed architecture offers 18% and 83% lower latency compared to similar architectures respectively. Note, in this delay calculation process, we consider 50 cycles as an cut-off threshold value for choosing the overshoot point (threshold cut-off point in considering packet delay).

IV. CONCLUSION

The CMesh-X2 and HLMGS topology were proposed to overcome the common pitfall of mostly accepted Mesh topology i.e. large packet delay in higher network dimension. CMesh-X2 and HLMGS-4 both lower the packet delay under light weight traffic but saturates early due to the limitation of CMesh architecture i.e. smaller bisection bandwidth. X-network comes with comparatively low packet delay and higher saturation load. Proposed MSM network is an attempt to further increase its performance keeping low area constraint. Comprehensive simulation studies shows that proposed architecture lower the packet delay compared with negligible area overhead and also higher the saturation point. In future, we are aiming to assess its power efficiency and scalability in large scale dimensional network.

978-1-5386-1357-3/17 $31.00 © 2017 IEEE

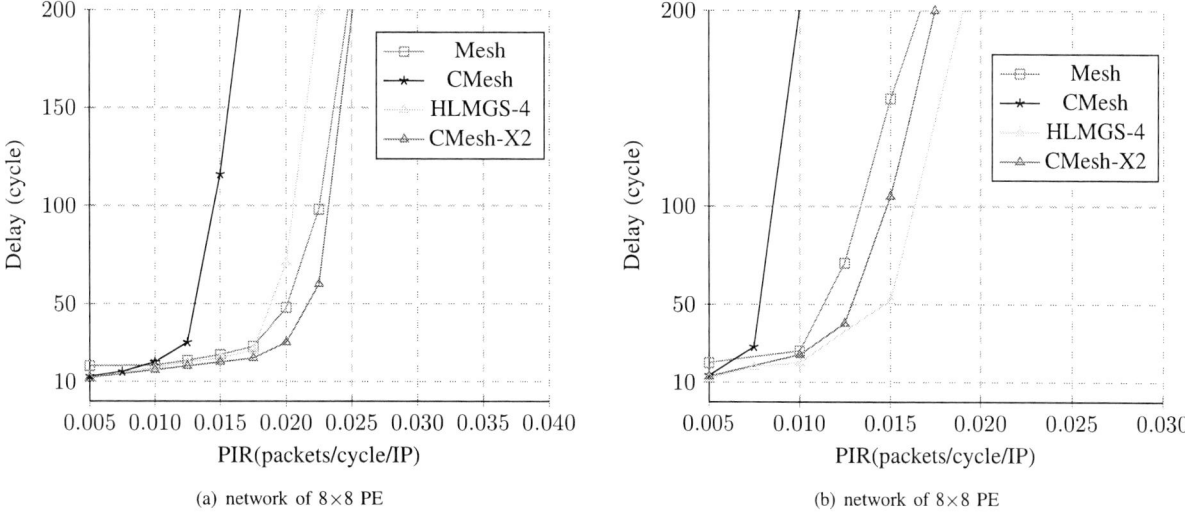

(a) network of 8×8 PE   (b) network of 8×8 PE

Fig. 3: Delay as a function of packet injection rate under (a) random (b) transpose traffic in NSND selection process.

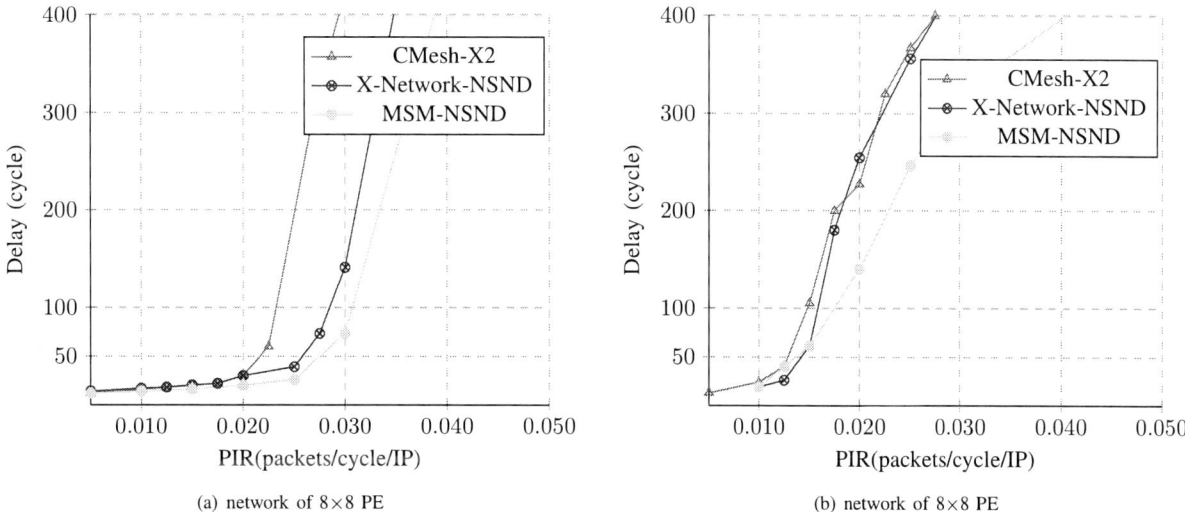

(a) network of 8×8 PE   (b) network of 8×8 PE

Fig. 4: Delay as a function of packet injection rate under (a) random (b) transpose traffic in XY/XY-Star routing in NSND selection.

REFERENCES

[1] S. Vangal, J. Howard, G. Ruhl, S. Dighe, H. Wilson, J. Tschanz, D. Finan, P. Iyer, A. Singh, T. Jacob, S. Jain, S. Venkataraman, Y. Hoskote, and N. Borkar. An 80-tile 1.28tflops network-on-chip in 65nm cmos. In *Solid-State Circuits Conference, 2007. ISSCC 2007. Digest of Technical Papers. IEEE International*, pages 98–589, Feb 2007.

[2] International Technology Roadmap for Semiconductors. Online.

[3] Shashi Kumar, Axel Jantsch, Juha-Pekka Soininen, Martti Forsell, Mikael Millberg, Johny Öberg, Kari Tiensyrjä, and Ahmed Hemani. A network on chip architecture and design methodology. In *VLSI, 2002. Proceedings. IEEE Computer Society Annual Symposium on*, pages 105–112. IEEE, 2002.

[4] James Balfour and William J Dally. Design tradeoffs for tiled cmp on-chip networks. In *Proceedings of the 20th annual international conference on Supercomputing*, pages 187–198. ACM, 2006.

[5] Samia Loucif. Concentration and its impact on mesh and torus-based noc performance. In *Parallel, Distributed and Network-Based Processing (PDP), 2015 23rd Euromicro International Conference on*, pages 361–364. IEEE, 2015.

[6] Tuhin Subhra Das and Prasun Ghosal. A provably good performance centric noc topology. In *Microelectronics and Electronics (PrimeAsia), 2013 IEEE Asia Pacific Conference on Postgraduate Research in*, pages 170–174. IEEE, 2013.

[7] Xiaofang Maggie Wang and Leeladhar Bandi. X-network: An area-efficient and high-performance on-chip wormhole interconnect network. *Microprocessors and Microsystems*, 37(8):1208–1218, 2013.

[8] Wen-Hsiang Hu, Seung Eun Lee, and Nader Bagherzadeh. Dmesh: a diagonally-linked mesh network-on-chip architecture. *Network on Chip Architectures*, page 14, 2008.

[9] Umit Y Ogras and Radu Marculescu. Prediction-based flow control for network-on-chip traffic. In *Proceedings of the 43rd annual design automation conference*, pages 839–844. ACM, 2006.

[10] Chen Sun, Chia-Hsin Owen Chen, George Kurian, Lan Wei, Jason Miller, Anant Agarwal, Li-Shiuan Peh, and Vladimir Stojanovic. Dsent-a tool connecting emerging photonics with electronics for opto-electronic networks-on-chip modeling. In *Networks on Chip (NoCS), 2012 Sixth IEEE/ACM International Symposium on*, pages 201–210. IEEE, 2012.

[11] Maurizio Palesi, Davide Patti, and Fabrizio Fazzino. Noxim-the noc simulator. *Online, http://noxim. sourceforge. net*, 2010.

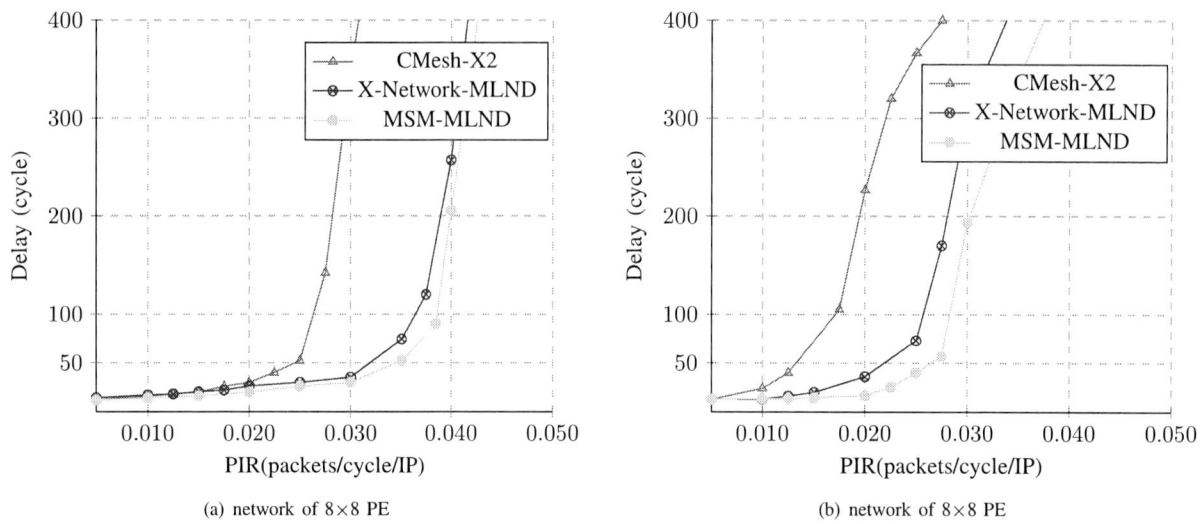

(a) network of 8×8 PE   (b) network of 8×8 PE

Fig. 5: Delay as a function of packet injection rate under (a) random (b) transpose traffic in quasi minimal adaptive routing in MLND selection.

Rapid prototyping IoT End Applications using Software Development Kits and Add on plugins

Manoj R
Connected MCU
Texas Instruments India Pvt. Ltd.
Bengaluru, India
manojr@ti.com

Adrian Fernandez
Connected MCU
Texas Instruments
Dallas, United States of America
adrianf@ti.com

Abstract— In any Internet of Things (IoT) related end equipment the most commonly used components are Sensors, Actuators. On the connectivity side technologies like Bluetooth Low Energy (BLE), Wi-Fi, Zigbee etc. are also used. Most common problems faced by developers while developing end applications is interfacing the sensors and actuators, working with IoT Software Development Kits (SDK) like Azure IoT SDK, Amazon AWS SDK and connectivity stacks. Many sensor vendors do provide sensor drivers, but user has to research and access multiple sources to get the necessary information and source code. Developer has to spend efforts in porting or interfacing the driver to his Micro Controller (MCU) platform. Portability also becomes important to protect software investments for future. This paper discusses about SDK for MCUs, Plugins for MCU SDKs, and provides details on how they simplify efforts for the developers in rapidly prototyping IoT end applications.

Keywords—Internet of Things, plugin, edge node

I. INTRODUCTION

The Internet of Things is the next major revolution in happening. Analysts predict that over 20+ billion devices will get connected to internet by 2020. IoT is a scenario where things, people and various cloud services connect via internet to enable new use cases and business models. IoT enables businesses to reduce expenses by making them smarter and leaner. It is becoming cheaper to add connectivity to MCUs, MCUs these days are more power efficient than ever which enables them to operate for years on a single coin cell battery. Further the increasing number of smartphone users is leading to the increasing adoption of IoT. The smartphones can get used as user interfaces for the connected things, it can act a gateway for the connected things, all of these are leading to a steady increase in the adoption of IoT. Right now we are at an IoT inflexion point.

The software challenges associated with embedded systems development have grown significantly in recent years. These systems continue to add advanced functionality with higher speed interfaces and multiple connectivity options, driving the need for more sophisticated scheduling algorithms, increased power consumption and code size. End-product needs are also changing rapidly as companies strive to serve multiple markets with different adaptations of their base product. As time-to-market windows shrink, developers need a robust software foundation with intuitive levels of abstraction and operating system support to enable faster creation of applications. This foundation should support a broad portfolio of devices that can easily reuse application code so that initial investments can be ported to a variety of products with varying system requirements.

The SimpleLink™ microcontroller (MCU) software development kit (SDK) is a complete set of validated, fully documented drivers, stacks and code examples that enable engineers to develop innovative and differentiated applications with the SimpleLink connected-MCU family from Texas Instruments (TI). This powerful SDK provides a cohesive and consistent experience by packaging essential software components and easy-to-use examples in one comprehensive package. Everything a customer needs to quickly and efficiently develop new applications using an ARM® Cortex®-M-based MCU from TI is included in the SDK, from the drivers and communication stacks to an OS kernel. The SDK is also well integrated into the development environment making advanced debug capabilities easily accessible.

SDK Plugins are intended to extend functionality of each individual base SDK to include specialized use-cases. Such as support for Sensors and Actuator software, Support for IoT SDKs etc.

Sensors and actuators are typically part of any IoT product. Sensors help the end equipment sense the surroundings or sometimes some parameters related to internals of the equipment, these may be sensors like light sensors, temperature sensors, Ultrasonic sensors etc. The sensor data sometimes gets consumed within the system, while most of the times the useful information read from the sensors gets stored in the cloud to perform actions such as predictive maintenance, user notifications etc. For Example: A gas leak detected in home can be notified to the user using gas leak sensor, in this particular use case the gas leak sensor data is also sent to the cloud so as to notify the end user.

Actuator is an electronic element, using which operations in end equipment can be controlled; these may be electronic components like solenoid valves, stepper motors, dc motors, buttons, etc. These actuators may be used to control things like gas valve, water valve, fan etc. these actuators may be controlled by some algorithm running in the cloud automatically (based on the various sensor information

978-1-5386-1357-3/17 $31.00 © 2017 IEEE

analyzed) or sometimes user may want to directly control them using an interface such as mobile application or web page.

The software used to interface sensor and actuator elements becomes one of the important pieces of software in the software to be designed for the IoT end equipment. In many cases these sensors, actuators may be from multiple vendors, there may be no software available from the vendor, in some cases, software even if available may be primitive, user will have to do considerable research to get the relevant software needed for his specific sensor. This specialized function is addressed by the **Sensor and Actuator Plugin for the SimpleLink SDK.**

Similarly we need specialized functions such as support for IoT SDKs which can enable us to connect to cloud. This is realized by IoT SDK Plugins such as by **Azure IoT plugin, AWS IoT plugin.**

The list of Plugins which will be supported for SimpleLink SDKs is growing day by day addressing various specialized use-cases.

This Paper specifically considers the SimpleLink™ MCUs and explains the concepts of SDK and later dives into the plugins for SDKs. The aim of the paper is to present a novel approach of providing software support for rapidly prototyping IoT end applications. Other contributions of the paper include

- Discussing important aspects of Software Development kit for SimpleLink MCUs.

- Showing how software abstraction enables portability across MCU platforms.

- Rapidly prototyping example applications using the Sensor and Actuator Plugin and IoT SDKs.

The remainder of this paper is organized as follows. We first discuss on some of the existing software development ecosystems for MCUs (Section II). We then discuss on the SimpleLink SDK and how it enables portability, in the same section we discuss about the plugin concept and also provide details on various plugins supported on SimpleLink SDKs. This section also provides a comparison of available development ecosystems and emphasizes why the approach using SimpleLink SDK and Plugins is appropriate (Section III), the Section IV discusses rapid prototyping IoT Applications using SimpleLink SDK and its plugins. Finally, we provide concluding remarks in section V.

II. SOLUTIONS AVAILABLE TODAY

Some of popular development ecosystems are Arduino, Mbed.

Arduino provides C/C++ framework for AVR, ARM etc. (based on Wiring). The software programs in Arduino are called as sketches. Arduino has good collection of libraries which can provide extra functionality to sketches such as reading a temperature sensor, controlling LED Matrix etc. Libraries for connectivity like BLE and Wi-Fi also exist for Arduino. A number of the libraries come as part of the default installation of Arduino IDE, but libraries can also be downloaded separately.

The Arm Mbed IoT Device Platform provides the operating system, cloud services, tools and developer ecosystem to make the creation and deployment of IoT based solutions.

The Mbed Software Development Kit (SDK) is a C/C++ microcontroller software platform.

It is built on the low-level ARM CMSIS APIs. In addition to RTOS, USB and Networking libraries, a cookbook of hundreds of reusable peripheral and module libraries have been built on top of the SDK by the Mbed Developer Community.

Mbed provides access to various libraries which are searchable from Mbed online IDE, many of these libraries are contributions from the open source community and third party developers there can be multiple libraries you can find for each sensor and there is no standardization of the sensor libraries. The coding convention, implementation of each sensor library differs from another.

The SimpleLink SDK with along with its Plugins provides extensive software support needed for prototyping of IoT Applications. This is a solution specific to Texas Instruments SimpleLink MCU family.

III. UNDERSTANDING THE SIMPLELINK SDK AND PLUGIN CONCEPT

To really understand what plugin to the SDK means we should first get a clear understanding of SDK.

The major components of the SDK include Drivers which have common API signature across MCUs in SimpleLink family these provide feature rich access to the MCU peripherals, OS/Kernel – Real-time Multitasking operating system, POSIX compliant API enable use of various RTOS, Middleware such as communication stacks, graphics library etc., examples and user documentation. Figure 1 shows the structure of SimpleLink MCU SDK.

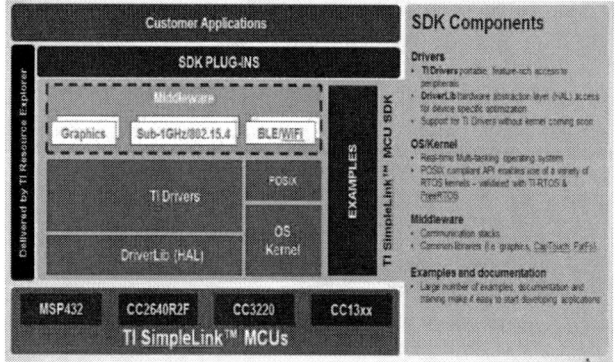

Figure 1. SimpleLink MCU SDK

The Paragraphs below provide detailed information on various components of the SimpleLink SDK

DriverLib hardware abstraction layer (HAL) consists of C functions that abstract away the details of the device's hardware registers. The TI Drivers and OS kernel support use the HAL to access hardware features. The HAL, built on top of device-specific header files, follows the ARM CMSIS standard, simplifying access to device modules beyond the register level. With the DriverLib HAL access, developers can peel back the layers of the TI drivers for greater control of their applications or to enhance the software for peripheral and device-specific optimization.

The integrated **TI-RTOS kernel** provides real-time, multitasking services such as timing and scheduling of tasks. TI-RTOS is a robust solution you can trust, already deployed in thousands of applications across various TI embedded solutions. The kernel is open source (BSD license) and was developed in lock-step with TI's silicon portfolio to enable very low latency in an efficient code footprint. Developers can optimize applications for power consumption, performance and code size to meet their needs. SimpleLink SDKs also support alternative RTOS kernels, such as FreeRTOS.

TI Drivers offer portable, feature-rich access to peripherals. The TI Drivers API exposes the functionality of the hardware-specific drivers in the same way across all TI SimpleLink devices, giving developers portable, feature-rich access to a variety of peripherals. TI drivers are open source [Berkeley Software Distribution license (BSD)] and built on the hardware abstraction layer, offering full access to the device's complete capability. This device-agnostic approach provides **easy portability of the application code across SimpleLink devices now and into the future**.

POSIX-compliant API offers support of additional third party kernels such as FreeRTOS. The POSIX layer abstracts the RTOS kernel functionality used by applications. POSIX is an IEEE industry standard for compatibility between operating systems. Requiring less than 2 KB of code in typical applications, the POSIX layer allows examples and user applications to be easily re-used and ported to a different kernel. POSIX-compatibility also allows TI third-party companies to interface with SimpleLink SDK devices to add support for their kernel. **This provides complete freedom to design with any OS, including FreeRTOS.**

Middleware adds functionality on top of drivers and includes communication stacks, graphics libraries, math libraries, an open source file system and more. TI has completed all testing and integration of the middleware for SimpleLink devices, making it fast and easy to integrate new technologies like Wi-Fi or Bluetooth low energy into an application.

The SDK provides a wide range of **free examples**. Using these examples, customers can quickly and easily start writing applications straight out of the box. Each example comes with its own documentation and project files, giving you everything you need to get started. Examples are provided using the supported RTOS kernels. For certain SDKs, examples that do not use an RTOS are also provided.

SDK Plugins are intended to extend functionality of each individual base SDK to include specialized use-cases. These specialized use cases can range anywhere from adding wireless functionality to extending a base SDK's example base. It is important to note that each plugin contains all of the necessary components to function fully alongside the base SDK. Plugins do not install inside of the SDK, but rather in a folder next to

the SDK. This is to simplify the maintenance model as well as provide customers with a genuine experience for updating and switching between plugin versions.

Today we have SDK plugins addressing various specialized use cases. Below is list of some of the plugins which are supported on the SimpleLink SDK.

TABLE I.   PLUGINS AVAILABLE TODAY FOR SIMPLELINK SDK

PLUGIN	Description
SIMPLELINK-SDK-BLUETOOTH-PLUGIN	Bluetooth Plugin for SimpleLink™ MCU SDK
SIMPLELINK-SDK-SENSOR-ACTUATOR-PLUGIN	Sensor and Actuator Plugin for SimpleLink™ MCU SDKs
SIMPLELINK-WIFI-CC3120-SDK-PLUGIN	SimpleLink™ WI-FI® CC3120 SDK Plugin
SIMPLELINK-SDK-VOICE-DETECTION-PLUGIN	Voice Detection plugin for the SimpleLink™ MCU Software Development Kit (SDK)
SIMPLELINK-CC3220-SDK-HOMEKIT-PLUGIN	SimpleLink(TM) Wi-Fi(R) SDK Plugin for Apple Homekit
SimpleLink CC32XX SDK Azure IoT Plugin	Azure IoT Plugin for CC3220 devices
SimpleLink CC32XX SDK AWS IoT Plugin	AWS IoT Plugin for CC3220 devices

Figure 2 shows how Sensor and Actuator Plugin works alongside the SimpleLink SDK.

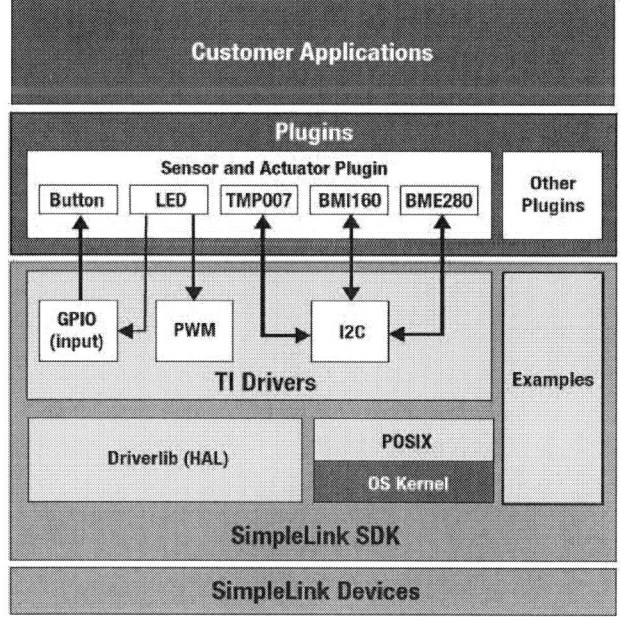

Figure 2. Sensor and Actuator Plugin

The modules of the Sensor and Actuator Plugin use TI Drivers to interface the sensors/actuators. Different sensors have different interface requirements. Some sensors use I2C interface, some use UART and some simple input and output devices like buttons and LEDs just need GPIO.

Similarly, other plugins work alongside the SimpleLink SDK and provide specialized functions.

Some of the Plugins like Azure IoT Plugin also bundle Sensor and Actuator Plugin within it. This enables complete access to all the sensors/actuators supported by Sensor and Actuator Plugin.

When it comes to building the real world embedded IoT systems, availability of RTOS is important. Frequently a system developer has to juggle with multitasking in his system and RTOS helps him achieve that easily.

Most of the IoT applications specifically the ones which act as the edge nodes need to consume low power. The edge nodes are typically battery powered. Low power design is needed is needed to ensure extended battery life of such nodes. SimpleLink MCU SDK includes power management framework to enable low power design of IoT applications.

End developer would like to use tool chain of his choice. Enabling multiple toolchains and providing example projects for multiple integrated development environments (IDE) like KEIL, IAR can help developers.

Cloud tool suite will enable developers to quickly evaluate examples and quickly enable a developer to get started with development on his interested platform.

Dynamically including interested library in the example project is useful for developers as it reduces manual effort of changing linker options and inclusion of header files.

TABLE II. is a comparison table which compares the important features of some of the available MCU development ecosystems

TABLE II.   COMPARAISION TABLE OF IMPORTANT FEATURES OF SOME OF THE DEVELOPMENT ECOSYSTEMS FOR MCU

Feature	Ecosystem		
	Arduino	Mbed Ecosystem	SimpleLink SDK & Plugins
Out of the Box RTOS Support	No	Yes (supports mbed OS)	Yes (validated on TIRTOS and FreeRTOS). Any customer POSIX compliant rtos can be used.
Power management capabilities	minimal	Available only for few platforms	Advanced power management capabilities on all SimpleLink platforms
Cloud support (for faster evaluation)	Extensive with cloud editor. Execution from cloud supported	Currently. Enables compilation on cloud IDE. Execution not supported.	Compilation and execution supported on cloud.
Tool chain support	gcc	Extensive. Default uses mbed compiler. But projects can be exported to IDE of user preference like IAR, KEIL etc.	Extensive. Example projects support multiple tool chains like TI,, IAR, GCC, KEIL
Dynamic library inclusion. (Reduces manual effort of modification of the linker and c files)	Yes	yes	No

IV. Rapid prototyping IoT Applications using SimpleLink SDK and Plugins

In this section we show how to rapidly prototype simple IoT Applications using the SimpleLink SDK and plugins.

The first example shows how to create a BLE enabled room temperature monitor which can be monitored using a Mobile phone having BLE support.

This example shows the implementation using MSP432 (where user is actually extending the feature set of existing implementation which does not have connectivity options like BLE, Wi-Fi etc.). Alternatively users can directly use BLE devices (CC26xx) from Texas Instruments, The SimpleLink SDK and the Sensor and Actuator Plugin are also supported on these devices.

The solution here uses the CC3220 the Single-Chip Wireless MCU Solution. It uses the SimpleLink SDK for CC3220 device and Azure IoT Plugin which includes sensor and actuator support within it.

Wide range of IoT Applications can be prototyped ground up using the SimpleLink SDK and its Plugins.

V. Conclusion

SimpleLink SDK and plugins simplify low power design of IoT applications, enables easier multitasking in end applications by providing opportunity to use any POSIX compliant RTOS. Scaling an existing design to make it connected is much simple as SimpleLink SDK is supported across entire portfolio of SimpleLink MCUs. With a single software architecture, modular development kits and free software tools for every point in the design lifecycle, the SimpleLink MCU platform allows 100% code reuse across the portfolio of microcontrollers; which supports a wide range of connectivity standards and technologies including RS-485, Bluetooth® low energy, Wi-Fi®, Sub-1 GHz with ZigBee® and Thread. There is still of scope in improving the overall experience but none the less it provides novel way of enabling rapid prototyping of the real world IoT Applications.

References

[1] http://www.ti.com/ww/en/internet_of_things/iot-overview.html

[2] http://www.ti.com/lit/SWSY004

[3] https://www.mbed.com/en/

[4] https://www.arduino.cc/

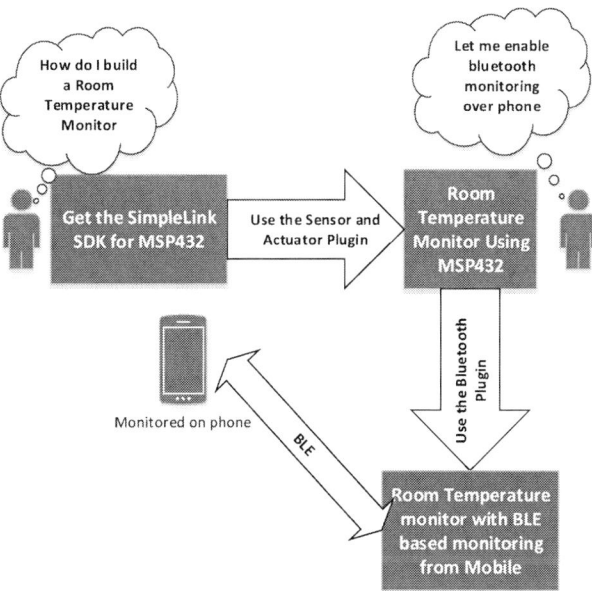

Figure 3. Room temperature monitor (BLE Enabled)

The second example shows implementation of room temperature monitor which is cloud connected, this enables remote monitoring.

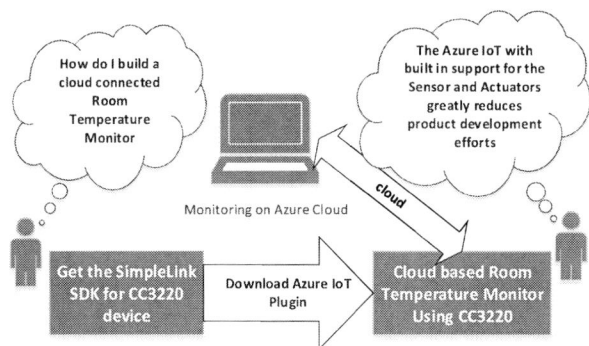

Figure 4. Cloud connected room temperature monitor

978-1-5386-1357-3/17 $31.00 © 2017 IEEE

Author Index

Aditya, Japa	90
Agrawal, Yash	34
Ahmed, Mohammed	184
Ahmed, Rekib Uddin	179
Akhilesh, Gangishetty	46
Anirudh, Grandhi Sai	56
Arora, Hardik	6
Aslam, Mohd.	190
B., Nithin Kumar Y.	129
Baghini, Maryam Shojaei	163
Bajpai, Shriya	195
Bansal, Shrestha	50
Bhargava, Lava	243
Bhattacharjee, Pritam	224
Bhattacharyya, Bidyut K.	206, 210
Božanić, Mladen	74
Brahma, Kaustav	61
Britt, Keith A.	123
Chakraborty, Susanta	145, 157
Chatterjee, Pratima	40
Chattopadhyay, Saranyu	61
Chaturvedi, Saurabh	74
Chowdhury, Tapan	157
D, Anoop	129
Das, Debasree	145
Das, Monalisa	210
Das, Tuhin Subhra	257
Deb, Sujay	50
Devulapalli, Arti	238
Fernandez, Adrian	263
Gade, Sri Harsha	50
Gamad, R. S.	220
Gambhira, Sashank	1
Gaur, Manoj Singh	243
Gautam, Shipra	28
Ghai, Garima	111
Ghosal, Prasun	40, 257
Gore, Ganesh R.	163
Gunti, Nagendra Babu	99, 105
Gupta, Yogendra	243
H., Vasantha M.	129
Humble, Travis S.	123
Hussain, Sarfraz	133, 139
J, Soumya	56
Joshi, Rathin	34
K, Sudeendra	252
K, Sudeendra Kumar	151
K, Sudeendra kumar	246
Kachave, Deepak	11
Khosla, Mamta	169
Kumar, Chintoo	28
Kumar, Dharmendra	28
Kumar, Pankaj	173, 200
Kumar, Rajesh	133, 139
Kumar, S. Dinesh	117
Kurzekar, Rahul	6
Laxmi, Vijay	243
Lingasubramanian, Karthikeyan	99, 105
Mahapatra, A.	252
Mahapatra, Abhishek	151, 246
Mahapatra, K.K.	151, 246, 252
Majudmer, Alak	206
Majumder, Alak	210, 224, 229
Mansore, S. R.	220
Meher, P. K.	23
Mishra, Akhilesh Chandra	238
Mishra, D. K.	220
Mishra, Vipul Kumar	66, 69
Mitra, Debasis	28
Mondal, Abir J.	206, 210
Mondal, Hemanta Kumar	50
Muhuri, Samya	145
Mukherjee, Arijit	157
Munoz-Coreas, Edgard	123
N.S., Murty	215
Nagateja, T.	90
Naskar, Mrinal Kanti	15
Nath, Bipasha	229
Neema, Shuba	11
Pande, Kirti S.	215
Panugothu, Sri Harsha	11
Parekh, Rutu	34
Patheja, Pushpinder Singh	111
Patil, Mahesh B.	163
Pattanaik, Manisha	234
Periasamy, C.	93

Author Index

Prasad, D. Prem	238
R, Manoj	263
Raj, Balwinder	169
Rajput, Amit Singh	234
Rathlavat, Santosh	15
Ray, Arkaprova	61
Ray, Ashok	173, 200
Rengrajan, Krishnan S.	163
Roy, Dipanjan	20
Roy, Soudip Sinha	78, 84
Roymohapatra, Sitansusekhar	163
Saha, Prabir	179
Sahoo, Rasmita	46
Sahoo, Sauvagya	151, 246
Sahoo, Sauvagya Ranjan	252
Sahoo, Subhendu Kumar	23, 46
Sahu, Chitrakant	93
Sarkar, Pallabi	15
Sarma, Manash Pratim	200
Sarma, Mansh Pratim	173
Sasikanth, M. Naga	1
Sengupta, Anirban	11, 15, 20, 66
Shafi, Nawaz	93
Sharad, Mrigank	1, 61
Sharma, Ashish	243
Sharma, Deepak G.	195
Sharma, Dheeraj	190, 195
Sharma, Neeraj	190, 195
Sharma, Sanjeev Kumar	169
Shrestha, Rahul	6
Singh, Jawar	93
Singh, Jeetendra	169
Sinha, Saurabh	74
Soni, Deepak	190, 195
Suthar, Rajani	215
Swain, A.K.	252
Swain, Ayas Kanta	151, 246
Thapliyal, Himanshu	117, 123
Tirkey, Sukeshni	195
Tiwari, R.K.	234
Trivedi, Gaurav	133, 139, 173, 200
Vaddi, Ramesh	90
Varun, T. S. S.	123
Vaseer, Gurveen	111
Yadav, Akanksha	163
Yadav, Dharmendra Singh	190, 195
Yadav, Shivendra	190
Yadav, Sonal	243
Zaman, Syed Samsuz	173, 200

IEEE Computer Society
Technical & Conference
Activities Board

T&C Board Vice President
Hausi Müller
University of Victoria, Canada

IEEE Computer Society Staff
Evan Butterfield, *Director of Products and Services*
Patrick Kellenberger, *Manager, Conference Publishing Services*

IEEE Computer Society Publications
The world-renowned IEEE Computer Society publishes, promotes, and distributes a wide variety of authoritative computer science and engineering texts. These books are available from most retail outlets. Visit the CS Store at *http://www.computer.org/portal/site/store/index.jsp* for a list of products.

IEEE Computer Society *Conference Publishing Services* (CPS)
The IEEE Computer Society produces conference publications for more than 300 acclaimed international conferences each year in a variety of formats, including books, CD-ROMs, USB Drives, and on-line publications. For information about the IEEE Computer Society's *Conference Publishing Services* (CPS), please e-mail: cps@computer.org or telephone +1-714-821-8380. Fax +1-714-761-1784. Additional information about *Conference Publishing Services* (CPS) can be accessed from our web site at: *http://www.computer.org/cps*

Revised: 18 January 2012

CPS Online is our innovative online collaborative conference publishing system designed to speed the delivery of price quotations and provide conferences with real-time access to all of a project's publication materials during production, including the final papers. The *CPS Online* workspace gives a conference the opportunity to upload files through any Web browser, check status and scheduling on their project, make changes to the Table of Contents and Front Matter, approve editorial changes and proofs, and communicate with their CPS editor through discussion forums, chat tools, commenting tools and e-mail.

The following is the URL link to the *CPS Online* Publishing Inquiry Form:
http://www.computer.org/portal/web/cscps/quote